Lecture Notes in Physics

Edited by H. Araki, Kyoto, J. Ehlers, München, K. Hepp, Zürich
R. Kippenhahn, München, D. Ruelle, Bures-sur-Yvette
H.A. Weidenmüller, Heidelberg, J. Wess, Karlsruhe and J. Zittartz, Köln
Managing Editor: W. Beiglböck

368

Luis Garrido (Ed.)

Statistical Mechanics of Neural Networks

Proceedings of the XIth Sitges Conference
Sitges, Barcelona, Spain, 3–7 June 1990

Springer-Verlag
Berlin Heidelberg GmbH

Editor

Luis Garrido
Facultad de Física
Departamento de Física Fundamental
Universidad de Barcelona
Diagonal 647, E-08028 Barcelona, Spain

ISBN 978-3-662-13785-7   ISBN 978-3-540-46808-0 (eBook)
DOI 10.1007/978-3-540-46808-0

© Springer-Verlag Berlin Heidelberg 1990
Originally published by Springer-Verlag Berlin Heidelberg New York in 1990
Softcover reprint of the hardcover 1st edition 1990

2153/3140-543210 – Printed on acid-free paper

ACKNOWLEDGEMENTS

I would like to take this opportunity to express my sincere thanks to all those who collaborated in the organization of this Conference.

Also, I wish to extend my warmest thanks to the Generalitat de Catalunya and to the Ministerio de Educación y Ciencia de Madrid for their economic support. To the City of Sitges I express my gratitude for allowing us again to use the Palace Maricel as a lecture hall.

Finally, I wish to thank my wife for her unremitting help and support.

Sitges, 7th June 1990   L. Garrido

CONTENTS

ON THE STATISTICAL-MECHANICAL FORMULATION OF NEURAL NETWORKS

Recently, we have witnessed a spectacular development of the theory of neural networks and its application to biology and computer science. Statistical mechanics, in particular spin-glass theory, has played an important role in the growth of this new area.

This chapter is not intended to be a complete review [1,2,3] of the subject but a short introduction in which some of the main points of the same, such as the analogy between neurons and Ising spins, the process of storing and retrieving memory, and the Hopfield model, are discussed.

From the biological point of view, neurons are the cellular constituents of the brain and were identified by Ramón y Cajal in the last century [4]. They are connected with each other by means of synapses and consequently can be viewed as a network. The complexity of such a system is the result of the great number of neurons which are present in the brain (about 10^9) as well as the number of synapses per neuron (about 10^5 in some cases). This large number of connections should be contrasted with that for electrical circuits (4 or 5 connections per element).

In the case of electric circuits some devices may become independent whereas for neural networks cooperative phenomena are favored.

Our present concept of a neural network is derived from several key developments. In 1943 McCulloch and Pitts [5] established the mathematical basis for modeling neural networks. In particular, they proposed a binary code for the outputs of the neurons. These authors were among the first to address the problem of pattern recognition. In 1949 Hebb [6] proposed his famous learning rule. The process of learning was then related to the modification of the synapses. One then may use Hebb's rule to store patterns.

One of the main problems of neural-network models is to find out the maximum number of patterns that may be stored with the smallest fraction of errors. In 1974 Little [7] pointed out the analogy between neural networks and magnetic systems. All these ideas led to the establishment of a model by Hopfield [8] in 1982 which would reproduce some of the most important characteristics of neural networks. The model's mean-field version was solved by Amit, Gutfreund, and Somplinsky [9] in 1987 using techniques from the spin-glass theory. The Hopfield model has been the starting point of most of the subsequent analyses.

The development of the spin-glass theory [10] has played an important role in the study of neural networks. Concepts such as the spin of a neuron or techniques such as the replica method have been used to formulate a statistical-mechanical theory of neural networks. To elucidate the close relationship between both theories, we must first of all introduce a mathematical model for neurons [11]. In this model, information is transmitted by the firing frequencies of neurons. These frequencies depend on the incoming pulses which arrive through the synapses in such a way that the total input is a linear function of such pulses. The output firing rate is a nonlinear function of the total inputs and of a threshold potential. This function is, in general, found to reduce to a Heaviside step function. In the process of carrying information, each neuron compares all inputs arriving at a certain time and the potential threshold. If the inputs exceed the value of the potential, the neuron fires; if not, it doesn't. Neurons are therefore elementary processors, which have also been called threshold automata or perceptrons.

The existence of a binary code makes it possible to assign a spin variable to each neuron. By convention, the value of the spin of the i-th neuron, s_i, is +1 if the neuron fires and -1 if it doesn't.

One may then formulate the discrete dynamic rule

where N is the total number of neurons, J_{ij} are the synapses (excitatory or inhibitory), and U_i is the potential threshold. This rule is deterministic and consequently may describe a situation in which the "temperature" is zero. In the non-zero temperature or noisy state, each spin will have a probability of pointing up or down. Statistical mechanics then gives the value for such probability, which is proportional to the corresponding Boltzmann factor.

Once we have established the close analogy between neurons and Ising spins, we will show how the existence of a great number of them may lead to collective behavior of the network. In fact, information can be stored and retrieved, giving rise to learning and memory functions. In "spin language" we have a need for the existence of stable firing patterns for which the spins $S_i(t)$ are not changed by the dynamic rule.

These patterns will constitute fixed points for attractors of the dynamics in such a way that information is encoded in them. The learning process was formulated mathematically by Hebb and is reflected in his rule

where i is a stored pattern or an attractor of the dynamics and is constant. As said before learning is associated with the modification of the synapses. One may then start from a "tabula rasa" situation (zero connection strengths) and by storing information through different firing patterns which are attractors of the dynamics. Of course on can imagine that the number of stored patterns cannot be infinite and there will exist a limit fixed by the storage capacity of the network. The problem of determining such limits will constitute one of the most important tasks in the study of neural networks. In this sense, Gardner [12] derived an upper limit for the storage capacity which is independent of the learning rule one employs.

Using once again the analogy with magnetic systems, Hopfield proposed his model by introducing the hamiltonian which is clearly a function of the pattern. The hamiltonian exhibits two contributions, namely exchange energy due to pairwise interactions between Ising spins and interaction with the external "magnetic field", $h_i$ext. Since the synaptic strengths may be positive (ferromagnetic) or negative (antiferromagnetic) the model is reminiscent of a spin glass and consequently exhibits properties such as frustration or the existence of many equilibrium configurations. It was shown that the stored patterns correspond to the minima of the hamiltonian, provided that Hebb's rule be satisfied. Using the replica approach, Amit, Gutfreund, and Sompolinsky [9] found the T-α phase diagram where T is the temperature and $\alpha =$ P/N, p being the number of stored patterns. From their analysis one infers the existence of two regimes. If P/N $\rightarrow$ 0, when N $\rightarrow \infty$ all the stored memories can be retrieved, provided that one starts from a pattern not too far from them; otherwise, one obtains spurious patterns. The regime P/N=α , when N $\rightarrow \infty$ is more interesting. There exist two phase transitions at $\alpha \simeq$ 0.05 and $\alpha \simeq$ 0.14. When $\alpha \leq$ 0.05 the replica symmetry is broken and ultrametricity holds. For $\alpha \geq$ 0.14 patterns are not recognized. On the other hand, retrieval is not possible in the paramagnetic phase defined above T=1.

L. Garrido and M. Rubi
Departamento de Física Fundamental
Universidad de Barcelona

References

1) H. Sompolinsky, Physics Today, December 1988, p.70
2) T. Geszti, Physical Models of Neural Networks, World Scient. 1990
3) W. Kinzel, Europhysics News, 21, 101 (1990)
4) S. Ramón y Cajal, Histologie du Systéme Nerveux, CSIC, Madrid
5) W.S. McCulloch and W. Pitts, Bull. Math. Biophys. 5, 115 (1943)
6) D.O. Hebb, The Organization of Behavior, Wiley, New York, 1949
7) W.A. Little, Math. Biosci. 19, 201 (1974)
8) J.J. Hopfield, Proc. Natl. Acad. Sci. USA, 79, 2554 (1982)
9) D.J. Amit, H. Gutfreund and H. Sompolinsky, Phys. Rev. A32 1007 (1985)
10) M. Mézard, G. Parisi, and M.A. Virasoro, Spin Glass Theory and Beyond, World Scientific, Singapore 1987 (and references therein)
11) A.V. Holden, Models of the stochastic activity of neurons, Lecture Notes in Biomathematics, Vol. 12, Springer-Verlag, 1976
12) E. Gardner, J. Phys. A, 21, 257 (1988)

MODEL NEURONS: FROM HODGKIN-HUXLEY TO HOPFIELD

L.F. Abbott* and Thomas B. Kepler**
Physics Department*, Biology Department** and
Center for Complex Systems
Brandeis University Waltham, MA 02254

Model neural networks are built of model neurons. While real biological neurons exhibit extremely complex and rich behavior, neuronal dynamics must be considerably simplified to make networks analytically and computationally tractable. The most complete and realistic descriptions of neuronal behavior are based on analytic fits of detailed voltage clamp measurements of the voltage and time dependence of cell membrane currents. Many different membrane currents may be involved and the resulting models can be extremely complex. The simplest model of this type, and the one which will be considered here, is the classic Hodgkin-Huxley model [1] of the squid giant axon. Even in this case the relative complexity of the model makes it rather intimidating to imagine studying large networks built from Hodgkin-Huxley cells. On the other extreme is the binary, on/off (McCulloch-Pitts, Little or Hopfield) model cell used in most studies of attractor neural networks [2]. The simplicity of the binary neuronal model makes extremely detailed studies of network behavior possible but leaves much of the biological complexity behind. Clearly a first priority in improving the biological accuracy of network modeling is to consider networks built from more realistic model neurons.

Neuronal models falling between the extremes represented by the Hodgkin-Huxley and binary, Hopfield-like models do of course exist. Most prominent are the integrate and fire models [3] and the FitzHugh-Nagumo model [4]. These were constructed phenomenologically to match the basic behavior displayed by the Hodgkin-Huxley model and to a lesser degree by real neurons. The absence of a more logical derivation of these models makes it difficult to see their exact relationship to more complete descriptions and to understand the nature of the approximations which have tacitly been used. Here we will present a step by step reduction of the classic Hodgkin-Huxley model based on work done in collaboration with Eve Marder. The advantage of this derivation is that it is mathematically explicit and that it provides a variety of descriptions of increasing simplicity as more and more simplifying approximations are made.

The starting point of the derivation is the Hodgkin-Huxley model. Through a reduction procedure the Hodgkin-Huxley equations are approximated by a system of two first-order differential equations. This results in a model similar in spirit to that of FitzHugh and Nagumo, but considerably different from it in detail. Our derivation provides missing links between the Hodgkin-Huxley description and more phenomenological models while at the same time producing a simplified model which is more accurate than the FitzHugh-Nagumo model. A reduction similar in some ways to ours has been given previously [5] but our methods are both more accurate and more general.

Once the two-dimensional model has been derived there are two possible approaches for further reduction and simplification. One of these leads to either a linear or nonlinear integrate and fire model while the other gives rise to a binary model with a time-dependent threshold [6,7]. Through our reduction procedure all of the parameters arising in these simplified models will be directly obtained from parameters of the original Hodgkin-Huxley description and all steps and approximations in the derivation will be explicit. In particular, we will arrive at a binary model of the general Hopfield type except that it includes hysteresis and a time-dependent threshold factor with dynamics completely determined by the underlying Hodgkin-Huxley description.

This work was supported by Department of Energy Contract DE-AC0276-ER03230, NIH Training Grant NIH-T32-HS0792 and NIMH grant NIMH-MH-46742.

THE HODGKIN-HUXLEY MODEL

The basic equation governing the dynamics of any neuron with spatially constant membrane potential V is given by conservation of electric charge,

$$C\frac{dV}{dt} = -F + I \qquad (2.1)$$

where C is the cell capacitance, F the membrane current and I the sum of external and synaptic currents entering the cell, each per unit of cell membrane area. If the potential is not spatially constant an additional Laplacian term enters this equation making the analysis much more complicated. We will restrict our attention to the spatially constant or space clamped case although our results can be applied to the more general situation as well. The membrane capacitance C is typically 1 μFarad per square centimeter. In the Hodgkin-Huxley model the membrane current F arises primarily from the conduction of sodium and potassium ions through voltage dependent channels in the membrane. In addition, other ionic currents are described by an Ohmic leakage contribution. F is a function of V and of three time- and voltage-dependent conductance variables m, h and n,

$$F(V, m, h, n) = g_L(V - V_L) + g_K n^4(V - V_K) + g_{Na} h m^3(V - V_{Na}) \qquad (2.2)$$

where g_L = 0.3 mmho/cm^2, g_K = 36 mmho/cm^2, g_{Na} = 120 mmho/cm^2, V_L = -54.402 mvolt, V_K = -77 mvolt and V_{Na} = 50 mvolt.

The conductance variables m, h and n are both voltage and time dependent and by definition take values between zero and one. They approach asymptotic values $\overline{m}(V)$, $\overline{h}(V)$ and $\overline{n}(V)$ with time constants $\tau_m(V)$, $\tau_h(V)$ and $\tau_n(V)$ respectively,

$$\tau_m(V)\frac{dm}{dt} = \overline{m}(V) - m \quad \tau_h(V)\frac{dh}{dt} = \overline{h}(V) - h \quad \tau_n(V)\frac{dn}{dt} = \overline{n}(V) - n. \qquad (2.3)$$

The six functions $\tau_m(V)$, $\tau_h(V)$, $\tau_n(V)$, $\overline{m}(V)$, $\overline{h}(V)$ and $\overline{n}(V)$ are given by fairly complicated formulas which are the result of detailed fits to experimental data. For all three variables,

$$\tau_{(m,h,n)} = \frac{1}{\alpha_{(m,h,n)} + \beta_{(m,h,n)}} \qquad (\overline{m}, \overline{h}, \overline{n}) = \frac{\alpha_{(m,h,n)}}{\alpha_{(m,h,n)} + \beta_{(m,h,n)}} \qquad (2.4)$$

where

$$\alpha_m = \frac{.1(V + 40)}{1 - \exp[-.1(V + 40)]} \qquad \alpha_h = .07\exp[-.05(V + 65)] \qquad (2.5)$$

$$\alpha_n = \frac{.01(V+55)}{1 - \exp[-.1(V+55)]} \qquad \beta_m = 4\exp[-.0556(V+65)]$$

$$\beta_h = \frac{1}{1 + \exp[-.1(V+35)]} \qquad \beta_n = .125\exp[-.0125(V+65)].$$

Here and in the following we suppress the units in formulas with the understanding that all potentials are in mvolts, all times in msec and all currents in μamps per cm^2. Although the Hodgkin-Huxley equations are quite complicated they mix in a rather minimal way. The cell potential V is affected by all three of the other dynamic variables m, h and n. However, m, h and n do not directly couple to each other, they only interact through V. This property will allow us to approximate the dynamics through the introduction of an auxiliary potential variable.

The functions $\overline{m}$, $\overline{h}$, $\overline{n}$ and all three τ's are plotted in Figure 1. Note that the time constant

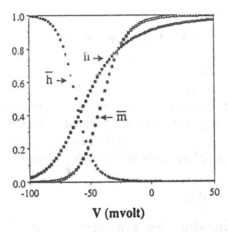
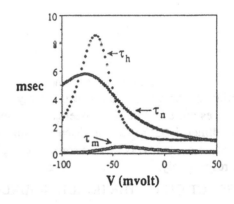

Figure 1

which governs the behavior of m is much smaller than those for h and n. Furthermore the time constants for h and n are roughly the same over most of the voltage range. These properties will be important for our reduction of the model.

The result of integrating the Hodgkin-Huxley equations are as follows. For $I = 0$, V remains at a resting potential of -65 mvolt. A positive current I of sufficient strength and duration will induce a sudden depolarization of the cell to about 50 mvolts followed rapidly by a hyperpolarization and then a slower recovery back to the resting level. This is of course an action potential. The time course of a typical action potential is shown in Figure 2. Note that the rise of the action potential is extremely rapid and that its fall has two parts, an initial downward ramp followed by a more rapid drop to a hyperpolarized potential. There is then a relatively long recovery back to the resting potential. The sudden rise of the action potential is caused by a rapid increase of the variable m which turns on a positive inward sodium current. The action potential is terminated as the variables h and n adjust more slowly to the change in membrane potential. The variable h decreases shutting off the sodium current which drove the upward swing of the action potential. At the same time an increase in n initiates a positive outward potassium current which hyperpolarizes the cell. The final recovery involves the readjustment of h and n back to their resting values.

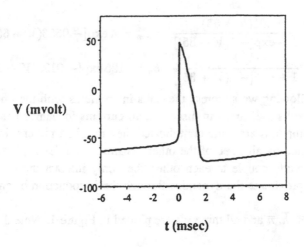

V (mvolt)

t (msec)

Figure 2

Three essential features are exhibited by the Hodgkin-Huxley model and these ideally should be present in any serious model of neuronal dynamics. They are 1) the action potential itself, 2) the refractory period after an action potential during which the slow recovery of the potassium and sodium conductance has a drastic impact on electrical properties and 3) the ability of the model cell to capacitively integrate incoming current pulses. These last two features are essential if a model cell is to react appropriately to synaptic inputs. The reduced two-dimensional model we discuss retains all three properties while the integrate and fire or binary models involve either some or complete loss of accuracy in describing property 2) or 3) respectively.

REDUCTION OF THE HOGDKIN-HUXLEY MODEL

The Hodgkin-Huxley model involves four time-dependent dynamical variables V, m, h and n. The four-dimensional nature of the phase space makes it difficult to visualize and intuitively understand the workings of this model. There is a tremendous conceptual advantage in reducing the model to two dimensions so that the phase space can be depicted in a straightforward manner. We therefore propose a procedure for reducing the number of dynamical variables from four to two. We do this in two steps. First, as discussed above the time scale associated with changes in m, τ_m, is much smaller than those associated with h and n. Thus, m will reach its asymptotic value $\overline{m}(V)$ much more rapidly than other changes in the model. If we are willing to give up some of the accuracy of the model over very short time scales we can replace m by its asymptotic value $\overline{m}(V)$ and ignore the differential equation for m entirely writing

$$m \approx \overline{m}(V) \tag{3.1}$$

and

$$F(V, m, h, n) \approx F(V, \overline{m}(V), h, n) \equiv F. \tag{3.2}$$

The last equivalence signifies that when the symbol F is used below it stands for the function

$F(V, \overline{m}(V), h, n)$. The instantaneous approximation for m reduces the number of dynamic variables from four to three.

It would simplify things considerably if we could also replace h and n by their asymptotic values. However, if we did this we would destroy the ability of the model to generate action potentials since h and n would terminate the action potential as quickly as m could initiate it. Instead because of their longer time constants, these variables should lag behind, reaching their asymptotic values more slowly. This effect can be simulated by introducing an auxiliary voltage variable U and replacing h and n by their asymptotic values not at the potential V but rather at U. The dynamics of the U variable will cause it to lag behind V but to approach it asymptotically. Thus we write,

$$h \approx \overline{h}(U) \qquad n \approx \overline{n}(U) \tag{3.3}$$

so that

$$F(V, m, h, n) \approx F(V, \overline{m}(V), \overline{h}(U), \overline{n}(U)) \equiv f(V, U). \tag{3.4}$$

Again the last equivalence indicates that f refers to the function $F(v, \overline{m}(V), \overline{h}(U), \overline{n}(U))$. Because both $\overline{h}$ and $\overline{n}$ are monotonic functions the replacement of either one of the variables h or n by its asymptotic value at U would correspond to a simple change of variables. Indeed our reduction procedure is exact in this case. However, because we are replacing both h and n by a single variable U the reduction is approximate. In order to minimize the impact of this approximation we must choose the variable U carefully. Clearly we want the time dependence of U in f to mimic the time-dependence induced into F in the full model by the changing values of h and n. Thus, we equate time derivatives of F at constant V in the full and reduced models,

$$\frac{\partial F}{\partial h}\frac{dh(V)}{dt} + \frac{\partial F}{\partial n}\frac{dn(V)}{dt} = \left(\frac{\partial f}{\partial \overline{h}}\frac{d\overline{h}(U)}{dU} + \frac{\partial f}{\partial \overline{n}}\frac{d\overline{n}(U)}{dU} \right)\frac{dU}{dt}. \tag{3.5}$$

We now use the original formulas for dh/dt and dn/dt incorporating the approximations $h \approx \overline{h}(U)$ and $n \approx \overline{n}(U)$ so that

$$\tau_h(V)\frac{dh}{dt} \approx \overline{h}(V) - \overline{h}(U) \qquad \tau_n(V)\frac{dn}{dt} \approx \overline{n}(V) - \overline{n}(U). \tag{3.6}$$

Using these result we can solve the equal time-derivative condition for dU/dt in terms of V and U. This gives us a reduced two-dimensional version of the Hodgkin-Huxley model,

$$C\frac{dV}{dt} = -f(V, U) + I \tag{3.7}$$

and

$$\frac{dU}{dt} = g(V, U) \tag{3.8}$$

where

$$g(V, U) = \frac{A}{B} \tag{3.9}$$

with

$$A = \frac{\partial F}{\partial h}\left(\frac{\overline{h}(V) - \overline{h}(U)}{\tau_h(V)} \right) + \frac{\partial F}{\partial n}\left(\frac{\overline{n}(V) - \overline{n}(U)}{\tau_n(V)} \right) \tag{3.10}$$

and

$$B = \frac{\partial f}{\partial \overline{h}} \frac{d\overline{h}(U)}{dU} + \frac{\partial f}{\partial \overline{n}} \frac{d\overline{n}(U)}{dU}.$$ (3.11)

where $\partial F/\partial h$ and $\partial F/\partial n$ are to be evaluated at $h = \overline{h}(U)$ and $n = \overline{n}(U)$. Note that when $U = V$, $g(V,U) = 0$ so U will approach V asymptotically if it can.

Simulations have shown that this reduced model is a good approximation of the full model. The easiest way to envision the dynamics given by Eqs. (3.7)-(3.11) is to plot the curves (isoclines) $dV/dt = -f + I = 0$ and $dU/dt = g = 0$ in the $U - V$ plane and to indicate the flows of U and V. This is done for $I = 0$ on the left side of Figure 3 below. The straight line corresponds to $U = V$ which makes $dU/dt = 0$. Along the curved line $dV/dt = 0$. The point $V = -65$ and $U = -65$ where these two curves intersect is the resting equilibrium point. The arrows indicate the flows toward these two curves. The right side of Figure 3 shows $dV/dt = 0$ isoclines for various positive values of the current I. The lowest curve corresponds to $I = 0$ and the left-hand portion of the curve rises as I is increased.

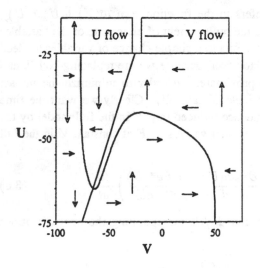
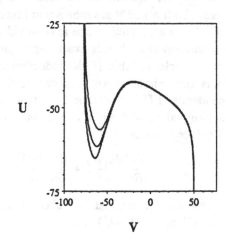

Figure 3

The model we have derived is, at least in spirit, similar to the FitzHugh-Nagumo model [4] which also involves two first-order differential equations. However, it has some essential and important differences. In the FitzHugh-Nagumo model g(V,U) is taken to be linear in both U and V while $f(V,U)$ is cubic in V and linear in U. The linear form for g is not such a bad approximation but the form assumed by FitzHugh and Nagumo for f considerably distorts the dynamics. In particular, as shown on the right side of Figure 3 the low voltage section of the $dV/dt = 0$ isocline moves up with positive current I. However, the high voltage portion of this curve is almost completely insensitive to external current. In the FitzHugh-Nagumo model the entire $dV/dt = 0$ isocline moves up by the same amount as a function of I and this causes some fairly severe misrepresentations of cell behavior. For example, the amplitude of the action potentials in the full Hodgkin-Huxley model and in our reduced model decrease when the cell is pushed to a high firing rate by large injected current. In the FitzHugh-Nagumo model the amplitude is fairly independent of firing frequency. Furthermore no direct relation between parameters of the FitzHugh-Nagumo model and the underlying Hodgkin-Huxley model exists.

To indicate the behavior of the two-dimensional model we first consider the response of the system to constant positive current. Figure 4 shows the phase plane for zero current (lower curve) and for a positive current large enough to cause repetitive firing (I=10). The circular point on the lower curve is the resting point of the cell with zero current. This stable equilibrium lies at the point where the two isoclines shown intersect. When positive current is applied the low-voltage portion of the $dV/dt = 0$ isocline moves up until the intersection of the two isoclines falls well within the portion of the $dV/dt = 0$ isocline with positive slope. This point is unstable so the system goes into a limit cycle as shown by the arrows in Figure 4. This produces a train of action potentials.

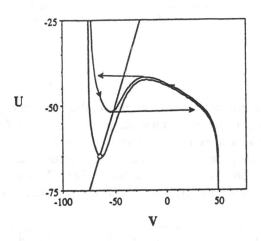

Figure 4

Another way to exhibit the dynamics of the reduced model is to subject a model cell which is initially at the resting potential $V = -65$ and $U = -65$ to (square) current pulses of various amplitudes and durations. Figures 5 and 6 show responses to such pulses. Along with the $dV/dt = 0$ and $dU/dt = 0$ isoclines, the phase-space trajectory is shown with data points marked at equal time intervals and arrows indicating the sense of the motion. The initial line of data points in all the figures shown indicates the behavior of V and U during the application of the current pulse while the curved trajectories give the subsequent response. The left side of Figure 5 shows the effect of a small pulse of current. This causes an upward shift in V that is insufficient to move V out of the region where $dV/dt < 0$ (compare with Figure 3). As a result V moves back down to its resting value and no action potential is fired. The loop in the return trajectory indicates that the U variable has also responded to the current pulse although this response is less dramatic than that for V. The right side of Figure 5 shows the effects of a stronger current pulse which is nevertheless still too weak to produce an action potential. Here, the dynamics of the U variable plays an essential role in preventing the firing of an action potential. Note that the current pulse is sufficiently strong to move V

into the region to the right of the $dV/dt = 0$ isocline where $dV/dt > 0$. However, no action potential results because the increase in the U variable causes the trajectory to curve back into the region where $dV/dt < 0$.

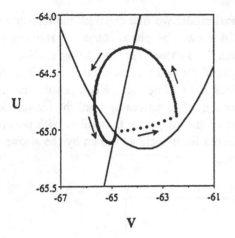
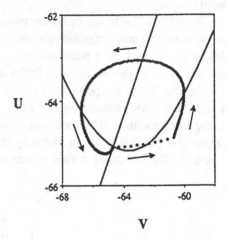

Figure 5

Figure 6 shows the effects of a current pulse sufficiently strong to produce an action potential. Here, V is pushed far enough into the unstable $dV/dt > 0$ region so that it climbs all the way up to the high-voltage portion of the $dV/dt = 0$ isocline. The initial trajectory is shown at left and the complete loop from the resting potential, through the action potential and back to the resting potential again is shown at right. It is interesting to note that the phase plane trajectory stays quite close to the $dV/dt = 0$ isocline except initially near the resting potential.

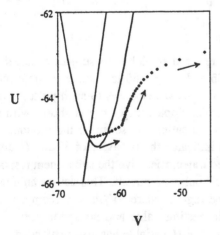
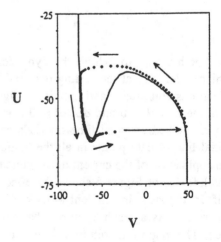

Figure 6

FURTHER REDUCTION OF THE MODEL

We will now show how the two-dimensional model can be further reduced to yield some even simpler models. Clearly with two differential equations in the model an obvious simplification is to ignore either one or the other. The differential equation for V reflects the capacitive properties of the cell, as the presence of the capacitance C would suggest. The differential equation for U on the other hand reproduces the time dependence of the membrane conductance. The capacitive behavior reflected in the V equation is responsible for the integrative behavior of the cell while the time-dependent conductances given by the U equation produce the refractory period. Thus, to proceed we must be willing to give up an accurate description of one or the other of these two essential features. As we will see, integrate and fire models ignore the U equation and must incorporate refractoriness in an ad hoc and crude manner. Binary models on the other hand ignore the V equation and thus do not simulate the integrative and time-delay behavior which capacitance provides. Nevertheless, each of these models rather accurately reproduces the remaining dynamics and thus provides an extremely useful if partial description of neuronal behavior.

A standard procedure for reducing a set of two nonlinear equations like (3.7)-(3.11) (if such a reduction is in fact possible) is to treat one of the variables as instantaneous. For example if dV/dt is much greater than dU/dt the V variable will rapidly move to a value which makes $dV/dt = 0$ and the longer time scale dynamics of the system will be governed by the constraint $dV/dt = -f + I = 0$ and the slower behavior given by the differential equation for U. Indeed this is the case through most of the dynamic range of V and U. This is evident in Figure 6 where we noted that the phase-space trajectory stays quite close the $dV/dt = 0$ isocline throughout the action potential. In fact throughout most of the phase plane $|f| >> |g|$, so that V is constrained to stay close to the $dV/dt = 0$ isocline most of the time. However, to accurately predict the firing pattern of a cell we are most interested in what happens near the resting potential and firing threshold and here the situation is more complex. Figure 7 shows $-f(V, -65)$ and $g(V, -65)$ in the relevant region of V. In this region f and

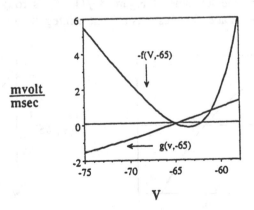

Figure 7

g are of comparable magnitude. Below the resting potential the magnitude of f tends to be bigger. Thus in this region V can be considered instantaneous and the dynamics is determined by the isocline constraint $-f + I = 0$ and the equation for dU/dt. However for potentials slightly above the resting potential f is actually smaller the g. This region is essential for triggering an action potential and we see that ignoring the capacitive effects incorporated in the equation for dV/dt here would be a mistake. A reduction of the two differential equations involves a trade-off. If the region between the resting potential and the firing threshold is of most interest then the V dynamics dominates. This is the approach taken by integrate and fire models as we discuss next. The approximation $dV/dt = -f + I = 0$ which removes V as an independent dynamical variable leaving only the differential equation for U is valid throughout most of the range of V except in the threshold region. This approximation gives rise to the binary models.

INTEGRATE AND FIRE MODELS

Integrate and fire models can approximately duplicate the capacitive behavior of a cell, but they treat the refractory properties very roughly at best. To derive an integrate and fire model from the reduced model of the last section we ignore the dynamics of the U variable. This is essentially a large C approximation of Eqs. (3.7)-(3.11). Since the time-dependence of the membrane conductance is not reproduced adequately the model cannot itself produce action potentials. Rather it is assumed that when the membrane potential exceeds some threshold value the result is the firing of an action potential. Likewise, since the U dynamics is ignored no refractory period arises in the model but this is sometimes included crudely by specifying that the V dynamics freezes for a certain time period following an action potential.

The elimination of the dynamics of the U variable can be done in two different ways. First we can set $U = V$ permanently making $dU/dt = 0$. The remaining dynamics is then just

$$C\frac{dV}{dt} = -f(V, V) + I. \tag{5.1}$$

As can be seen from the left side of Figure 8 $f(V, V)$ is roughly linear and this further approximation is often made. The result is the linear integrate and fire model. To write it we

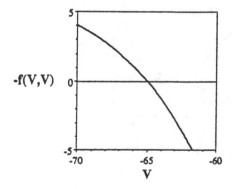
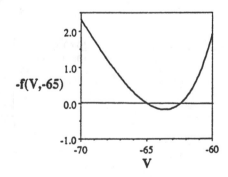

Figure 8

re-express V as

$$V = v - 65 \tag{5.2}$$

so that

$$C\frac{dv}{dt} = -\frac{v}{R} + I \tag{5.3}$$

where by fitting the curve $f(V, V)$ we find $R = 0.8$ kOhm-cm^2.

The approximation $U = V$ is actually a pretty terrible one to make. As can be seen in the phase-space curves for the two-dimensional model, U is never very close to V except at the resting value of -65. A more sensible approximation is to set $U = -65$ and leave it there. Then, V is governed by the equation

$$C\frac{dV}{dt} = -f(V, -65) + I. \tag{5.4}$$

The curve $f(V, -65)$ is given in the right side of Figure 8. This curve is quite far from linear, in particular it passes through zero twice. The higher value of V for which $f = 0$ defines the threshold potential since for all higher potentials up to the peak of the action potential $dV/dt > 0$. Fitting this curve in the relevant region gives a nonlinear integrate and fire model again written in terms of v,

$$C\frac{dv}{dt} = -.250v + .083v^2 + .008v^3 + I \tag{5.5}$$

with $v_{threshold} = 2.5$. Note that the parameters for both these integrate and fire models are given directly by the underlying Hodgkin-Huxley model through the reduction procedure and simple curve fitting.

BINARY MODELS

As discussed above the membrane potential V tends to stay quite close to the $dV/dt = 0$ isocline most of the time throughout most of the phase space. Thus, another approach to further reduction is to assume that the neuron has small enough capacitance C so that the differential equation for V can be replaced with the constraint equation

$$f(V, U) = I. \tag{6.1}$$

In other words, we assume that the capacitance is small enough so that changes of V caused by changes in the current I and in U can be approximated as instantaneous. We can then solve the constraint equation for V as a function of U. The resulting values of V versus U are plotted at the left in Figure 9. Note that V is a multi-valued function of U. To correctly define the inverse we can introduce a binary variable S which keeps track of which branch of the curve is being used. In Figure 9 an upper region given by $S = +1$ and a lower region given by $S = -1$ are indicated. For this figure $I = 0$. The correspondence between this S variable and the binary variable in Hopfield-type models should be obvious. $S = +1$ when the membrane potential takes high values characteristic of an action potential while $S = -1$ when the cell is not firing and near its resting potential. The region between those indicated by $S = +1$ and $S = -1$ does not need to be considered because V will never enter this region. As the arrows indicate V will jump instantaneously between the $S = \pm 1$

regions. This is an instantaneous approximation to the beginning and end (specifically the second rapidly dropping part of the end) of an action potential and is a result of our ignoring the cell capacitance. Since these processes are extremely rapid this is actually quite a good approximation.

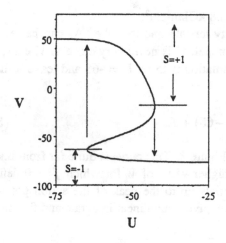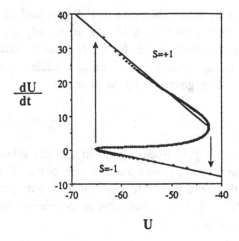

Figure 9

From the curve shown on the left side of Figure 9, V can be eliminated in favor of S and U. Then dU/dt depends on U and S rather than on U and V. This is shown as the hatched line in the right-hand panel of Figure 9. Also shown by solid lines in this figure are linear fits to dU/dt in the regions $S = +1$ and $S = -1$ which are used to further simplify the model. As can be seen these linear fits are quite good. A non-zero current can be included by noting that for $S = +1$ current has no appreciable impact while for $S = -1$ it effectively shifts the U variable (see the right side of Figure 3). In particular we find that to a good approximation

$$\frac{dU}{dt} = -.3(U + 65 - .6I) \tag{6.2}$$

for $S = -1$ and

$$\frac{dU}{dt} = -1.3U - 50 \tag{6.3}$$

for $S = +1$. The worst part of this approximation is the fact that the differential equation for U is linear in I when in fact the response becomes nonlinear if the current is too large.

The final binary model involves two equations, one to maintain the value of the S variable and the other a differential equation for U obtained by combining the two equations for dU/dt given above. We will keep track of S by demanding that it flip from -1 to +1 when U goes through the value -66 from above and from +1 to -1 when U passes through -43 from below. To simplify the final formulas we make the change of variables

$$U = 11.5u - 65 \qquad i = .03I \tag{6.4}$$

Then the binary model is given by

$$S = sign[S + .9 - u + (1 - S)i] \tag{6.5}$$

with

$$\frac{du}{dt} = -(.8 + .5S)u + 1.5(1 + S) + .3(1 - S)i \qquad (6.6)$$

The equation (6.5) is similar to that of a binary, Hopfield model with three important extra properties. 1) The presence of the S variable in the sign function provides hysteresis. 2) The u variable acts as a time-dependent threshold determined by the differential equation (6.6). 3) The current i only couples when $S = -1$. The addition of these three features lifts the binary model to a level where it is accurately reflecting the underlying dynamics of the full Hodgkin-Huxley model except of course that capacitive effects are not included. The additional complications introduced to achieve this are really quite minimal. In particular the differential equation for u is linear so it can be solved easily and analytic results can be obtained. A model similar to (6.5) and (6.6) has been studied extensively [7]. It is possible to derive analytic expressions for firing rates, phase response curves, firing delays, phase locking regions and many other responses of the cell to constant and time varying external currents.

A discrete time version of the model can be generated by turning Eq. (6.5) into a dynamic map and by integrating Eq. (6.6) over one time step Δt,

$$S(t + \Delta t) = sign[S(t) + .9 - u(t) + (1 - S(t))i(t)] \qquad (6.7)$$

with

$$u(t + \Delta t) = \overline{u} + (u(t) - \overline{u}) \exp[-(.8 + .5S(t))\Delta t] \qquad (6.8)$$

where

$$\overline{u} = \frac{1.5(1 + S(t)) + .3(1 - S(t))i(t)}{.8 + .5S(t)}. \qquad (6.9)$$

Systems of binary model cells can be constructed by using the standard expression

$$i = \frac{1}{2} \sum_{j=1}^{N} J_j(S_j + 1) \qquad (6.10)$$

for the synaptic current where the sum is over other cells coupled to the neuron being studied and the J_j are synaptic weights. This has also been done in a related model [7].

CONCLUSIONS

Following a well-defined reduction procedure we have constructed three types of reduced models. The two-dimensional model incorporates both the effects of cell capacitance and of time-dependent membrane conductances and so is the most accurate of the three. Integrate and fire models can be derived only at the expense of weakening the treatment of time-dependent conductances and most importantly refractoriness. Nevertheless these provide quite a good description near the resting and threshold potentials. By ignoring cell capacitance we obtained an interesting binary model which incorporates quite accurately aspects of the time-dependent conductances through hysteresis and through a time-dependent threshold factor. An addition difference between this derived binary model and those more commonly in use is that synaptic current only couples to the model cell when it is in the non-firing $S = -1$ state. In all three cases, parameters of the reduced models were obtained directly from the underlying full Hodgkin-Huxley model. An advantage of our reduction approach is that it allows us to include the effects of other membrane currents besides those included in the original Hodgkin-Huxely description. For example we can determine how additional currents affect the time-dependent threshold of the binary model. This is presently being done.

REFERENCES

1. Hodgkin, A.L. and Huxley, A.F. (1952) *J. Physiol. (London)* 117. 500.
2. McCulloch, W.S. and Pitts, W. (1943) *Bull. Math. Biophys.* 5, 115; Little, W.A. (1975) *Math Biosci.* 19, 101; Hopfield, J.J. (1982) *Proc. Natl. Acad. USA* 79, 2554.
3. For a review see: Tuckwell (1988) H. *Introduction to Theoretical Neurobiology* (Cambridge University Press, Cambridge).
4. FitzHugh, R. (1961) *Biophys. J.* 1, 445; Nagumo, J.S., Arimoto, S. and Yoshizawa, S. (1962) *Proc. IRE* 50, 2061.
5. Krinskii, V.I. and Kokoz, Y.M. (1973) *Biofizika* 18, 533 and 937.
6. Horn, D. and Usher, M. (1989) *Phys. Rev.* A40, 1036.
7. Abbott, L.F. (1990) *J. Phys. A* (in press); Proc. Natl. Acad. Sci USA (submitted).

Statistical Mechanics for Networks of Analog Neurons

R. Kühn

Sonderforschungsbereich 123, Universität Heidelberg
Im Neuenheimer Feld 294, D-6900 Heidelberg

Well, it's a matter of continuity.
Thomas Pynchon [1]

Foreword

The present paper is a report on work that has been done in collaboration with S. Bös and J.L. van Hemmen. With me, they will sign responsible for the hard results in what is presented below, though perhaps not for some of the metaphysics that may have sneaked into the text. It is a pleasure to thank them for their help and advice.

1. Introduction

In attempts to understand the workings of the central nervous system, two opposite starting points come to mind. One might be to begin by studying the physico-chemical mechanisms that govern the dynamical behaviour of *single neurons*, the other to concentrate on *collective phenomena* emerging as the result of the interaction of many simplified neuron-like elements — the assumption being that fine details at the single neuron level will not be all-important for the dynamical behaviour of the whole.

Both approaches may be associated with exisisting research programs. The former has been highlighted by the work of Hodgkin and Huxley [2] on mechanisms underlying spike generation in giant axons of the squid. The latter — more precisely its current phase — has been initiated by the work of Little [3] and Hopfield [4], and it has of late become quite popular in the physics community.

The Hodgkin-Huxley equations have provided an impressively detailed microscopic description of the dynamical behaviour of single neurons. On the other hand, they are so complicated that any attempt at studying collective behaviour of large networks of Hodgkin-Huxley neurons would turn out unwieldy to the point of losing its heuristic value. There have, however, been recent attempts to simplify the Hodgkin-Huxley approach to the end of making a dynamical description of networks of such neurons feasible again, while keeping essential details underlying spike generation at the single neuron level. The reader will find an account of such work in the contribution of L. Abbott to these proceedings [5].

In the approach of Hopfield [4] and his followers (see, e.g. [6-9]), the single neuron is stripped of all its details, and is represented as a simple threshold automaton, capable of attaining only two states: firing at maximum rate, and non-firing. An interpretation of the "on" and "off" states as presence or absence of a *single* spike has also been proposed [6].

It appears not widely appreciated though that devising a faithful description of the *dynamics* of a collection of interacting formal two-state neurons is still a delicate task. To mention only one major complication, there are various transmission delays due to different axonal and dendritic distances that signals must travel. This will cause neurons to respond to firing states of other neurons, some of which are already outdated, some not. Details of the resulting haphazard anarchy are difficult to capture in general, or by formal rules; they will depend on specific characteristics of the network architecture. The favourite choice of physicists has been to replace this anarchy by another one: asynchronous dynamics with a random order of updatings. While it is easy to state what has been gained by this desperate trick — analytic tractability — it appears very difficult to assess what has been lost. One of the main undesirable features of asynchronous dynamics, for instance, is that neurons do not update their firing-state any time their postsynaptic potential crosses a given threshold, but rather when the order of updatings is on them. The other popular alternative — parallel dynamics — must be regarded as even more unrealistic from a biological point of view, globally synchronized operation being an unlikely feature of natural nerve nets.

Networks of formal neurons with symmetric synaptic couplings have been analyzed very successfully by a statistical mechanical approach pioneered by Peretto [7] and by Amit et al. [8]; for recent comprehensive reviews of this approach, the reader may consult Refs. 6 and 9. Most of the results so far obtained along this line pertain to networks of two-state neurons, equipped with a stochastic dynamics of Glauber or heat bath type — asynchronous or parallel. There are a few exceptions regarding the set of possible output states of individual neurons[10-16], though not the underlying dynamics. Kanter [10] has considered networks of q-state Potts spins, and finds that the storage capacity scales as $\alpha_c(q) \sim \frac{1}{2}q(q-1)\alpha_c^H$ with the number q of Potts states, $\alpha_c^H \simeq 0.138$ denoting the capacity of the Hopfield model. A numerical study by Meunier et al. [11] of a system of three-state (spin-1) neurons, with a threshold dynamics adapted to the three state nature of the neurons, indicated that the storage capacity of such a system might be markedly lower than that of the standard model, but exhibited also some interesting dynamical features arising from the null-state of the neurons. The system has been studied analytically in the highly dilute limit by Yedidia [12]. Its capacity to store a collection of unbiased three-state patterns $\{\xi_i^\mu \in \{0, \pm 1\}; 1 \le i \le N\}$ increases with decreasing activity, i.e., fraction of non-zero ξ's in the stored patterns, and in the low-activity limit the system performance approaches that of a sparsely connected network of 0-1 neurons [12,17]. The approach of Ref. 12 was extended by Rieger to cover more general input-output (I/O) relations [13]. His finding is, in particular, that the capacity of networks of graded response neurons to store a collection of unbiased binary random patterns exhibits a pronounced increase as one moves away from the infinite gain limit [13], a result that has also been obtained by Mertens [14]. We shall see below that this feature does not survive the removal of the sparse-connectivity limit.

Fully connected networks of graded response neurons with a parallel *discrete-time* iterated map dynamics have been studied by Marcus et al. [18]. They were able to show that, in case of synaptic symmetry, the only attractors of the dynamics are fixed points and period-two limit cycles, as in the Little model [3]. Moreover, it was established that for neurons with sufficiently low analog gain only fixed point attractors survive, and self-consistency equations describing these stationary attractors were proposed [19]. For systems with Hopfield synaptic couplings, these equations agree with those describing the stochastic Ising spin network, provided that the neurons have a hyperbolic tangent input-output relation and analog gain is identified with inverse temperature [19]. As a consequence, there are stationary retrieval and spin-glass like phases as in the Hopfield model, and the boundary between them coincides with the corresponding spin-glass transition line found by Amit et al. [8]. At high analog gain, the stability criterion for fixed point attractors is violated and period two limit cycles prevail. For the system with pseudo-inverse couplings, no phase analogous to the spin-glass phase in its stochastic counterpart [20] was found. The total number of fixed points was shown to decrease with decreasing analog gain [21].

Very recently, graded response neurons with stochastic asynchronous dynamics have been considered by Treves [22], and by Rieger [23]. In many respects, the performance of these networks turns out to be similar to that of their two-state-neuron counterparts.

The present contribution, too, is devoted to networks of analog neurons. Our aim is to improve upon some of the above-mentioned undesirable features of the asynchronous or parallel dynamics that has been underlying all modeling so far, while keeping an analysis of collective behaviour feasible. More precisely, we shall demonstrate that existing improvements [24] can be analyzed to a level of detail that does hardly fall short of what has been accomplished for stochastic asynchronous neural networks. In so doing, we shall also be able to drop another constraint on the type of networks that can be handled, viz. the *homogeneity* constraint: There will be no need to assume that all neurons of the network behave identically. A preliminary account of this work can be found in Ref. 25.

2. Continuous-Time Dynamics and Graded Response Neurons

We study networks of *graded response* (analog) neurons with a *deterministic* continuous-time dynamics described by a set of differential RC charging equations. Such a description of neural network dynamics models the single neuron as an iso-potential object which has a capacitance that is charged by other neurons connected to it or by external sources, and whose instantaneous output can be described by some continuous function of its transmembrane voltage. The network equations read

$$C_i \frac{dU_i}{dt} = \sum_{j=1}^{N} J_{ij} V_j - \frac{U_i}{R_i} + I_i \quad , \tag{1.a}$$

$$V_j = g_j(\gamma_j U_j) \quad , \tag{1.b}$$

They have been proposed by Hopfield [24] in an attempt to capture the influence of capacitive input delays, of transmembrane leakages and graded, i.e. finite gain, input-output (I/O) characteristics that would always be present in a system of real neurons. In (1), C_i denotes the input capacitance of the i–th neuron, R_i its transmembrane resistance, U_i its postsynaptic potential and V_i its instantaneous output. The I/O characteristics of a neuron is described by its gain or transfer function g_j as in (1.b), where γ_j is a gain parameter. The I_i represent external (current) sources, and the synaptic weights are as usual denoted by J_{ij}.

In neural network modeling, the V_i should be interpreted as neuronic firing rates. The graded-response property of the neurons as described by (1.b) then implies that Eqs. (1) retain at least some of the features of frequency coding that is presumably used in biological neural systems — though perhaps still in a rather crude manner. It should, however, be borne in mind that a description of neural network dynamics in terms of *spike rates* as in (1) is not impartial with respect to one of the fundamental undecided questions of neurobiology, i.e., the question of the *neural code*: As yet, it is not clear whether a description in terms of spike rates does indeed cover the essentials of the global network dynamics, or whether on the contrary the presence or absence of *individual spikes* carries information that is crucial for the dynamical evolution of the whole. As to neurobiology, the results presented below should therefore be read with this proviso in mind.

Alternatively, Eqs. (1) can be interpreted as providing a rather realistic description of the dynamics of a network of resistively coupled operational amplifiers, in which case the V_j are output voltages of amplifiers with gain functions g_j. Networks of such devices can be used, for instance, for solving hard optimization problems [26], and a theoretical analysis of their performance could thus be used as a guide for improving the design of such devices.

Our approach to analyzing the collective behaviour of networks of graded response neurons described by (1) is in the spirit of statistical mechanics. The necessary condition for our method to be applicable is the existence of a Lyapunov function for (1) — a condition that is satisfied for networks whose synaptic matrix is symmetric and whose neurons have monotone increasing I/O relations [24]. The I/O characteristics may be *otherwise arbitrary*, and may, in particular, vary from neuron to neuron. Non-*decreasing* I/O relations are covered as limiting cases by a continuity argument.

Apart from the symmetry constraint, the connectivity may also be arbitrary — at least in principle. In particular, self-connections are allowed, in contrast to the situation in asynchronous stochastic models. In practice, of course, the demands of analytic tractability do impose restrictions on the type of connectivity that can be handled. Just as in the stochastic models, it should be such that a mean field analysis becomes exact (or a good approximation).

Up to stability, the fixed points of the iterated map and of the continuous time dynamics should be identical, and in some respects, our findings confirm the results of Marcus et al. [19]. We will, however, also come across a number of differences and discrepancies. They will be discussed as we go along.

The Lyapunov function governing (1) is given [24] by

$$\mathcal{H}_N = -\frac{1}{2}\sum_{i,j=1}^{N} J_{ij}V_iV_j + \sum_{i=1}^{N}\frac{1}{\gamma_i R_i}G_i(V_i) - \sum_{i=1}^{N}I_iV_i \quad , \tag{2}$$

where G_i is the integrated inverse input-output relation of neuron i,

$$G_i(V) = \int_{V_{i0}}^{V} g_i^{-1}(V')dV' \quad , \tag{3}$$

and V_{i0} denotes the output voltage of the neuron at zero transmembrane potential U, i.e., $V_{i0} = g_i(0)$. To wit, compute

$$\begin{aligned}
\frac{d\dot{\mathcal{H}}_N}{dt} &= \sum_{i=1}^{N}\frac{\partial\mathcal{H}_N}{\partial V_i}\frac{dV_i}{dt} \\
&= -\sum_{i=1}^{N}\left(\sum_{j=1}^{N}J_{ij}V_j - \frac{U_i}{R_i} + I_i\right)C_i\frac{dU_i}{dt}\frac{\gamma_i}{C_i}g'(\gamma_iU_i) \quad .
\end{aligned} \tag{4}$$

In (4), we have used the symmetry of the synaptic matrix. By monotonicity of the g_i and Eq. (1) the assertion follows, i.e. $\frac{d\mathcal{H}_N}{dt} \leq 0$, with equality only at stationary points of (1).

A typical example of a gain function is shown in Fig. 1. One may, for instance, think of the sigmoidal I/O relation $g(x) = \tanh(x)$. The corresponding inverse I/O relation $g^{-1}(x) = \text{arctanh}(x)$ has integrable singularities at $x = \pm 1$, so that $G(x) = x\tanh(x) - \log\cosh(x)$ remains finite at $x = \pm 1$.

The existence of a Lyapunov function entails that the dynamics of networks described by (1) always converges to fixed points, which are global or local minima of $\mathcal{H}_N$. This observation immediately tells us how our analysis of the collective behaviour of neural nets described by (1) should proceed. We have to compute the deterministic zero-temperature ($\beta \to \infty$) limit of the free energy

$$\begin{aligned}
f_N(\beta) &= -(\beta N)^{-1}\log\text{Tr}_V\exp(-\beta\mathcal{H}_N) \\
&= -(\beta N)^{-1}\log\int\prod_i d\rho(V_i)\exp(-\beta\mathcal{H}_N) \quad ,
\end{aligned} \tag{5}$$

and analyze the nature of its stable and metastable phases. Here $d\rho(V)$ denotes an a priori measure of the output voltage of the individual neurons, which — guided by Laplace's principle of insufficient reason — we take to be *uniform* on its support. It will turn out that, as long as we are interested in zero-temperature properties of our system, the support of $d\rho$ is indeed all that matters. We will have occasion to return to this point later on.

Note that the input capacitances C_i of the neurons do not enter $\mathcal{H}_N$ and thus do not occur in the free energy (5). That is, the characteristic processing times $\tau_i = R_iC_i$ of individual neurons do *not* affect the nature or the number of fixed points of (1). They must, however, be expected to determine the size and shape of basins of attraction, convergence times, the way in which the fixed points of (1) are approached, and other intrinsically dynamic properties of the network.

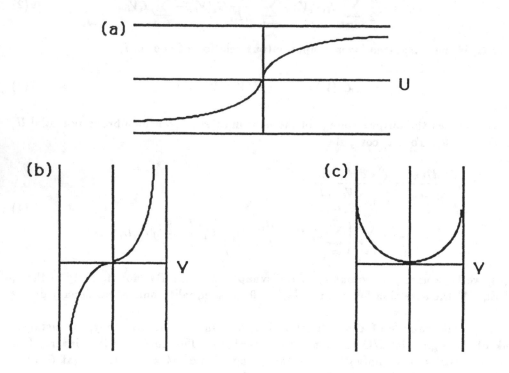

Fig. 1: (a) Example of a sigmoidal gain function. (b) Corresponding inverse I/O relation. (c) Integrated inverse I/O relation.

We now proceed to specify the details of our system. To facilitate comparison with the perhaps most well-known and best understood model, our first choice will be a soft-neuron version of the Hopfield network. Thus, we assume that the couplings are given by

$$J_{ij} = \frac{1}{N} \sum_{\mu=1}^{q} \xi_i^\mu \xi_j^\mu \quad , \quad i \neq j \quad , \quad J_{ii} = 0 \quad , \tag{6}$$

designed to store a set of q unbiased binary random patterns $\{\xi_i^\mu, 1 \leq i \leq N\}, 1 \leq \mu \leq q$. For simplicity and as a first step, we shall take the network to be homogeneous: All neurons are assumed to attain output voltages V_i in the interval $[-1, 1]$, and $R_i = C_i = 1$ (in suitable units) throughout the network. Moreover, the input-output relation will be taken to be the same for all neurons. The mean-field analysis below can be carried out without specifying the gain function g, and we will therefore not restrict generality by choosing a specific input-output relation until it comes to the numerical solution of the fixed point equations describing the attractors of (1).

The above homogeneity assumptions are by no means necessary to keep the analysis feasible. They can and will be relaxed later on. In particular, non-zero self-interactions and input-output relations varying from neuron to neuron are easily dealt with. They introduce nothing but an extra element of on-site disorder, the analytic description of which presents no additional difficulties of principle. All this will eventually allow us to study, for instance, networks consisting of several types of neurons.

For now, let us however stick to homogeneous networks in the sense outlined above, and to Hopfield's synaptic couplings (6). For such systems, the free energy (5) may be expressed as

$$f_N(\beta) = -(\beta N)^{-1} \log \int \prod_i d\tilde{\rho}(V_i) \exp\left(\frac{N\beta}{2} \sum_{\mu=1}^q m_\mu^2\right) \quad , \tag{7}$$

where we have introduced the overlaps

$$m_\mu = \frac{1}{N} \sum_{i=1}^N \xi_i^\mu V_i \quad , \quad 1 \le \mu \le q \quad , \tag{8}$$

and where the integrated inverse I/O relation as well a correction term taking account of the vanishing self-interactions in (6) have been absorbed in the single site measure for the V_i,

$$d\tilde{\rho}(V) = d\rho(V) \exp[-\alpha\beta V^2/2 - \beta\gamma^{-1}G(V)] \quad . \tag{9}$$

There are, as $N \to \infty$, two essentially different limits to investigate, the limit of *finitely* many patterns, and the limit of *extensively* many patterns, $q = \alpha N$. It will be instructive to consider both.

The evaluation of the free energy and the derivation of an associated set of fixed point equations closely follows Amit et al. [8], with a few minor modifications to cope with the continuous nature of our fundamental variables. To deal with these, the large deviations techniques outlined e.g. in Ref. 27, though not essential, do come in handy.

3. Finitely Many Patterns

If the number of stored patterns remains finite, as $N \to \infty$, so that $\alpha \to 0$ in (9), we get

$$f(\beta) \equiv \lim_{N \to \infty} f_N(\beta) = \frac{1}{2} \sum_{\mu=1}^q m_\mu^2 - \beta^{-1} \left\langle \log \int d\tilde{\rho}(V) \exp\left(\beta V \sum_{\mu=1}^q m_\mu \xi^\mu\right) \right\rangle \quad , \tag{10}$$

where the m_μ must be chosen to satisfy the fixed point equations

$$m_\mu = \langle \xi^\mu [V]_\xi \rangle \quad . \tag{11}$$

Here the angular brackets denote an average over the ξ's, and we have introduced the short-hand notation

$$[V]_\xi = \frac{\int d\tilde{\rho}(V) V \exp\left(\beta V \sum_\mu m_\mu \xi^\mu\right)}{\int d\tilde{\rho}(V) \exp\left(\beta V \sum_\mu m_\mu \xi^\mu\right)} \quad . \tag{12}$$

Let us recall at this point that for the description of the attractors of (1) we have to investigate these equations in the limit $\beta \rightarrow \infty$. In this limit the average (12) is easy to compute. Provided that the a priori measure $d\rho(V)$ is sufficiently smooth on its support, we find, using (9), that the effective probability measure used to evaluate $[V]_\xi$ in (12) converges, as $\beta \rightarrow \infty$, to a Dirac measure at the point $\hat{V} = \hat{V}(\xi)$ where $\exp[-\gamma^{-1}G(V) + V\sum_\mu m_\mu \xi^\mu]$ is maximal. This point is given by $\hat{V}(\xi) = g(\gamma \sum_\mu m_\mu \xi^\mu)$ so that

$$f(\beta) \rightarrow f = \frac{1}{2}\sum_\mu m_\mu^2 - \gamma^{-1}\left\langle (\gamma \sum_\mu m_\mu \xi^\mu) g(\gamma \sum_\mu m_\mu \xi^\mu) - G\left(g(\gamma \sum_\mu m_\mu \xi^\mu)\right)\right\rangle, \quad (13)$$

and the fixed point equations reduce to

$$m_\mu = \left\langle \xi^\mu g(\gamma \sum_\mu m_\mu \xi^\mu)\right\rangle . \quad (14)$$

Among the various solutions of (14), only those giving rise to dynamically stable phases are of interest for the study of the attractors of (1). For these, the matrix of second derivatives of f, $\partial^2 f/\partial m_\mu \partial m_\nu = \delta_{\mu,\nu} - \gamma\langle \xi^\mu \xi^\nu g'(\gamma \sum_\mu m_\mu \xi^\mu)\rangle$, must be positive definite.

For generic sigmoid input-output relations g with $\lim_{x\rightarrow\pm\infty} g(x) = \pm 1$, the large γ behaviour will always be like the large β behaviour known from the stochastic version of the standard Hopfield model. Moreover, if we consider standardized g's, having $g(0) = 0$ and $g'(0) = 1$, then the paramagnetic solution $m_\mu = 0$ will loose local stability, as the gain γ increases above $\gamma_c = 1$.

Details of the emergent collective behaviour, in particular the nature and order of phase transition in the vicinity of $\gamma = 1$, will strongly depend on the gain-function g of the model. Consider for instance retrieval solutions $m_\mu = m\delta_{\mu,\nu}$, for which (14) reads $m = g(\gamma m)$. The nature of the transition from $m = 0$ to non-zero m's is determined by the sign of the first non-zero higher order derivative of g at $x = 0$. If the sign is negative, the transition will be continuous, with critical exponent $\beta = 1/(k-1)$ if k is the order of the derivative in question, and it will occur *exactly* at $\gamma = \gamma_c = 1$. If the sign is positive, the transition will be discontinuous or first order. In this case, the precise location of the (equilibrium) transition point depends on global properties of the gain function g, and there will quite generally be hysteresis effects.

For the special choice $g(x) = \tanh(x)$, it is not difficult to show that the system is formally equivalent with its stochastic Ising model counterpart at inverse temperature $\beta = \gamma$. In contrast, however, to what has been claimed for the iterated map network [19], this formal equivalence does *not* persist in the limit of extensively many stored patterns.

4. Extensively Many Patterns

In this limit, the free energy per neuron is evaluated by the replica method (see e.g. [8,9,27]). For states which have macroscopic overlaps with at most finitely many, say p, of the $q = \alpha N$ stored patterns, the replica symmetric approximation gives

$$f(\beta) = \frac{1}{2} \sum_{\mu=1}^{p} m_\mu^2 + \frac{\alpha}{2} \left\{ \beta^{-1} \log[1 - \beta(q_0 - q_1)] + (q_0 - q_1)\tilde{r} + \beta(q_0 - q_1)r \right\}$$

$$- \beta^{-1} \Big\langle\!\Big\langle \log \int d\tilde{\rho}(V) \exp \Big\{ \beta[(\sum_{\mu=1}^{p} m_\mu \xi^\mu + \sqrt{\alpha r} z)V + \frac{1}{2}\alpha\tilde{r}V^2] \Big\} \Big\rangle\!\Big\rangle \quad . \tag{15}$$

Here, the double angular brackets represent a combined average over the ξ^μ associated with the (at most) p macroscopically condensed patterns, and the Gaussian random variable z with zero mean and unit variance. The measure $d\tilde{\rho}(V)$ has been defined in (9), and

$$r = \frac{q_1}{[1 - \beta(q_0 - q_1)]^2} \quad , \quad \tilde{r} = \frac{1}{1 - \beta(q_0 - q_1)} \quad .$$

The m_μ, q_0 and q_1 in (15) must be chosen so that they satisfy the fixed point equations

$$m_\mu = \langle\!\langle \xi^\mu [V]_{\xi,z} \rangle\!\rangle \quad ,$$

$$q_0 = \langle\!\langle [V^2]_{\xi,z} \rangle\!\rangle \quad , \tag{16}$$

$$q_1 = \langle\!\langle [V]_{\xi,z}^2 \rangle\!\rangle \quad ,$$

where now

$$[F(V)]_{\xi,z} = \frac{\int d\tilde{\rho}(V) F(V) \exp\left[\beta(\sum_\mu m_\mu \xi^\mu + \sqrt{\alpha r} z)V + \frac{\alpha\beta}{2}\tilde{r}V^2\right]}{\int d\tilde{\rho}(V) \exp\left[\beta(\sum_\mu m_\mu \xi^\mu + \sqrt{\alpha r} z)V + \frac{\alpha\beta}{2}\tilde{r}V^2\right]} \quad . \tag{17}$$

Again, we have to investigate these equations in the limit $\beta \to \infty$, where the effective probability measure used to evaluate $[F(V)]_{\xi,z}$ in (17) converges to a Dirac measure at the point $\hat{V} = \hat{V}(\xi, z)$ where $\exp[-\alpha V^2/2 - \gamma^{-1} G(V) + (\sum_\mu m_\mu \xi^\mu + \sqrt{\alpha r} z)V + \frac{\alpha}{2}\tilde{r}V^2]$ is maximal. In contrast to the case of finitely many patterns, $\hat{V}$ can no longer be specified explicitly in terms of ξ and z. It is determined *implicitly* as solution of the fixed point equation

$$\hat{V} = g\left(\gamma\left[\sum_\mu m_\mu \xi^\mu + \sqrt{\alpha r} z + \alpha(\tilde{r} - 1)\hat{V}\right]\right) \tag{18}$$

on the support of $d\rho$. This is the main source of differences between our results and those of Ref. 19.

For the purposes of a numerical solution one rewrites (16) in terms of the variables m_μ, q_1 and $C \equiv \lim_{\beta \to \infty} \beta(q_0 - q_1)$, so as to obtain, as $\beta \to \infty$,

$$m_\mu = \langle\!\langle \xi^\mu \hat{V}(\xi, z) \rangle\!\rangle \quad ,$$

$$C = \frac{1}{\sqrt{\alpha r}} \langle\!\langle z\hat{V}(\xi, z) \rangle\!\rangle \quad , \tag{19}$$

$$q_1 = \langle\!\langle \hat{V}^2(\xi, z) \rangle\!\rangle \quad ,$$

with $\hat{V} = \hat{V}(\xi, z)$ determined by (18), and $r = q_1/(1 - C)^2$ while $\tilde{r} = 1/(1 - C)$.

As $\gamma \to \infty$ (recall that we have to take the $\beta \to \infty$ limit first), Eqs. (19) reduce to a set of fixed point equations which are formally equivalent with those describing the zero-temperature limit of the Hopfield model [8], provided that $g(x)$ is monotone increasing from -1 to +1 as x increases from $-\infty$ to $+\infty$. This equivalence does *not* persist as one moves away from the infinite gain limit, not even if the I/O relation is taken to be $g(x) = \tanh(x)$; see the phase diagram Fig. 2. In particular, the "paramagnetic" null-state at high inverse gain becomes locally unstable against spin-glass type ordering with $q_1 \neq 0$, as the inverse gain is decreased below $\gamma_g^{-1} = 1 + 2\sqrt{\alpha}$. This curve is *universal* among standardized I/O functions with $g(0) = 0$ and $g'(0) = 1$. It should be contrasted with the line $\beta_g^{-1} = 1 + \sqrt{\alpha}$ separating the paramagnetic and spin-glass phases in the stochastic Hopfield model [8]. Here, our results agree with findings of Marcus et al. [19].

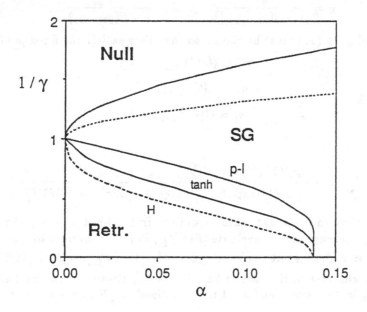

Fig. 2: Phase diagrams for soft neuron versions of the Hopfield model. Phase boundaries between the retrieval and spin-glass (SG) phase are marked tanh and p-l for the models with gain-functions $g(x) = \tanh(x)$ and the piecewise-linear function $g(x) = \text{sgn}(x) \times \min\{|x|, 1\}$, respectively. The boundary between the "paramagnetic" null-state and the SG-phase is the same for both models. Dashed lines give stochastic Ising model results for comparison; for these, the inverse gain scale should be identified with the temperature scale.

Upon closer inspection, however, there are several differences. For instance, the self-consistency equations of Marcus et al. [19] which characterize the stationary attractors of the iterated map network (if they are stable), are formally identical with the fixed point equations describing the stochastic Hopfield network, provided that $g(x) = \tanh(x)$ and analog gain is identified with inverse temperature. They would thus predict phase boundaries as in the Hopfield model, in particular that spin-glass type order ceases to exist as the inverse gain is increased above $1 + \sqrt{\alpha}$, leaving only the "paramagnetic" null-solution. On the other hand, the null-state was demonstrated to lose local stability at inverse analog gains as high as $1 + 2\sqrt{\alpha}$. The explanation of this paradox is twofold. First, Marcus et al. failed to introduce correction terms that would take the absence of self-interactions into account. Second the analysis of Ref. 19 is de facto a signal-to-noise ratio analysis, and therefore only approximate. In passing we note that pseudo-inverse couplings [20] also give results [28] different from those of Ref. 19.

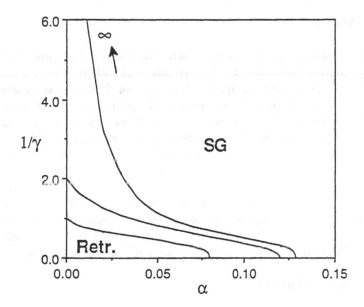

Fig. 3: Phase boundaries between retrieval and spin-glass phases for a network in which half of the neurons have a piecewise-linear, the other half a hyperbolic tangent I/O relation. For various values of the gain parameter of the piecewise-linear neurons the critical inverse gain of the tanh-neurons has been plotted as a function of the storage ratio α. From top to bottom, we have $\gamma_{p-l} = 2.0$, $\gamma_{p-l} = 1.5$, and $\gamma_{p-l} = 1.0$. As $\alpha \to 0$, the critical inverse gain of the tanh-neurons approaches $(2 - \gamma_{p-l})^{-1}$, if $\gamma_{p-l} < 2$, and it diverges for every $\gamma_{p-l} \geq 2$.

Equations (15)-(19) are for systems which are homogeneous in the sense described above. If we have non-zero self-interactions, $J_{ii} \neq 0$, and input-output relations $V_i = g_i(\gamma_i U_i)$ as well as transmembrane resistances R_i varying from site to site, the single site measure $d\tilde{\rho}(V)$ in (9) must be replaced by the site dependent measure

$$d\tilde{\rho}_i(V) = d\rho(V) \exp[\beta(J_{ii} - \alpha)V^2/2 - \beta\lambda_i^{-1}G_i(V)] \quad , \tag{20}$$

where $\lambda_i = \gamma_i R_i$, and where G_i denotes the integrated inverse input-output relation for g_i as in (3). It is then fairly straightforward to see that, if the J_{ii}, the R_i, the γ_i and the g_i are randomly selected according to some distribution satisfying rather mild regularity conditions (finitenes of the family of possible I/O relations suffices), Eqs.

(15)-(19) remain formally unaltered, except that the double angular brackets in (15), (16) and (19) now imply an *additional average* over the random measure $d\tilde{\rho}_i(V)$.

Fig. 3 shows boundaries between the retrieval and the spin-glass phase for a soft-neuron version of the Hopfield model in which half of the neurons are described by a hyperbolic tangent I/O relation,the other half by a piecewise linear I/O characteristics. The phase boundaries depend *strongly* on the gain parameter of the piecewise linear neurons.

5. Noisy Systems

The theory presented above characterizes the attractors of the deterministic dynamics described by the evolution equations (1). The dynamics of biological neural networks or artificial devices, however, is never completely free of noise. If a local stochastic driving force $\zeta_i(t)$ is added to the differential equations (1.a), the invariant distribution of the resulting Markov dynamics will in general *not* be the Gibbs distribution generated by $\mathcal{H}_N$, because in the V_i representation the noise turns out to be of a multiplicative nature. Under certain conditions on the nature of the noise [29] one can nevertheless work out invariant distributions as Gibbs-type measures, albeit generated by a modified energy function [30]. In this case, details of the a-priori measure $d\rho(V)$ beyond its support do become relevant. Luckily, however, the analysis of this case *reveals* what the a priori measure should be. Perhaps not surprisingly, one finds that $d\rho(V)$ should be homogeneous (though not normalized) on its support, i.e., $d\rho(V) = dV$, which confirms our choice above. The case of noisy dynamics is currently being evaluated.

6. Discussion and Outlook

The main idea of the present contribution was to show that neural networks with a (deterministic) continuous-time dynamics as described by (1) are amenable to analysis by the tools of statistical mechanics. In many respects, the nature of the fixed points of (1) turns out to be like that encountered in the stochastic asynchronous models — finite gain in networks of analog neurons playing a role similar to thermal noise in networks with stochastic dynamics. Details will, of course, depend on the choice of the gain function.

While the results presented above may thus perhaps not be surprising, we do believe that our ability to handle a large class of gain functions as well as networks consisting of several types of neurons, might carry some potential for the integration of further neurophysiological detail into neural network models. The problem of low local firing rates [31], for instance, can find an explanation in terms of neuronal input-output relations [30] which is perhaps more natural than those previously considered [32,33].

On the other hand, synaptic symmetry remains as one of the major unrealistic features of the systems that can be handled by our approach, though experience with the stochastic models suggests that the reintroduction of synaptic asymmetry does not discontinuously upset the performance of networks such as considered in the present paper [34,35]. At a more fundamental level, however, there is the still undecided question of the neural code alluded to above, which strictly needs clarification before the neurophysiological relevance of our results can be assessed at all.

For networks of operational amplifiers our analysis is *quantitative*. It might thus be used in the process of designing such devices. In particular, if networks of this type are used for performing hard optimization tasks, analog gain is one of the means to control the appearance or disappearance of spurious sub-optimal solutions. Indeed, for the travelling salesman problem, networks operating at moderate gain were found to perform consistently better than those operating in the high gain limit [26].

We have as yet not studied the stability of our results with respect to replica symmetry breaking (RSB). However, in the retrieval phase, we expect RSB to occur only at fairly high gains, where the system is already almost fully ordered, so that the effects should be small.

References

1. T. Pynchon, *Gravity's Rainbow*, (The Viking Press, New York, 1973) p 664.
2. A.L. Hodgkin and A.F. Huxley, J. Physiol. **117**, 500 (1952).
3. W.A. Little, Math. Biosci. **19**, 101 (1974).
4. J.J. Hopfield, Proc. Natl. Acad. Sci. USA **79**, 2554 (1982).
5. L.F. Abbott, *Model Neurons: From Hodgkin-Huxley to Hopfield*, these proceedings.
6. D.J. Amit, *Modeling Brain Function — The World of Attractor Neural Networks*, (Cambridge University Press, Cambridge, 1989).
7. P.Peretto, Biol. Cybern. **50**, 51 (1984).
8. D.J. Amit, H. Gutfreund, and H. Sompolinsky, Phys. Rev. A **32**, 1007 (1985); Phys. Rev. Lett. **55**, 1530 (1986); Ann. Phys. (N.Y.) **173**, 30 (1987).
9. E. Domany, J.L. van Hemmen, and K. Schulten (eds.), *Physics of Neural Networks*, (Springer, Heidelberg, 1990).
10. I. Kanter, Phys. Rev. A **37**, 2739 (1988); see also the contribution of D. Bollé, these proceedings.
11. C. Meunier, D. Hansel, and A. Verga, J. Stat. Phys. **55**, 859(1989).
12. J. Yedidia, J. Phys. A **22**, 2265 (1989).
13. H. Rieger, Thesis, Köln, (1989), unpublished.
14. S. Mertens, *Storing p-state Patterns in a Network with Analog Neurons*, this conference, and private communication.
15. J. Cook, J. Phys. A **22**, 2257 (1989).
16. J. Stark and P. Bressloff J. Phys. A **23**, 1633 (1990).
17. M.V. Tsodyks, Europhys. Lett. **7**, 203 (1988).
18. C.M. Marcus and R.M. Westervelt, Phys. Rev. A **40**, 501 (1989).
19. C.M. Marcus, F.M. Waugh, and R.M. Westervelt, Phys. Rev. A **41**, 3355 (1990).
20. I. Kanter and H. Sompolinsky, Phys. Rev. A **35**, 380 (1987).
21. F.M. Waugh, C.M. Marcus, and R.M. Westervelt, Phys. Rev. Lett. **64**, 1986 (1990).
22. A. Treves, *Graded-Response Neurons and Information Encodings in Auto-Associative Memories*, Phys. Rev. A (1990), in press; see also *Neural Networks in the Hippocampus Involved in Learning*, these proceedings.
23. H. Rieger, *Properties of Neural Networks with Multi-State Neurons*, these proceedings.

24. J.J. Hopfield, Proc. Natl. Acad. Sci. **81** 3088 (1984).
25. R. Kühn, S. Bös, and J.L. van Hemmen, *Statistical Mechanics for Networks of Graded Response Neurons*, submitted to Phys. Rev. A (1990).
26. J.J. Hopfield and D.W. Tank, Biol. Cybern. **52**, 141 (1985); D.W. Tank and J.J. Hopfield, IEEE Transactions on Circuits and Systems Cas-**33** 533 (1986).
27. J.L. van Hemmen and R. Kühn, *Collective Phenomena in Neural Networks*, in Ref. 9.
28. S. Bös, to be published.
29. R. Graham and H. Haken, Z. Phys. **243** 289 (1971).
30. R. Kühn et al., to be published.
31. M. Abeles *Local Cortical Circuits*, (Springer, Berlin, Heidelberg, New York, 1982); M. Abeles, E. Vaadia, and H. Bergman, Network **1**, 13 (1990).
32. D. Amit and A. Treves Proc. Natl. Acad. Sci. USA **86**, 7871 (1989); A. Treves and D.J. Amit, J. Phys. A **22**, 2205 (1989).
33. N. Rubin and H. Sompolinsky, *Neural Networks with Low Local Firing Rates*, Europhys. Lett. (1990), in press.
34. H. Sompolinsky, in: *Heidelberg Colloquium on Glassy Dynamics*, J.L. van Hemmen and I. Moegenstern (eds.), (Springer, Heidelberg, 1987) p 485.
35. W. Kinzel, in: *Heidelberg Colloquium on Glassy Dynamics*, J.L. van Hemmen and I. Morgenstern (eds.), (Springer, Heidelberg, 1987) p 529.

Properties of Neural Networks with Multi-State Neurons [1]

H. Rieger

Institut für Theoretische Physik
Universität zu Köln
FRG

1 Introduction

The original constituents of neural network models were neurons that can take on two values, corresponding to the two states "firing" and "not firing" and synaptic efficacies varying over a continuous range [1-3]. The introduction of neurons with graded response [4] does not drastically change the retrieval properties of the networks, but the description of the effects in a more quantitative manner is more difficult.

A convenient way to investigate neural networks with analog neurons is to define the dynamics by an iterated map or a coupled set of differential equations [4-6]. There are some similarities between the analog gain of these neurons and the temperature in networks with two-state neurons. Therefore one can improve the retrieval process in this deterministic network by varying the gain parameter instead of temperature in a probabilistic net. In the case of symmetric couplings the analytical investigation of the dynamics of analog neural networks makes use of the existence of a Lyapunov function (not a Hamiltonian) whose minima should describe the fixed-point properties of them.

A further generalization of models with two-state neurons is the Potts-glass model of neural networks [7], which is reminiscent of a Potts spin glass, and the clock neural network [8], which is reminiscent of a clock spin glass, but the relevance of these models to biological systems or technical applications is not clear. In addition phasor neural networks have attracted some attention [9], where neurons are considered to be unit-length 2-vectors. Another model incorporating P-state neurons is the three-state net [10,11] which is similar to the S-Ising spin glass [12]. By utilizing the third state — the zero state — of the neurons it is possible to provide this network with some new interesting features.

The aim of this paper is to set up a general framework in which simultanously the dynamics of analog and P-state neurons (to be defined in section 2) can be investigated [13]. With the help of the formalism, developped in section 3, the properties of three different models (concerning the synaptic efficacies) are analyzed: the fully asymmetric SK-model [14,15], the asymmetrically and extremely diluted Little-Hopfield model [16] and the fully

---

[1]Work performed within the research program of the Sonderforschungsbereich 341 Köln-Aachen-Jülich supported by the Deutsche Forschungsgemeinschaft.

connected Little-Hopfield model [1–3,17]. The most interesting result, presented in section 4, is the fact that in the second model the storage capacity generally increases with the introduction of multi-state neurons, whereas in the third model it decreases. These results are summarized and discussed in section 5.

2 Definitions

The model that I shall consider consists of N neurons interacting via $N \cdot (N-1)$ synaptic couplings J_{ij}. The neurons are updated in a random sequential manner. After updating the i-th neuron the output σ_i depends on the local field

$$h_i = \sum_{j(\neq i)} J_{ij}\sigma_j + b_i \, , \tag{1}$$

where b_i is an external field which is in some cases related to the postsynaptic threshold. The output depends on the input h_i in the deterministic case via an IO-function $g(h)$:

$$\sigma_i^{\text{new}} = g(h_i) \, . \tag{2}$$

In Fig. 1 a typical sigmoid-shaped IO-function is shown.

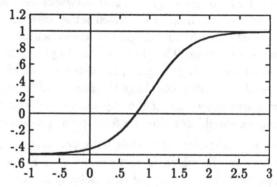

Figure 1:
Typical IO-function
of an analog neuron.
x-axis: input h,
y-axis: output σ.

One way to introduce stochastic noise into the dynamics is to define a transition probability according to

$$w(\sigma_i \rightarrow \sigma_i') = \frac{1}{\varepsilon\sqrt{2\pi}} \exp\left[-\frac{[\sigma_i' - g(h_i)]^2}{2\varepsilon^2}\right] \, , \tag{3}$$

where ε is the noise parameter. For $\varepsilon \rightarrow 0$ one gets the noiseless case because in this limit the r.h.s. of (3) is a delta-function:

$$w(\sigma_i \rightarrow \sigma_i') = \delta(\sigma_i' - g(h_i)) \, , \tag{4}$$

i.e. the probability for the i-th neuron to take on a state where the output σ_i is within the interval $[\sigma_i - d\sigma_i, \sigma_i + d\sigma_i]$ is one for $g(h_i)$ belonging to that interval and zero otherwise. The special form of the transition probabilities is not relevant for what follows, except in the case where one has to fulfil detailed balance, a point to which I shall return later. Only the input-output relation in the noiseless case is of essential importance. The appropriate tool to describe the dynamics of the system (which is probabilistic even in the noiseless case because of the random sequential updating) is a Master equation which will be discussed later.

To make contact with McCulloch-Pitts neurons one should observe that P-state neurons can be characterized in the noiseless case by a discontinuous IO-function as depicted in Fig. 2.

Figure 2:
IO-function for
the P-state
neuron (here is
$P = 9$).

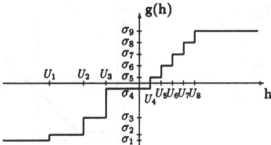

A P-state neuron can have only P $(< \infty)$ different output values $\sigma_1 < \sigma_2, \quad < \sigma_P$. In the absence of noise the neuron switches into the state σ_k if the local field is within the interval $[U_{k-1}, U_k[$ $(k = 1, \ldots, P)$:

$$w(\sigma_k) = \begin{cases} 1 & \text{for} \quad h \in [U_{k-1}, U_k[\\ 0 & \text{otherwise.} \end{cases} \tag{5}$$

The thresholds are ordered in magnitude: $-\infty = U_0 < U_1 < \ldots < U_{P-1} < U_P = +\infty$. Two-state neurons are now described by the choice of the parameters $P = 2$, $U_1 = 0$, $\sigma_1 = \tau$, $\sigma_2 = +1$, where conveniently $\tau = 0$ or $\tau = -1$. In the latter case the transition probabilities in the presence of noise are defined by

$$w(\sigma_k) = \frac{e^{\beta h \sigma_k}}{e^{\beta h} + e^{-\beta h}} \quad , \quad k = 1, 2. \tag{6}$$

The parameter β is called the inverse temperature $\beta = 1/T$ and $T \to 0$, $(\beta \to \infty)$, reproduces (5). This choice is motivated by the existence of a Hamiltonian $H = -\frac{1}{2}\sum_{i \neq j} J_{ij}\sigma_i\sigma_j$ if the synaptic couplings are symmetric ($J_{ij} = J_{ji}$). Since detailed balance holds for the choice (6) with respect to the equilibrium distribution $\exp(-\beta H)/Z$, one is able to calculate static properties like overlap functions and storage capacity. Therefore it is also advantageous to consider the Hamiltonian

$$H\{\underline{\sigma}\} = -\frac{1}{2}\sum_{i \neq j} J_{ij}\sigma_i\sigma_j + U \sum_{i=1}^{N} \sigma_i^2 \, , \tag{7}$$

which was used in spin glass theory of classical spins that are able to take on the values $-S, -S+1, \ldots, S-1, S$ [12]. In this case the transition probabilities are

$$w(\sigma_k) = \frac{e^{\beta(h\sigma_k - U\sigma_k^2)}}{\sum_{m=1}^{P} e^{\beta(h\sigma_m - U\sigma_m^2)}} \, , \tag{8}$$

and for vanishing noise one gets an IO-function as in Fig. 2 with the thresholds

$$U_k = U \cdot (\sigma_k + \sigma_{k+1}) \quad \text{for} \quad k = 1, \ldots, P-1 . \tag{9}$$

If the output-values can vary continuously over an interval $[\sigma_{\min}, \sigma_{\max}]$ one gets a piece-wise linear, continuous IO-function, that is $\sigma_{\min}$ for $h < 2U\sigma_{\min}$, increases linearly with a slope $1/2U$ for $h \in [\sigma_{\min}, \sigma_{\max}]$ and equals $\sigma_{\max}$ for $h > 2U\sigma_{\max}$ (see Fig. 3).

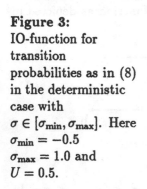

Figure 3:
IO-function for transition probabilities as in (8) in the deterministic case with $\sigma \in [\sigma_{\min}, \sigma_{\max}]$. Here $\sigma_{\min} = -0.5$ $\sigma_{\max} = 1.0$ and $U = 0.5$.

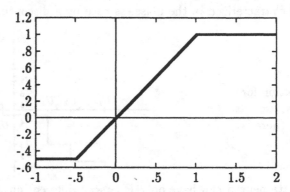

Thus one can construct a network that consists of neurons with a smooth, sigmoid shaped IO-function, where equilibrium properties can easily be calculated in the presence of symmetric couplings because the dynamics is described by a Hamiltonian.

Until now I have presented the type of neurons that will be considered in this paper. Still I have to define the synaptic couplings and for further reference I list the models that will be used later:

Model A: The fully asymmetric SK-model [14,15]:

Every synaptic coupling J_{ij} is an independently distributed random variable with

$$P(J_{ij}) = \sqrt{\frac{N}{2\pi J^2}} \exp\left(-\frac{J_{ij}^2 N}{2J^2}\right) . \tag{10}$$

Since there are no correlations between J_{ij} and J_{ji} the dynamics of this network can be solved exactly [14,15].

Model B: The asymmetrically and extremely diluted Little-Hopfield model [16].

$$J_{ij} = c_{ij} \cdot \frac{1}{K} \sum_{\nu=1}^{p} \xi_i^\nu \xi_j^\nu , \tag{11}$$

where $\underline{\xi}^\nu = (\xi_1^\nu, \ldots, \xi_N^\nu)$ are $p = \alpha K$ random patterns with

$$P(\xi_i^\nu) = \frac{1}{2}\delta(\xi_i^\nu - 1) + \frac{1}{2}\delta(\xi_i^\nu + 1) . \tag{12}$$

The variable c_{ij} is responsible for the dilution of the network

$$P(c_{ij}) = (1 - c) \cdot \delta(c_{ij}) + c \cdot \delta(c_{ij} - 1) , \tag{13}$$

the parameter K is the mean connectivity of the neurons so that the bond-concentration is $c = K/N$. Note that the dilution is fully asymmetric (that means c_{ij} and c_{ji} are uncorrelated). If $K \ll \log N$ and $K \to \infty$ in the limit $N \to \infty$ this model becomes completely equivalent to the fully asymmetric SK-model, the parameter J^2 has to be replaced by α and the external field b by $b = \underline{m} \cdot \underline{\xi}$, where

$$m^\nu = \frac{1}{N} \sum_{i=1}^N \overline{\xi_i^\nu \langle \sigma_i \rangle} \tag{14}$$

is the overlap of the networks state with the ν-th pattern.

Model C: The fully connected Little-Hopfield model near saturation [3].

Here the synatic couplings are defined to be

$$J_{ij} = \frac{1}{N} \sum_{\nu=1}^p \xi_i^\nu \xi_j^\nu \quad , \quad J_{ii} = 0 . \tag{15}$$

While the two previous models are fully asymmetric, and therefore solvable, this model has symmetric couplings (that means $J_{ij} = J_{ji}$) and is again solvable if one defines the transition probabilities as in (6) or (8). Again the number of patterns is $p = \alpha N$ and $\xi_i^\nu = \pm 1$ with equal probabilities. It should be mentioned that, if one has more than two neuron-states, it should be possible to store more complex patterns where ξ_i^ν can take on any possible neuron-output. In that case the probability distribution of the random variable ξ_i^ν is not confined to two values (see e.g. [10]).

One of the main purposes of this paper is to show that the introduction of multi-state neurons leads to complete different behaviour in models B and C.

3   Formalism

The starting point of my investigation of networks with multi-state neurons is the Master equation for the time evolution of the probabilities of the configuration $\underline{\sigma} = (\sigma_1, \dots, \sigma_N)$ of the network. As mentioned in the last section the output σ_i of the i-th neuron can vary over a certain parameter range, either discrete or continous. In the former case it is called a P-state neuron, in the latter an analog neuron.

I consider random sequential updating and the transition probabilities $w(\sigma_i \to \sigma_i')$ do not depend on the initial state σ_i of the i-th neuron: $w(\sigma_i \to \sigma_i') = w(\sigma_i')$. Therefore the Master equation can be written as

$$\frac{\partial}{\partial t} P\{\underline{\sigma}; t\} = -\sum_{i=1}^N \mathrm{Tr}_{\sigma_i'} \left[w(\sigma_i') P\{\underline{\sigma}; t\} - w(\sigma_i) P\{(\sigma_1, \dots, \sigma_i', \dots, \sigma_N); t\} \right] , \tag{16}$$

where the operator Tr_{σ_i} means the trace over all possible output-values σ_i'. In the case of P-state neurons $\mathrm{Tr}_{\sigma'}$ is a sum over $\sigma_1', \dots, \sigma_P'$ and in the case of analog neurons its $\mathrm{Tr}_{\sigma'} = \int d\sigma'$. Normalizing the transition probabilities by requiring

$$\mathbf{Tr}_{\sigma_i}[\mathbf{w}(\sigma_i)] = 1 . \tag{17}$$

and defining the Liouville-operator

$$\hat{\mathcal{L}}_i(t) = 1 - \mathbf{w}(\sigma_i)\mathbf{Tr}_{\sigma_i} , \tag{18}$$

one can formally write down the solution of the Master equation (16) as

$$\mathbf{P}\{\underline{\sigma}; t\} = \hat{\mathbf{T}} \exp\left(-\sum_{i=1}^{N} \int_{t_0}^{t} d\tau \hat{\mathcal{L}}_i(\tau)\right) \mathbf{P}\{\underline{\sigma}; t_0\} , \tag{19}$$

where $\hat{\mathbf{T}}$ is the time-ordering operator and $\mathbf{P}\{\underline{\sigma}; t_0\}$ is the probability distribution of the network configurations at the initial time t_0. The transition probabilities $\mathbf{w}(\sigma_i)$ occuring in the operator $\hat{\mathcal{L}}_i$ are generally nonlinear functions of the local fields $h_i = b_i + \sum_{j(\neq i)} J_{ij}\sigma_j$ and by this of the synaptic couplings J_{ij} which have to be averaged out. It is therefore convenient to write (19) in the following way [18], incorporating functional integrals over real fields $h_i(t)$, $\hat{h}_i(t)$, $\lambda_i(t)$ and $\hat{\lambda}_i(t)$:

$$\mathbf{P}\{\underline{\sigma}; t\} = \int \mathbf{D}(\underline{h}, \hat{\underline{h}}, \underline{\lambda}, \hat{\underline{\lambda}}) \exp\left(-\int_j i\hat{h}_j(h_j - b_j) - \int_j i\hat{\lambda}_j\lambda_j + \int_{i,j} J_{ij}i\hat{h}_i\lambda_j\right) \prod_{i=1}^{N} \hat{\mathbf{P}}_i(\sigma_i, t) , \tag{20}$$

where $\hat{\mathbf{P}}(\sigma_i, t)$ obeys the differential equation

$$\frac{\partial}{\partial t}\hat{\mathbf{P}}_i(\sigma_i, t) = -[\hat{\mathcal{L}}_i(t) - i\hat{\lambda}_i\sigma_i]\hat{\mathbf{P}}_i(\sigma_i, t) \quad ; \quad \hat{\mathbf{P}}_i(\sigma_i, t_0) = \mathbf{P}_i(\sigma_i, t_0) . \tag{21}$$

One has to assume that at the initial time the probabilty distribution factorizes according to $\mathbf{P}\{\underline{\sigma}, t_0\} = \prod_{i=1}^{N} \mathbf{P}_i(\sigma_i, t_0)$. In what follows the initial conditions are taken to be homogeneous and therefore I can drop the neuron index i from now on.

The advantage of this formulation is that now the quenched disorder, the couplings J_{ij}, occur linearly in the exponent and that the average over their distribution can be done easily in certain models, e.g. the models A–C defined in the last section. This procedure does not change in the presence of multi-state neurons. What is new is hidden behind the differential equation (21) for the pseudo distribution $\hat{\mathbf{P}}(\sigma, t)$. Defining its normalization by

$$\mathcal{Z}_0(t) = \mathbf{Tr}_{\sigma}[\hat{\mathbf{P}}(\sigma, t)] \tag{22}$$

and its moments by

$$\mathcal{M}_{\nu}(t) = \frac{1}{\mathcal{Z}_0(t)} \cdot \mathbf{Tr}_{\sigma}[\sigma^{\nu}\hat{\mathbf{P}}(\sigma, t)] , \tag{23}$$

one gets from (21) the following system of coupled differential equations:

$$\frac{d}{dt}\mathcal{Z}_0(t) = i\hat{\lambda}(t)\mathcal{Z}_0(t)\mathcal{M}_1(t) , \tag{24}$$

$$\frac{d}{dt}\mathcal{M}_\nu(t) = -\mathcal{M}_\nu(t) + \mathbf{W}_\nu[h(t)] + i\hat{\lambda}(t)[\mathcal{M}_{\nu+1}(t) - \mathcal{M}_1(t)\mathcal{M}_\nu(t)] , \quad \nu \geq 1 ,$$

where $\mathbf{W}_\nu(h)$ are the moments of the transition probabilities (which depend on the local field h by definition, see (3-6), (8)).

$$\mathbf{W}_\nu(h) = \mathrm{Tr}_\sigma[\sigma^\nu \mathbf{w}(\sigma)] . \tag{25}$$

The normalization, which is according to (24) given by

$$\mathcal{Z}_0(t) = \exp\left(\int_{t_0}^t d\tau\, i\hat{\lambda}(\tau)\, \mathcal{M}_1(\tau)\right) , \tag{26}$$

is identical with the generating functional for completely decoupled neurons ($J_{ij} = 0$) and correlation functions can be calculated via

$$\langle\sigma(t_1)\cdots\sigma(t_n)\rangle = \hat{\delta}(t_1)\cdots\hat{\delta}(t_n)\,\mathcal{Z}_0(t)|_{\hat{\lambda}=0} , \quad t > \max\{t_1,\ldots,t_n\} , \tag{27}$$

where I used the short-hand notation $\hat{\delta}(t)$ for the functional derivative $\delta/\delta i\hat{\lambda}(t)$. With the help of (24) one can prove that $\hat{\delta}(t)^\nu \mathcal{Z}_0|_{\hat{\lambda}=0} = \mathcal{M}_\nu(t)|_{\hat{\lambda}=0}$ and therefore, as one would expect:

$$\langle\sigma^\nu(t)\rangle = \langle\sigma^\nu(t_0)\rangle \cdot e^{-(t-t_0)} + \int_{t_0}^t d\tau\, e^{-(t-\tau)} \cdot \mathbf{W}_\nu(h(\tau)) . \tag{28}$$

Performing the average over the couplings J_{ij} leads in the limit $N \to \infty$ again to a single neuron dynamics whose generating functional is renormalized. This renormalization yields for the models A–C mentioned in the last section the following results:

Model A: Here the generating functional for correlation and response function is given by [15]

$$\mathcal{Z}[\hat{\lambda}, b] = \left\langle \mathcal{Z}_0[\hat{\lambda}, b + \phi(\tau)]\right\rangle_{\{\phi(\tau)\}} , \tag{29}$$

where $\mathcal{Z}[\hat{\lambda}, b] = \mathcal{Z}_0(t)$ is given in (26) and $\langle\cdots\rangle$ means an average over a Gaussian stochastic process $\phi(\tau)$ with correlations

$$\langle\phi(t)\rangle = 0 , \quad \langle\phi(t)\phi(t')\rangle = J^2 C(t, t') \tag{30}$$

and the averaged autocorrelation function $C(t, t')$ has to be determined self-consistently via

$$C(t, t') = \frac{1}{N}\sum_{i=1}^N \overline{\langle\sigma_i(t)\sigma_i(t')\rangle} = \hat{\delta}(t)\hat{\delta}(t')\,\mathcal{Z}[\hat{\lambda}, b]|_{\hat{\lambda}=0} . \tag{31}$$

Model B is the same as model A, but one has to replace J^2 by α and the external field b by $\underline{m}\cdot\underline{\xi}$, where the overlaps $m^\nu(t)$ (see (14)) have to be determined selfconsistently:

$$m^\nu(t) = \left\langle \xi^\nu\, \hat{\delta}(t)\,\mathcal{Z}_0[\hat{\lambda}, \underline{m}\cdot\underline{\xi} + \phi(\tau)]|_{\hat{\lambda}=0}\right\rangle_{\underline{\xi},\{\phi(\tau)\}} . \tag{32}$$

Model C: In this case the renormalization is more troublesome but formally the same as in the (± 1)–neuron case (see [17]). Since I shall only need the time-persistent order parameters in equilibrium to calculate for example the storage capacity, I restrict the formulas to statics. Suppose the transition probabilities are defined as in (8). Then the statics is determined by an effective Hamiltonian incorporating a static random field and a modification of the local quadratic term (see e.g. [19]):

$$H_z^{\text{eff}} = h_z \sigma + \tilde{U}\sigma^2 \tag{33}$$

with

$$h_z = \underline{m} \cdot \underline{\xi} + z\sqrt{\alpha r} , \tag{34}$$

where z is a Gaussian random variable with mean zero and variance one, m^ν is the overlap of a condensed pattern with the networks state, r is the mean square random overlap of the non-condensed patterns and $\mathcal{K}$ is the mean square of the output of the neurons, which is a measure of the mean activity of the network. Defining furthermore the susceptibility $\chi = \beta(\mathcal{K} - q)$ and the EA-order parameter q one has to solve the following self-consistency equations for the order parameters:

$$
\begin{aligned}
\underline{m} &= \langle\!\langle \underline{\xi}\, \mathbf{W}_1(h_z) \rangle\!\rangle_{\xi,z} & , && \tilde{U} &= U + \frac{\alpha}{2} \cdot \frac{\chi}{1-\chi} , \\[2mm]
\mathcal{K} &= \langle\!\langle \mathbf{W}_2(h_z) \rangle\!\rangle_{\xi,z} & , && \chi &= \langle\!\langle \tfrac{\partial}{\partial h_z} \mathbf{W}_1(h_z) \rangle\!\rangle_{\xi,z} , \\[2mm]
q &= \langle\!\langle \{\mathbf{W}_1(h_z)\}^2 \rangle\!\rangle_{\xi,z} & , && r &= \frac{q}{(1-\chi)^2} ,
\end{aligned}
\tag{35}
$$

where $\mathbf{W}_\nu(h_z)$ is defined as in (25) to be the ν-th moment of the transition probability $w(\sigma)$ (see (8)) within a field h_z and the renormalized threshold-parameter $\tilde{U}$:

$$\mathbf{W}_\nu(h_z) = \frac{\text{Tr}_\sigma[\sigma^\nu e^{\beta(h_z\sigma - \tilde{U}\sigma^2)}]}{\text{Tr}_{\tilde\sigma}[e^{\beta(h_z\tilde\sigma - \tilde{U}\tilde\sigma^2)}]} \tag{36}$$

and $\langle\!\langle \cdots \rangle\!\rangle_{\xi,z}$ means an average over ξ and z.

4   Results

Model A

The self-consistency equation for the autocorrelation function $C(t,t') = \langle \sigma(t)\sigma(t') \rangle$ and the activity $\mathcal{K}(t) = \langle \sigma^2(t) \rangle$ that have to be solved in model A are given by:

$$C(t_1, t_2) = \mathcal{K}(t_2)e^{-(t_1 - t_2)} + \int_{t_2}^{t_1} d\tau \int_{t_0}^{t_2} d\tau' e^{-(t_1 - \tau)} e^{-(t_2 - \tau')} F[C(\tau, \tau')] . \tag{37}$$

$$\frac{d}{dt}\mathcal{K}(t) = -\mathcal{K}(t) + \langle\!\langle\mathbf{W}_2(zJ\sqrt{\mathcal{K}(t)})\rangle\!\rangle_z \,, \tag{38}$$

where $\langle\!\langle\cdots\rangle\!\rangle_z$ means an average over the Gaussian random variable z with mean zero und variance one. The function $F[C(\tau,\tau')]$ is defined to be

$$F[C(\tau,\tau')] = \sum_{k=0}^{\infty} \frac{[C(\tau,\tau')]^k}{k!}\,\langle\!\langle\mathbf{W}_1^{(k)}(zJ\sqrt{\mathcal{K}(\tau)})\rangle\!\rangle_z \cdot \langle\!\langle\mathbf{W}_1^{(k)}(zJ\sqrt{\mathcal{K}(\tau')})\rangle\!\rangle_z \,. \tag{39}$$

When the system has reached its stationary state (i.e. $t_0 \to -\infty$) the activity takes on a time-independent value determined by

$$\mathcal{K} = \langle\!\langle\mathbf{W}_2(zJ\sqrt{\mathcal{K}})\rangle\!\rangle_z \,, \tag{40}$$

the correlation function depends only on time differences $C(t_1,t_2) = C(t_1 - t_2)$ and obeys the differential equation

$$\ddot{C}(t) = C(t) - F[C(t)] \,, \tag{41}$$

with $C(0) = \mathcal{K}$ and $\lim_{t\to\infty} = q < \mathcal{K}$, where the EA order parameter $q = \overline{\langle\sigma\rangle^2}$ has to be determined via $q = F[q]$. If $\mathbf{W}_1(h)$ is an odd function of h (i.e. $\mathbf{W}_1(-h) = -\mathbf{W}_1(h)$), then $q = 0$ is the stable solution. In case of an asymmetric function $\mathbf{W}_1(h)$ (for example in the case of $(0,1)$–neurons [20]) generally it is $q \neq 0$.

The autocorrelation function decays asymtotically as $C(t) \propto \exp(-t/\tau_\infty)$ with

$$\tau_\infty^{-1} = \sqrt{1 - F'[q]} \,. \tag{42}$$

In the case of analog neurons with transition probabilities as defined in (3) one gets

$$\mathbf{W}_1(h) = g(h) \,,$$
$$\mathbf{W}_2(h) = [g(h)]^2 + \varepsilon \tag{43}$$

and therefore

$$\mathcal{K} = \varepsilon + \langle\!\langle[g(zJ\sqrt{\mathcal{K}})]^2\rangle\!\rangle_z \,,$$
$$F[x] = \sum_{k=0}^{\infty} \frac{x^k}{k!} \cdot \langle\!\langle J^k g^{(k)}(zJ\sqrt{\mathcal{K}})\rangle\!\rangle^2 \,. \tag{44}$$

In what follows I assume that the IO-function $g(h)$ is odd (i.e. $g(-h) = -g(h)$) and approaches ± 1 in the limit $h \to \pm\infty$. If furthermore $g(h)$ is analytic in the vicinity of $h = 0$ one gets

$$\mathcal{K} = (g_1 J)^2 \cdot \mathcal{K} + O(\mathcal{K}^3) + \varepsilon \quad, \quad g_1 = g'(0) \,, \tag{45}$$

that means $\mathcal{K} > 0$ for $g_1 J > 1$ or $\varepsilon > 0$ and $\mathcal{K} = 0$ for $g_1 J \leq 1$ and $\varepsilon = 0$. Thus, in the noiseless case ($\varepsilon = 0$) all neurons are in the zero-output state if the gain g_1 is lower than J^{-1}. This has drastic consequences for the relaxation-time of the autocorrelation function in the vicinity of $g_1 = J^{-1}$: since $g(h)$ is odd, q vanishes and (42) yields

$$\tau_\infty^{-1} = \sqrt{1 - a_1} = \sqrt{1 - \langle\langle J g'(zJ\sqrt{\mathcal{K}})\rangle\rangle_z^2} \,. \tag{46}$$

One sees easily that a_1 approaches one for $g_1 J \searrow 1$ and therefore we observe critical slowing down

$$\lim_{g_1 J \searrow 1} \tau_\infty = \infty \,. \tag{47}$$

In Fig. 4 the activity $\mathcal{K}$ and the asymptotic relaxation time τ_∞ is shown for three different IO-functions ($\varepsilon = 0$):

$$\begin{aligned}
g_A(h) &= \operatorname{erf}(g_1 h \sqrt{\pi}/2) \,, \\
g_B(h) &= \tanh(g_1 h) \,, \\
g_C(h) &= \begin{cases}
-1 & \text{for} & h < -g_1^{-1} \\
g_1 h & \text{for} & h \in [-g_1^{-1}, g_1^{-1}] \\
+1 & \text{for} & h > +g_1^{-1}
\end{cases}
\end{aligned} \tag{48}$$

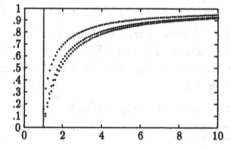 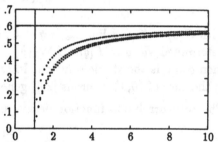

Figure 4: Activity $\mathcal{K}$ (left) and inverse asymptotic relaxation time τ_∞^{-1} (right) as a function of the gain $g_1 J$. The lower curve corresponds to $g_B(h)$, the central to $g_A(h)$, the upper to $g_C(h)$.

The error function is defined to be $\operatorname{erf}(x) = 2/\sqrt{\pi} \int_0^x dz\, e^{-z^2}$. All functions approach the sign–function in the limit $g_1 \to \infty$ (i.e. (± 1)–neurons) and then $\mathcal{K} = 1$ and $\tau_\infty = \sqrt{\pi/(\pi - 2)}$ (compare with ref. [15]).

In the case of P-state neurons with transition probabilities as defined in (8) (i.e. in the noiseless case) one gets

$$\mathcal{K} = \sum_{k=1}^p \sigma_k^2 \int_{\tilde{U}_{k-1}}^{\tilde{U}_k} \frac{dz}{\sqrt{2\pi\mathcal{K}}} e^{-z^2/2\mathcal{K}} \quad \text{with} \quad \tilde{U}_k = U/J \,. \tag{49}$$

Suppose there is a $k \in \{1, \ldots, p\}$ with

$$0 \in\,]U_{k-1}, U_k[\quad \text{and} \quad \sigma_k = 0 \,, \tag{50}$$

then $\mathcal{K} = 0$ is a stable solution of the self-consistency equation (49). Generally there is a second stable solution of (49), depending on the choice of the parameters $\{(\sigma_k, U_k) | k = 1, \ldots, p\}$. If (50) is not fulfilled, i.e. the value $h = 0$ is not exactly within the interval

$]U_{k-1}, U_k[$ where $g(h) = 0$, then the activity never vanishes. In neural networks with P-state neurons, where (50) holds (e.g. $P = 3$ and $\sigma_k = -1, 0 + 1$ and U_k as in (9), see [11]) the configuration in which all outputs are simultanously zero is a stable state, similar to the analog-neurons with $g_1 J < 1$. The difference is, that in the former case a stable state with a nonvanishing activity exists and has mostly a greater basin of attraction. Thus, in which state the network is in equilibrium depends on the initial conditions.

If $\mathbf{W}_1(h)$ is an odd function of h the asymptotic relaxation time is given by

$$\tau_\infty^{-2} = 1 - \frac{1}{2\pi\mathcal{K}} \left\{ \sum_{k=1}^{P-1} (\sigma_{k+1} - \sigma_k) \exp\left(\frac{-\tilde{U}_k^2}{2\mathcal{K}}\right) \right\} \tag{51}$$

and one can observe that because of

$$\frac{\partial}{\partial U_k} \tau_\infty^{-2}\bigg|_{U_1=\ldots=U_{P-1}=0} = \frac{2}{\pi} \frac{\partial}{\partial U_k} \mathcal{K}\bigg|_{U_1=\ldots=U_{P-1}=0} < 0 \tag{52}$$

the asymptotic relaxation time is increased by the introduction of new output-states of the neurons. But a critical slowing down as in networks of analog-neurons with a gain q_1 approaching J^{-1} cannot occur in the case of P-state neurons.

Model B

If the synaptic couplings are defined according to (11-13) one is interested in the existence of retrieval states with nonvanishing, macroscopic overlap (see (14)) of the networks state with one of the patterns. In this context one has to look for the critical storage capacity $\alpha = p/K$ and how it changes in dependence of the IO-function of the neurons.

As mentioned in the previous section (see (32)) the self-consistency equation for the overlap reads

$$\frac{d}{dt} m^\nu(t) = -m^\nu(t) + \langle\!\langle \xi^\nu \mathbf{W}_1(z\sqrt{\alpha\mathcal{K}(t)} + \underline{\xi} \cdot \underline{m}) \rangle\!\rangle_z \tag{53}$$

and for the calculation of the activity $\mathcal{K}(t)$ one has to replace in (38) the random field $zJ\sqrt{K}$ by $z\sqrt{\alpha\mathcal{K}(t)} + \underline{\xi} \cdot \underline{m}$. In retrieval states the vector $\underline{m}$ has only one non-vanishing component m. The stationary values of $\mathcal{K}$ and m in the case of analog neurons with odd IO-function $g(h)$ and vanishing noise ($\varepsilon = 0$) are now given by the two coupled equations

$$\mathcal{K} = \langle\!\langle g^2(z\sqrt{\alpha\mathcal{K}} + m) \rangle\!\rangle_z ,$$
$$\tag{54}$$
$$m = \langle\!\langle g(z\sqrt{\alpha\mathcal{K}} + m) \rangle\!\rangle_z .$$

One can recognize that the Gaussian noise with variance $\alpha\mathcal{K}$ has the tendency to diminish the overlap m. If the mean activity decreases, which is the case, if new neuron-outputs lower than one in absolute value are available, the overlap should grow. Simultaneously lowering the gain has a similar effect as increasing the temperature in the two-state net and therefore competes with the former effect. As we shall see the latter effect is dominated by the first one. Therefore the critical storage capacity should be greater than in the (± 1)-case. Let us make these statements more quantitative:

Solutions of (54) with $m \neq 0$ exist for $\alpha < \alpha_c(g_1)$. If $\alpha \nearrow \alpha < \alpha_c(g_1)$ the overlap vanishes continuously for IO-functions like g_{A-C} in (48). In the vicinity of $\alpha = \alpha_c(g_1)$ the equation for m reads

$$m = m \cdot \underbrace{\langle\!\langle g'(z\sqrt{\alpha K})\rangle\!\rangle_z}_{=:\rho} + O(m^3) . \tag{55}$$

The critical storage capacity is determined by $\rho = 1$. For $\rho \leq 1$ the stable solution of (54) is $m = 0$. Suppose that in the limit $g_1 \to \infty$ the IO-function approaches the sign-function, i.e. $g'(h) \to 2\delta(h)$, in such a way that for $g_1 \gg 1$ the function $g_1(h)$ looks like a smeared 2δ-function of the width $\epsilon = g_1^{-1}$:

$$g'(h) \approx \frac{2}{\epsilon\sqrt{2\pi}} e^{-h^2/2\epsilon^2} \quad \text{für} \quad \epsilon \ll 1 . \tag{56}$$

In this case it is

$$\rho = \sqrt{\frac{2}{\pi}(\alpha K + \epsilon^2)^{-1}} , \tag{57}$$

and therefore the critical storage capacity is given by

$$\alpha_c(\epsilon) = \left(\frac{2}{\pi} - \epsilon^2\right) \cdot K^{-1}(\epsilon) , \quad (\epsilon \ll 1) . \tag{58}$$

For $g_1 \to \infty$ one gets the well-known result for two-state neurons [16] $\alpha_c = 2/\pi$ (because $K(g_1 \to \infty) = 1$). From

$$\left.\frac{\partial \alpha}{\partial \epsilon}\right|_{\epsilon=0} = -\frac{2}{\pi} K'(0) > 0 \quad (\text{since} \quad K'(0) < 0) \tag{59}$$

one concludes that the storage capacity in the asymmetrically, extremely diluted Little-Hopfield-model can be increased by the introduction of neurons with IO-functions of the type with $g_1 < \infty$ but not too small. If the gain-parameter g_1 is decreased further, α_c reaches its maximum in $g_1 = 1$. If $g_1 < 1$ no retrieval states exist and if $g_1\sqrt{\alpha} < 1$ the network is in the zero-state. In Fig. 5 the overlap m as function of the gain g_1 for fixed values of α is plotted (with the IO-function $g_A(h)$ in (48)). One observes the occurence of a second order phase transition in dependence of g_1 at $g_1 = 1$.

Figure 5:
Overlap m as function of the neural gain g_1. From top to bottom it is $\alpha = 0.3$, 0.5, 0.6, 0.7, 0.8.

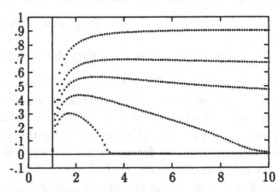

One can calculate the EA order parameter by solving

$$q = \langle\!\langle\,\langle\!\langle g(z\sqrt{\alpha(\mathcal{K}-q)} + y\sqrt{\alpha q})\rangle\!\rangle_y^2\,\rangle\!\rangle_z \tag{60}$$

and observes that for $m \to 0$ also $q \to 0$ and therefore (see (46)) critical slowing down occurs in the case $g_1 \searrow 1$ and $\alpha \nearrow 1$.

Concerning P-state neurons one can identify the same phenomena. The critical storage capacity is determined by

$$1 = \frac{1}{\sqrt{2\pi\alpha_c\mathcal{K}}} \sum_{k=1}^{p-1} (\sigma_{k-1} - \sigma_k) \exp\left[\frac{-\tilde{U}_k^2}{2\alpha_c\mathcal{K}}\right] \tag{61}$$

and therefore in the case of $\sigma_1 = -1$, $\sigma_P = +1$:

$$\frac{\partial}{\partial U_k}\alpha_c\bigg|_{U_1,\ldots,U_{p-1}=0} = -\frac{2}{\pi}\frac{\partial}{\partial U_k}\mathcal{K}\bigg|_{U_1,\ldots,U_{p-1}=0} > 0, \tag{62}$$

which means that again the storage capacity increases for $U_k \neq 0$ $(k = 1,\ldots,P-1)$. The special case of three-state neurons considered in [11] confirms exactly this general result.

Model C

Here one has to solve the self-consistency equations (36) for the order parameters. It should be mentioned that the solution (36) was obtained by neglecting the anomalous response (see [17]) and corresponds therefore to the replica-symmetric solution in statics. Nevertheless, replica-symmetry-breaking effects are expected to be as weak as in the original Little-Hopfield model with two-state neurons [3].

As the simplest example let us look for the case $T = 0$ (vanishing noise) of the three-state net $(\sigma = -1, 0, +1)$ that was extensively simulated on a computer in [10]. Then we have for $\underline{m} = (m, 0, 0, \ldots)$:

$$\begin{aligned}
m &= \frac{1}{2}\left\{\mathrm{erf}\left(\frac{m+\tilde{U}}{\sqrt{2\alpha r}}\right) + \mathrm{erf}\left(\frac{m-\tilde{U}}{\sqrt{2\alpha r}}\right)\right\}, \\
\mathcal{K} &= q = 1 - \frac{1}{2}\left\{\mathrm{erf}\left(\frac{m+\tilde{U}}{\sqrt{2\alpha r}}\right) - \mathrm{erf}\left(\frac{m-\tilde{U}}{\sqrt{2\alpha r}}\right)\right\}, \\
\chi &= \frac{1}{\sqrt{2\pi\alpha r}}\left\{\exp\left(\frac{-(m+\tilde{U})^2}{2\alpha r}\right) + \exp\left(\frac{-(m-\tilde{U})^2}{2\alpha r}\right)\right\}.
\end{aligned} \tag{63}$$

Solving these equations numerically and searching for the maximum value α_c of the storage capacity up to which retrieval states $(m \neq 0)$ exist, one gets the curve $\alpha_c(U)$ which is shown in Fig. 6. One can immediately see that at all values of the threshold U the critical storage capacity is lower than in the original two-state Little-Hopfield model, where one gets $\alpha_c = 0.138$. There is also no enhancement of the overlaps m for intermediate values of α (≈ 0.05) in contrast to results of simulations presented in [10]. But this can be a consequence of a slightly different definition of overlaps together with the storage of patterns incorporating neurons in the zero-state, with other words: the probability distribution (12) for the random variable ξ was changed in [10] into $\mathbf{P}(\xi) = p_+\delta(\xi-1) + p_0\delta(\xi) + p_-\delta(\xi+1)$.

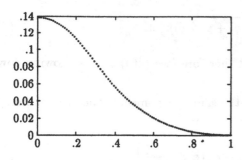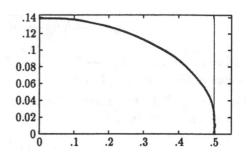

Figure 6: The critical storage capacity α_c in dependence of the threshold-parameter U, on the left side for three-state neurons, on the right side for the neurons with a piecewise linear IO-function as depicted in Fig. 3.

In the case of a piecewise linear IO-function as depicted in Fig. 3 the resulting critical storage capacity shows a very different behaviour: Up to $U \approx 0.03$ the storage capacity increases a little bit from $\alpha_c = 0.1383$ to $\alpha_c = 0.1386$, that means only 0.3%, and decreases then monotonically to zero at $U = 0.5$, where the slope of the IO-function is equally one. In contrast to the three-state net the critical storage capacity vanishes at the critical gain rapidly (with an algebraic singularity). This latter behaviour is similar to that of model B with analog neurons, although the storage capacity is not essentially enhanced.

5   Conclusions

I have studied the properties of several neural network models that use neurons characterized by an input-output relation that can be either continuous or discrete. In models, where the synaptic strengths are completely uncorrelated, one can easily solve the dynamics and observes different behaviour for nets with P-state neurons and analog neurons: only the latter show critical slowing down for the neural gain approaching a certain value and always settle down into the zero state for smaller gains. Furthermore an increase of the storage capacity can be observed in the asymmetrically and extremely diluted Little-Hopfield model. In the case of analog neurons it increases up to $\alpha = 1$ for the critical gain and in the case of P-state neurons it reaches its maximum at certain values of the threshold parameters. The reason for this effect is the ability of the network to reduce the random noise that has the tendency to destroy the overlap. This reduction is achieved by lowering the mean activity of the neurons.

This is in contrast to the behaviour of the storage capacity of the fully connected Little-Hopfield model, where it is lowered in all cases considered up to now. Also here the random noise is decreased, but because of the symmetry of the couplings more complicated effects are incorporated. For example the threshold value in the three-state net is renormalized in such a way that the mean energy is increased. This complicated interplay between avoiding frustrations and lowering the overlap in the high storage regime is still an open problem and requires a more detailed analysis. Recently it was shown that in the deterministic case the three state net can be rewritten as a two-state net with twice the number of neurons, and it was claimed that therefore always an increase in performance should be expected [21]. As mentioned above this holds for the asymmerically and extremely diluted Little-Hopfield model, but for the fully connected Little-Hopfield model this statement is wrong.

Finally it should be mentioned that it is still not clear whether the optimal storage of patterns, which leads to a storage capacity of $\alpha = 2$ for uncorrelated patterns in two-state nets [22], can be enhanced by the introduction of multi-state neurons. Perhaps one should search for optimal neurons for a given task.

References

1. Little, W. A.: Math. Biosci. **19** 101 (1974).

2. Hopfield, J. J.: Proc. Natl. Aca. Sci. USA. **79** 2554 (1982).

3. Amit, D. J., Gutfreund, H., Sompolinsky, H.: Phys. Rev. **A32** 1007 (1985), Phys. Rev. Lett. **55** 1530 (1985), Annals of Physics **173**, 30 (1987).

4. Hopfield, J. J.: Proc. Natl. Aca. Sci. USA. **81** 3088 (1984).

5. Marcus C.M., Westervelt, R.M.: Phys. Rev. **A39** 347 (1989), Phys. Rev. **A40** 501 (1989). Marcus C.M., Waugh F.R., Westervelt, R.M.: Phys. Rev. **A41** 3355 (1990). Waugh, F.R., Marcus C.M., Westervelt, R.M.: Phys. Rev. Lett. **64** 1986 (1990).

6. Sompolinsky, H., Crisanti, A., Sommers, H. J.: Phys. Rev. Lett. **61** 259 (1988).

7. Kanter, I.: Phys. Rev. **A37** 2739 (1988).

8. Cook, J.: J. Phys. **A22** 2057 (1989).

9. Noest, A. J.: Europhys. Lett. **6** 469 (1988).

10. Meunier, C., Hansel, D., Verga, A.: J. Stat. Phys. **55** 859 (1989).

11. Yedidia, J. S.: J. Phys. **A22** 2265 (1989).

12. Ghatak, S. K., Sherrington, D.: J. Phys. **C10** 3149 (1977).

13. Rieger, H.: Ph.D. Thesis, University of Cologne, unpublished (1989).

14. Crisanti, A., Sompolinsky, H.: Phys. Rev. **A37** 4865 (1988).

15. Rieger, H., Schreckenberg, M., Zittartz, J.: Z. Phys. **B74** 527 (1989).

16. Derrida, B., Gardner, E., Zippelius, A.: Europhys. Lett. **4** 167 (1987). Kree, R., Zippelius, A.: Phys. Rev. **A36** 4421 (1987).

17. Rieger, H., Schreckenberg, M., Zittartz, J.: Z. Phys. **B72** 523 (1988).

18. Sommers, H. J.: Phys. Rev. Lett. **58** 1268 (1987).

19. Sompolinsky, H., Zippelius, A.: Phys. Rev. **25** 6860 (1982). Horner, H.: Z. Phys. **B57** 29 (1985).

20. Tsodyks, M. V., Feigel'man, M. V.: Europhys. Lett. **6** 101 (1988). Horner, H., Bormann, D., Frick, M., Kinzelbach, H., Schmidt, A.: Z. Phys. **B76** 381 (1989).

21. Stark, J., Bressloff, P.: J. Phys. **A23** 1633 (1990).

22. Gardner, E.: J. Phys. **A21** 257 (1988).

Adaptive Recurrent Neural Networks And Dynamic Stability

B. Schürmann, J. Hollatz, U. Ramacher

Corporate Research & Development, Siemens AG

D-8000 München 83, West Germany

Abstract

We discuss properties of stable recurrent artificial neural systems. In particular, we focus on the "detailed balance" net which is a stable recurrent asymmetric net for Hebbian learning, and on its applications. Furthermore, we summarize results obtained for a hybrid approach containing Hebbian as well as back propagation terms.

1 Introduction

Two major classes of models for artificial neural networks (ANNs) have emerged. These are on the one hand stable recurrent ANNs (Hopfield (82, 84), Cohen & Grossberg (83), Kosko (87, 88), Schürmann (89)), and multi layer feedforward ANNs (Werbos (74), le Cun (85), Parker(85), Rumelhart et al. (86)) on the other. The natural application domain of the first class is pattern association and/or completion, whereas for the second it is pattern classification.

The mechanisms of storing and recalling patterns are qualitatively different for the two classes. For stable recurrent ANNs, patterns are stored as local minima of an energy, or more generally, of a Lyapunov function in activation state space. In the strongly nonlinear ("high gain") limit, i.e. for a steep signal (transfer, threshold, squashing) function, the dominant term of the Lyapunov function is quadratic in the activation state signals of the formal neurons (nodes, units, processing elements) with a coefficient matrix whose elements are the connection weights between the nodes. To guarantee stability of the net, the weight matrix $W = (w_{ij})$ has to satisfy certain constraints, the most widely used of which is the symmetry property $w_{ij} = w_{ji}$ (c.f., e.g., Hopfield (82)). Less restrictive is the "detailed balance" condition (Almeida (87), Schürmann (89)) $\mu_j w_{ij} = \mu_i w_{ji}$ with positive constant factors μ_i. A third possibility for maintaining stability is to obey an upper bound for the weights (Atiya (87)).

In many instances, the weight matrix is assumed to be not only symmetric, but also to be constant over time. Such ANNs possess no learning phase, and hence may be termed "nonadaptive". Their weight matrices are constructed explicity from the patterns to be stored, often in outer product form (Hopfield (82), Kosko (88)).

Under certain conditions, stability may be preserved also for *adaptive* ANNs, i.e. for systems with a learning phase (Kosko (87)). Sufficient conditions for global stability, the connection of the gradients of the Lyapunov function with the update rule for the neurons' activation states and for the weights, respectively, and the adiabatic ("fixed point") approximation are discussed in section 2. In contrast to Kosko, who assumed a symmetric weight matrix, Schürmann (89) studied the more general case of a net with a "detailed balance" weight matrix. The latter approach is reviewed in section 3 together with its "fixed point" approximation. By obtaining an approximate analytic expression within the fixed point approximation for the "detailed balance" weight matrix, Schürmann et al. (90) were able to exhibit the connection of adaptive stable models with non-adaptive Hopfield-type models. This is described in section 4. An important task within the framework of the "detailed balance" net is to determine the numerical values of its constants which are connected to the passive decay of the activation states, to forgetting during learning, and to the degree of asymmetry in the weight matrix. Fixing these constants by use of performance studies for a small size model problem is the subject of section 5. In section 6, the universality of the constants' numerical values is tested on a more realistic, larger size task.

Hebbian learning has several disadvantages, one of which is its inability to handle non-linearly separable learning tasks. This poses no problem for multi layer perceptrons with a backpropagation learning rule. For this reason and others it is desirable to consider a hybrid approach containing both Hebbian and backpropagation elements. The results of this approach are summarized in section 7. The paper closes with a brief summary.

2 Globally Stable Artificial Neural Networks

Stability implies dynamical control of the net which is a feature desired for both theoretical studies and hardware implementations. Pioneering theoretical work on the question of stability in the area of neural network research has been performed by Amari (77), Hopfield (82), Cohen & Grossberg (83), Marcus & Westerfelt (89), and Hirsch (89).

Before turning to specific models we discuss the issue of global stability for ANNs in a general framework. We consider an artificial neural system whose activation state vector at time t we denote by $\vec{z}(t)$, and whose weight matrix at time t we denote by $\vec{W}(t) = (w_{ij}(t))$. The weights will vary with time during the learning phase only. If learning is terminated at time t_f, then $\vec{W}(t) = \vec{W}(t_f) = constant$ for $t \geq t_f$.

2.1 Criteria for Dynamic Stability

In generalization of a theorem by Cohen & Grossberg (83), an ANN is asymptotically stable if there exists a cost function $L[\vec{z}(t), \vec{W}(t)]$, bounded from below, with the property

$$\frac{dL}{dt} < 0. \tag{1}$$

Such a cost function is termed "Lyapunov function". It controls the net dynamics in the sense that only cost decreasing state transitions are possible, with the final states being local minima of L. Writing (1) as

$$\frac{dL}{dt} = grad_z L \cdot \frac{d\vec{z}}{dt} + grad_W L \cdot \frac{d\vec{W}}{dt} < 0, \tag{2}$$

a sufficient condition for an ANN to be stable is

$$grad_z L = -\alpha^2 \frac{d\vec{z}}{dt}, \tag{3}$$

and

$$grad_W L = -\beta^2 \frac{d\vec{W}}{dt}. \tag{4}$$

The proportionality factors α^2 and β^2 may still be functions of $\vec{z}$ and $\vec{W}$.

If L is known, the activation state update rule (3) and the learning rule (4) completely determine the dynamics of the adaptive stable ANN. If on the other hand, the net dynamics is given by explicit expressions for $d\vec{z}/dt$ and $d\vec{W}/dt$, and if it is possible to construct from (3) and (4) a function L with the property (1), the equations for $d\vec{z}/dt$ and $d\vec{W}/dt$ describe the dynamics of a stable ANN. Failure of constructing a Lyapunov function is no proof, however, that the ANN at hand is unstable, since (1), (3), and (4) are only sufficient but not necessary conditions for stability. We are currently investigating the possibilities of constructing stable adaptive ANNs by use of methods of symbolic computation (Schürmann & Wang (90)).

In practice, it is useful to assume that the activation states change much faster with time than the weights. This assumption is central to our investigations and is discussed in the next subsection.

2.2 The "Adiabatic" or "Fixed Point" Approximation

Let us assume the dynamics of a stable ANN to be given by the differential equations

$$\frac{d\vec{z}}{dt} = \vec{F}(\vec{z}, \vec{W}) \tag{5}$$

and

$$\frac{d\vec{W}}{dt} = \vec{G}(\vec{z}, \vec{W}),$$ (6)

and the activation state vector $\vec{z}$ to change on a much faster time scale than the weight matrix $\vec{W}$. In analogy to similar situations encountered in the behavior of many particle systems in physics, we term such an assumption "adiabatic" approximation. The activation state vector $\vec{z}(t)$ will then approach its constant asymptotic limit $\vec{z}^F \equiv \vec{z}(t \to \infty)$ much faster than does the weight matrix $\vec{W}(t)$. In the extreme case, one may replace eq. (5) by

$$\frac{d\vec{z}}{dt} = 0 = F(\vec{z}^F, \vec{W}),$$ (7)

thereby turning the differential equation (5) into an algebraic aquation for the fixed point(s) $\vec{z}^F$. Accordingly, approximating (5) by (7) is termed "fixed point" approximation.

Eqs. (6) and (7) constitute an important concept for the learning phase which can be characterized as follows: (i) start with an initial (for instance randomly chosen) set of weights, (ii) determine $\vec{z}^F$ from equation (7) for this set of weights, (iii) insert the fixed point $\vec{z}^F$ into eq. (6), (iv) calculate the weights for the next time step, (v) insert this new set of weights in (7) to obtain a new fixed point, and so on, until a fixed point $\vec{W}^F$ (determined from eq. (6) by the condition $d\vec{W}/dt = 0$) is reached. When this happens, learning is terminated.

In describing this procedure, we did not refer explicitly to the patterns to be stored during the learning phase. Eq.(5) and its approximation (7) will depend on these, and the weight matrix $\vec{W}^F$ at the end of the learning phase will be a function of all of these patterns. It is beyond the scope of this paper to discuss the various forms of input patterns (e.g. continuous or discontinuous in time, deterministic or stochastic) as well as the various ways of pattern presentation (e.g. random draws, sweep through all patterns). For the sake of clarity, we assume for the moment that only a single static pattern is to be learnt. Generalization of the results obtained to several patterns is straightforward and will be discussed in section 4. In the next section , an example for an adaptive stable ANN is given, for the general case as well as for the fixed point approximation.

3   The "Detailed Balance" Net

It was shown by Kosko (87) that neural nets may be stable for certain Hebbian type learning rules if symmetric connection weights are used. As has been shown by Schürmann (89), stability still holds if symmetry is replaced by the more general "detailed balance" condition introduced below.

The essential features of the "detailed balance" net are briefly summarized. We consider a fully connected, unstructured artificial neural system with Q nodes, whose dynamics is described by a set of coupled differential equations for the nodes z_k,

$$\dot{z}_k = f_k(z_k)\left[-g_k(z_k, K_k) + \sum_{s=1}^{Q} w_{sk}S_s(z_s)\right], \quad k = 1,\ldots,Q, \tag{8}$$

and for the weights w_{sk},

$$\dot{w}_{sk} = -\lambda_{sk}w_{sl} + \varepsilon_{sk}S_s(z_s)S_k(z_k). \tag{9}$$

The functions $f_k > 0$ and g_k are continuous, $\vec{K}$ is a static external pattern vector, and the functions S_s are signal functions, assumed to be monotonically increasing, i.e. $dS_s(z_s)/dz_s \equiv S'_s > 0$. The forgetting constants $\lambda_{sk} > 0$ are symmetric, and the asymmetry constants $\varepsilon_{sk} > 0$ are defined by $\varepsilon_{sk} = \mu_{max(s,k)}/\mu_k$ with positive μ_s and μ_k. The weights obey the "detailed balance" condition

$$\mu_k w_{sk} = \mu_s w_{ks}. \tag{10}$$

The dynamical system described by eqs. (8)-(10) is globally stable. This can be shown by specifying a Lyapunov function $L(\{z_k\}, \{w_{sk}\})$ for the system,

$$L(\{z_k\}, \{w_{sk}\}) = \sum_{k=1}^{Q} \int_0^{z_k} \mu_k g_k(\xi_k, K_k) S'_k(\xi_k) d\xi_k$$

$$- 1/2\sum_{s=1}^{Q}\sum_{k=1}^{Q} \mu_k w_{sk} S_s(z_s) S_k(z_k) + 1/4 \sum_{s=1}^{Q}\sum_{k=1}^{Q} \lambda_{sk}\mu_k w_{sk}^2/\varepsilon_{sk}, \tag{11}$$

i.e. L is bounded from below and satisfies $dL/dt < 0$ (Schürmann (89)). We note that

$$\dot{z}_k = -\frac{f_k}{\mu_k}\frac{1}{S'_k}\frac{\partial L}{\partial z_k} = -\frac{f_k}{\mu_k}\frac{\partial L}{\partial S_k}, \tag{12}$$

and

$$\dot{w}_{sk} = -\frac{\varepsilon_{sk}}{\mu_k}\frac{\partial L}{\partial w_{sk}}. \tag{13}$$

According to 2.1, $\dot{z}_k = 0$ for $t \to \infty$ and $\dot{w}_{sk} = 0$ for $t \to \infty$. As in 2.2, the activation states z_k will be assumed to change much more rapidly with time than the weights w_{sk}. In the fixed point approximation, eqs. (8), (9) and (11) become

$$g_k(z_k^F, K_k) = \sum_{s=1}^{Q} w_{sk}S_s(z_s^F), \tag{14}$$

$$\dot{w}_{sk} = -\lambda_{sk}w_{sl} + \varepsilon_{sk}S_s(z_s^F)S_k(z_k^F), \tag{15}$$

and

$$L^F(\{w_{sk}\}) = \sum_{k=1}^{Q} \int_0^{z_k^F} \mu_k g_k(\xi_k, K_k) S'_k(\xi_k) d\xi_k$$

$$- \frac{1}{2}\sum_{s=1}^{Q}\sum_{k=1}^{Q} \mu_k w_{sk} S_s(z_s^F) S_k(z_k^F) + \frac{1}{4}\sum_{s=1}^{Q}\sum_{k=1}^{Q} \lambda_{sk} \mu_k w_{sk}^2/\varepsilon_{sk} \qquad (16)$$

with $\kappa_{sl} = \mu_{max(s,k)}/\mu_k$.

In the sections to follow, we specifically choose

$$f_k(z_k) = 1, \quad g(z_k, K_k) = C_k z_k - K_k \qquad (17)$$

so that eq. (8) becomes

$$\dot{z}_k = -C_k z_k + \sum_{s=1}^{Q} w_{sk} S_s(z_s) + K_k \qquad (18)$$

with the passive decay constants $C_k > 0$. The constants C_k^{-1} and λ_{sk}^{-1} determine the time scales of the changes of the activation states and of the weights, respectively. The adiabatic approximation discussed in 2.2 is equivalent to

$$C_k/\lambda_{sk} \gg 1. \qquad (19)$$

4 The "Detailed Balance" Learning Matrix in the Strongly Nonlinear Limit

In this section, we discuss an analytic result for the "detailed balance" connection strengths at the end of the learning phase. To begin with, we define the setting in which our investigations take place. During the learning phase, for a time interval $\delta t_\alpha^{(\nu)}$, in the α-th presentation cycle and for the ν-th pattern distribution, sample pattern vectors $\vec{K}^{(\nu)}$ from the pattern distribution P_ν are offered to the system, a new sample at each time step within $\delta t_\alpha^{(\nu)}$. Altogether, there are m distributions of different patterns and c cycles. The system learns the centroids of the distributions which are the "clean" patterns $K_{clean}^{(\nu)}$ (Schürmann (90)).

4.1 An Approximate Analytic Expression

We derive an analytic result for the connection strengths at the end of the learning phase. For transparency, we make the simplifying assumption that in each cycle α, $\alpha = 1, \ldots, c$, the presentation periods for samples from a specific distribution P_ν, $\nu = 1, \ldots, m$, are equal, i.e. $\delta t_\alpha^{(\nu)} \equiv \delta t^{(\nu)}$. In addition we assume the forgetting constants to be independent of node numbers, i.e. $\lambda_{sk} \equiv \lambda$. Moreover, we introduce the duration time $\tau = \sum_\nu \delta t^{(\nu)}$ for one learning cycle. Because of (19), at the end of sufficiently large time intervals $\delta t^{(\nu)}$, $sgn(z_s) \simeq sgn(K_s^{(\nu)clean}) \equiv \xi_s^{(\nu)}$ so that

$$\dot{w}_{sk}^{(\nu)} \simeq -\lambda w_{sk}^{(\nu)} + \varepsilon_{sk} \xi_s^{(\nu)} \xi_k^{(\nu)}, \tag{20}$$

with the solution

$$w_{sk}^{(\nu)} = w_{sk}^{(\nu-1)} e^{-\lambda t} + (\varepsilon_{sk}/\lambda)\xi_s^{(\nu)}\xi_k^{(\nu)}(1 - e^{-\lambda t}), \quad \nu = 1, \ldots, m. \tag{21}$$

By use of this recurrence relation, an analytic expression for the connection strengths at the end of the learning phase is obtained:

$$w_{sk}^{learn}(w_{sk}^{(0)}, \{\vec{K}_{clean}^{(\nu)}\}; \lambda; \varepsilon_{sk}; \{\delta t^{(\nu)}\}, \tau, c) =$$

$$w_{sk}^{(0)} e^{-\lambda \tau c} + \frac{1 - e^{-\lambda \tau c}}{1 - e^{-\lambda \tau}} \sum_{\nu=1}^{m} \varepsilon_{sk} \xi_s^{(\nu)} \xi_k^{(\nu)} \frac{e^{\lambda \delta t^{(\nu)}} - 1}{\lambda} e^{-\lambda \sum_{\mu=\nu}^{m} \delta t^{(\mu)}}, \tag{22}$$

To simplify the discussion, we assume $w_{sk}^{(0)} = 0$. It is seen that the pattern learnt last is weighted the most and the one learnt first is weighted the least. Hence the system tends to forget its past during the course of learning. If desired, forgetting can be remedied by offering the earlier patterns for a longer time period $\delta t^{(\nu)}$ than the later ones.

4.2 The Hopfield limit

In this subsection, the Hopfield limit of (22) is discussed. For the relation to the marginalist learning scheme of Mézard et al. (86), we refer to Schürmann et al (90).

The forgetting constant λ plays a crucial role in (22). If it becomes too large, the system will not learn at all. For small λ, the Hopfield limit (Hopfield (82)) is obtained for $c = 1$, $\varepsilon_{sk} = 1$, $\delta t^{(\nu)} = 1$, and $\lambda \to 0$.

The approximate analytic expression discussed in this section has led to valuable insight in the process of learning and in the relation of the "detailed balance" to other models. In applications to specific tasks, it can however, not replace precise numerical calculations. We turn to these in the next section.

5 Fixing the Constants of the "Detailed Balance" Net by Performance Studies

In the description of the dynamics of the "detailed balance" neural network by eqs. (8)-(18), we have to fix the passive decay constants C_k, the forgetting constants λ_{sk}, and the asymmetry factors ε_{sk}. We will address this task on a qualitative level first.

5.1 Qualitative Considerations

From the fixed point approximations to eqs. (15) and (18) one finds (Hollatz & Schürmann (90))

$$C_k \approx 1 + (1/|K_k|) \sum_s (\varepsilon_{sk}/\lambda_{sk}). \tag{23}$$

This equation relates the constants with one another as well as with the pattern size $|K_k|$. Eq. (23) tells us that $C_k > 1$. Furthermore, one observes that in order to make C_k independent of the net size, the pattern component size $|K_k|$ should scale with the net size, i.e. $|K_k| = \alpha Q$. In the following we choose $\alpha = 1$. To constrain the magnitude of the constants further, we make use of the adiabatic approximation in the form of eq. (19). A reasonable choice for the passive decay constants is $\lambda_{sk} \simeq 0.1\,C_k$. If λ_{sk} becomes too small, the "detailed balance" model approaches the Hopfield limit, c.f. 4.2. This is a feature we do not desire. To narrow down C_k further, we consider the discretized version of the differential equation (18). This is

$$z_k(t + \Delta t) = (1 - C_k \Delta t)z_k(t) + \Delta t \sum_s w_{sk}(t)S_s[z_s(t)] + K_k \Delta t. \tag{24}$$

In order for (24) to converge, one has to require $C_k < 1/\Delta t$.

Concerning the asymmetry factors ε_{sk}, they are useful for a fine-tuning of the net (Schürmann (89)). For moderate asymmetry of the weights, the assumption $\varepsilon_{sk} \in [0.5, 1.5]$ is reasonable. For large asymmetry, the usefulness of the "detailed balance" net is anyhow unclear (Hollatz & Schürmann (90)). We summarize the constraints obtained for the passive decay constants C_k, the forgetting constants λ_{sk}, and the asymmetry factors ε_{sk}:

$$1 < C_k < 1/\Delta t, \quad \lambda_{sk} \simeq 0.1 \cdot C_k, \quad 0.5 \le \varepsilon_{sk} \le 1.5. \tag{25}$$

5.2 Extraction of the Constants from Computer Simulations

To check and quantify the above qualitative considerations, we now consider a specific "detailed balance" net employed for a specific problem. The setting is the same as the one discussed in section 4. Learning a specific pattern during a cycle is automatically terminated if the overlap function (direction cosine, correlation function; c.f. Kinzel (85) for a corresponding definition for 2-state models)

$$a^\nu(t) = \frac{\sum_k S_k[K_k^{(\nu)clean}]S_k[z_k(t)]}{\sqrt{\sum_k S_k^2[K_k^{(\nu)clean}]}\sqrt{\sum_k S_k^2[z_k(t)]}} \simeq 1. \tag{26}$$

This determines the time intervals $\delta t_\alpha^{(\nu)}$. The number c of cycles is fixed when $\partial L/\partial w_{ij} = 0$. We divide the net into 2 layers with 7×7 formal neurons each, with connections only between layers and not within layers. The sample vectors are noisy pairs of letters, the

λ	C	$\ll a^{re} \gg$	$\ll a^{as} \gg$
1.0	2.0	0.90	0.66
2.0	4.0	0.91	0.62
0.1	10.0	0.09	0.07
0.4	4.0	0.39	0.04
0.8	4.0	0.91	0.71
0.8	10.0	0.94	0.80

Table 1: Average overlaps and numerical values for the model constants

number of distributions is 10. One letter is the input to the bottom layer, and the other one of the pair is the input to the top layer.

In the simulations we use the same signal function $S(z) = tanh(z/T)$ for all neurons, and a steepness parameter $T = 0.02$. As step size Δt in the numerical implementation of the differential equations we employ $\Delta t = 0.05$. The estimations (25) then yield $1 < C_k < 20$, $0.1 < \lambda_{sk} < 2$, $0.5 \leq \varepsilon_{sk} \leq 1.5$. We have made an extensive search for the optimal set of constants, guided by these confinements. For simplification we have assumed the constants to be independent of the neuron number, except for ε_{sk} where we have to take into account that $\varepsilon_{sk} = 1$ for $s \leq k$, but $\varepsilon_{sk} \neq 1$ for $s > k$ in general. Only a few typical results will be reported here. These are summarized in Table 1. The quantities $\ll a^{re} \gg$ and $\ll a^{as} \gg$ are the overlap functions (26) averaged over 10 randomly chosen initial conditions for eq. (18), and over the 10 letter pairs in the recall phase. The overlap $\ll a^{as} \gg$ denotes the average for the task retrieving clean pairs of letters, if only one noisy letter of each pair is presented to the net (association). The latter task is much harder than the former, and hence its sensitivity to different choices for the constants is larger. For the symmetry factors we found the best choice to be $\varepsilon_{sk} = 0.8$ for $s > k$ if only one letter at the bottom layer was presented, and the reciprocal value $\varepsilon_{sk} = 1.25$ if only the letter at the top layer was presented.

Table 1 displays remarkable results for the passive decay and forgetting constants C and λ, respectively. For similar values of C and λ, the performance of the net is satisfactory for retrieval but only marginal for association. For too small λ, the performance of the net on the average becomes disastrous. This is because, c.f. 4.2, $\lambda \to 0$ corresponds to the Hopfield net which does very well for a few patterns but breaks down if a critical number of patterns is exceeded (phase transition). The best performance for the "detailed balance" net we find for $C/\lambda \gg 1$ as anticipated in section 3. An optimal choice is $C = 10$, $\lambda = 0.8$, and $\varepsilon_{sk} = 0.8$ (1.25) for $s > k$. For the symmetric case we find $C = 10$ and $\lambda = 0.8$ again to be optimal. The performance of the symmetric net for retrieval is the same as in the asymmetric case but is a few percent worse for association, $\ll a^{as}_{sym} \gg = 0.75$ as compared to $\ll a^{as}_{asym} \gg = 0.80$.

So far, we have discussed averages over *all* patterns. The averages $< a^{re} >$ and $< a_{as} >$ over randomly chosen initial conditions in dependence of the *individual* patterns are displayed in Fig.1 for our optimal choice of the constants. For retrieval, we witness a rather smooth behavior of the average overlap as function of the pattern num-

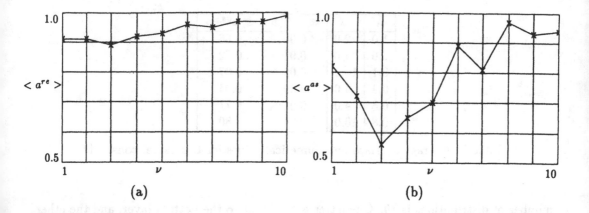

Figure 1: Average overlaps as function of the pattern number ν. (a) Retrieval, (b) Association. ($\lambda = 0.8$, $C = 10$, $\varepsilon_{sk} = 0.8$ for $s > k$)

ber ν. This is in contrast to the corresponding average overlap for association. Here, the forgetting effect is apparent.

Regarding the Hopfield limit $\lambda \rightarrow 0$, we have studied the average overlap as function of the number m of stored patterns (Hollatz & Schürmann (89)). For the hard task of pattern association, the Hopfield model did very well for the first 4 patterns but broke down after $m > 4$ patterns. The crucial difference between the "detailed balance" and the Hopfield model hence is the sudden collapse of the latter after exceeding a crucial number of patterns and the gradual decline of the former.

6   Work pieces Recognition

Having fixed the constants of the "detailed balance"net in the preceding section, we now test the numerical values obtained on a more realistic problem. This consists in the learning and recognition of six different work pieces. Offered to the system are pictures with a resolution of 256×256 pixels and 256 different grey levels. Each work piece presented in the recall phase is either to be reconstructed to a previously learnt one or merely to be classified.

Preprocessing plays a significant part in applications of neural nets. In our task, this consists in bringing translated and rotated work pieces in a standard position. This is achieved by calculating invariant moments up to 3th order of distribution functions representing the corresponding grey level pictures. For details, c.f. Hollatz (90).

To confine the number of nodes and weights to a reasonable size, the resolution of the picture is reduced to $32 \times 32 = 1024$ pixels. If we use a two-layer architecture like in section 5, with 512 nodes in each layer and connections only between layers, we have more than

5×10^5 connections. Offering during the learning phase one half of each picture to the bottom-up and the top-down layer, respectively, learning takes about 5 hours. The recall phase lasts about 30 seconds.

In many cases, *classification* of the object is sufficient. To save learning time, we use the "detailed balance" net as a classifier. The architecture is again that of a two-layer net with connections between the layers only. The lower layer has 1024 nodes, with a noisy and incomplete picture as input in the learning phase. The upper layer now functions as an output layer only, with 6 nodes serving as classifiers of the six classes of work pieces. The winning class is the one with the highest output value. Note that the net now possesses $2 \times 6 \times 1024 = 12288$ weights only and that it is highly asymmetric. As in the model case, the net functions again as an associative memory. After having been brought in standard position, pictures of varying quality of illumination are presented in the learning phase.

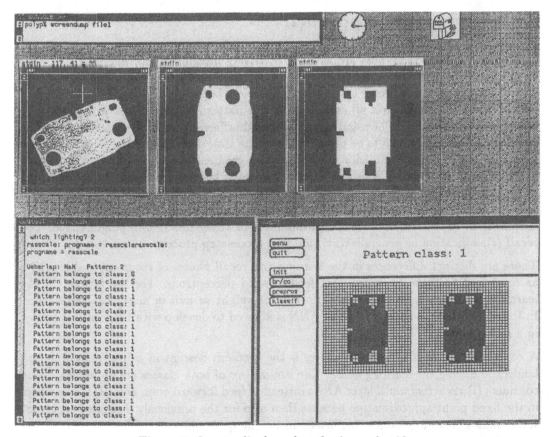

Figure 2: Screen display of work pieces classifier

In Fig. 2, a typical screen display is shown, with different processing steps. The window in the upper left is the original picture, the one in the upper middle is its binary form in standard position, and in the upper right window the original resolution is reduced to 32×32. This picture is presented to the net (lower right). In the lower left window, the protocol shows how the net first incorrectly classifies the picture as belonging to class 2, then as belonging to class 5, and finally as belonging to the (correct) class 1.

It is remarkable that the same constants which led to optimal performance in the model problem, also lead to optimal performance in this case. This indicates that the values of the constants are independent of the size of the net.

To close, we comment on a performance comparison of the "detailed balance"net with other models with regard to work pieces recognition. These models were: single layer perceptron (slp) with a perceptron learning rule, slp with a generalized delta learning rule, slp with a generalized delta rule applied to moments up to the 7th order, multi layer perceptron, and a conventional correlation calculation. All models yielded a recognition rate about 80%. For more details, we refer to Hollatz et al. (1990).

7 The Hybrid Approach

Multi layer feed forward nets are almost exclusively used in connection with problems where the training data consists of well identified input-output pairs. During the training phase, a function consisting of the squared differences between the actual and the desired output signals summed over all output units and patterns is minimized with respect to the connection weights. The negative gradient of this "error" function with respect to the weights at time t is defined to be proportional to the time derivative of the weights at time t and yields the generalized delta rule. When this gradient becomes smaller than a small prefixed error, learning is finished. Hence, for multi layer perceptrons, in the learning phase gradient descent in weight space is performed. The activation states are considered to be functions of the weights and are updated according to a prescribed recipe. Learning is equivalent to the search of the global minimum of a cost function in weight space. The recall (classification or generalization) phase is a one-step process.

There are distinct differences in the learning and recall phases of recurrent stable ANNs as compared to those for multi layer feed forward perceptrons. For the former, in the learning phase gradient descent is performed in weight *as well in activation state space*. In the recall phase, a recurrent stable ANN is allowed to develop with time until it settles in a stable state.

A common feature of both ANN classes is the gradient descent in a Lyapunov function landscape during the learning phase. The similarities of both classes are increased, if one considers (i) *recurrent* multi layer ANNs instead of feed forward ones, and (ii) if one works in the fixed point approximation because then also for the previously discussed recurrent stable nets gradient descent is performed in weight space only.

Recurrent multi layer neural systems have been investigated by Almeida (87) and Pineda (88) with emphasis on the generalized delta ("back propagation") learning rule. Almeida (87) discusses the issue of stability but employs a Cohen & Grossberg type Lyapunov function which is not consistent with the backpropagation learning rule. Pineda(88) ignores stability problems altogether.

The present approach differs from that of Almeida, Pineda, and others in that (i) a hybrid cost function containing the Hopfield type as well as error function type terms is considered, (ii) a learning rule is derived from this cost function which contains Hebbian as well as backpropagation terms, and (iii) the question of stability is addressed.

7.1 The Hybrid Cost Function

We devise an objective function L which combines elements of Hebbian learning with recurrent backpropagation.

Inspired by eq. (16) and by the form of the error function for backpropagation, we propose an objective function of the form

$$L^F(\{w_{sk}\}) = \sum_{k=1}^{Q} \int_0^{z_k^F} \mu_k g_k(\xi_k, K_k) S'_k(\xi_k) d\xi_k - \frac{1}{2} \sum_{s=1}^{Q} \sum_{k=1}^{Q} \mu_k w_{sk} S_s(z_s^F) S_k(z_k^F)$$

$$\left| \quad \frac{1}{4} \sum_{s=1}^{Q} \sum_{k=1}^{Q} \lambda_{sk} \mu_k w_{sk}^2 / \kappa_{sk} + \frac{1}{2} \sum_{k \in \Omega} \alpha_k [S(K_k) - S(\gamma_k^F)]^2 \right. \quad (27)$$

with $\kappa_{sk} = \mu_{max(s,k)}/\mu_k$.

The last term of (27) is the recurrent backpropagation term. Ω is the set of indices which denote input and output nodes. For hidden nodes, the last term is zero.

7.2 The Hybrid Learning Rule

Our aim now is, to derive a learning rule from (27) by calculating dL/dw_{ij} and assuming this to be proportional to $-\dot{w}_{ij}$. Learning is terminated ($\dot{w}_{ij} = 0$) when $dL/dw_{ij} = 0$.

By use of

$$\frac{dL}{dw_{ij}} = \sum_k \frac{\partial L}{\partial z_k^F} \frac{dz_k^F}{dw_{ij}} + \frac{\partial L}{\partial w_{ij}} \quad (28)$$

and defining in analogy to the "detailed balance" model (c.f. eq. 13)

$$-\frac{\kappa_{ij}}{\mu_j} \frac{dL}{dw_{ij}} \equiv \dot{w}_{ij} \quad (29)$$

we obtain in the general case, with $S_s^F \equiv S_s(z_s^F)$ and $S'^F_k \equiv S'_k(z_k^F)$,

$$\dot{w}_{ij} = -\frac{1}{2} \frac{\kappa_{ij}}{\mu_j} v_{ij}^F + S_i^F S_j^F \left[\kappa_{ij} - \frac{1}{2} \frac{\kappa_{ij}}{\mu_j} (\mu_j w_{ij} - \mu_i w_{ji}) \right]$$

$$- \frac{1}{2} \left[\lambda_{ij} w_{ij} + \lambda_{ji} (\mu_i/\mu_j)^2 w_{ji} \frac{dw_{ji}}{dw_{ij}} \right] + \frac{\kappa_{ij}}{\mu_j} y_j^F S_i^F, \quad (30)$$

and in the "detailed balance" limit $\mu_j w_{ij} = \mu_i w_{ji}$

$$\dot{w}_{ij} = -\lambda_{ij} w_{ij} + \varepsilon_{ij} S_i^F S_j^F + \frac{1}{2} \varepsilon_{ij} (\frac{1}{\mu_j} y_j^F S_i^F + \frac{1}{\mu_i} y_i^F S_j^F). \tag{31}$$

The activation states $z_k^F \in (-\infty, +\infty)$ are solutions of the fixed point equations

$$z_k^F = \frac{1}{C_k} \Big[\sum_s w_{sk} S_s^F + K_k \cdot \Phi_k \Big], \tag{32}$$

The error signals $y_r^F \in (-\infty, +\infty)$ and the quantities $v_{ij}^F \in (-\infty, +\infty)$ obey the corresponding fixed point equations

$$y_r^F = \frac{1}{C_r} S_r'^F \cdot (\sum_j w_{rj} y_j^F + J_r \cdot \Phi_r), \tag{33}$$

and

$$v_{ir}^F = \frac{1}{C_r} S_r'^F \sum_j \Big[w_{rj} v_{ij}^F + S_i S_j (\mu_r w_{jr} - \mu_j w_{rj}) \Big], \tag{34}$$

with

$$J_r = \alpha_r [S(K_r) - S_r^F], \quad \text{and} \quad \Phi_j = \begin{cases} 1 & \text{if } j \text{ is an output unit} \\ 0 & \text{otherwise} . \end{cases} \tag{35}$$

Eqs. (30)-(34) constitute the fundamental equations of the hybrid approach. Details of their derivation and the relevance of these equations for hardware realizations are discussed in Ramacher & Schürmann (90).

8   Summary

In this article we have discussed adaptive recurrent artificial neural systems with an emphasis on the question of stability. In particular, we have investigated the properties of a stable asymmetric net for Hebbian learning, the "detailed balance" net, including applications to model as well as to "real world" problems. Being aware of certain weakness of Hebbian learning, we have derived a learning rule containing Hebbian as well as back propagation type terms. Though the utility of such an hybrid approach has yet to be proven with regard to performance measures, the advantages for devising a neuro simulator or a neural chip are obvious.

References

[1] Almeida, L.B., Proc. of the IEEE 1st Int. Conf. on Neural Networks, San Diego, California, Vol. II (1987) 609-618.

[2] Amari, S.-I., Biol. Cybern. 26 (1977) 175-185.

[3] Atiya, A.F., in: Neural Information Processing Systems, D.Z. Anderson (Editor), American Institute of Physics (1988) 22-30.

[4] Cohen, M.A., Grossberg, S., IEEE Transactions on Systems, Man and Cybernetics, SMC-13 (1983) 815-826.

[5] Hirsch, M., Neural Networks 2 (1989) 331-349.

[6] Hollatz, J., Diploma Thesis, Techn. Univ. Munich (1990).

[7] Hollatz, J., Schürmann, B., Proc. of the IEEE Int. Conf. on Neural Networks, San Diego (1990), to appear.

[8] Hollatz, J., Knerr, S., Kuner, P., Leuthäusser, I., Schürmann, B., Proc. of the Conf. on Neural Networks for Computing, Snowbird, Utah (1990), to appear.

[9] Hopfield, J.J., Proc. Natl. Acad. Sci. USA 79 (1982) 2554-2558.

[10] Hopfield, J.J., Proc. Natl. Acad. Sci. USA 81 (1984) 3088-3092.

[11] Kinzel, W., Zeitschr. für Physik B 60 (1985) 671-680.

[12] Kosko, B., Applied Optics 26 (1987) 4947-4960.

[13] Kosko, B., Proc. of the IEEE 2nd Int. Conf. on Neural Networks, San Diego, Vol.I (1988) 141-152.

[14] le Cun, Y., Proc. of Cognitiva (1985) 599-604.

[15] Marcus, C.M., Westervelt, R.M., Phys. Rev. A39 (1989) 347-359.

[16] Mezard, M., Nadal, J.P., Toulouse, G., J. Physique 47 (1986) 1457-1462.

[17] Parker, D.B., Technical Report TR-47, Massachusetts Institute of Technology (1985).

[18] Pineda, F.J., Journal of Complexity 4 (1988) 216-245.

[19] Ramacher, U., Schürmann, B., in: VLSI Design of Neural Networks, U. Ramacher (Editor), Kluwer Academic Publ. (1990), to appear.

[20] Rumelhart, D.E., Hinton, G.E., Williams, R.J., in: Rumelhart, D.E., McClelland, J.L., Parallel Distributed Processing, Vol. I, MIT Press (1986) 318-362.

[21] Schürmann, B., Phys. Rev . A 40 (1989) 2681-2688.

[22] Schürmann, B., Proc. of the IJCNN Conf., Washington,D.C., Vol.I (1990) 459-462.

[23] Schürmann, B., Leuthäusser, I., Hollatz, J., in: Parallel Processing in Neural Systems and Computers, R. Eckmiller, G. Hartmann, G. Hauske (Editors), North-Holland (1990) 213-216.

[24] Schürmann, B., Wang, D., Proc. of the Int. Neural Network Conf., Paris (1990), to appear.

[25] Werbos, P.J., PhD thesis, Harvard University (1974).

Neuronal Oscillators: Experiments and Models

Carme Torras i Genís

Institut de Cibernètica (CSIC-UPC)

Diagonal 647, 08028-Barcelona. SPAIN.

1 Introduction

The Neural Networks field has traditionally focussed on the *spatial* aspects of both stimulation and neural activity. This is sufficient for tasks like pattern recognition, associative memory and optimization, as long as they involve only spatial patterns. However, in the study of other tasks, such as motor control, dynamical system modelling, and the recognition and association of *temporal* patterns, the dominant trend is clearly not sufficient.

The widespread use of logical neuron models is a consequence of the fact that meaning is attributed only to the instant-by-instant states, disregarding evolution. Moreover, these models consider neurons to be merely passive stimulus-transducers, while there is a large body of experimental evidence of neurons that fire spontaneously.

Even further, if neurons are logic transducers, then learning has to be confined to changes in synaptic efficacy. This presupposes that learning modifies not the autonomous functioning of neurons, but only the influence that some have upon others or that the stimuli exert on them, thus only implying structural changes in the network. However, there is evidence that long-lasting changes in the spontaneous firing rate of a neuron can be obtained through specific stimulation strategies.

In this paper, we focus on neurons that show spontaneous oscillatory activity, i.e. that fire autonomously at a given rate, which are called *pacemaker neurons*. In particular, we review existing work on these neurons both in the experimental and modelling domains. On the one hand, experimental data reveal many interesting features in the behaviour of some pacemaker neurons. On the other hand, computational requirements force a progressive simplification of the models as the behaviour to analyse gets more complex. The ideal goal is to retain as many interesting features as possible in carrying out the simplifications, so as to obtain the simplest model that permits analysing the desired behaviour.

2 Experimental Data on Pacemaker Neurons

Evidence of neurons that fire spontaneously with a given frequency has been collected from the nervous systems of several invertebrates (e.g. squid, sea snail, oyster, watersnake, sea lobster, crayfish, medicinal leech, limax, and land lobster). The following four subsections are devoted to describing the four features of their behaviour that we are interested in modelling.

2.1 Spontaneous Interspike Interval Fluctuations

Pacemaker neurons are not ideal oscillators. Instead, their spontaneous interspike intervals vary from cycle to cycle. The histograms of these intervals are usually unimodal and slightly positively skewed [Junge and Moore 1966; Holden and Ramadan 1980, 1981b; Kohn et al. 1981].

Concerning the quantitative aspects of this variability, average firing frequencies between 2 and 15Hz, with standard deviations between 0.02 and 0.1, have been reported for the tonic stretch receptor in the last thoracic segment of crayfish [Buno and Fuentes 1984], and for neurons of the abdominal ganglion of Aplysia [Junge and Moore 1966] and of the watersnake [Holden and Ramadan 1980].

2.2 Response to Occasional Perturbations

Let the *phase* ϕ of a pacemaker neuron be the fraction of the average spontaneous interspike interval elapsed since the last discharge. Occasional perturbations usually have only an ephemeral effect on the spontaneous firing frequency, but a permanent effect on the phase. This permanent effect is graphically represented by means of the *Phase Response Curve (PRC)*, which plots the delay (or advancement) of phase provoked by an input impulse, as a function of the phase of the pacemaker just before the arrival of the impulse.

Figure 1 shows the PRC's obtained experimentally for neurons in the abdominal ganglion of Aplysia and the tonic stretch receptor in the last thoracic segment of crayfish. The two PRC's for excitatory stimuli can be approximated by two straight regression lines in the form of a "V", while the ones for inhibitory stimuli can be roughly approximated by a single straight line of slope between 0 and 1. Several researchers have reported qualitatively similar PRC's for many other invertebrate pacemaker neurons [Beltz and Gelperin 1980; Pinsker and Ayers 1983].

The parameters characterizing the "V"-shape of the PRC vary usually as a function of the amplitude of the stimulation. Essentially, the inflexion point in the "V"-shape moves to the left and down as the stimulation amplitude increases [Pinsker 1977b].

2.3 Entrainment

Entrainment —or *phase-locking*— is the relationship between an oscillator and a periodic input where the sequence of phases of the oscillator at the times of arrival of the impulses is periodic. $(s : r)$-entrainment refers to the fact that s cycles of the stimulus correspond to r cycles of the response. *Synchronization* is a particular case, since it requires the coincidence of both null phases.

The most outstanding characteristics of entrainment in invertebrate pacemaker neurons are the following:

(a) The entrainment is usually synchronized, for excitatory inputs, and non-synchronized, for inhibitory ones [Ayers and Selverson 1979].

(b) The attainment of stable entrainment does not depend on the initial phase, but only on the ratio between the stimulation and firing frequencies [Pinsker 1977b].

(c) The bands of stimulation frequencies that cause entrainment widen as the amplitude of the stimulation increases [Pinsker 1977b; Pinsker and Ayers 1983].

(d) Simple entrainment ratios (1:1, 1:2, 2:1, 1:3, etc.) are predominantly observed [Holden and Ramadan 1981a].

(e) Some authors maintain that entrainment always arises [Kohn et al. 1981], while others claim to have observed aperiodic responses [Holden and Ramadan 1981a].

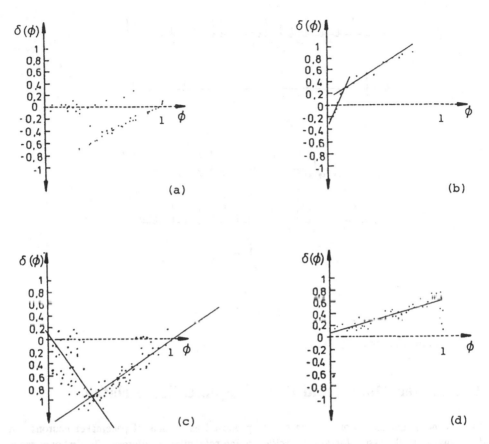

Figure 1. Phase response curves: (a) and (c) for excitatory stimuli; (b) and (d) for inhibitory stimuli; (a) and (b) for neurons of the abdominal ganglion of Aplysia [from Pinsker 1977a; Pinsker and Ayers 1983]; (c) and (d) for the stretch receptor of crayfish [from Buño and Fuentes, Centro Especial 'Ramón y Cajal', Madrid; Kohn et al. 1981]. $\delta(\phi)$ is the phase-delay provoked by an input impulse, as a function of the phase ϕ of the pacemaker just before the arrival of that impulse. Thus, negative values in the ordinate axis mean phase-advancement.

2.4 Plasticity in the Firing Rate

Neuronal plasticity refers to the consistent changes of certain properties of the neuron as a function of the stimulation and of its own previous activity. Von Baumgarten (1970) reports that pacemaker neurons of the abdominal ganglion of Aplysia, submitted to excitatory stimulation at frequencies slightly higher or lower than the spontaneous one, for a time interval ranging from a few minutes to an hour, continue firing at the imposed frequency and phase, even after the stimulation has stopped. This behaviour may last only a few cycles (Fig. 2) or go on for as long as 20min. Similar results have been obtained with bursting neurons of the same ganglion [von Baumgarten 1970; Parnas et al. 1974].

Further experimental support for the consistent modification of a pacemaker's firing rate, through external stimulation, comes from the use of conditioning paradigms [Kristan 1971; Kandel 1976] and phase-contingent stimulation [Pinsker and Kandel 1977].

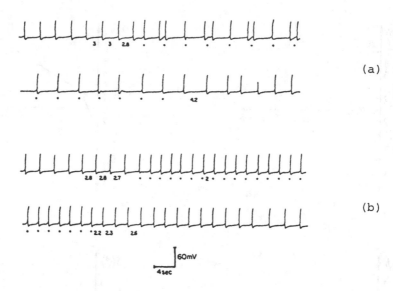

Figure 2. Plasticity in the firing pattern: (a) lenghtening, and (b) shortening of the inter-spike interval, resulting from the application of rhythmic excitatory stimuli to neurons in the abdominal ganglion of Aplysia [from von Baumgarten 1970]. Dots indicate the instants of presentation of input impulses.

3 Pacemaker Models and their Simplification Relations

The different nonlinear oscillators that have been proposed as models of pacemaker neurons can be joined into two general classes: *limit-cycle oscillators* and *relaxation oscillators*. The first ones require a minimum of two state-variables and are characterized by the existence of a closed convergent trajectory in their phase-plane. The second ones presuppose a variable that increases (or decreases) monotonically and that, when it surpasses a threshold, is automatically reset to its initial value.

Many pacemaker models come from the modification of some parameter conditions in the models previously established for non-pacemaker neurons; in the same way as many models of the relaxation oscillator class have, in turn, been obtained by simplification of models of the limit-cycle type. The triplet Hodgkin-Huxley (1952) / BVP (1961) / Perkel et al. (1964) constitutes a paradigmatic instance of this simplifying process.

The *Hodgkin-Huxley model* has become a true landmark in the evolution of neural modelling, since it reproduces with such exactness the flow of sodium and potassium ions underlying the action potential, that has been used as reference in the validation of many later models. The great value of this model, which incorporates four state-variables, resides in having established a correspondence between experimental data about the same phenomenon, obtained at two very different levels (evolution of the action potential and changes to the permeability to sodium and potassium).

In the two subsections that follow, we will describe one model of the limit-cycle class —the BVP— and one model of the relaxation class —that initially proposed by Perkel et al. (1964) and later modified by Torras (1985a). The reader interested in pacemaker neuron models can find additional information in Pavlidis (1973), Winfree (1980) and Pinsker and Ayers (1983).

3.1 Limit-cycle Oscillators

The *BVP (Bonhoeffer-Van der Pol) model*, proposed by Fitzhugh (1961), results of combining Bonho-effer's model of the excitability of iron wire in nitric acid with van der Pol's model of the heartbeat. Its analytic expression is:

$$\dot{x} = c(y + x - \frac{x^3}{3} + z); \quad \dot{y} = -\frac{(x - a + by)}{c};$$ (1)

where the variables x and y are approximately linear functions of two different pairs of state-variables of the Hodgkin-Huxley model, z is the input, and the parameters $a, b, c \in \Re$ are subject to the following constraints:

$$1 - \frac{2b}{3} < a < 1; \quad 0 < b < 1; \quad b < c^2.$$ (2)

These constraints ensure the existence of a unique singular point (where the lines $\dot{x} = 0$ and $\dot{y} = 0$ cross) in the phase-plane (Fig. 3(a)). The system remains in this point until an input impulse moves it along the dotted line of the previous figure. The state then follows the trajectory that begins in the corresponding point of that line, to return eventually to the resting point. The action potential occurs only for long translations along the dotted line.

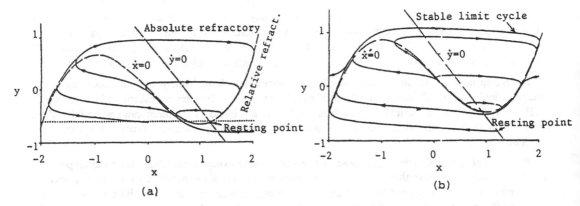

Figure 3. Phase-plane representation of the BVP model: (a) in the absence of stimulation [from Harmon and Lewis 1966]; (b) when submitted to a constant depolarizing current.

A sufficiently-intense constant depolarizing current makes the singular point unstable and surrounded by a stable limit-cycle (Fig. 3(b)). It is then that the model becomes an oscillator.

This kind of models has not been used to study the behaviour of pacemakers submitted to periodic stimulation, because of the high computational cost the numeric integration of their differential equations requires and the difficulty of obtaining a clear vision of their dynamics under these stimulation conditions. Then, simpler models have been used, which we will next describe.

3.2 Relaxation Oscillators

In the Hodgkin-Huxley model and its many derivatives, the state-variables vary slowly, except at the time of firing, when the variation becomes extremely quick, without losing continuity. This has favoured the development of a large number of models which substitute a discontinuous jump for this quick transition. All these models adhere to the same interpretation of neuronal functionig: The membrane potential increases as a result of the entrance of sodium ions into the cell body, while the threshold for a sudden increment of the permeability to sodium remains constant or decreases gradually from the time of firing; as soon as the values of the two coincide, an action potential is generated and both are discontinuously reset to widely separate values, from which they resume their approximation towards equality. Depending on whether the increment in membrane potential is spontaneously driven or requires external stimulation to surpass the threshold, the modelled neuron will or will not be a pacemaker.

The general expression of the models that incorporate a discontinuity as described above, called *integrate-and-fire models*, is:

$$\begin{cases} \frac{dP}{dt} = F_1(P, x(t)) \text{ and } \frac{dH}{dt} = F_2(H), & \text{while } P < H \\ P = P_0 \text{ and } H = H_0, & \text{when } P = H \end{cases} \tag{3}$$

where P is the membrane potential, H is the threshold, and $x(t)$ is the input.

The simplest instance of non-oscillatory model of this kind —the *linear integrator*— is that in which $F_1 = x(t)$ and $F_2 = 0$ [Sokolove 1972; Knight 1972]. Two slightly more sophisticated instances are the *leaky-integrator* [Rescigno et al. 1970; Knight 1972; Stein et al. 1972] and the *integrator with variable time-constant* [Fohlmeister et al. 1974, 1977]. In both cases, $F_1 = -\gamma P + x(t)$, with γ constant in the former and a function of time in the latter. The frequency response of the above models to different intensities of stimulation has been widely studied and characterized in the mentioned papers, as has also the one obtained with discrete versions of those and other related models [Nagumo and Sato 1972; Sato 1972; Yoshizawa 1982; Yoshizawa et al. 1982], to which we will come back in Section 4.

Focussing now on the intrinsically oscillatory models that have been proposed according to the interpretation of neuronal functioning given above, let us say that they are necessarily of the relaxation oscillator type and differ basically in the functional form attributed to the membrane potential and the threshold evolution with respect to time. Linear, sinusoidal and exponential functions are the most commonly used.

Perkel et al.'s model (1964), proposed in relation to some experimental data about the response of the stretch receptor of crayfish and the abdominal ganglion of Aplysia to periodic stimulation, postulates exponential evolutions in opposite directions for the spontaneous membrane potential and the threshold (Fig. 4). The postsynaptic potentials are assumed to be punctate and are instantaneously added to the spontaneous potential, to afterwards drop exponentially to zero. When there is no stimulation, the model behaves as a leaky integrator with exponential threshold, subjected to a constant input.

Thus, in terms of equation (3), the model is expressed as follows:

$$\begin{cases} \frac{dPb}{dt} = -\tau_b(Pb - Pb_1) \text{ and } \frac{dPs}{t} = -\tau_s Ps + x \text{ and } \frac{dH}{dt} = -\tau(H - H_1), & \text{while } (Pb + Ps) < H \\ Pb = Pb_0 \text{ and } Ps = 0 \text{ and } H = H_0, & \text{when } (Pb + Ps) \geq H \end{cases} \tag{4}$$

where Pb is the spontaneous membrane potential, Ps is the postsynaptic potential, Pb_0 and H_0 are the values of the membrane potential and the threshold after firing, Pb_1 and H_1 are the asymptotic limits of the spontaneous potential and the threshold, respectively, and the τ's are time-constants.

Note that this model supposes a further simplification of Hodgkin-Huxley model beyond that of the BVP model, since now the phase characterizes the system uniquely, so that any perturbation is translated into a phase-shift, as in all relaxation oscillators.

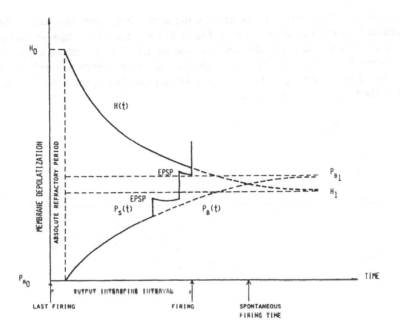

Figure 4. Dynamics of Perkel et al.'s pacemaker neuron model.

We have extended the above model to account for the experimental data described in Section 2. In particular, we have introduced stochasticity and plasticity into the model by:

1. Considering Pb_1 a stochastic process, of constant value in the interspike interval and of Gaussian probability density —with mean μ_{Pb_1} and standard deviation σ_{Pb_1}— at the time of firing.

2. Considering the parameter μ_{Pb_1} as a variable, subject to the following learning rule that accounts for the modification of the firing rate according to the stimulation received:

When $(Pb + Ps) \geq H$:

$$\mu_{Pb_1} = \begin{cases} \min \ (\mu_{Pb_1} + c, \overline{\mu_{Pb_1}}), & \text{if } Ps \geq P^* \\ \mu_{Pb_1}, & \text{otherwise} \end{cases}$$

When $x > 0$:

$$\mu_{Pb_1} = \begin{cases} \max \ (\mu_{Pb_1} - c, \underline{\mu_{Pb_1}}), & \text{if } T \leq T^* \\ \mu_{Pb_1}, & \text{otherwise} \end{cases} \qquad (5)$$

where $\overline{\mu_{Pb_1}}$ and $\underline{\mu_{Pb_1}}$ are the upper and lower bounds, respectively, for μ_{Pb_1}, c is the learning rate, P^* is the amount of postsynaptic potential required for accelerative learning to take place, T is the time elapsed since the last discharge, and T^* is the amount of time after firing open to decelerative learning.

The experimental justification of these two modifications can be found in Torras (1985a & b). Essentially, experimental studies indicate that changes in the slope of the membrane potential underlie both stochasticity and plasticity in the firing rate.

By using this model, we were able to replicate the experimental data described in Section 2. The interested reader is referred again to Torras (1985a & b) for details, since here we will only highlight the concordance between the PRC's obtained with the model (Fig. 5) and the experimental ones, as well as the finding that the postulated learning rule favours the emergence of simple entrainment ratios to the detriment of the more complex ones (Fig. 6), which provides an explanation for the experimental evidence described in point (d) of Section 2.3.

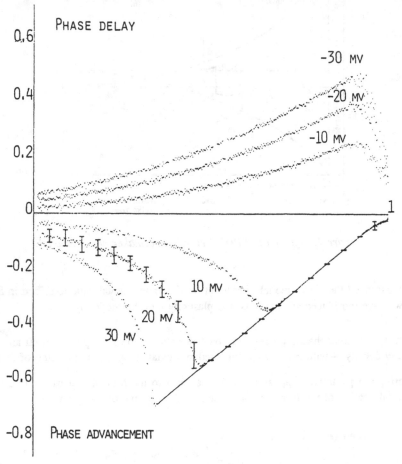

Figure 5. PRC's for three amplitudes of excitatory stimulation (10 mv, 20mv and 30mv) and three amplitudes of inhibitory stimulation (-10mv, -20mv and -30mv).

4 A Further Simplification: Modelling the Response to Impulses

Even the simplest models described in the preceding section result still too complex when one is interested in characterizing the response of a pacemaker to all possible frequencies of stimulation. In this case, namely when one is not interested in the internal functioning of pacemaker neurons, but only in their response to rhythmic stimulation, more simplified models that characterize neurons by their PRC's have been used.

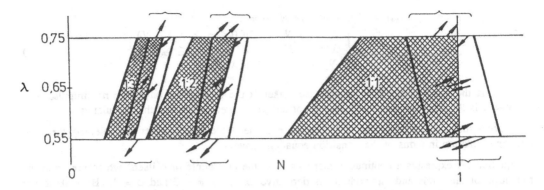

Figure 6. Vectors that characterize learning in the different regions of the (N, λ)-parameter space where it takes place (between brackets), in relation to the entrainment areas (shaded).

In particular, the PRC of the pacemaker neurons dealt with in this paper has been approximated through a piecewise linear function [Segundo 1979; Segundo and Kohn 1981; Torras 1986], as follows:

$$\delta_\lambda(\phi) = \begin{cases} -((1 - \lambda)/\lambda)\phi, & \text{if } 0 \leq \phi < \lambda \\ -1 + \phi, & \text{if } \lambda \leq \phi < 1 \end{cases} \tag{6}$$

where, as before, ϕ is the phase of the pacemaker at the time of arrival of an input impulse, $\delta_\lambda(\phi)$ is the phase delay produced by that input impulse, and λ is the minimum phase for which the arrival of an input impulse triggers the immediate discharge of the pacemaker (see Fig. 7). Thus, by leaving λ as a parameter, the PRC just described can be thought to model the response of the pacemaker to an input impulse of any amplitude.

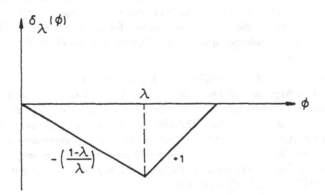

Figure 7. Piecewise linear "V"-shaped PRC.

Assuming that: (a) the phase delay produced by an input impulse is independent of the number of input impulses applied within an interspike interval, and (b) the arrival of an input impulse modifies the duration of the present interval, but does not affect subsequent intervals, the following transition equation has been derived:

$$\phi_{n+1} = \ < \ \phi_n - \delta(\phi_n) + N \ >_1 \ =$$
$$= \begin{cases} \lambda^{-1}\phi_n + N, & \text{if } 0 \leq \phi_n < \lambda(1-N) \\ \lambda^{-1}\phi_n + N - 1, & \text{if } \lambda(1-N) \leq \phi_n < \lambda \\ N, & \text{if } \lambda \leq \phi_n < 1 \end{cases} \qquad (7)$$

where ϕ_n is the phase of the postsynaptic pacemaker at the time of arrival of the nth impulse, and $0 \leq N < 1$ is the ratio between the frequency of the pacemaker and that of the stimulation.

Figure 8 shows two instances of (2:1)-entrainment, one stable and the other unstable, together with their representation in terms of the transition equation above.

Equation (7) expresses a continuous mapping from the unit circle onto itself, which preserves the orientation of the circle and has continuous derivative except at $\phi = 0$ and $\phi = \lambda$. By making use of the properties of this kind of mappings [Coddington and Levinson 1955; Bernussou 1977], it can be concluded that *stable entrainment arises always, but for a set of measure zero of values of N* [Torras 1986]. This is concordant with the experimental evidence mentioned in point (e) of Section 2.3. Furthermore, *the entrainment ratio* $\rho_{g_\lambda}(N)$ *is an extended Cantor function*[1] *of N, whose steepness increases as the value of λ decreases* (see Fig. 9). Finally, there exists a recursive procedure to determine the input/output entrainment patterns that emerge.

Recently, a curious relationship between these entrainment results and those obtained by Nagumo and Sato (1972) has been discovered. Nagumo and Sato modelled the behavior of a *non-pacemaker* neuron under a periodic input of *constant frequency* and *variable amplitude*, and found that stable entrainment arises always, but for a set of measure zero of input *amplitudes*. Furthermore, the entrainment ratio was shown to be an extended Cantor function of those *amplitudes*.

Although parallelism in the phrasing of these results and those above could suggest an equivalence between the stable entrainment behaviours of the two models, this is not the case, and it has been proven that the true equivalence is between stable entrainment for Nagumo and Sato's model and unstable entrainment for our model. To prove this, a *transformation* has been defined that relates the transition equations characterizing both models [Torras 1987]. Thus, through the results of Nagumo and Sato, a characterization of the unstable entrainment for our model has also been attained. On the other hand, there is no unstable entrainment for Nagumo and Sato's model corresponding to stable entrainment in our model.

Three neurophysiological facts underlie these mathematical results. First, non-pacemaker neurons (the ones modelled by Nagumo and Sato) submitted to a pulsed stimulus can only fire synchronously with input impulses, while pacemaker neurons (the ones we have modelled) can also fire in the interval between consecutive input impulses. Therefore, entrainment of non-pacemaker neurons is always synchronous, while pacemakers can give rise to the more general phenomenon of *phase-locking*, which encompasses the possibility of non-synchronous entrainment as well. Second, for pacemakers that are reset after firing, synchronous entrainment is stable. Third, non-synchronous entrainment in pacemakers characterized by a "V"-shaped PRC is unstable.

---

[1]Because of its resemblance to the Cantor function, a detailed description of which can be found in Gelbaum and Olmsted (1964), we say that $f : [a,b] \rightarrow \Re$ is an extended Cantor function if it satisfies the following conditions:

1. f is continuous and monotonically increasing.

2. $f(a) = a$ and $lim_{x \rightarrow b} f(x) = b$.

3. $f'(x) = 0, \forall x \in [a,b] \setminus C$, where C denotes a set of measure zero.

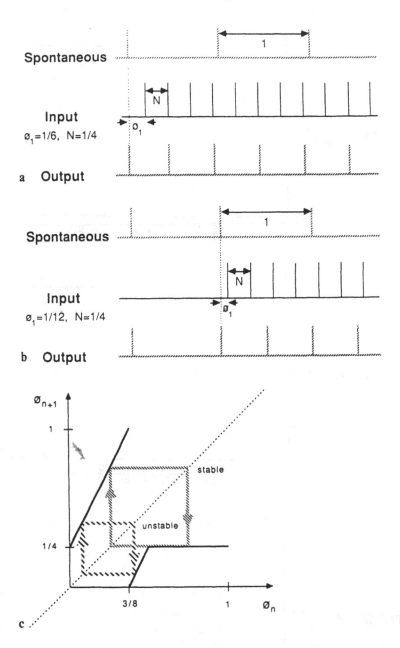

Figure 8. (2:1)-entrainment for $\lambda = 1/2$, in response to an input of frequency four times higher than the pacemaker's spontaneous one. (a) Turning the stimulus on when the pacemaker is at phase $\phi_1 = 1/6$ results in stable synchronized entrainment, while (b) turning the stimulus on when the pacemaker is at phase $\phi_1 = 1/12$ results in unstable non-synchronized entrainment. (c) Graphical representation of the transition equation (7) —thick continuous lines— together with the periodic solutions corresponding to the two instances of (2:1)-entrainment shown in (a) —big cycle marked with arrows— and (b) —small cycle marked with arrows.

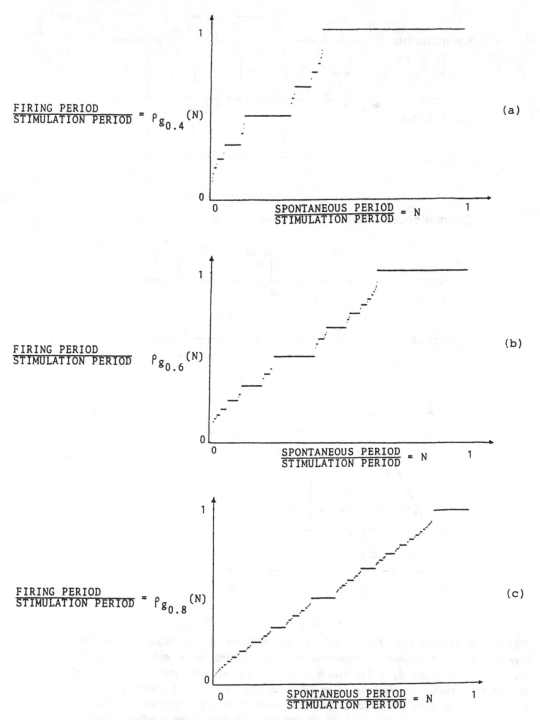

Figure 9. Graphical representation of the entrainment ratio for the values:
(a) $\lambda = 0.4$, *(b)* $\lambda = 0.6$, *(c)* $\lambda = 0.8$.

5 Open Research Issues

Related to the work described in this paper are some results by Herman (1977) and Keener (1980). Herman proved that for paths of class C^1 of diffeomorphisms of the circle of class $C^r (r \geq 3)$, irrational rotation occurs on a Cantor set of positive measure. Keener showed that the same result holds for some strictly-increasing piecewise-continuous mappings from the circle onto itself, while the measure of the Cantor set is zero for other one-to-one non-onto mappings of the same kind. The diffeomorphism in equation (7) is of class C^0, but with non-continuous derivative, falling thus somewhere in between the two types mentioned above; it yields irrational rotation on a Cantor set of measure zero. These facts raise the theoretical question of what factors determine the measure of the Cantor set where irrational rotation takes place.

In this direction, and taking into account its being obtained experimentally through regression, it would be of interest to analyse the implications of introducing slight modifications in the piecewise linear PRC studied, such as for example admitting that $\delta(0)$ and $\delta(1)$ could adopt values different from zero or that the slope of the segment corresponding to the upper phases could differ from 1. The modified PRC's would give place to phase transition curves (on the circle) discontinuous at the origin, of rotation number greater than 1, or decreasing in certain segments. This perhaps would originate aperiodic responses in a deterministic way, thus explaining the second type of evidence described in point (e) of Section 2.3.

Another line of future research is the analysis of the interaction between entrainment and learning. A first result along this line has been displayed in Fig. 6, where learning has been interpreted as a gradual deformation of the PRC.

Let us also mention that there is a demand, from the experimental domain, that a similar study of entrainment to the one reported here be carried out for inhibitory stimulation.

To end this list of open research issues, the analytic study of the interaction between several pacemakers would be very much appreciated, especially when there are simulation results available [Torras 1986] for a network of 100 pacemakers, which are difficult to analyse.

Let us conclude by saying that the research here reported corresponds to one of the research lines on Neural Networks pursued at the Institute of Cybernetics, namely that dealing with the analysis of temporal aspects of neural activity. We are at present pursuing two other research lines [Bofill et al. 1990; Torras 1989], namely the use of neural relaxation algorithms to solve combinatorial optimization problems [Torras and Bofill 1989] and the design of reinforcement learning algorithms for robot path finding [Millán and Torras 1990a & b].

REFERENCES

Ayers J.L. and Selverson A.I. (1979): "Monosynaptic entrainment of an endogenous pacemaker network: a cellular mechanism for von Holt's magnet effect". *J. Comp. Physiol.* A 129, 5-17.

Beltz B. and Gelperin A. (1980): "Mechanisms of peripheral modulation of salivary burster in Limax maximus: a presumptive sensorimotor neuron". *J. of Neurophysiol.* 44, 675-686.

Bernussou, J. (1977): *"Point Mapping Stability"*. Oxford: Pergamon Press.

Bofill P., Millán J. del R. and Torras C. (1990): "Short-term and long-term optimization in neural networks: Two applications". In *"Cellular Automata and Neural Networks"*, edited by E. Goles, Plenum Press, to appear.

Buño W. and Fuentes J. (1984): "Coupled oscillators in an isolated pacemaker-neuron?". *Brain Res.* 303, 101-107.

Coddington E.A. and Levinson N. (1955): *"Theory of Ordinary Differential Equations"*. New York: McGraw-Hill.

Fitzhugh R. (1961): "Impulses and physiological statesin theoretical models of nerve membrane". *Biophys. J.* 1, 445-466.

Fohlmeister J.F., Poppele R.E. and Purple R.L. (1974): "Repetitive firing: Dynamic behavior of sensory neurons reconciled with a quantitative model". *J. of Neurophysiol.* 37, 1213-1227.

Fohlmeister J.F., Poppele R.E. and Purple R.L. (1979): "Repetitive firing: A quantitative study of feedback in model encoders". *J. Gen. Physiol.* 69, 815-848.

Gelbaum B.R. and Olmsted J.M.H. (1964): *"Counterexamples in Analysis"*. San Francisco: Holden-Day.

Harmon L.D. and Lewis E.R. (1966): "Neural Modelling". *Physiol. Rev.* 46, 513-591.

Herman M.R. (1977): "Measure de Lebesgue et nombre de rotation". In *"Geometry and Topology"*, edited by J. Palis and M. de Carmo, Lect. Notes in Math. 597, 271-293, Berlin Heidelberg New York: Springer.

Hodgkin A.L. and Huxley A.F. (1952): "A quantitative description of membrane current and its application to conduction and excitation in nerve". *J. Physiol.* 117, 500-544.

Holden A.V. and Ramadan S.M. (1980): "Identification of endogenous and exogenous activity in a molluscan neurone by spike train analysis". *Biol. Cybern.* 37, 107-114.

Holden A.V. and Ramadan S.M. (1981): "The response of a molluscan neurone to cyclic input: entrainment and phase-locking". *Biol. Cybern.* 41, 157-163.

Junge D. and Moore G.P. (1966): "Interspike-interval fluctuations in Aplysia pacemaker neurons". *Biophys. J.* 6, 411-434.

Kandel E.R. (1976): *"Cellular basis of behavior: An introduction to behavioral neurobiology"*. Freeman.

Keener J.P. (1980): "Chaotic behavior in piecewise continuous difference equations". *Trans. Am. Math. Soc.* 261(2), 589-604.

Knight B.W. (1972): "Dynamics of encoding in a population of neurons". *J. Gen. Physiol.* 59, 734-766.

Kohn A.F., Freitas da Rocha A. and Segundo J.P. (1981): "Presynaptic irregularity and pacemaker interactions". *Biol. Cybern.* 41, 5-18.

Kristan W.B. (1971): "Plasticity of firing patterns in neurons of Aplysia pleural ganglion". *J. of Neurophysiol.* 34, 321-336.

Millán J. del R. and Torras C. (1990a): "Reinforcement learning: Discovering stable solutions in the robot path finding domain". *Proc. 9th European Conf. on Artif. Intell. (ECAI'90)*, Stockholm, Aug.

Millán J. del R. and Torras C. (1990b): "Reinforcement learning in robot path finding: A comparative study". *Proc. 3rd. COGNITIVA*, Madrid, Nov.

Nagumo J. and Sato S. (1972): "On a response characteristic of a mathematical neuron model". *Kybernetic* 10, 155-164.

Parnas I., Amstrong D. and Strumwasser (1974): "Prolonged excitatory and inhibitory synaptic modulation of a bursting pacemaker neuron". *J. of Neurophysiol.* 7, 594-608.

Pavlidis T. (1973): *"Biological Oscillators: Their Mathematical Analysis"*. Academic Press.

Perkel D.H., Schulman J.H., Bullock T.H. Moore, G.P. and Segundo J.P. (1964): "Pacemaker neurons: Effects of regularly spaced synaptic inputs". *Science* 145, 61-63.

Pinsker H.M. (1977a): "Aplysia bursting neurons as endogenous oscillators. I - Phase response curves for pulsed inhibitory synaptic input". *J. Neurophysiol.* 40, 527-543.

Pinsker H.M. (1977b): "Aplysia bursting neurons as endogenous oscillators. II - Synchronization and entrainment by pulsed inhibitory synaptic input". *J. Neurophysiol.* 40, 544-556.

Pinsker H.M. and Ayers J. (1983): "Neuronal Oscillators". Chapter 9 in *"Neurobiology"*, edited by W.D. Willis, Churchill Livingstone Inc.

Pinsker H.M. and Kandel (1977): "Short-term modulation of endogenous bursting rhythms by monosynaptic inhibition in Aplysia neurons: effects of contingent stimulation". *Brain Res.* 125, 51-64.

Rescigno A., Stein R.B., Purple R.L. and Poppele R.E. (1970): "A neuronal model for the discharge patterns produced by cyclic inputs". *Bull. Math. Biophys.* 32, 337-353.

Sato S. (1972): "Mathematical properties of responses of a neuron model. A system as a rational number generator". *Kybernetic* 11, 208-216.

Segundo J.P. (1979): "Pacemaker synaptic interactions: Modelled locking and paradoxical features". *Biol. Cybern.* 35, 55-62.

Segundo J.P. and Kohn A.F. (1981): "A model of excitatory synaptic interactions between pacemakers. Its reality, its generality and the principles involved". *Biol. Cybern.* 40, 113-126.

Sokolove P.G. (1972): "Computer simulation of after-inhibition in crayfish slowly-adapting stretch receptor neuron". *Biophys. J.* 12, 1423-1451.

Stein R.B., Frech A.S. and Holden A.V. (1972): "The frequency response, coherence, and information capacity of two neuronal models". *Biophys. J.* 12, 295-322.

Torras C. (1985a): "Pacemaker neuron model with plastic firing rate: Entrainment and learning ranges". *Biol. Cybern.* 52, 79-91.

Torras C. (1985b): *"Temporal-pattern learning in neural models"*. Lect. Notes Biomath. 63, Berlin Heidelberg New York: Springer.

Torras C. (1986): "Entrainment in pacemakers characterized by a V-shaped PRC". *J. Math. Biol.* 24, 291-312.

Torras C. (1986): "Neural network model with rhythm-assimilation capacity". *IEEE Trans. on Syst., Man, and Cybern.* 16 (5), 680-693.

Torras C. (1987): "On the relationship between two models of neural entrainment". *Biol. Cybern.* 57, 313-319.

Torras C. (1989): "Relaxation and neural learning: Points of convergence and divergence". *J. of Parallel and Distributed Computing* 6, 217-244.

Torras C. and Bofill P. (1989): "A neural solution to finding optimal multibus interconnection networks". *Proc. IX Conf. of the Chilean Computer Science Society and XV Latinoamerican Conference on Informatics* 1, 446-454.

von Baumgarten R. (1970): "Plasticity in the nervous system at the unitary level". In *"The Neurosciences: Second Study Program"*, edited by F.O. Schmitt, Rockefeller University Press, 206-271.

Winfree A.T. (1980): "The Geometry of Biological Time". Springer-Verlag.

Yoshizawa S. (1982): "Periodic pulse sequences generated by an analog neuron model". In *"Competition and Cooperation in Neural Nets"*, edited by S. Amari and M.A. Arbib, Lect. Notes in Biomath. 45, Springer-Verlag.

Yoshizawa S., Osada H. and Nagumo J. (1982): "Pulse sequences generated by a degenerate analog neuron model". *Biol. Cybern.* 45, 23-33.

NEURONAL NETWORKS IN THE HIPPOCAMPUS INVOLVED IN MEMORY

Alessandro Treves and Edmund T Rolls

Department of Experimental Psychology, University of Oxford
South Parks Road, Oxford OX1 3UD, U K

1 Introduction: the hippocampus and memory

The peculiar features of the hippocampal formation have fascinated brain scholars for a long time [28]. In recent years there has been an accumulation of evidence from different disciplines, which has shed some light on the role of the hippocampus in learning and memory [24]. The purpose of this paper is to suggest how the analysis of formal network models may contribute to this enquiry.

The hippocampus is located deep inside the temporal lobe, and via the adjacent parahippocampal gyrus and entorhinal cortex it receives inputs from virtually all association areas in the neocortex, including those in the parietal, temporal and frontal lobes [32,23]. Therefore the hippocampus has available highly elaborated multimodal information, which has already been processed extensively along different, and partially interconnected, sensory pathways. Additional inputs come from the amygdala and, via a separate pathway, from the cholinergic and other regulatory systems. An extensively divergent system of output projections enables the hippocampus to feed back into most of the areas from which it receives inputs [32].

Lesion studies indicate that the hippocampus is important in the encoding of a type of memory which has been described by various authors as declarative, episodic, or 'working' memory [28,26]. Human patients who have suffered hippocampal damage tend to show, especially if the lesion is bilateral, anterograde amnesia, an inability to recognise new stimuli or remember events occurred after the trauma [32]. An example is the failure to recognize the doctor who sees the patient daily. Previously encoded memories tend to be spared [40], and this seems to indicate that the hippocampus is crucial during the process of memory acquisition and consolidation, but that long-term storage occurs elsewhere (i.e., in neocortex). Hippocampectomized monkeys tend to be impaired in recognition tasks such as identifying a stimulus seen previously from a list of novel ones [24].

A way to describe hippocampal processing is to note that it involves a snapshot type of memory, in which one whole scene must be remembered [26]. In vivo recordings from primates performing recognition tasks yield data compatible with this picture, in that some hippocampal neurons are found which tend to respond selectively to the combination of a particular object and a particular location in which the object is shown [20]. The fact that there are cells which learn relatively rapidly to respond to specific associations between, for example, visual stimuli and spatial responses supports the conclusion that one of the functions of the hippocampus is the rapid learning of such associations.

Neurophysiological evidence also directly indicates that many of the synapses within the hippocampus can become modified as a result of experience. This has been shown in many studies on long-term potentiation, whereby presynaptic activity concurrent with strong postsynaptic depolarization can result in a strengthening of the synaptic efficacy [19,7]. Evidence has also been claimed for a mechanism of long-term depression [33]. These synaptic modifications appear to be involved in the types of learning for which the hippocampus is necessary, as shown in studies in which the blocking of long-term potentiation with drugs results in the impairment of these types of learning [21].

2 The networks in the hippocampus

Information is processed within the hippocampus along a distinctly unidirectional path, consisting of three major stages. Input fibers reach the granule cells in the dentate gyrus via the perforant path, and also proceed to make synapses on the apical dendrites of pyramidal cells in the next stage, CA3. There are about $6 \cdot 10^5$ to $1 \cdot 10^6$ dentate granule cells in the rat, and approximately $9 \cdot 10^6$ in man (more detailed anatomical studies are available for the rat) [32]. They project to CA3 cells via the mossy fibers, which form a relatively sparse but powerful synaptic matrix, with a marked topographic organization; each fiber makes, in the rat, about 14 synapses onto the proximal dendrites of CA3 pyramidals. As there are some $1.6 \cdot 10^5$ CA3 pyramidals in the rat ($2.3 \cdot 10^6$ in man [30]), each of them receives no more than 52-87 mossy synapses, and the probability of a CA3 cell receiving input from a given granule cell would be, if it were uniform, of the order of 10^{-4}. By contrast, there are many more (and probably weaker) direct perforant path inputs onto each CA3 cell, in the rat of the order of $2 \cdot 10^3$ [25].

Far more than in the dentate gyrus or in the third processing stage, CA1, in CA3 there is a very extensive system of intrinsic axon collaterals, which allows exchange of signals between cells in the same stage. As CA3 pyramidals are large cells, with about 10^4 spines in the rat, the majority (around 3/4) of their excitatory inputs, occupying the middle portion of their apical dendrites, comes from fellow CA3 cells. In the rat these associational projections, which show only a faint topographic organization, yield a contact probability within CA3 of about 5% [25,32], and a similar figure is expected to hold in man. Other input and output projections run through the fimbria, while intrinsic connections exist with a variety of (mainly inhibitory) interneurons. Extrinsic axonal projections from CA3, the Schaffer collaterals, provide the major input to CA1

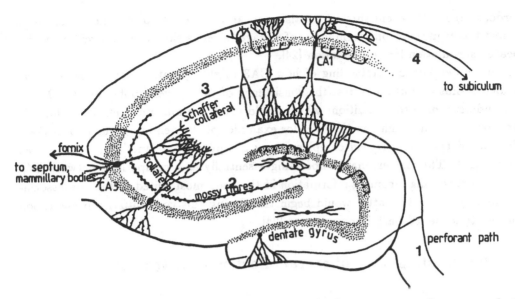

Figure 1: Representation of connections within the hippocampus. Inputs reach the hippocampus through the perforant path (1), which makes synapses with the dendrites of the dentate granule cells and also with the apical dendrites of the CA3 cells. The dentage granule cells project via the mossy fibers (2) to the CA3 pyramidal cells. The well-developed recurrent collateral system of the CA3 cells is indicated. The CA3 pyramidal cells project via the Schaffer collaterals (3) to the CA1 pyramidal cells, which in turn have connections (4) to the subiculum.

pyramidals, of which there are about $2.5 \cdot 10^5$ in the rat, and $4.6 \cdot 10^6$ in man. The output of CA1 returns via the subiculum to the entorhinal cortex, from which it is redistributed to neocortical areas.

The most salient feature of this architecture is that information gathered from all over higher processing areas in the neocortex can be funnelled through and converge onto a network, CA3, which is small enough as to allow a relatively high contact probability via axon collaterals between its cells, almost irrespective of their topographical location within CA3 itself. This information can then be returned through diverging projections to the neocortex. This, and the fact that most of the excitatory connections within the hippocampus are thought to be modifiable with learning, has led to the hypothesis [17,24] that CA3 acts as an autoassociation matrix memory. The association between the various elements of an event would be represented by the pattern of simultaneous firing activities of the CA3 pyramidals, and a memory trace of this pattern could then be encoded in the synaptic efficacies of the collateral system.

An additional hypothesis is that the subsequent stimulation of a subset of the cells active in a particular pattern should be sufficient to spread the activation to the rest and thus elicit the full representation of the event. This mechanism could then be used both in order to recall the event through partial cues, and in order to direct the formation of memory traces in more long-term forms of storage, where a slower process might require repeated stimulation for consolidation [24].

A more detailed understanding of how CA3 might operate as an autoassociative memory is thus important to test the consistency of the above hypotheses [26]. Moreover, understanding the conditions which would allow the maximum efficiency for this operation could shed light on the role, for example, of the dentate gyrus and of CA1, which would then be assigned complementary computational functions, as suggested by Rolls [24]. The former, maybe operating essentially as a competitive network [24], could help preparing incoming information in the code most suited to the autoassociative memory, while the latter would begin translating this code into a representation appropriate to be fed back into neocortex [24].

3 Understanding the collateral effect with attractors

Marr's theory of archicortex [17] represents an early and ambitious attempt to produce a formal, predictive and detailed model of the hippocampus. Later studies [12] along the same line have concentrated more specifically on the role of axon collaterals. A weak point of this approach is that it is difficult to test the generality and robustness of the results as some of the assumptions that it makes, such as those involved in describing single fibers, the time course of their activation, or the plasticity of synaptic efficacies, are modified to be made more realistic.

The statistical mechanics formalism that has been developed [3,4] to analyse Hopfield's model of autoassociative memory [14], on the other hand, can be extended to try to account for one element of hippocampal operation, the collateral effect. By focusing on the structure of the attractor states emerging in the long-time limit of the dynamics, this formalism deals with features that can be hoped to be relatively robust to changing the details, in particular of the dynamics itself. In order to apply it to the CA3 network, though, it is necessary to examine some of the basic ingredients of attractor models [2] as originally formulated.

One common way to model synaptic modification as effected by learning, is in terms of a discrete set of p firing patterns, memorised with equal strength. The additive modification produced by each pattern is proportional to the product of the deviations of the pre- and post-synaptic activities from their average values, according to what has been termed a 'covariance' learning rule [29]. This description of synaptic plasticity involves a number of untested assumptions, concerning for example the coexistence of, and delicate equilibrium between, long-term potentiation and depression, and therefore should not be seen as anything more than a simple arbitrary ansatz. In view, however, of the lack of clarity concerning the neurophysiological mechanisms underlying the phenomenon, not to mention their quantitative description, this simple rule has appealing features that make it interesting for use in higher level modelling. One merit of the covariance rule is that it can be conceived as the result of one-shot learning, in line with the fast learning the hippocampus is thought to subserve. Other advantages can be argued *a posteriori* from studies of specific models, which indicate [5,39,31] that the associative behaviour induced by this rule is very robust to a variety of perturbating effects.

A special consequence of the covariance rule is that if it supposed to express the total modification to a pair of reciprocal synaptic connections between two pyramidal neurons, both connections being initially present and with equal efficacies, the efficacies remain equal after learning. To assume this symmetry in the connections makes it possible to analyze the long-time behaviour of densely connected networks. Comparable studies of non-symmetric cases have had to be restricted to very highly diluted connectivities [9], with the fan-in to each neuron of the order of the logarithm of the number of cells, i.e. 10-15 in the case of CA3. There are some indications [36] that the situation where a degree of asymmetry results only from an incomplete, but still dense connectivity may be approximated better by a fully connected than by a very strongly diluted network. This, and the fact the aim here is to model the collateral effect among pyramidal neurons, rather than, for example, to capture oscillatory behaviours due to excitatory-inhibitory feedbacks, suggests the opportunity to maintain the simplying assumptions of a covariance rule, full connectivity and symmetry of synaptic connections.

The main concern here, by contrast, is with the statistics of the patterns of firing activity encoded in the synaptic efficacies. To examine the statistical properties of the behaviour of the network, these patterns are drawn at random from a probability distribution, which is taken to be the product of independent equal distributions P_η for the firing activity η of each cell in each pattern. In the Hopfield and related models P_η has been assumed bimodal, in fact consisting of two delta functions of equal weight, one centered at zero activity and the other at an assumed saturation firing rate. From the point of view of a (spatial) population distribution, this implies that at any given moment half of the neurons would be quiescent, and half firing at saturation. Note that due to the gauge symmetry of the interaction, the η value corresponding to saturation could be rescaled independently for each cell, thereby allowing a distribution of physical saturation rates. Still, this situation hardly corresponds to experimentally observable firing rate distributions, either in the hippocampus or in neocortex [1]. This fact has stimulated consideration of the case in which much less than half of the cells are active in each pattern, which has led to uncovering the very interesting properties of *sparsely* coded autoassociative nets [38,8].

From the point of (temporal) single cell activity distributions over different patterns, however, the awkward assumption remained of a bimodal distribution, with each cell firing at saturation in some of the patterns. Attempts to dispose of part of the problem have been made [37,6,27,13], by describing models in which the firing level of active neurons is not determined by single cell saturation, but rather by reaching an equilibrium between overall excitation and inhibition. In these models, P_η still consists of two delta functions, and neurons are represented formally as McCullough-Pitts binary units [18]. To go further, and avoid also the *bimodality* of single cell distributions by considering alternative forms for P_η, a modification is required in the way one models single cell responses. In fact, assuming a non-bimodal P_η makes little sense, if then binary neurons are used, producing inherently bimodal response.

Realistic representations of the way particular classes of neurons respond by firing to different modes of stimulation can be reconstructed from neurophysiological data, and expressed in simplified form by input-output relationships of typically sigmoidal shape. Such relatively detailed descriptions of single cell behaviour greatly complicate, on the other hand, the quantitative analysis of collective network properties, and in fact not much progress has been done in this direction beyond arguing that the type of behaviour produced by 'sigmoidal' neurons is similar to that produced by binary neurons [15], if the sigmoid itself is similar enough to a binary Heaviside function.

Threshold-linear units have been suggested [34], instead, as a possible formal model of neuronal graded response in attractor neural networks. The point being made here is that by more realistically representing graded response at the single neuron level, one is allowed to explore the effects of encoding information in firing patterns of different, non-binary, structure. This has implications, in particular, for understanding the best way in which the CA3 network could operate as an autoassociative memory. Would it be better for it to receive information in a binary code? Could the dentate gyrus have, as one of its functions, the task to 'binarize' incoming inputs to CA3? Or would CA3 operate just as efficiently with non-binary codes?

One possible measure of performance of an autoassociative memory is its storage capacity. It appears a biologically relevant measure in light of the hypothesized role of the hippocampus as a temporary memory store, which has to remember learnt associations for at least all the time needed for consolidation of the memory in neocortex. In a wide class of formal models, the maximum number of stored patterns p is proportional to the number of collateral inputs to each cell, which in practice is limited by the physical size of the cell. Thus achieving a high proportionality factor would seem an advantage for CA3. Although the precise way in which the capacity is defined in this formal framework may be very simplistic, it is reasonable to assume that it bears some relation with a biologically meaningful feature. The issue of how the form in which information is encoded affects the storage capacity of a network of threshold-linear neurons has therefore been studied [35] by considering the limits on memory overloading as a function of varying P_η. Here the model used and the main results are reviewed.

4 Modelling graded response

The firing rates of N pyramidal cells are represented by positive variables V_i, with $i = 1, \ldots, N$. They are determined by the membrane potentials h_i, as

$$V = \begin{cases} g(h - T_{hr}) & h > T_{hr} \\ 0 & h < T_{hr}, \end{cases} \qquad (1)$$

where g is a gain parameter, and T_{hr} a threshold. h_i is a real-valued variable summarizing the influence on unit i of all other units, as well as of external stimulation and of additional units not explicitly represented, such as inhibitory interneurons. The threshold-linear form is conceived as the simplest way to model the nonlinearity due to the threshold as well as graded-response above threshold. As the regulation of firing activity is ascribed to global inhibition, saturation effects are not represented, for simplicity, at this single cell level (in contrast with sigmoidal representations).

The encoded firing patterns, labelled $\mu = 1, \ldots, p$, are specified by associating with each unit a positive number η_i^μ, proportional to the activity of that unit in the encoding of that pattern. The η's are drawn, independently for each μ and i, from a distribution P_η, and are normalized by setting their average over P_η to a value a. Imposing that also

$$\int_0^\infty P_\eta(\eta)\eta^2 d\eta \equiv a \qquad (2)$$

turns a into a parameter that measures the sparseness of the coding scheme. Specific cases for P_η are considered later on.

The network is supposed to operate under external stimulation. The initial stimulation is taken to determine essentially the initial configuration $\{V_i\}$. The subsequent, persistent, external stimulation is split into a uniform part, plus pattern-specific terms. The latter come as contributions to the membrane potential, in the form [5]

$$\sum_\mu s^\mu \frac{\eta_i^\mu}{a}, \qquad (3)$$

where the s's give the relative strength of one or more patterns embedded in the external stimulus.

The positive strength of excitatory-excitatory connections is taken to result from an original, uniform, positive average strength, with superimposed incremental and decremental modifications as expressed by the covariance rule, i.e. $J_{ij} = J_{ij}^0 + J_{ij}^c$, with

$$J_{ij}^c = \frac{1}{Na^2} \sum_{\mu=1}^p (\eta_i^\mu - a)(\eta_j^\mu - a). \qquad (4)$$

The remaining uniform external stimulation, the contribution to EPSPs from the baseline excitatory strength J_{ij}^0, and PSPs from cells that are not explicitly represented, including inhibitory ones, are all lumped together into an additional term in the membrane potential of each cell. The approximation is made, to consider this additional term as uniform over different units, and depending only on the average network activity. In conclusion, the membrane potential is written

$$h_i = \sum_{j \neq i}^{N} J_{ij}^c V_j + b(\sum_j \frac{V_j}{N}) + \sum_\mu s^\mu \frac{\eta_i^\mu}{a}. \tag{5}$$

The second term contributing to h_i thus represents a variety of effects, whose common feature is that they are uniform with respect to the pattern structure encoded in the J_{ij}^c's.

Independently of the details of the dynamical evolution, as long as it leads the network towards attractor states that can be described microscopically by Eqs. 1-5, a static analysis can determine the range of existence of the various types of attractors in terms of the different parameters of the model. This analysis was used to find the range of storage values $\alpha \equiv p/N$ for which the network can operate as an associative memory, i.e. the range of existence of *retrieval* states [3,2], or attractors macroscopically correlated with a single pattern.

It was shown in Ref. [35], with a calculation using replica symmetry, that the maximum α for which retrieval states exist for a given P_η is independent of the form of the b term in Eq. 5, and of the threshold T_{hr}. There is a dependence on the gain g, which is itself renormalized by the interaction to a new α-dependent value. The highest capacity α_c occurs for intermediate values of the gain. It can be found [35] as the value α for the which the function

$$E_1(w, v) \equiv (A_1 + \delta A_2)^2 - \alpha A_3 \tag{6}$$

has a maximum at $E_1 = 0$ in the w, v plane. Here w and v are two parameters measuring signal-to-noise ratios, and A_1, A_2, A_3 are averages over P_η and over a noise variable z:

$$A_1(w, v) = \int_0^\infty P_\eta(\eta) d\eta \frac{(\eta - a)}{v(1-a)} \int^+ \frac{e^{-z^2/2} dz}{2\pi} (w + v\frac{\eta}{a} - z) - \int_0^\infty P_\eta(\eta) d\eta \int^+ \frac{e^{-z^2/2} dz}{2\pi}$$

$$A_2(w, v) = \int_0^\infty P_\eta(\eta) d\eta \frac{(\eta - a)}{v(1-a)} \int^+ \frac{e^{-z^2/2} dz}{2\pi} (w + v\frac{\eta}{a} - z) \tag{7}$$

$$A_3(w, v) = \int_0^\infty P_\eta(\eta) d\eta \int^+ \frac{e^{-z^2/2} dz}{2\pi} (w + v\frac{\eta}{a} - z)^2$$

where the superscript $+$ indicates that the z-average has to be carried out only in the range where $w + v\eta/a - z > 0$. δ is a further parameter that measures the fraction of the correlation of the retrieval state with the retrieved pattern, that is directly elicited by the pattern-specific component of the external stimulus [35]. Eq. 6 thus determines, for any firing distribution encoded in the synaptic efficacies, the factor α_c, whose value will depend on δ as well as on P_η.

5 Sparse and graded information encodings

Distributions P_η can be chosen in a vast space, and one could try to model actually observed distributions of firing activities. In order to gain insight as to the effect of sparse and graded information encodings on the capacity of the network, it is useful to start by constrasting the effect of a binary distribution with that of a *ternary* one. That is, ternary patterns can be used to explore what happens when memories become increasingly structured. As P_η becomes more structured, more and more parameters are required to describe it, and this would make it hard to obtain any general insight. Therefore, the simplest non-binary structure seems a natural choice for a first analysis.

The binary P_η will be

$$P_\eta(\eta) = (1 - a)\delta(\eta) + a\delta(\eta - 1), \tag{8}$$

with $\delta(x)$ Dirac's function. A possible choice for a ternary distribution is

$$P_\eta(\eta) = (1 - \frac{4}{3}a)\delta(\eta) + a\delta(\eta - \frac{1}{2}) + \frac{a}{3}\delta(\eta - \frac{3}{2}), \tag{9}$$

Both these distributions have been expressed in terms of the sparse coding parameter a introduced earlier. One can see that, although in general this parameter is defined in terms of the moments of the distribution P_η, in these particular cases a is simply proportional to the fraction of cells with non-zero activity in the encoding phase. The qualification 'sparse coding' refers to the region $a \ll 1$, which has attracted attention in the literature also for reasons unrelated to biological plausibility. In fact it was found [8,38] that the information content that can be stored in binary networks endowed with covariance imprinting rules approaches, for $a \to 0$, the upper bound derived for binary networks and for any choice of synaptic strengths[11]. In contrast, when e.g. $a = 0.5$, the use of a covariance rule allows the storage of only a small fraction of the information storable with other choices of the synaptic strengths[4,16,11]. Moreover, the stable states of the fully connected system (those studied here) have been shown [10] to be equivalent to the ones obtained as fixed points of the dynamical evolution of the strongly diluted system. This would suggest that capacity estimates obtained with static methods might be expected to be valid also for diluted systems, provided the storage capacity is defined as $\alpha \equiv p/C$, where C, the connectivity, replaces N, which is the connectivity of a fully connected system.

The resulting maximum storage capacities are plotted in Fig. 2 as functions of a and for two values of δ. The main feature of the graphs, both for binary and ternary distributions, is the strong increase of α_c as $a \to 0$, to first approximation as a^{-1}. This effect has already been well understood by studying networks of binary neurons [8,38]. Indeed the modelling of single neurons as threshold-linear rather than binary unit does not alter significantly the capacity of the network, if a binary form is still used for P .

As for the ternary case, one notes that the storage capacity is roughly twice than for binary patterns, for corresponding values of a. Thus, the number of patterns that

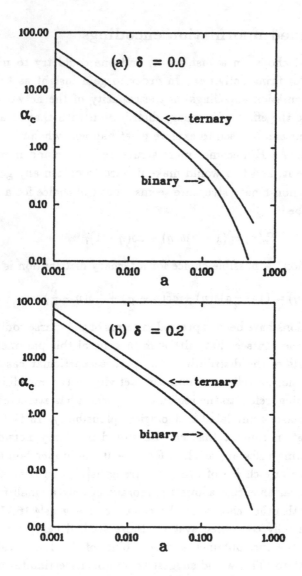

Figure 2: Storage capacity ($\alpha_c \equiv p/N$, with p the maximum number of stored patterns) vs. the sparse coding parameter a, for binary and ternary pattern distributions (Eqs. 8-9). a) No pattern-specific persistent input ($\delta = 0$); b) $\delta = 0.2$.

can be stored in the auto-associative memory appears to increase, as they become more structured. It should be remarked, however, that a has been defined in terms of the distribution of rates in the encoding phase, i.e., for present purposes, in terms of the distributions of the synaptic modifications. An examination of the distributions of activities in the retrieval phase reveals that most cells encoded as having intermediate activity in the ternary distribution, tend to have zero activity during retrieval. This amounts to effectively making the code more sparse (only cells with high activity tend to sustain the collateral effect), and thus explains the enhanced storage capacity.

These results are shown not to depend on the value of δ. When some degree of pattern-specific persistent input contributes to elicit a certain pattern, the storage capacity is higher (see Fig. 2b and Ref. [4]). However, the increase appears essentially decoupled from the dependence on P_η. Therefore, while it is interesting to consider a non-zero amount of specific external stimulation, in order to make contact with realistic situations, an examination of the case $\delta = 0$ will already reveal the effect of the structure of the encoding.

While one can compare storage capacities, it has been noted that another interesting measure of performance is the total amount of information which can be stored *and* retrieved. This quantity depends on the number of patterns stored as well as on the information stored in each of them, minus the fraction of this information content that gets lost due to noise in retrieval. A formal expression for the information, I, stored in a network of threshold-linear neurons has been given elsewhere [35]. That expression can be maximized over α and the maximum is in general for $\alpha < \alpha_c$ and a specific value of the gain g. The maximum is quite flat, however, so that I does not depend crucially on the precise value of g in a certain range.

Fig. 3 shows the amount of information retrievable in the two cases considered, as a function of the sparse coding parameter a. Note that, because α_c grows, as $a \to 0$, approximately as $\alpha_c \approx 1/(a \ln(1/a))$, what is actually plotted is $\alpha_c a$ versus $1/\ln(1/a)$, also to facilitate a comparison with Fig. 2 of Ref. [8]. As noted above, the capacity for retrieval of binary patterns of the threshold-linear model is broadly similar to that of the binary model, the main difference, i.e. the marked advantage of the threshold-linear model in the region far from the $a \to 0$ limit, being due to the fact that the choice of threshold adopted in Ref. [8] is sub-optimal in that region [22]. The behaviour is the same also in terms of information content, showing that the storage of binary patterns is not affected sustantially by replacing binary processing units with threshold-linear ones.

Considering now the ternary case, Fig. 3b, it is clear that while the storage capacity has increased, the retrievable information is less than in the binary case. Interestingly, this happens despite the fact that, for a given a, each ternary pattern contains more information than its binary counterpart. In other words, of the three factors that contribute in determining I one, the retrieval quality, plays a dominant role in decreasing

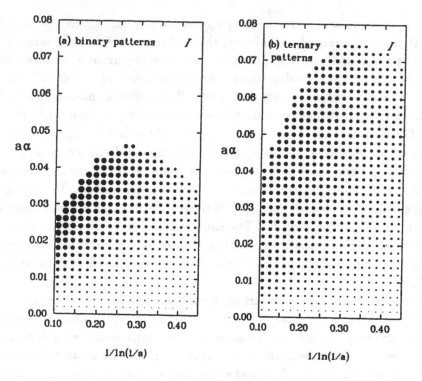

Figure 3: Storage capacity times a vs. $1/\ln(1/a)$. Below α_c (the dotted region) the linear size of the dots is proportional the maximal information content I achievable for given a, α. a) Patterns distributed according to Eq. 8 (the largest dot corresponds to $I = 0.338$); b) ternary patterns, Eq. 9 (largest dot, $I = 0.209$).

the efficiency (as measured by I) of storing ternary patterns rather than binary ones. The greater amount of information lost during retrieval more than compensates for the increased number of patterns retrievable and for the larger amount of information each one carries.

One can check that this behaviour is not a peculiar feature of the particular choice, Eq. 9, used as a ternary P_η [35]. Such an analysis reveals that most ternary distributions are broadly equivalent to binary ones, both in terms of pattern capacity and of information content. When they are not, which happens essentially when the fraction of cells with higher activity is small with respect to the fraction of those with intermediate activity, then the behaviour is as exemplified by the specific choice given by Eq. 9. To the extent to which the ternary case can be extrapolated to a more general situation, these results then suggest that binary encoding of information is optimal in the sense measured by I, whereas the number of stored pattern can increase considerably by using more structured forms of encoding.

6 Conclusions

Is the above formal analysis, using threshold-linear units and a covariance rule, indicative of general features arising from the possibility of encoding structured activity patterns in autoassociative networks in which neurons have graded response? If that is the case, what are the implications for the operation of the hippocampal network?

The answer to the first question seems to rest mainly on a fuller understanding of two problems: the effects of noise, and the relevance of the assumed covariance rule. Various types of noise source can be incorporated in a formal memory of the type treated here. Some of the ways to model noise have been employed extensively in the study of nets of binary units, and it would be interesting to carry on these approaches in order to appreciate how noise affects sparsely coded memories. Ultimately, however, what is needed is an evaluation of the importance of specific noise sources as they operate in real neuronal networks, which at the moment is just being guessed. Similarly and more importantly, the plausibility of the covariance rule is an important matter, in view of the crucial role that it seems to play in allowing extensive memory storage. Thus it seems that formal modellers will have to watch closely the rapid developments in the study of synaptic plasticity by neurophysiologists, for their discoveries will have a strong impact on the biological meaningfulness of present models.

Within these limits, the indications emerging from the analysis are the following. The crucial factor in determining the capacity of autoassociative nets is the sparseness of the coding scheme. A moderate enhancement may also arise from having a more structured coding, which makes effective use of the ability of neurons to produce graded response. This enhancement may be counterbalanced by a decrease in other indicators of performance, such as the information content. There is in no case a drastic advantage, however, to binary encoding. This suggests that if CA3 operates along the lines hypothesised, preprocessing stages such as the dentate gyrus would have no special task of rendering in binary form the information it receives. It may be that such stages make the coding more sparse [24]. Further work on the functional implications of the singular anatomical connections which precede the CA3 cells in the dentate gyrus will, at any rate, be of great interest.

References

[1] M Abeles, E Vaadia and H Bergman, *Network* **1** 13 (1990)

[2] D J Amit, *Modelling Brain Function* (Cambridge Univ Press, NY, 1989)

[3] D J Amit, H Gutfreund and H Sompolinsky, *Phys Rev* **A 32** 1007 (1985)

[4] D J Amit, H Gutfreund and H Sompolinsky, *Ann Phys* **173** 30 (1987)

[5] D J Amit, G Parisi and S Nicholis, *Network* **1** 75 (1990)

[6] D J Amit and A Treves, *Proc Natl Acad Sci USA* **86** 7871 (1989)

[7] T H Brown, A H Ganong, E W Kairiss, C L Keenan and S R Kelso, in *Neural Models of Plasticity*, J H Byrne and W O Berry eds, (Academic Press, San Diego, 1989)

[8] J Buhmann, R Divko and K Schulten, *Phys Rev* **A 39** 2689 (1989)

[9] B Derrida, E Gardner and A Zippelius, *Europhys Lett* **4** 167 (1987)

[10] M R Evans, *J Phys* **A 22** 2103 (1989)

[11] E Gardner, *J Phys* **A 21** 257 (1988)

[12] A R Gardner-Medwin, *Proc Roy Soc* **B 194** 375 (1976)

[13] D Golomb, N Rubin and H Sompolinsky, *Phys Rev* **A 41** 1843 (1990)

[14] J J Hopfield, *Proc Natl Acad Sci USA* **79** 2554 (1982)

[15] J J Hopfield, *Proc Natl Acad Sci USA* **81** 3088 (1984)

[16] I Kanter and H Sompolinsky, *Phys Rev* **A 35** 380 (1987)

[17] D Marr, *Phil Trans Roy Soc* **B 262** 23 (1971)

[18] W S McCullough and W Pitts, *Bull Math Biophys* **5** 115 (1943)

[19] B L McNaughton, in *Neurobiology of the Hippocampus*, W Seifert ed, (Academic Press, London, 1983)

[20] Y Miyashita, E T Rolls, P M B Cahusac, H Niki and J D Feigenbaum, *J Neurosci* **61** 669 (1989)

[21] R G M Morris, E Anderson, G S Lynch and M Baudry, *Nature* **319** 774 (1986)

[22] C J Perez Vicente and D J Amit, *J Phys* **A 22** 559 (1989)

[23] E T Rolls, in *The Scientific Basis of Clinical Neurology*, M Swash and C Kennard eds, (Churchill Livingstone, London, 1985)

[24] E T Rolls, in *Neural Models of Plasticity*, J H Byrne and W O Berry eds, (Academic Press, San Diego, 1989)

[25] E T Rolls, in *The Computing Neuron*, R Durbin, C Miall and G Mitchison eds, (Addison-Wesley, Wokingham, England, 1989)

[26] E T Rolls, in *Neurobiology of Comparative Cognition*, D S Olton and R P Kesner eds, (Erlbaum, Hillsdale, N J, 1990)

[27] N Rubin and H Sompolinsky, *Europhys Lett* **10** 465 (1989)

[28] W Seifert (ed), *Neurobiology of the Hippocampus* (Academic Press, London, 1983)

[29] T Sejnowski, *J Math Biol* **4** 303 (1977)

[30] L Seress, *J Hirnforsc* **29** 335 (1988)

[31] H Sompolinsky, *Phys Rev* **A 34** 2571 (1986)

[32] L R Squire, A P Shinamura and D G Amaral, in *Neural Models of Plasticity*, J H Byrne and W O Berry eds, (Academic Press, San Diego, 1989)

[33] P K Stanton and T J Sejnowski, *Nature* **339** 215 (1989)

[34] A Treves, *J Phys* **A 23** in press (1990)

[35] A Treves, *Phys Rev* **A** in press (1990)

[36] A Treves and D J Amit, *J Phys* **A 21** 3155 (1988)

[37] A Treves and D J Amit, *J Phys* **A 22** 2205 (1989)

[38] M V Tsodyks and M V Feigel'man, *Europhys Lett* **6** 101 (1988)

[39] K Y M Wong and D Sherrington, *J Phys* **A 23** L175 (1990)

[40] S Zola-Morgan, L R Squire and D G Amaral, *J Neurosci* **6** 2950 (1986)

BASINS OF ATTRACTION AND SPURIOUS STATES IN NEURAL NETWORKS

by L. Viana, E. Cota and C. Martínez,
Lab. de Ensenada, Instituto de Física, UNAM, A.Postal 2681,
22800 Ensenada B.C., México.

Abstract. A long range Ising Neural Network is considered, where p patterns have been stored according to either the Hebb rule or a modification of it which stores patterns with different weights. By performing computer simulations, the size of the basins of attraction of pure and spurious memories is evaluated in the absence of noise, as a function of the load parameter of the system. It is found that the use of the modified Hebb prescription, decreases considerably the percentage of configuration space occupied by spurious memories, which translates into an improvement in the retrieval capabilities of the network.

1. Introduction.

In the last few years, there has been a growing interest in studying the global behaviour of Neural Networks (NN) due to their features as associative fault-tolerant memories. In this way, we find professionals of various fields such as neurobiologists, computer scientists, psychologists, physicist, etc, working on common grounds and trying to find results from their various points of view. The works presented in this conference reflect some of the interests within this wide field of current research.

Statistical physicists became interested in this problem after the pioneering work of Hopfield [1], who made a mathematical analogy between an assembly of neurons and certain disordered magnetic materials called Spin Glasses [2]. This connection allowed the use of methods developed in Statistical Mechanics to study some general properties of NN in the thermodynamic limit. In particular, when the interactions between elements are symmetric, it is possible to assign an energy function to the system.

The retrieval capabilities of NN are a natural consequence of their dynamics, as by reducing their energy, they evolve spontaneously towards a minimum of the free energy which has a large 'overlap' [3] with the initial state. Therefore the final state of the network will depend on both the initial state and the energy 'landscape' which is determined by the connection coefficients $\{J_{ij}\}$. In this way, learning is related to a suitable choice of the specific set $\{J_{ij}\}$ which favours particular configurations by making them minima of the Hamiltonian. However, it has been found that in the process of storing information, other undesired minima appear; these minima also act as attractors and therefore have a negative influence on the retrieval capabilities of the system. We will call these states 'spurious' as opposed to the 'pure' states we stored on purpose.

For symmetrical NN $(J_{ij}=J_{ji})$, the analytical methods of statistical mechanics give us important information about the existence of equilibrium states (both pure and spurious). In this way, we know that spurious states are related to mixtures of pure memories and that their number grows very rapidly with the number of stored patterns. However, more relevant than the number of spurious minima is the total percentage of configuration space occupied by their domains of attraction, as this parameter will directly affect the retrieval capabilities of the network. If we are interested in obtaining a more complete picture of the configuration space, it is necessary to use other complementary techniques such as computer simulations.

2. Procedure

In this paper we carry out computer simulations for a long range Ising NN composed by N neurons-like elements, whose dynamics, in the absence of noise, is governed by an energy function given by:

$$H = -\left(\frac{1}{2}\right)\sum_{i,j} J_{ij} S_i S_j, \tag{1}$$

where S_i denotes the state of the ith neuron with $S_i = +1(-1)$ for a firing (quiescent) neuron; and $J_{ij}=J_{ji}$ represents the 'synaptic strength' between neurons i and j given by a modified Hebb rule, designed to reflect training [4]. This rule assumes that J_{ij} will be given by the sum of p random patterns $\{\xi_i^\mu\}$, according to

$$J_{ij} = \left(\frac{1}{N}\right)\sum_\mu J_\mu \xi_i^\mu \xi_j^\mu, \tag{2}$$

for $i \neq j$ and $J_{ii} = 0$; where $\{\xi_i^\mu = \pm 1\}$ with $\mu = 1, \ldots, p$ are 'quenched' variables which correspond to p unbiased stored patterns, and J_μ is the weight assigned to the μth pattern. In particular, the case with all $J_\mu = 1$ reduces to the Hebb rule. Both learning prescriptions (with equal or different J_μ's) assume that synaptic strengths change in response to experience in a way proportional to the correlation between the firing of the pre and post synaptic neurons.

The equilibrium properties of this model have been studied analytically, in the thermodynamical limit ($N \to \infty$), for the Hebb rule with a finite [5] and an infinite number of stored patterns [6], and for the modified Hebb with p finite [4]. These studies were done as a function of the load parameter α defined as the ratio of the number p of stored memories, to the number N of elements in the net. The main results, for various noise levels, are as follows:

Hebb rule, $\alpha = 0$ [5]. For $0 < T < 1$, all pure memories are related to an energy ground state (where $T = 1$ is the critical level of noise below which ordered states appear). For $0.46 \lesssim T < 1$ these 'pure' or 'Mattis states' are the only minima; however, as the noise level decreases, spurious stable states show up. Spurious states consist of a symmetric mixture of $2s + 1$ pure memories [$s = 1, 2, \ldots, (p-1)/2$] and are equidistant, according to the Hamming distance [7], to each of the pure memories they are related with. As $T \to 0$, the number of such states grows exponentially with p, and $f_1 < f_3 < \ldots < f_{2s+1}$, where f_{2s+1} is the free energy of a mixture state of $2s + 1$ pure memories.

Hebb rule, $\alpha \neq 0$ [6]. For a finite ratio α, it is found that as α grows, the sum of the random overlaps among patterns becomes an important contribution to the energy function, until it eventually unstabilizes the stored patterns and there is a discontinuous breakdown for $\alpha \gtrsim 0.138$. Beyond this point, retrieval of stored patterns is no longer possible.

Modified Hebb, $\alpha = 0$ [4]. A modified Hebb rule given by (2), in which it is possible to reflect some degree of training has also been studied analytically [4]. In that case, strictly different weights J_μ were assigned to each of a finite number of stored patterns ($1 = J_1 > J_2 > \ldots > J_p > 0$). Due to the lack of symmetry between stored patterns, this model presents the following properties: At high temperatures ($0 < T < 1$), not all the states $\{\xi_i^\mu\}$ are stable. As T is decreased some of them become minima, the lower the value of J_μ the lower the value of T at which they become stable. As $T \to 0$ all pure states become local minima of the energy with an energy given by $f_{J_\mu} = -J_\mu/2$. Therefore it is no longer true that all minima related to 'pure memories' are global minima of the energy. Moreover, some spurious stable states can have energies higher than those of pure states. Another consequence is that symmetric spurious states do not exist. Instead, all spurious minima involve an unequal overlap of an odd number of memories, whose relative value depends on the noise level T.

In order to complement the previous information we analyzed numerically the relative importance of pure and spurious memories, for both learning prescriptions, through the relatives size of their domains of attraction. This is given simply by the relative number of times the system retrieves such states starting from configurations generated at random. In the case of the modified Hebb rule, we considered weights J_μ such that $<J_\mu> = 0.7$. The analysis was performed as a function of $\alpha = p/N$.

We considered values of N ranging from 32 to 512, and stored p nonoverlapping patterns, with p from 3 to 7, in order to eliminate some of the effects produced by the finite size of the sample [8]. Therefore, although our results could, in principle, be extrapolated to the thermodynamic limit, for both $\alpha = 0$ or $\alpha \neq 0$, our results are better suited to the first case since we considered nonoverlapping patterns in our problem. We know that for finite α, the sum of the random overlaps among patterns, which we are neglecting, plays an important part in the stability of the stored memories.

3. Results

Fig.1 shows a typical run consisting of 3000 trials on a sample with $N = 320$ elements and $p = 5$ stored patterns, for Hebb (Fig.1.a), and modified Hebb (Fig.1.b) with $<J_\mu> = 0.7$. The abscissas show, with open symbols, the relative number of times each 'pure memory' was retrieved, and full circles correspond to the percentage of times any of the 'spurious memories' was recalled. We can see from Fig.1.b that, as expected, the higher the value of J_μ the better the retrievability of the corresponding memory. It can also be appreciated that the relative importance of spurious memories decreases in case (b) with respect to case (a). In other words, if one introduces "training" into the model via the modified Hebb rule, one can expect overall improvement in the retrieval capabilities of the network, since the importance of some pure memories over others can be stressed and the percentage of configuration space occupied by spurious memories is significantly diminished.

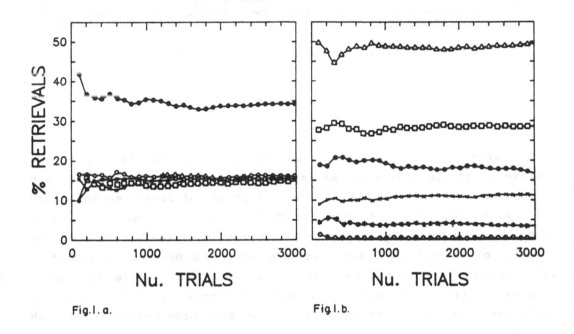

Fig.I. a.   Fig I. b.

Fig.2 shows, as a function of $\alpha = p/N$ ($p = 5$ and N ranging from 32 to 512), the percentage of times each memory was retrieved; full circles corresponding to spurious states. The cases with equal weights ($J_\mu = 1$), and different weights J_μ such that $<J_\mu> = 0.7$, are shown in parts (a) and (b), respectively. Again we observe, for the range of values of α considered, that the percentage of configuration space occupied by spurious memories decreases in case (b) with respect to (a). In fact, we may extrapolate from the figure to the $\alpha = 0$ limit to obtain approximate values of 32% and 16% for cases (a) and (b), respectively, for this value of p.

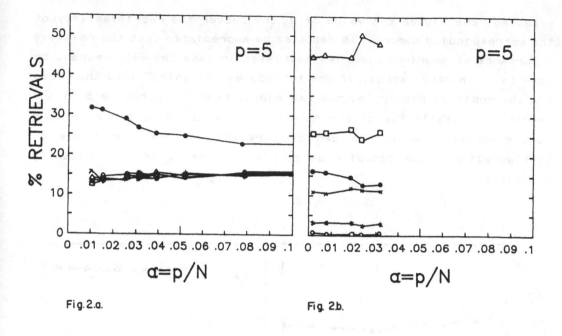

Fig. 2.a.

Fig. 2.b.

This remark is more clearly seen in Fig.3, where the percentage retrieval of spurious memories is shown as a function of α, for both equal and different weights. These two groups of cases are indicated as (pA) and (pB) respectively, for $p = 3,4,5,6$ and 7. Here, it can be seen that the total 'area' of the configuration space occupied by the domains of attraction of spurious memories, is considerably higher if all memories are stored with equal weight. Also, notice that as p increases, the percentage of configuration space occupied by spurious memories apparently tends to a limit, for both Hebb and modified Hebb rules.

From Figs.1.b and 2.b, it can be observed that, for the modified Hebb rule, memories with larger weights have a higher percentage of retrieval. This means that by varying the specific values of J_μ it is possible to modulate the percentage of times each memory is retrieved, i.e., it is possible to simulate training. This is shown explicitly in Fig.4, for the case $p = 7, N = 192$ and $<J_\mu> = 0.7$.

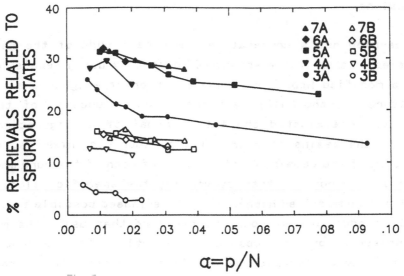

Fig. 3.

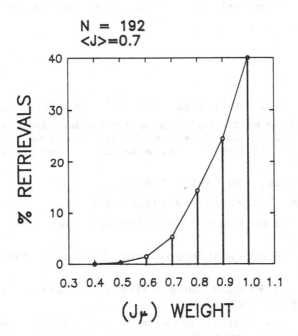

Fig. 4.

4. Conclusions.

We have carried out a comparative numerical study of the retrieval capabilities of a NN in the absence of noise, by using Hebb's learning rule and a modified Hebb's rule where different weights are assigned to the stored (orthogonal) patterns. As a function of the load parameter α, we evaluated the total 'area' of configuration space occupied by the basins of attraction of spurious memories, and the relative size of the domains of attraction of each of the pure memories. We find from our results that by varying the specific values of the weights J_μ in the modified Hebb's rule it is indeed possible to simulate training. In addition, we consistently find that using the modified Hebb's prescription reduces significantly the percentage of configuration space occupied by spurious memories, which translates into an improvement in the retrieval capabilities of the network.

Acknowledgements. One of the authors (LV) wishes to thank M.A. Virasoro for useful comments and Prof. L. Garrido for his hospitality during the XI Sitges conference.

References.

[1]   JJ Hopfield; Proc. Natl. Acad. Sci. USA 79 (1982) 2554.

[2]   For recent reviews on Spin glasses see K Binder, AP Young, Rev.Mod. Phys. **58** (1987) 801, and M Mezard, G Parisi and MA Virasoro "Spin Glass Theory and Beyond", Lecture Notes in Physics Vol 9, World Scientific 1987.

[3]   W Kinzel, Z. Phys. **B60** (1985) 205.

[4]   L Viana, J. Phys. (France) **49** (1988) 205.

[5]   DJ Amit, H Gutfreund and H Sompolinsky, Phys. Rev. **A32** (1985) 1007.

[6]   DJ Amit, H Gutfreund and H Sompolinsky, Phys. Rev. Lett. **55** (1985) 1530.

[7]   JA Hertz and RG Palmer; Duke University lecture notes to be published (1989), Secc. 2.2.1.

[8]   it approximates a feature present in an infinite NN with $\alpha = 0$, where the overlaps among the stored patterns are of order zero. However some other effects are not removed, since our sample still has $\sum \xi_i^\mu \neq 0$.

TAILORING THE PERFORMANCE OF ATTRACTOR NEURAL NETWORKS

K Y M Wong and D Sherrington
Department of Theoretical Physics, Oxford University,
1 Keble Road, Oxford OX1 3NP, U.K.

Abstract:

First, we study the effects of introducing training noise on the retrieval behaviours of dilute attractor neural networks. We found that, in general, training noise enhances associativity, but also reduces the attractor overlap. At a narrow range of storage levels, however, the system exhibits re-entrant retrieval behaviour on increasing training noise.

Secondly, we consider optimization of network performance, and subsequently the storage capacity, in the presence of retrieval noise (temperature). This is achieved by adapting the network to an appropriate training overlap, which is determined self-consistently by the optimal attractor overlap. The maximum storage capacity deviates from the storage capacity of the maximally stable network on increasing temperature, and in the high temperature regime ($T \geq 0.38$ for Gaussian noise), the Hebb-rule network yields the maximum storage capacity. Our analysis demonstrates the principles of specialization and adaptation in neural networks.

1. Introduction

Learning in neural network models can be considered as an optimization process [1-4]. This usually involves adjusting the synaptic interactions, or other adjustable parameters of the network, so that a certain performance requirement, such as the attractor overlaps or the size of the basins of attraction of a set of stored patterns, is optimized. Since neural networks may operate in different retrieving environments, their performance optimization may vary accordingly. For instance, networks performing well in noisy retrieving environments may not do so in less noisy environments, and vice versa. We are therefore interested in "tailor-making" neural networks which have optimal performances in specified retrieving environments.

In this article, we are interested in the effects of ambient noise on neural networks. In particular, we address two questions:

(i) What are the effects of introducing noises during the *training* stage of a neural network?

(ii) How can we optimize the performance of a neural network when it is used to retrieve patterns in the presence of noise?

Through our discussion, we shall also introduce the concept of adaptation which allows us to optimize the network performance in the presence of ambient noise. In fact, the principle of adaptation is very general, and can be applied to optimize the network performance in an arbitrarily specified environment. To emphasize the concepts, we shall keep the mathematical aspects in this article minimal. Interested readers are referred to [3-4] for further details.

2. The model

More concretely we shall use a particular model to illustrate the underlying principles. Consider a randomly diluted asymmetric neural network, in which the neuronal state at node i ($i = 1, \ldots, N$), takes the possible values ± 1, and is updated at each time step according to the sign of the local field, i.e.

$$S_i(t+1) = \mathrm{sgn} h_i(t); \quad h_i(t) = \frac{1}{\sqrt{C}} \sum_{j=i_1}^{i_C} J_{ij} S_j(t) \tag{1}$$

Here $h_i(t)$ is the (normalized) local field at node i, with C being the connectivity of a node, and $j = i_1, \ldots, i_C$ the nodes feeding node i. The interactions J_{ij} satisfy the spherical constraint $\sum_j J_{ij}^2 = C$, and we shall be interested in the case of large connectivity $C \gg 1$. We shall also restrict our discussion to dilute networks satisfying $C \ll \ln N$ whose retrieval dynamics, as we shall see, is completely determined by the *retrieval mapping* [5], rendering the problem solvable.

The network is assigned to store p patterns $\{\xi_i^\mu\}$ ($\mu = 1, \ldots, p$). If the network configuration has a macroscopic overlap m with only one of the stored patterns, say pattern 1, then the dynamics of the dilute network is completely determined by the retrieval mapping f(m), where

$$m(t+1) = f(m(t)); \quad m(t) = \frac{1}{N} \sum_{i=1}^{N} \xi_i^1 S_i(t) \tag{2}$$

The attractor overlap corresponds to the stable fixed points $m^* = f(m^*)$ of the retrieval mapping, and the basin boundary of the attractor is determined by its unstable fixed points $m_B = f(m_B)$. The storage capacity of the network is reached when the stable fixed point m^* corresponding to pattern retrieval coalesces with one of the unstable fixed points m_B as the storage level is varied.

It is well known that retrieval in attractor neural networks is very robust against noisy input data [5-7]. When the initial network configuration is a stored pattern messed up by a moderate amount of noise, it can still drift towards the neighbourhood of the stored pattern. This capability to process noisy input data will be referred to as *associativity.* High associativity of a network implies a large basin of attraction for its retrieval attractors.

An effective approach to optimize associativity, called "training with noise" [8], is to encode the synaptic interactions according to the perceptron learning rule [9-10], but with the input configuration of the stored patterns slightly distorted by random noise. The rationale is that if the network learns only clean patterns, it will retrieve clean patterns, but not necessarily noisy patterns, i.e. its associativity is low. But if the network learns noisy patterns, it may possibly retrieve noisy as well as clean patterns, i.e. its associativity is high.

Simulation results of Gardner, Stroud and Wallace [8] have demonstrated the success of this approach. Indeed, training noise enhances associativity, but on the other hand, excessive training noise also causes confusion, reducing the retrieval quality of the stored patterns.

3. Statistical mechanics of learning

Statistical mechanics has been successfully applied to study the *retrieving* stage of neural networks [6]. In this approach, one typically fixes the synaptic interactions $\{J_{ij}\}$ according to some prescription, say the Hebb rule, and then looks for network configurations $\{S_i\}$ that minimize an energy function, e.g. $E = -\frac{1}{2}\sum_{(ij)} J_{ij}S_iS_j$. In this approach, the averaged free energy is given by

$$-\beta F = \langle\langle \ln \text{Tr}_{\text{config space}} e^{-\beta E}\rangle\rangle_{\text{interaction space}} \tag{3}$$

In the usual language of spin glass theory, the free energy is *annealed averaged* over the configuration space, and *quenched averaged* over the interaction space.

Statistical mechanics of learning was first successfully formulated by Gardner [11] and Gardner and Derrida [1]. It can be viewed as an inverse problem of retrieving. Here one typically fixes the set of stored patterns $\{\xi_i^\mu\}$, and looks for interaction configurations $\{J_{ij}\}$ that minimize an energy function, e.g.

$$E = -\sum_\mu (\text{output overlap for correct input of pattern } \mu)$$

In this approach, the averaged free energy is given by

$$-\beta F = \langle\langle \ln \text{Tr}_{\text{interaction space}} e^{-\beta E}\rangle\rangle_{\text{pattern space}} \tag{4}$$

so that the free energy is annealed averaged over the interaction space, and quenched averaged over the pattern space. We shall concentrate on the statistical mechanics of learning.

Hence learning is equivalent to finding a network whose interactions minimize a certain energy function. By this specification of the learning process we are, at least for the moment, not interested in practical algorithms of learning. Instead, our optimization procedure will provide an upper bound for the performance of actual "training with noise" algorithms.

4. Statistical mechanics of noisy training

The interesting question, now, is whether the Gardner-Derrida approach can also be successfully applied to learning situations other than the learning of clean patterns, e.g. stochastic situations where random noise is present in the training. Our answer to this question is to consider input patterns $\{R_j^\mu\}$ being drawn from a *training ensemble* which consists of noisy versions of the patterns $\{\xi_j^\mu\}$. This means that

$$\Pr(R_j^\mu) = \frac{1}{2}(1+m_t)\delta(R_j^\mu - \xi_j^\mu) + \frac{1}{2}(1-m_t)\delta(R_j^\mu + \xi_j^\mu) \tag{5}$$

where we call m_t the *training overlap*. The corresponding energy function is now given by

$$E = -\sum_{\text{patterns}} \begin{pmatrix} \text{probability of a} \\ \text{noisy input pattern} \end{pmatrix} \begin{pmatrix} \text{output overlap of the} \\ \text{input noisy pattern} \end{pmatrix} \tag{6}$$

As remarked in Section 2, the retrieval dynamics of a dilute asymmetric network is completely determined by the retreival mapping. It turns out that the retrieval mapping $f(m)$ is in turn completely determined by the distribution $\rho(\Lambda)$ of the normalized field Λ of the clean input patterns (in the aligning direction of the output pattern bit) [12-14], i.e.

$$\Lambda_i^\mu = \xi_i^\mu \left(\frac{1}{\sqrt{C}} \sum_j J_{ij}\xi_j^\mu \right) \tag{7}$$

The more $\rho(\Lambda)$ shifts towards the positive, the more robust is the retrieval, and hence the better the associativity.

The retrieval properties of the network optimizing the noisy training energy function has been studied in [3]. The major results were that:

(i) When there is no training noise, all the aligning fields Λ are positive, but only marginally stable, i.e. the minimum Λ is zero. This agrees with the original results of Gardner and Derrida [1].

(ii) When an infinitesimal amount of training noise is introduced, all Λ's are *automatically* greater than the so-called maximal stability K, which is the maximal lower bound on the Λ's that can be achieved by iterative procedures like the perceptron learning rule [11] or the minover algorithm [15]. (This maximally stable network will be referred to as MSN hereafter.) This confirms the usefulness of training noises in enhancing memory associativity.

(iii) When the training noise $d_t \equiv \frac{1}{2}(1-m_t)$ becomes of the order $(\ln N)^{-1}$, errors at the retrieval attractor become inevitable. As the training noise increases, it causes confusion and the attractor overlap m^* continues to deteriorate, but the basin boundary m_B also decreases, signalling a widening of the basins of attraction, or alternatively an increase in associativity. For sufficiently high training noise, the boundary overlap m_B drops to zero, and the system is said to be in the *wide retrieval* phase.

(iv) In the extremely noisy limit, the system becomes a Hebb-rule network with $J_{ij} \sim \sum_\mu \xi_i^\mu \xi_j^\mu$.

Fig. 1 shows the training noise dependence of the aligning field distribution $\rho(\Lambda)$ from the maximally stable limit $(m_t = 1^-)$ to the Hebb-rule limit $(m_t = 0^+)$. Fig. 2(a-c) shows the training noise dependence of the attractor overlap m^* and the boundary overlap m_B for sufficiently low storage.

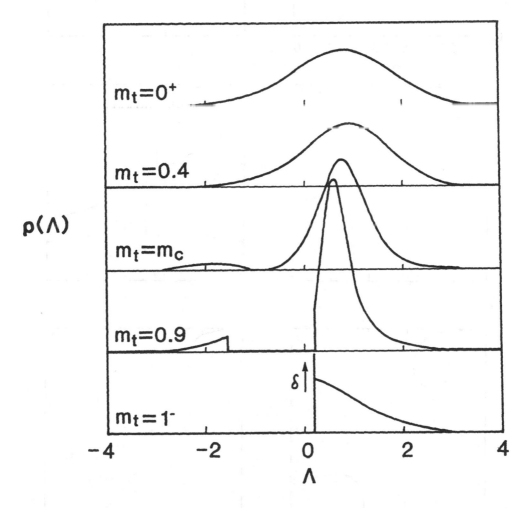

Fig. 1: The aligning field distribution $\rho(\Lambda)$ for different training overlaps from the maximally stable limit to the Hebb-rule limit, at $\alpha = 1.5$. Starting from the bottom, the curves correspond to $m_t = 1^-, 0.9, m_c, 0.4$ and 0^+. At $m_c = 0.78$ the two-band distribution becomes one-band. The curve for $m_t = 1^-$ has a delta function peak at $K = 0.19$. The vertical scale has units displaced by 0.6 for each curve.

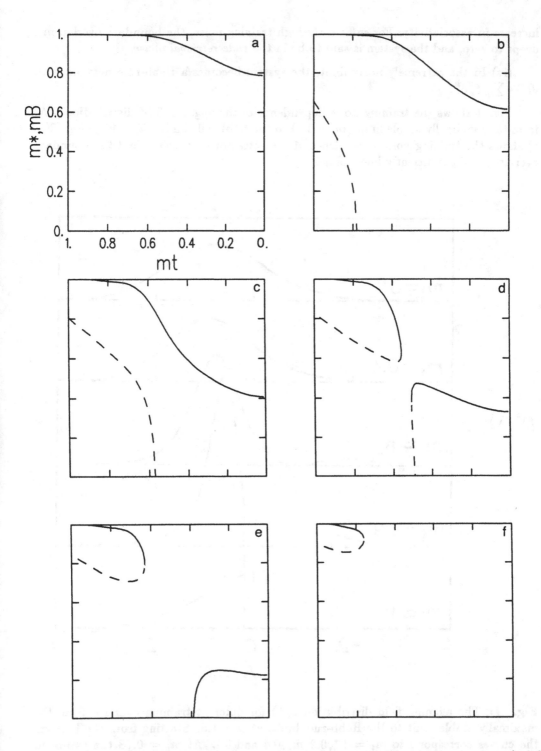

Fig. 2: The training noise dependence of the attractor and boundary overlaps (solid and dashed lines respectively) at storage levels $\alpha =$(a)0.4, (b)0.5, (c)0.58, (d)0.6, (e)0.62, (f)0.7.

5. New results

Recently, however, we have found some intriguingly surprising results when the storage level $\alpha \equiv p/C$ is varied beyond the low storage regime [16]. In Fig. 2(d) a *re-entrant* behaviour appears. At low training noise, the stored patterns have narrow basins, As the training noise increases, the attractor and boundary overlaps coalesce and the patterns are no longer retrievable. But as the training noise further increases, a pair of stable and unstable fixed points begins to appear again and bifurcate. Thus the attractors of the stored patterns appear again with narrow basins of attraction, which becomes wide basins on further increase of training noise.

Fig. 2(e) shows a similar re-entrant behaviour at a higher storage level, but the re-entrant retrieval phase has a wide basin of attraction for all the training noise levels where it exists. The extent of the re-entrant retrieval phase shrinks with increasing storage level.

For high storage levels (Fig. 2(f)), the re-entrant retrieval phase disappears. Narrow retrieval is possible at low training noise, and no retrieval is possible at high training noise.

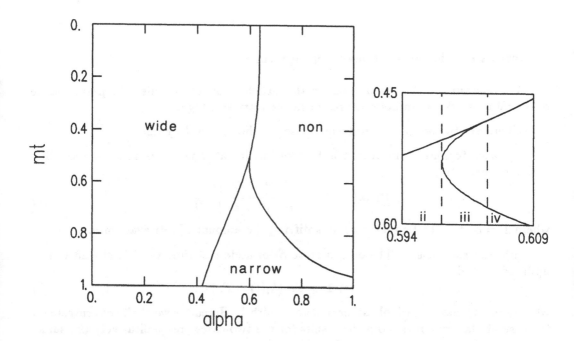

Fig. 3: (a) The phase diagram of retrieval behaviours in the $m_t - \alpha$ space. The lower right curve extends to the point of maximum storage $(\alpha, m_t) = (2, 1)$. (b) The amplified phase diagram around the tricritical point, showing three transition behaviours. (Horizontal magnification = 10 times vertical magnification.)

Fig. 3 shows the phase diagram in the $m_t - \alpha$ space. Generally speaking, wide retrieval phase exists at low storage level, above which narrow retrieval phase exists at high training overlap, and non-retrieval phase exists at low training overlap. The wide retrieval phase is bounded by boundaries of continuous transition to either narrow retrieval or non-retrieval, and the transition between narrow retrieval and non-retrieval phases is discontinuous. The phase transition lines meet at a tricritical point with a common slope.

The re-entrant behaviour can be observed around the tricritical point. This is indicated by a bend of the discontinuous transition line before arriving at the tricritical point. As training noise increases, the transition behaviours are:
(i) Only wide retrieval for α below 0.42.
(ii) Narrow $\rightarrow$ wide retrieval for α between 0.42 and 0.599.
(iii) Narrow $\rightarrow$ non $\rightarrow$ narrow $\rightarrow$ wide retrieval for α between 0.599 and 0.604.
(iv) Narrow $\rightarrow$ non $\rightarrow$ wide retrieval for α between 0.604 and 0.64.
(v) Narrow $\rightarrow$ non-retrieval for α above 0.64.

The explanation and implication of re-entrant retrieval has not been fully explored yet. Does it indicate that networks corresponding to high training noises (e.g. Hebb-rule net) and low training noises (e.g. MSN) have different mechanism for retrieval? Is the same phenomenon observed in non-dilute networks? Surely these interesting questions deserve further attention.

6. Retrieval noise and retrieval temperature

We now turn to the second issue in this article: how to optimize the performance of neural networks when noise is present in the retrieval stage?

There are two ways of introducing noises to the retrieval dynamics:

(i) Discrete noise – The output state S_i of a node i at time $t+1$ is updated according to the probability

$$\Pr(S_i(t+1)) = \frac{e^{\beta h_i(t) S_i(t+1)}}{e^{\beta h_i(t)} + e^{-\beta h_i(t)}} \qquad (8)$$

where $T = \beta^{-1}$ is the temperature, quantifying the amount of retrieval noise.

(ii) Gaussian noise – The output state S_i of node i at time $t + 1$ is stochastically updated according to

$$S_i(t+1) = \text{sgn}(h_i(t) + Tz) \qquad (9)$$

where z is a Gaussian variable of mean 0 and width 1. Hereafter we shall concentrate on the case of Gaussian noise, but the results for the two cases are qualitatively the same.

Recent work by Amit et al [7] has shown that the Hebb-rule net has a higher storage capacity than the MSN in the high temperature regime. This shows that although the MSN has the maximum storage capacity at zero temperature, its performance is far from optimal at high temperatures. Thus we are interested in performance optimization at non-zero temperatures here.

Intuitively, we may expect that training noises are required to improve the retrieval performance in noisy networks. In the presence of retrieval noises, the input signals to a node can never be perfect even in the attractor of a stored pattern. Optimizing the performance of the network therefore requires its adaptation to imperfect input signals. The notion of training noise adaptation is therefore relevant.

7. The principle of adaptation

To determine the amount of training noise to be introduced, we make use of *the principle of adaptation*. This principle has a very strong biological flavour. As a biologist may expect, a species of living organism will survive best in a given environment if it adapts to that environment. Interestingly, a similar principle of adaptation exists in the world of neural networks:

When a system optimizes its performance in a training environment, then its performance is optimized, among other systems, in the same retrieval environment.

Thus we can optimize the network performance by adjusting the training noise to be at the same level as the error in the retrieval attractor. Since the retrieval error again depends on the training noise, they can only be determined self-consistently.

More concretely, let us define the appropriate energy function to be minus the output overlap of the stored patterns at temperature T and an input overlap m_t, which will be determined self consistently later.

$$E = - \sum_{patterns} \begin{pmatrix} \text{probability of a} \\ \text{noisy input pattern} \end{pmatrix} \begin{pmatrix} \text{output overlap of the input} \\ \text{noisy pattern at temperature T} \end{pmatrix} \quad (10)$$

To determine the training overlap m_t which gives the optimal performance for a constant temperature T and storage level α, we invoke the principle of adaptation which, in this case, relates the training overlap m_t and the retrieval overlap m:

(i) If we consider a fixed retrieval overlap m and search the space of interactions, the network which gives the best output overlap is the one corresponding to the training overlap $m_t = m$.

(ii) Conversely, if we consider a network with fixed training overlap m_t and search the space of state configurations, the output overlap which is better than those of any other networks is found at the retrieval overlap $m = m_t$.

Thus we can envisage a family of retrieval curves $f_{m_t}(m)$ each being enveloped by the curve $f_m(m)$ above them (see Fig. 4). Mathematically, this is equivalent to

$$\frac{d}{dm_t} f_{m_t}(m)\Big|_{m_t=m} = 0 \quad (11)$$

Now if we consider the training overlap m_t as an adjustable parameter, the principle of adaptation implies that the stable fixed point of the envelope $f_m(m)$ would give the best attractor overlap. Hence for a constant temperature T and storage level α, we would choose the training overlap m_t to be the fixed point m of the envelope $f_m(m)$:

$$f_m(m) = m \qquad (12)$$

Once we obtain the fixed point of the envelope, the retrieval mapping of the corresponding training overlap will touch the envelope at the same point, and its fixed point will be identical. As schematically shown in Fig. 4, it will give a greater fixed point overlap than any other network for the particular T and α. Hence the fixed point of the envelope gives, on one hand, the training overlap required to optimize the network performance and, on the other hand, the attractor overlap during retrieval.

8. Results for the optimally adapted retrievers

We compare the behaviours of the optimally adapted network (hereafter referred to as OAN) and the MSN in Fig. 5(a-d). First consider $T = 0$ (Fig. 5(a)). Since no noise is present during retrieval, the MSN is already optimal in its retrieval overlap. Hence the stable fixed points of the retrieval mapping of the MSN and the retrieval envelope are identical up to the maximum storage capacity $\alpha = 2$. The unstable fixed point of the MSN, however, is always greater than that of the retrieval envelope.

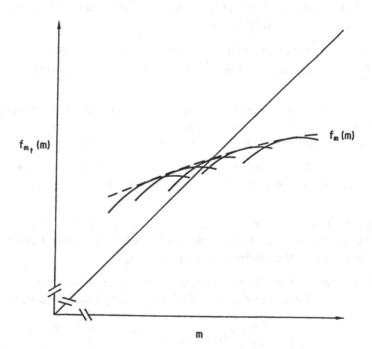

Fig. 4: A schematic diagram of the family of retrieval curves $f_{m_t}(m)$ and their envelope $f_m(m)$. The fixed point of the envelope optimizes the network performance.

Note that for α between 0.60 and 0.64, there exists a second smaller stable fixed point of the retrieval envelope, indicating the existence of a weaker retriever having a locally maximal performance when compared with other networks in its neighbourhood of the interaction space. In this range of α, the retrieval envelope has two convex regions in the range $0 \leq m \leq 1$, and hence there exist two non-zero stable fixed points, in contrast to the retrieval mapping of the MSN, which has at most one non-zero stable fixed point in the same range (see Fig. 6). In fact, this weak retriever corresponds to the local optimum in the re-entrant retrieval phase discussed in Fig. 2.

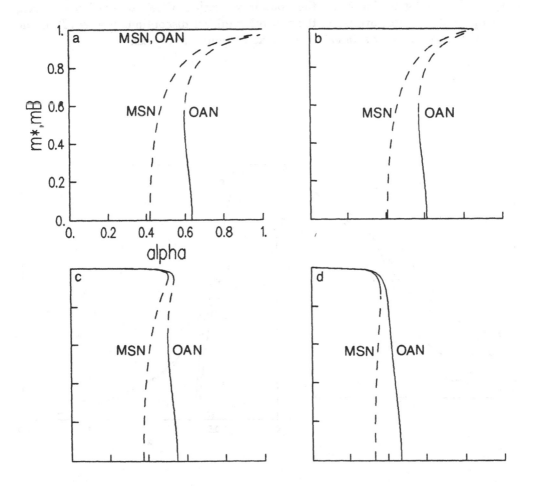

Fig. 5: The dependence of the fixed point overlap of the OAN on the storage level for $T =$(a)0, (b)0.2, (c)0.4 and (d)0.6. The stable fixed point of $f_m(m)$ is shown in solid curve, and the unstable in dashed curve. The retrieval behaviour of the MSN is also shown for comparison. In (a) the three curves extend to the point of maximum storage $(\alpha, m) = (2, 1)$.

The unstable fixed point of an individual retrieval mapping yields the basin boundary of attraction. On the other hand, the retrieval envelope describes the behaviour of a self-adaptive process, which involves continually optimizing the network performance in the (adiabatically evolving) environment created by its own retrieval stage, so that the attractor performance of the network is eventually optimized. Hence the unstable fixed point yields the basin boundary of self-adaptation.

At low but non-zero T (Fig. 5(b)), the optimal training noise is also low, and the improvement in both the attractor overlap and the storage capacity from the MSN to the OAN is only marginal. The attractor overlap of the strong retriever, however, is no longer perfect, but drops with both T and α. At some $\alpha = \alpha_c(T)$, the attractor overlap of the strong retriever and the boundary overlap of self-adaptation coalesce, and the optimal attractor overlap of the network vanishes discontinuously. $\alpha_c(T)$ is the maximum storage capacity attainable by attractor neural networks.

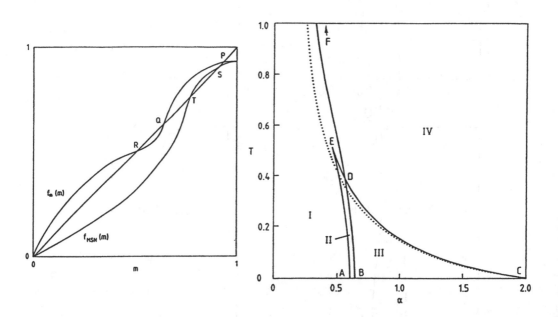

(Left)**Fig. 6:** A schematic plot of the retrieval envelope $f_m(m)$ having two non-zero stable fixed points in the range $0 \leq m \leq 1$. The retrieval curve $f_{MSN}(m)$ of the MSN, which has only one non-zero stable fixed point, is also shown for comparison.

(Right)**Fig. 7:** The retriever phase diagram of the optimal network in the temperature-storage space. The retrieval phase boundary of the MSN (dotted line) is also shown for comparison.

As T increases (Fig. 5(c)), the improvement in performance from the MSN to the OAN is increasingly marked. On the other hand, the strong retriever of high training overlap is less and less favourable when compared with the weak retriever of lower training overlap. At sufficiently high temperature, the strong retriever vanishes before the weak on increasing α. Thus the Hebb-rule network gives the maximum storage capacity for attractor neural networks, and the optimal attractor overlap undergoes a continuous transition.

As T further increases (Fig. 5(d)), the attractor overlaps and the storage capacities of both the MSN and OAN drop. The distinction between strong and weak retrievers disappears, indicating that re-entrant behaviours no longer exist, and the retriever phase vanishes continuously when α is increased from 0 to $\alpha_c(T)$.

Fig. 7 shows the phase diagram in the $T - \alpha$ space. The curve AE corresponds to the appearance of the weak retriever, CE corresponds to the discontinuous vanishing of the strong retriever, and BF to the continuous vanishing of the weak retriever. Hence the regions I to IV are respectively the single retriever, strong and weak retrievers, strong retriever, and non-retriever phases. The maximum storage capacity of attractor neural networks is given by the discontinuous transition line CD for $T \leq 0.38$, and the continuous transition line DF for $T \geq 0.38$. Note the increase in storage capacity from the MSN to the OAN when an optimal amount of training noise is introduced.

The retrieval performance and phase diagrams do *not* correspond to those of networks with *fixed* synaptic prescriptions, say the MSN or the Hebb-rule network. Instead, we are searching the entire space of interactions for the optimal performance. The network for each value of T and α corresponds to a unique interaction configuration, or a unique *retriever*. To emphasize this distinction, we call a phase diagram such as Fig. 7 a *retriever*, rather than *retrieval*, phase diagram, Alternatively, this phase diagram can be interpreted as one of *self-adaptation*, rather than one of *attraction*. The retrieval behaviour of a network of fixed synaptic prescriptions is determined by α, but not T. the retrieval behaviour of a self-adapted network is optimally determined by both α and T. If we consider networks self-adapted at a fixed temperature T_1, the corresponding phase diagram of attraction will have the phase boundaries touching those of self-adaptation at T_1, but not necessarily at other temperatures.

9. How to optimize associativity?

We have found the optimal attractor overlap as a function of T and α, and the maximum storage capacity as a function of T. A related issue is to optimize the associativity of the attractor neural network (i.e. to maximize its basins of attraction). Similar arguments lead us to conclude that the optimal basin boundary for a constant T and α is given by the unstable fixed points of the envelope $f_m(m)$, and a training overlap equal to this basin boundary should be introduced to attain this optimum.

Take, for example, the low T plot in Fig. 5(b). As α increases from 0, the only unstable fixed point of the retrieval envelope $f_m(m)$ is m=0. Here we have the wide retriever phase, and the Hebb-rule net assoicates best. For intermediate values of α, the retrieval envelope has one zero and one non-zero unstable fixed point, corresponding respectively to the existence of both a wide and a narrow retriever. As α further increases, only the non-zero unstable fixed point remains, and we have the narrow retriever phase. Above $\alpha_c(T)$, we have the non-retriever phase.

Since the stable and unstable fixed points coalesce at the phase boundaries, the phase diagram in this case is identical to Fig. 7, except that the regions I to IV are now respectively the wide retreiver, wide and narrow retrievers, narrow retriever, and non-retriever phases. This suggests that the maximum storage capacity $\alpha_c(T)$ can be attained by either maximizing the retrieval overlap or the basin of attraction; the two requirements are equivalent at the phase boundary. In regions I and II, the Hebbian synaptic prescription gives the maximum possible associativity, although it may not give the best retrieval overlap.

10. Conclusion

In this article, we have answered the two questions posed in Section 1:

(i) For sufficiently low storage levels, training noise enhances memory associativity, making possible the retrieval of very noisy patterns; but at the same time, training noise reduces the quality of the retrieved patterns. At higher storage levels, however, the network exhibits re-entrant retrieval behaviour.

(ii) By introducing an appropriate amount of training noise, we can optimally improve the performance of a neural network in the presence of retrieval noise ($T \neq 0$). The amount of training noise to be introduced is determined self-consistently by equating it with the amount of errors in the retrieval attractor.

Besides these specific conclusions, our work has demonstrated some general principles which can be applied to other networks.

Our work has demonstrated the principle of specialization for different environments in neural networks, namely, networks optimizing the performance in one environment do not optimize at another environment. The effect of temperature is a good example. At $T = 0$ the MSN stores most, but at higher temperatures the networks with the maximum storage capacity correspond to training overlaps less than 1^-. In the high temperature regime ($T \geq 0.38$) the Hebb-rule network yields the maximum storage capacity. Therefore one cannot attain the best storage at all temperature ranges using a single network. In other words, networks are *specialized*.

The notion of specialization also applies to different performance requiremants in neural networks. The best retrieval overlap and associativity are generally given by a higher and lower training overlap, which correspond respectively to stable and unstable fixed points of the retrieval envelope. One cannot attain both the best retrieval overlap and associativity using a single network, except at the phase boundaries. Again, networks can be described as specialized. The implication to network design is that separate or modular networks have to be considered in order to achieve both objectives.

Our work has also demonstrated the principle of adaptation, namely, to optimize the performance P of a network to be operated in the retrieval environment E, the network should adapt P to the training environment $E_t = E$ during training. In our case, the performance function is the averaged output overlap, the training environment is specified by the training overlap, and the retrieval environment by the attractor overlap. In general, these parameters can be replaced by other performance functions, e.g. chaoticity (fraction of non-frozen spins), or selectivity (ability to differentiate similar patterns) etc, and other environmental parameters, e.g. temperature, or thresholding function etc. The important point is that the principle of adaptation still applies.

Our studies have also hinted at the possibility of using self-adaptation as a basis for *practical adaptive algorithms* to optimize the network performance in the presence of noise (or in other retrieval environment). It may be possible that the learning procedure involves relaxing the network to an attractor, and then updating the synaptic interactions to optimize the performance at the retrieval attractor. This surely deserves further exploration.

References

[1] Gardner E and Derrida B 1988 J Phys A: Math Gen **21** 271

[2] Wong K Y M and Sherrington D 1988 Europhys Lett **7** 197

[3] Wong K Y M and Sherrington D 1990 J Phys A: Math Gen **23** L175

[4] Wong K Y M and Sherrington D 1990 "Optimally adapted attractor neural networks in the presence of noise" submitted to J Phys A: Math Gen

[5] Derrida B, Gardner E and Zippelius A 1987 Europhys Lett **4** 167

[6] Amit D, Gutfreund H and Sompolinsky H 1987 Ann Phys (NY) **173** 30

[7] Amit D, Evans M, Horner H and Wong K Y M 1990 J Phys A: Math Gen in press

[8] Gardner E, Stroud N and Wallace D 1989 J Phys A: Math Gen **22** 2019

[9] Rosenblatt F 1962 *Principles of Neurodynamics* (New York: Spartan)

[10] Minsky M and Papert S 1969 *Perceptrons* (Cambridge, MA: MIT Press)

[11] Gardner E 1988 J Phys A: Math Gen **21** 257

[12] Kepler T B and Abbott L F 1988 J Physique **49** 1657

[13] Krauth W, Nadal J-P and Mézard M 1988 J Phys A: Math Gen **21** 2995

[14] Gardner E 1989 J Phys A: Math Gen **22** 1969

[15] Krauth W and Mézard M 1987 J Phys A: Math Gen **22** 2019

[16] Wong K Y M and Sherrington D 1990 "Retrieval properties of noise-optimal attractor neural networks" to be published

LEARNING AND OPTIMIZATION

J. Bernasconi

Asea Brown Boveri Corporate Research

CH-5405 Baden, Switzerland

Abstract

Many learning paradigms for neural networks are based on the optimization of some measure of performance. The learning behavior of these networks then depends on the topography of the corresponding "fitness landscape" and on the chosen optimization method. In this paper, we apply different learning strategies to a number of Boolean problems and analyze how the topography of the fitness landscape is affected by the complexity of the problem and by the network architecture. We further discuss the application of stochastic optimization strategies to problems with delayed performance estimation and present results for a neurocontrol solution to the "lunar lander" problem.

1. Introduction

An important property of neural networks is their ability to learn, where learning is often defined as a process which optimizes the performance of the network with respect to a given task (see Ref. [1] for a review of learning procedures). In supervised learning, performance is usually measured in terms of the output error, and the majority of the proposed learning procedures, e.g., "error backpropagation" [1-3] or "Boltzmann machine learning" [1,3,4], use gradient descent techniques to find an optimal set of weights for the connections in the network. For networks with hidden units, the error landscape may contain many local minima which make it difficult for a gradient descent algorithm to find a good solution. Even more serious appears to be the presence of long flat valleys which often lead to excessively long learning times. An obvious way to deal with these problems is to use more efficient optimization techniques, and several authors have successfully applied stochastic optimization strategies to train a neural network [5-8].

An additional complication occurs in the application of neural networks to control problems. Here, the quality of the output cannot be rated directly, and the performance of the control network can only be estimated after a certain time delay, i.e., after many control cycles. Several strategies have been proposed to solve this problem of learning over time, e.g., backpropagation through time [9,10], or adaptive reinforcement prediction [9,11].

In this paper, we discuss several aspects of learning as an optimization problem. In particular, we analyze the "fitness" landscape of feedforward networks for some Boolean learning tasks, and we demonstrate that stochastic optimization strategies represent an interesting alternative to gradient descent learning. Sections 2 and 3 are concerned with supervised learning from examples. In section 2, we briefly review gradient descent procedures, such as error backpropagation [1-3], and then discuss several stochastic learning strategies. In section 3, the different learning procedures are applied to a number of small Boolean problems. From an analysis of the respective learning behaviors, we are able to obtain information about the topography of the error landscape over the space of weights, and about how this topography is affected by the network architecture and by the complexity of the learning problem. In section 4, we turn to the problem of delayed performance estimation in neurocontrol applications and discuss the use of stochastic strategies for optimization over time. In section 5, finally, a stochastic learning procedure is applied to train a neural network which is supposed to control a "lunar lander".

2. Supervised Learning from Examples

We restrict ourselves to networks of the feedforward type (multilayer perceptrons). These consist of a layer of input units, of one or more layers of hidden units, and of a layer of output units, and only feedforward connections are allowed. The state of unit i is described by a scalar variable, S_i, which is determined by the total weighted input to that unit,

$$S_i = f \left(\sum_j W_{ij} S_j - \Theta_i \right). \tag{1}$$

W_{ij} denotes the weight associated with the connection from unit j to unit i, and the nonlinear activation function f is usually a threshold function or a sigmoid-type function such as $f(x) = (1 - exp(-x))/(1 + exp(-x))$. By introducing an extra input unit whose state is held fixed, $S_0 = 1$, the thresholds Θ_i can be treated like ordinary weights, $\Theta_i = -W_{i0}$.

In a feedforward network, the states of the units are determined in a single forward pass through the network, and if the state variables, S_i, of the input and output units are explicitly identified as I_i and O_i, respectively, we have

$$O_i = O_i \left(\{I_i\}, \{W_{ij}\} \right). \tag{2}$$

For a fixed input pattern $\{I_i\}$, the output of the network is thus determined by the weights W_{ij}, and learning can be defined as a process which adjusts these weights in such a way that the performance of the network is optimized. In supervised learning, performance is often measured in terms of the squared output error, F, summed over a set of learning examples,

$$F = \sum_{\mu=1}^{N} F^\mu \quad , \quad F^\mu = \frac{1}{2} \sum_i \left(D_i^\mu - O_i^\mu\right)^2 , \tag{3}$$

where $\{D_i^\mu\}$ is the desired output pattern for the learning example with input $\{I_i^\mu\}$.

Gradient descent learning

Most learning procedures use gradient descent to find a set of weights which minimizes F, and one distinguishes between "batch learning" and "example-by-example learning". In the latter case, the learning examples μ are presented in random order, and each time the W_{ij}'s are changed by an amount proportional to the respective negative gradient of the error for the chosen example,

$$\Delta W_{ij} = -\eta \frac{\partial F^\mu}{\partial W_{ij}}. \tag{4}$$

If η is small enough, this procedure minimizes the total error, $F = \sum_\mu F^\mu$, to an accuracy proportional to η [12]. In batch learning, F^μ in Eq.(4) is replaced by F, i.e., the errors are accumulated over all learning examples before the weights are actually changed.

Error backpropagation [1-3] refers to a particularly efficient implementation of gradient descent minimization for feedforward networks. It represents the probably most widely used supervised learning procedure and is applied in the example-by-example as well as in the batch version.

On complicated error landscapes, gradient descent methods suffer from a number of problems. The probably most serious one is the presence of flat valleys which often leads to extremely long learning times. Large values of the learning parameters η, on the other hand, cause oscillations which prevent the algorithm from converging. The error surface, moreover, may contain many local minima, so that it may be very difficult to find a good solution with a gradient descent algorithm which is only guaranteed to converge to the nearest local minimum.

To overcome these problems, one has to look for more efficient optimization procedures, and in this paper we shall, in particular, discuss the use of stochastic strategies. These have the additional advantage that they can also be applied if the weights are restricted to a discrete set of values, or if the activation function f is non-differentiable.

Stochastic learning strategies

As described above, the objective of supervised learning is to minimize an error function of the form

$$F(\mathbf{W}) = \sum_{\mu=1}^{N} F^{\mu}(\mathbf{W}),$$ (5)

where $F^{\mu}(\mathbf{W})$ is the output error for the μ-th learning example, e.g. given by Eq. (3), and where $\mathbf{W} = \{W_{ij}\}$ denotes a point in the space of weights. This space is usually assumed to be continuous, i.e., $W_{ij} \in \mathbf{R}$. For certain applications, however, it may be desirable to restrict the weights to a discrete set of values, e.g., $W_{ij} = \pm 1$.

Stochastic learning procedures are strategies which direct a random search in weight-space towards a good minimum of the error function $F(\mathbf{W})$. At each iteration step, the current weight vector $\mathbf{W}$ is changed by some random amount $\Delta \mathbf{W}$, and the new weight vector, $\mathbf{W}' = \mathbf{W} + \Delta \mathbf{W}$, is either accepted or rejected, according to some criterion. This criterion is usually formulated in terms of the associated change of the output error, $\Delta F = F(\mathbf{W}') - F(\mathbf{W})$.

A basic, and frequently used, strategy is *iterative improvement*, i.e. the new weight vector $\mathbf{W}'$ is only accepted if $\Delta F \leq 0$. This procedure represents the stochastic equivalent of gradient descent, and it suffers from the same problems with local minima, at least if the magnitude of the weight changes is bounded. Under certain assumptions for the probability distribution of $\Delta \mathbf{W}$, however, it can be proved that the procedure eventually converges to the global minimum, provided that the search is restricted to a compact region in weight space [13]. We have also found (see below) that the problems with local minima sometimes can be avoided by using an example-by-example version of the iterative improvement procedure, i.e., the new vector $\mathbf{W}'$ is accepted if $\Delta F^{\mu} \leq 0$, where μ refers to a randomly selected learning example. A more sophisticated strategy to prevent the search algorithm from becoming trapped in a bad local minimum is *simulated annealing* [14,15]. Here, the randomly created trial vector $\mathbf{W}'$ is accepted with probability $p(\Delta F)$, where

$$p(\Delta F) = \begin{cases} 1 & \text{if } \Delta F \leq 0 \\ exp(-\Delta F/T) & \text{if } \Delta F > 0. \end{cases}$$ (6)

The efficiency of the simulated annealing procedure depends crucially on the choice of the "cooling" schedule for the control parameter T ("temperature") [15,16].

Another promising class of optimization strategies is represented by the so-called *genetic algorithms* [5,6,17] which operate on a population of different weight vectors.

3. Error Landscape Topography and Learning Behavior

The two main factors that determine the learning behavior of a neural network are the topography of the "fitness landscape" on which the learning process takes place and the chosen optimization strategy. For supervised learning, "fitness" is measured in terms of the output error, and we would expect that the topography of the corresponding error surface over the weight space depends on the complexity of the learning task as well as on the details of the network architecture.

A convenient measure for the complexity of a learning problem is the maximum possible information gain during a learning process. This can be calculated by interpreting a neural network as a thermodynamic system whose states are the weight-configurations, $\mathbf{W}$, and whose energy function is the total output error, $F(\mathbf{W})$, for a given set of learning examples [18, 19]. The information gain is then equal to the entropy decrease which occurs during learning, where the entropy S, as a function of temperature T, is given by

$$S(T) = \frac{d}{dT} \left(T log Z \right), \quad Z = \sum_{\mathbf{W}} e^{-F(\mathbf{W})/T}. \tag{7}$$

The maximum possible entropy decrease, ΔS, i.e., our complexity measure, is obtained if we assume that we have no information before learning and maximum information after learning, so that

$$\Delta S = S(T = \infty) - S(T = 0). \tag{8}$$

If the weights are restricted to a finite set of discrete values (e.g., $W_{ij} = \pm 1$), this reduces to the simple expression [18]

$$\Delta S(\text{bits}) = log_2 \frac{n}{n_P}, \tag{9}$$

where n denotes the total number of weight-configurations, and n_P the number of weight-configurations which solve the given learning problem P.

To study the connections between problem complexity, error surface topography, and learning behavior, we consider four different 2 input/1 output Boolean problems:

Input		Output			
I_1	I_2	XOR	I_1	AND	CONST
1	1	-1	1	1	1
1	-1	1	1	-1	1
-1	1	1	-1	-1	1
-1	-1	-1	-1	-1	1

We first calculate the complexity of these problems with respect to a feedforward network with one layer of hidden units and binary weights ($W_{ij} = \pm1$). The activation function f, see Eq. (1), for the hidden units and for the output unit is chosen as $f(x) = sign(x)$, and there are no direct input-output connections. The results are summarized in the following table.

Complexity ΔS (bits)

# hidden units	XOR	I_1	AND	CONST
2	5.0	5.0	4.0	2.7
3	5.1	5.1	4.1	2.6
4	4.6	4.6	3.9	3.1

For this type of networks, the XOR problem thus has the same complexity as the problem of reproducing one of the inputs at the output. The constant output problem has the lowest complexity, and the AND problem is inbetween the two. We further note that the complexity appears to depend only weakly on the number of hidden units, an observation which has also been made by Carnevali and Patarnello [18].

We now turn our attention to the learning probability, i.e., to the probability of finding a solution, which obviously depends on the chosen learning strategy (optimization strategy). The complexity ΔS of a given learning problem P, see Eq. (9), is directly related to the learning probability p_L for a purely random search in weight space for which $p_L = n_P/n$. In the next table, these learning probabilities are compared with those obtained by applying the batch version an iterative improvement strategy based on single weight changes.

Learning Probability p_L

# hidden units	Problem	Random search	Iterative improvement
2	XOR	0.03	0.15
2	I_1	0.03	0.48
2	AND	0.06	1.00
2	CONST	0.16	1.00
3	XOR	0.03	0.95
3	I_1	0.03	0.99
3	AND	0.06	0.95
3	CONST	0.17	0.97
4	XOR	0.04	0.99
4	I_1	0.04	1.00
4	AND	0.07	0.99
4	CONST	0.11	1.00

There does not appear to exist an obvious relation between the complexity of the problem and the p_L-values for the iterative improvement strategy. In particular, the iterative improvement learning probabilities increase very strongly when the number of hidden units is increased, while the problem complexity, i.e., the fraction of solutions, remains approximately constant. We also applied a simulated annealing strategy which, for the small problems considered here, always leads to a solution.

From these results we obtain some information about the topography of the error landscape over the space of weights. The random search results directly give the fraction of solutions, and the p_L-values for iterative improvement measure the relative size of the corresponding basins of attraction. The dependence of these learning probabilities on the number of hidden units indicates that the number of local minima, or the size of their basins of attraction, is strongly reduced when the number of hidden units is increased.

We have further found that the problems with local minima are much less severe if we apply an example-by-example version of the iterative improvement procedure. Corresponding results for the XOR-problem with 2 hidden units are shown in Figure 1. At each iteration step, one of the four learning examples was selected at random. If the network output for this example was incorrect, one of the (binary) weights was flipped, otherwise the weights were left unchanged. We observe that the learning probability approaches one after about 300 example presentations. The

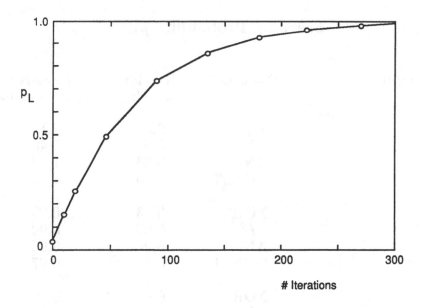

Fig. 1. Learning probability vs. number of iterations for an example-by-example version of the iterative improvement strategy. The results refer to the XOR-problem and to a network with two hidden units and binary weights.

batch version, on the other hand, converges much faster, but has a probability of 0.85 of becoming trapped in a local minimum.

If we allow the weights to take arbitrary real values ($-\infty < W_{ij} < \infty$), and if we choose a sigmoid activation function for the output and for the hidden units, the networks considered in this section can also be trained with the error backpropagation procedure, i.e., by using gradient descent. For the corresponding analysis of our four Boolean problems we have chosen a rather steep activation function, $f(x) = (1 - exp(-3x))/(1 + exp(-3x))$, and we have set the learning parameter η equal to 0.1. This choice is certainly not optimal, but our aim was to obtain a meaningful comparison of the different learning probabilities, and not to optimize the performance of the learning procedure. Some of our results are shown in Figs. 2 and 3. In these simulations, the initial weights were assumed to be uniformly distributed between -1 and +1, and the output error was considered zero if it was smaller than 0.2. We have used the batch version of the backpropagation procedure, i.e., the weights are updated after a complete cycle through the learning examples. From Fig. 2 we conclude that also for the backpropagation procedure, the XOR remains by far the hardest problem to learn. The I_1-problem, on the other hand, becomes much easiser than the AND. We have tried to obtain some information on the topography of the error landscape by analyzing the shape of the

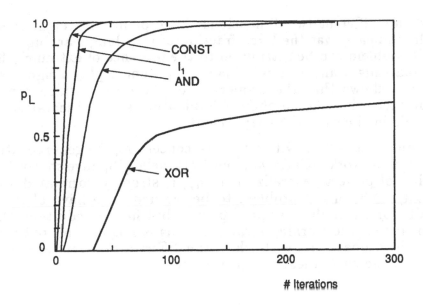

Fig. 2. Learning probability vs. number of iterations for the batch version of the error backpropagation procedure, applied to the four Boolean problems discussed in the text. All curves refer to a network with two hidden units.

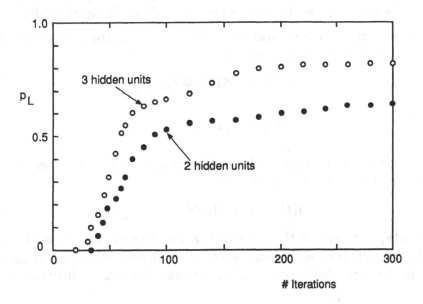

Fig. 3. Comparison of the XOR learning probability for networks containing two and three hidden units, respectively (error backpropagation learning, batch version)

error surface for different two-dimensional hyperplanes in weight space. These results indicate that the large fraction of very long learning times for the XOR-problem can be attributed to the presence of extremely flat valleys and plateaus, while local minima do not seem to play a significant role. Figure 3 shows that the properties of the error surface can be improved by increasing the number of hidden units, an observation we have also made in the case of binary weights.

We conclude this section with a remark concerning the generalization ability of neural networks. Carnevali and Patarnello [18] have shown that the probabiliy of perfect generalization, p_G, is strongly correlated with the complexity, ΔS, of the problem to be learned. We have obtained similar results for our multilayer perceptrons, but find, in addition, that p_G also depends on the learning strategy. This is not unexpected if we observe that the probabilities with which the different solutions are found may depend on the chosen learning procedure.

4. Delayed Performance Estimation

Many interesting applications of neural networks are concerned with the learning of a strategy, e.g., of a control strategy. In such cases, the desired output is usually not known, and the performance of the network can only be estimated in the future, i.e., only after a large number of control decision have been made. The error minimization procedures discussed in the previous sections are thus not directly applicable.

Consider a system whose state a time t is described by $\mathbf{x}(t)$, and whose dynamical evolution is given by

$$\mathbf{x}(t+1) = s(\mathbf{x}(t), u(t)), \tag{10}$$

where

$$u(t) = u(\mathbf{x}(t)) \tag{11}$$

represents the control strategy to be optimized. If the controller is realized by a neural network, we can write

$$u(t) = u(\mathbf{x}(t), \mathbf{W}), \tag{12}$$

where $\mathbf{W} = \{W_{ij}\}$ refers to the adjustable weights in the network. The criterion with which the performance of the control actions is measured often involves a summation over contributions referring to future times,

$$G(t) = \sum_{\tau=t}^{T-1} g(\mathbf{x}(\tau), u(\tau)) + H(\mathbf{x}(T)) = min, \tag{13}$$

and sometimes even depends only on the final state of the system, i.e., $G = H(\mathbf{x}(T))$. The learning problem for the control network thus refers to an optimization over time.

Several strategies have been proposed to deal with such a situation. One approach is to use a second neural network, often called the "critic", which learns to predict the future performance of the system [9,11]. The weights of the control network are then adjusted in such a way that the predicted performance is optimized. The corresponding learning procedures are variants of a heuristic treatment of the Bellman, or dynamic programming recursion for $G_{min}(t)$ [9].

Another possibility is to carry out the entire control process with the same set of weights, to evaluate the corresponding performance measure, $G(\mathbf{W})$, and to determine a set of weights which optimizes $G(\mathbf{W})$. If a so-called "emulator" neural network [10] is trained to simulate the response of the system, the control process can be represented as a sequence of control and emulator networks. This sequence can be regarded as a single large neural network, and the weights of the control network can then, e.g., be determined by "error backpropagation through time" [9,10]. As an alternative to this gradient descent approach, we propose the use of stochastic optimization strategies, and in the next section we shall apply such a procedure to find neural networks which are able to control the landing of a space vehicle on the moon.

5. A Neurocontrol Solution to the Lunar Lander Problem

We shall consider a simplified version of the lunar lander problem which is implemented in a number of computer games. The descent of the vehicle is described by the equations

$$x(t + \Delta t) = x(t) - v(t)\Delta t + \frac{1}{2}(f(t) - g)\Delta t^2 \tag{14a}$$

$$v(t + \Delta t) = v(t) - (f(t) - g)\Delta t, \tag{14b}$$

and

$$F(t + \Delta t) = F(t) - f(t)\Delta t, \tag{14c}$$

where x denotes the height above the surface of the moon, v the downward velocity, F the remaining fuel, and g the acceleration of gravity near the moon's surface. The deceleration force per unit mass, f, is assumed to be proportional to the rate at which the fuel is burned, and the decrease in vehicle mass is neglected. In regular time steps, the controller is given information about x, v, and F, and has to decide about the rate of fuel consumption, f, for the next time interval Δt. His task is to find a strategy,

$$f(t) = f(x(t), v(t), F(t)), \quad f(t) \leq f_{max}, \tag{15}$$

such that the "crash velocity" v_c, i.e., the velocity at $x = 0$, is sufficiently small to guarantee a smooth touchdown.

In continuous time, this problem can be solved exactly [20]. With the addtional condition that the total amount of fuel consumption should be a minimum, the solution becomes unique and consists of two phases. The first phase is a free fall, i.e., $f(t) = 0$, to a height $\tilde{x}$ given by

$$\tilde{x} = \frac{1}{2 f_{max}}(v_0^2 + 2gx_0),\tag{16}$$

where x_0 and v_0 denote the initial values of the height and downward velocity, respectively. From that point on, deceleration is maximal, $f(t) = f_{max}$, until touchdown. Landing takes place with $v = v_c = 0$, and the amount of fuel consumed during the deceleration phase, ΔF, is given by

$$\Delta F = \left[\frac{f_{max}}{f_{max} - g}(v_0^2 + 2gx_0)\right]^{1/2}.\tag{17}$$

We have attempted to realize the control of such a lunar lander by a neural network, and Fig. 4 shows one of the simple feedforward architectures we have analyzed. In our simulations we have used $g = 5\ feet/sec^2$, $f_{max} = 5g$, and $\Delta t = 0.5\ sec$, and we have applied a simulated annealing strategy to find an optimal set of weights and thresholds for the neural controller. After each landing simulation, a single weight was changed by a random amount, uniformly distributed between $-\delta$ and $+\delta$, where δ was chosen in the range $0.5 \leq \delta \leq 1.0$. The annealing temperature, T, was either decreased according to an exponential cooling schedule, or kept constant at a value close to the freezing temperature (see, e.g., Ref.[16]). Good performance was typically reached after about 500 landing simulations.

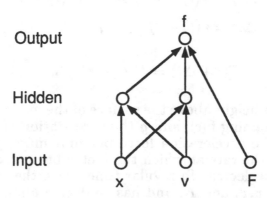

Fig. 4. Neural network architecture for the control of a lunar lander.

We have experimented with different forms for the performance criterion, $G(\mathbf{W}) = min$, and our results can be summarized as follows. If G represents the squared crash velocity for a single starting point (x_0, v_0), the algorithm finds many different types of solutions. Some of these neurocontrollers are very "cautions", i.e., they use the deceleration rockets during the entire landing approach, others wait until the last possible moment before they start to slow down the vehicle. If the performance criterion, however, not only involves the crash velocity, v_c, but also the amount of fuel consumed, the neurocontrol solutions represent very close approximations to the optimal strategy. The results shown in Figs. 5 and 6 refer to a criterion of the form

$$G = <v_c^2 + \alpha \int_0^{t_c} dt f(t) > = min, \qquad (18)$$

where $< ... >$ denotes an average over three different starting points in the (x, v)-plane. In Figure 5, one of the resulting landing approaches is

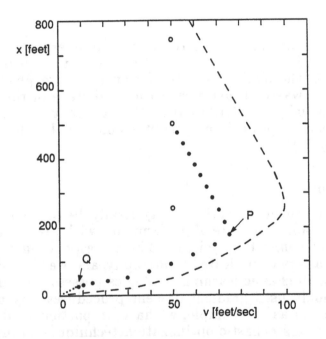

Fig. 5. Neurocontrol solution to the lunar lander problem. The landing approach from a height of 500 feet with an initial downward velocity of 50 feet/sec is plotted in steps of $\Delta t = 0.5$ sec. The three starting points used in the learning process are marked by open circles, and the region of starting points for which $v_c < 5$ feet/sec is indicated by the dashed line. The available fuel was limited to 150 units.

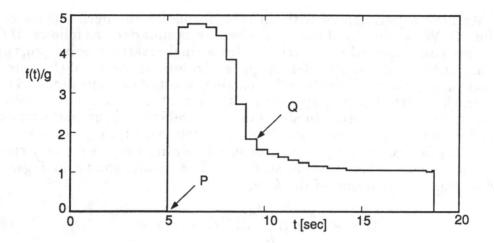

Fig. 6. Deceleration force f vs. time t for the landing approach shown in Fig. 5.

reproduced in detail, and the corresponding deceleration vs. time diagram is shown in Fig. 6. If we observe that the neurocontroller operates in discrete time, the similarity with the optimal two phase solution is remarkable. The waste of fuel during the final stage of the landing manoeuvre can partially be attributed to the fact that the requirement of minimal fuel consumption is not strictly enforced by the performance criterion of Eq.(18).

6. Conclusions

Learning in neural networks can generally be viewed as a process which optimizes some measure of performance with respect to the choice of the adjustable connection weights. The corresponding "fitness" landscape can have a very complicated topography, and we have shown that in such situations, stochastic learning strategies possess several advantages over the commonly used gradient descent procedures. By using the lunar lander problem as an example, we have, in particular, demonstrated the applicability of stochastic optimization techniques to problems with delayed performance estimation.

Acknowledgements

I have benefited from many fruitful discussions with Dr. H.J. Wiesmann, Dr. P. Wild, A. Kilchenmann, and Prof. K. Gustafson. This work was partially supported by the NFP23 program of the Swiss National Science Foundation.

REFERENCES

[1] G.E. Hinton, Artificial Intelligence 40, 185 (1989).

[2] D.E. Rumelhart, G.E. Hinton, and J.R. Williams, Nature 323, 533 (1986).

[3] "Parallel Distributed Processing: Explorations in the Microstructure of Cognition", vol. 1: "Foundations", ed. by D. E. Rumelhart and J. L. McClelland (MIT Press, Cambridge, MA, 1986).

[4] D. H. Ackley, G. E. Hinton, and T. J. Sejnowski, Cognitive Science 9, 147 (1985).

[5] D.J. Montana and L. Davis, in "Proc. 11th Int. Joint Conf. on AI" (Morgan Kaufmann, San Mateo, CA, 1989), pp. 762-767.

[6] H. Mühlenbein and J. Kindermann, in "Connectionism in Perspective", ed by R. Pfeifer, Z. Schreter, F. Fogelman-Soulié, and L. Steels (North-Holland, Amsterdam, 1989), pp. 173-198.

[7] N. Baba, Neural Networks 2, 367 (1989).

[8] H. Köhler, S. Diederich, W. Kinzel, and M. Opper, Z. Phys. B 78, 333 (1990).

[9] P.J. Werbos, Neural Networks 3, 179 (1990); and references therein.

[10] D. Nguyen and B. Widrow, in "Proc. IJCNN 1989" (IEEE Service Center, 1989), pp. II-357 - II-363.

[11] A.G. Barto, R.S. Sutton, and C.W. Anderson, IEEE Trans. on Systems, Man, and Cybernetics SMC-13, 834 (1983).

[12] B. Widrow and J.M. McCool, IEEE Transactions on Antennas and Propagation, AP-24, 615 (1976).

[13] F.J. Solis and R.J.B. Wets, Mathematics of Operations Research 6, 19 (1981).

[14] S. Kirkpatrick, C.D. Gelatt Jr., and M.P. Vecchi, Science 220, 671 (1983).

[15] R.H.J.M. Otten and L.P.P.P. van Ginneken, "The Annealing Algorithm" (Kluwer, Boston, 1989).

[16] J. Bernasconi, in "Chaos and Complexity", ed. by R. Livi, S. Ruffo, S. Ciliberto, and M. Buiatti (World Scientific, Singapore, 1988), pp. 245-259.

[17] L.B. Booker, D.E. Goldberg, and J.H. Holland, Artifical Intelligence 40, 235 (1989).

[18] P. Carnevali and S. Patarnello, Europhys. Lett. 4, 1199 (1987).

[19] N. Tishby, E. Levin, and S.A. Solla, in "Proc. IJCNN 1989" (IEEE Service Center, 1989), pp. II-403 - II-409.

[20] R. Bassein, The American Mathematical Monthly 96, 721 (1989).

STATISTICAL DYNAMICS OF LEARNING

J. A. Hertz

Nordita, Blegdamsvej 17, 2100 Copenhagen Ø

Abstract

These lectures review the basic aspects of learning in simple neural networks and outlines how one can study learning as a statistical dynamical process, stressing the role of phase transitions.

1   Introduction

A neural network consists of a collection of units or processors with very simple programs: each processor computes a simple weighted sum of the input data from other units and then outputs a single real number V_i (which we call the "state" or "activation" of the unit) which is a fixed, generally nonlinear, function of this weighted sum. This number is then sent to other units to be used in their own updatings of their states. The equation describing this is

$$V_i(t+1) = g(h_i(t)) = g\left(\sum_j w_{ij} V_j(t)\right) \tag{1}$$

The input data can be read in as an initial state of some subset of the units ("input units") and the result of the computation read off a possibly different subset of units ("output units") at a later time. The programming of such a network amounts to the choice of weights or *connections* w_{ij}, since it is these parameters which determine the dynamics of the system. Thus, this style of computation is called "connectionist". The activation values V_i can be either discrete or continuous; here we will generally work with continuous ones.

The general mathematical problem of neural networks has two principal aspects: how the network evolves in performing a given computation (the "retrieval problem" in an associative memory net) and the determination of the connection coefficients w_{ij} which determine what computation the network will carry out ("learning"). The former of these has been studied extensively in the last five years or so with the help of statistical mechanical methods (see, e.g. [1]. These lectures will describe some more recent results on the latter.

The "learning" process is generally an iterative one. One makes successive adjustments in the connection strengths until the network performs the desired computation with a certain accuracy. Thus learning is intrinsically a dynamical problem. Nevertheless (just as in the retrival problem), *static* statistical mechanical methods can also be very useful in studying learning. The application of equilibrium statistical mechanics, notably

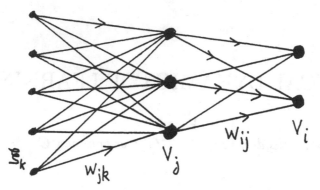

Figure 1: A perceptron.

techniques from spin glass theory, in this area was pioneered by Gardner and Derrida [2] and is still under active development (see [3] for an indication of the state of the art a year ago). In such work, one typically asks about the capacity of a particular net — how many different examples of a given class of associations it can perform, for example — independent of how it is "taught". Other lectures here will cover the latest developments in this paradigm.

We will be concerned here, on the other hand, with the actual dynamics of learning. In this paradigm one studies a particular learning algorithm in a particular network as a dynamical process. Methods from the statistical dynamics of disordered materials can again be brought to bear, as one has in general to average over noise in the process itself ("thermal" noise) and a statistical ensemble of computations to be learned ("quenched disorder").

There are two general kinds of learning, distinguished according to whether the changes are made directly in response to a comparison with correct examples of the computation or according to some other externally fixed criterion. The first kind is called *supervised learning*, the second *unsupervised*.

In these lectures we will discuss aspects of both kinds of learning. There will be more material on supervised learning because more work has been done on it. In both problems our focus will be the identification of phase transitions (broadly speaking), especially their dynamical aspects. We will be able to exploit a number of analogies with magnetic systems to give us insight.

feed-forward networks

In general, the retrieval and learning problems are intricately coupled. We want to focus on learning, so we will restrict our attention to layered feed-forward networks or *perceptrons* [4], where retrieval is very simple. Here the units are arranged in layers, each unit receives inputs only from the units in the preceding layer and send its output only to units in the next layer (Fig. 1). Input patterns ξ_k are read in at the input units k, and the result of the computation is read out from the activations of the units in the last layer. Because there is no feedback, the output of a perceptron can be written as an explicit function of its input, e.g. for a single intermediate layer

$$V_i = g(\sum_j w_{ij} g(\sum_k w_{jk} \xi_k)). \tag{2}$$

Thus the retrieval part of the problem is essentially trivial. This frees us to concentrate on the learning part.

The computational power of such a network depends on how many layers of connections it has. If it has only one, it is quite limited. The reason for this [5] is that it must discriminate solely on the basis of a *linear* combination of its inputs. In contrast, even one extra layer (if big enough) is sufficient to do any continuous function [6].

2 Supervised learning

The general nature of the task to be learned is a set of associations of input patterns ξ_k^μ with output patterns ζ_i^μ: When the input layer units are put in the configuration ξ_k^μ the output units should produce the corresponding ζ_i^μ. We call ζ_i^μ the *target patterns*. Associative memory task is one example of a task that can be described this way; in it the input and output patterns are the same. Another general kind of such task is categorization into particular output categories. In any case, in supervised learning we will want to choose the connection strengths so that the output V_i^μ (2) produced in response to input pattern number μ is equal to ζ_i^μ.

A way to iteratively compute the necessary connection strengths is based on changing them little by little so that the total error measure or cost function

$$H[w] \equiv \tfrac{1}{2} \sum_{i,\mu} (V_i^\mu - \zeta_i^\mu)^2 \tag{3}$$

decreases at each step. This can be guaranteed by taking the change in w proportional to the negative gradient of H with respect to w (sliding downhill in w-space on the surface $H[w]$):

$$\delta w_{ij} = -\eta \frac{\partial H}{\partial w_{ij}}. \tag{4}$$

This derivative can be computed using (2).

This prescription has been successfully used in a wide range of applications — text-to-phoneme conversion[7], sonar target identification [8], backgammon[9], extraction of the shape of objects from the shading in images[10], and protein secondary structure[11,12] — to name a few. But our emphasis here will be on the generic statistical properties of learning. That is, we will deal with ensembles of random patterns ζ_i^μ and ξ_k^{mu} and calculate measures of the learning process, averaged over this ensemble.

We will discuss two problems in supervised learning. The first is that of learning a set of p associations $\xi_k^\mu \to \zeta^\mu$, where the inputs and targets are independent binary random patterns. The questions one can ask here are how many such random associations can a network of a given size learn, how long does it take to learn them, and how is the learning affected by noise and constraints on the connections. The second is that of *generalization*. Here we assume there is a particular function from the input to the output space that the network is learning to implement by training (following (4) on a fixed number of examples of that function). The questions we ask are then how well can the network approximate the function when tested on inputs that were not among the training examples and how long does it take to achieve a given performance level in this task. The mathematical descriptions of these two problems are closely related, but it is important to understand the difference between them clearly.

2.1 learning random associations

We will treat this problem for a simple model with a single *linear* output unit and no hidden units. This is so simple as to appear trivial, but the averaging over the patterns makes it nontrivial. However, it is still soluble analytically, so it is very useful for theoretical insight. (It also has practical application: this kind of net is used in long-distance telephone lines to reduce noise.)

The cost function is much like (3) except that V_i^μ is a linear combination of the inputs instead of as in (2):

$$H[w] = \tfrac{1}{2} \sum_{i,\mu} [\zeta_i^\mu - N^{-\frac{1}{2}} \sum_j w_{ij} \xi^\mu]^2 + \tfrac{1}{2}\lambda \sum_{ij} w_{ij}^2 \tag{5}$$

The last term places a cost on large values of the connection weights. At the end of the calculation, we will fix the parameter λ by requiring that the mean square value of the w_{ij} (which will depend on λ) be equal to a fixed number S^2. Such a condition might be dictated (at least approximately), by constraints on resistance values or their precision in some hardware implementation. For example, suppose the resistor elements on a chip can only take on a single value. Thus (with the help of inverters), the w_{ij}'s can be, say, $\pm K$. The cost function (5) is then like the energy of an Ising spin system. From long experience with real spin systems we know that although the full Ising model may be insoluble, valuable insight can be gained from the corresponding spherical model, where the strong Ising constraint on the magnitude of every spin is replaced by a weaker one on the sum of the squares of the spin variables. Thus here we are doing the spherical model for this problem.

We are particularly interested in the effects of noise on the learning process. So we model an "unreliable teacher" by the stochastic differential equation (Langevin equation)

$$\frac{\partial w_{ij}}{\partial t} = -\frac{\partial H}{\partial w_{ij}} + \eta_{ij}(t), \tag{6}$$

where $\eta_{ij}(t)$ is white Gaussian random noise of covariance

$$\langle w_{ij}(t)w_{kl}(t')\rangle = 2T\delta_{ij,kl}\delta(t-t'). \tag{7}$$

(In physical systems modelled by such equations, T is the temperature, but, again, here it is just a measure of the noise in our process.) The formal task now involves calculating correlation functions of the w_{ij}'s, averaged over the noise and over the distribution of random input patterns. The time dependence of these correlations will tell us about the dynamics of the learning process.

Differentiating the cost function (5), we can write the equation of motion (6) in the form

$$\dot{w}_j = B_j - \sum_{j'} A_{jj'} w_{j'} - \lambda w_j + h_j(t) + \eta_j(t),$$ (8)

where the coefficients

$$B_j = \frac{1}{\sqrt{N}} \sum_{\mu} \zeta^{\mu} \xi_j^{\mu}$$ (9)

and

$$A_{jj'} = \frac{1}{N} \sum_{\mu} \xi_j^{\mu} \xi_{j'}^{\mu}$$ (10)

are random because they are functions of the random patterns ξ_j^{μ} and ζ^{μ}. We have dropped the output unit index i because the problem separates into independent problems for different i. We have also introduced a field $h_j(t)$ for formal purposes in the calculation as described below.

To physicsts, (8) looks like a model for a random medium, with $A_{jj'}$ the hopping amplitude or probability, depending on the problem. We solve it using the same kinds of methods one would use on these problems. All quantities of interest can be calculated from the average Green's function

$$G(t - t') = \left[\frac{\partial \langle w_i(t) \rangle}{\partial h_i(t')} \right]_{\zeta\xi},$$ (11)

where the notation $[\cdots]_{\zeta\xi}$ means the average over the binary random patterns ζ^{μ} and ξ_j^{μ}. The problem is a little bit different from the usual physical models because $A_{jj'}$ and B_j do not have such simple distributions as one would usually assume, and this difference is essential to the results one obtains. (This is also what differentiates mathematically between the retrieval problem in the Hopfield model and a spin glass.)

The physics of the problem is the competition between the learning of the desired associations $\xi_j^{\mu} \to \zeta_i^{\mu}$ and the constraint on the size of the connection strengths, and how this is affected by the noise. The details of the solution have been published elsewhere [13], so we will just give highlights and results necessary for the next subsection here.

Fourier transforming the equation of motion (8), we find we can write the Green's function as a Born series in A_{ij}:

$$G(\omega) = \left[G_0(\omega) - G_0^2(\omega) A_{ij} + G_0^3(\omega) A_{ik} A_{kj} - \cdots \right]_{\xi\zeta},$$ (12)

where

$$G_0(\omega) = \frac{1}{\lambda - i\omega}.$$ (13)

The summation of this series can be done by diagrammatic methods. Defining the self energy Σ by

$$G^{-1} = G_0^{-1} + \Sigma,$$ (14)

the diagrams for Σ that survive in the large-N limit are shown Fig. 2. Their evaluation

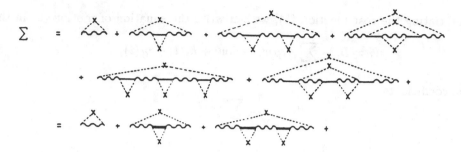

Figure 2: self-energy diagrams. The thick line stands for the full averaged local Green's function

involves averages of the form

$$\left[N^{-q} \sum_{\mu} \sum_{i_1 i_2 \dots i_{q-1}} \xi_i^{\mu} \xi_{i_1}^{\mu} \dots \xi_i^{\mu} \right]_{\xi} = \alpha \tag{15}$$

where we have defined the load parameter $\alpha \equiv p/N$. The series for the self energy, Σ, then becomes

$$\Sigma = \alpha - \alpha G + \alpha G^2 - \dots = \frac{\alpha}{1+G}, \tag{16}$$

and the response function is

$$G^{-1} = z + \frac{\alpha}{1+G}, \tag{17}$$

where $z = G_0^{-1}(\omega) = \lambda - i\omega$. We regard G as a function of z.

Solving (17) for G gives

$$G(z) = \frac{1 - \alpha - z + \sqrt{(z + \alpha - 1)^2 + 4z}}{2z}. \tag{18}$$

When $\alpha < 1$, G has a pole with of residue $1 - \alpha$ at $z = 0$.

This pole has the following meaning: For $\lambda = 0$ (i.e. unconstrained learning) the dynamics (8) take place only in the subspace spanned by the patterns. Any component of the initial state in the complementary subspace will not relax. The residue of the pole in G at $z = 0$ is just the number of these non-relaxing components. The capacity of the unconstrained network is $\alpha = 1$; this model has been analyzed in detail by Opper [14].

Returning to our problem, it remains to fix the chemical potential λ so that the constraint

$$\left[\langle w_i^2 \rangle \right]_{\xi\zeta} = S^2 \tag{19}$$

is satisfied. Using the fluctuation-response theorem, we can write $\left[\langle w_i^2 \rangle \right]_{\xi\zeta}$ as

$$\left[\langle w_i \rangle^2 \right]_{\xi\zeta} + \left[\langle (w_i - \langle w_i \rangle)^2 \rangle \right]_{\xi\zeta} = q + TG(\lambda), \tag{20}$$

where

$$q = [\langle w_i \rangle^2]_{\xi\zeta} \tag{21}$$

is a sort of Edwards-Anderson order parameter. Now it can be shown [13] that q can be expressed simply in terms of the zero-frequency Green's function $G(\lambda)$, with the result that the constraint (19) becomes

$$\frac{\partial}{\partial\lambda}(\lambda G(\lambda)) + TG(\lambda) = S^2 \tag{22}$$

Combining this equation with (18) and eliminating λ gives the required Green's function. Doing this, we find that there is a critical value of the load parameter α:

$$\alpha_c = \frac{S^2}{1 + S^2} \tag{23}$$

Below α_c, the solution in the low-noise limit $T \to 0$ has $\lambda \to 0$. That is, in this limit the learning is just that which would happen if there were no constraint; all p patterns are learned perfectly. Above α_c, a nonzero λ is found even for zero noise. Thus the learning of the patterns is compromised by the necessity of satisfying the constraint. Loading more patterns increases the size of the weights, and at α_c their minimum mean square size reaches the limit S^2 even without added noise.

In the presence of noise, this sharp transition is smeared out: The learning is never perfect at any α, but there is a sharp crossover from good to markedly inferior learning as α is increased through α_c.

From the Green's function we can calculate the relaxation time

$$\tau = \frac{\int_0^\infty tG(t)dt}{\int_0^\infty G(t)dt} = G(0) \lim_{\omega \to 0} \frac{\partial G^{-1}(\omega)}{\partial(-i\omega)}. \tag{24}$$

There are three interesting cases. When $\alpha < \alpha_c$ we find

$$\tau = \frac{1}{\gamma_0 T} \frac{S^2(1-\alpha) - \alpha}{(1-\alpha)^2}, \qquad (T \ll \alpha_c - \alpha \ll 1). \tag{25}$$

In the critical region ($|\alpha - \alpha_c| \ll T \ll 1$) we get

$$\tau = \left(\frac{2S^2(1 + S^2)}{T}\right)^{1/2} \frac{1}{\gamma_0(1-\alpha)}. \tag{26}$$

Finally when $\alpha > \alpha_c$ we obtain

$$\tau = \left[\gamma_0 x_0 \left(1 + \frac{T(1-\alpha_c)^2}{2x_0^2 \alpha_c \sqrt{\alpha_c/\alpha}}\right)(1 - \alpha_c)\left(1 + T\frac{1-\alpha_c}{x_0}\right)\right]^{-1} \tag{27}$$

for $T \ll \alpha - \alpha_c \ll 1$.

These equations also tell us something interesting about the learning process. From eq. (25) we see that although the patterns can in principle be learned perfectly for $\alpha < \alpha_c$, it takes forever to do so. At finite temperature the learning time is finite but at the cost

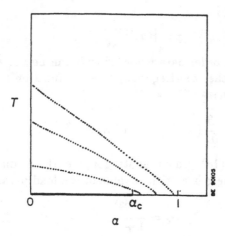

Figure 3: Phase diagram of the constrained linear learning model. The dotted lines are contours of constant learning time.

of imperfect learning (because then λ is nonzero). Note also that for fixed T the learning time shrinks as α approaches α_c, but this does not mean that the learning is improving. Rather the asymptotic performance is getting worse, so it takes less time to reach this state. Right at α_c the learning time does not diverge as rapidly as $T \to 0$ (like $T^{-\frac{1}{2}}$ instead of T^{-1}), but this advantage is of course offset by a correspondingly more severe degradation of the asymptotic performance by small amounts of noise. Above α_c the learning time is always finite even at zero temperature (though it diverges as $(\alpha - \alpha_c)^{-1}$). Correspondingly the asymptotic performance is never perfect. The situation is shown in Fig. 3.

Thus we have found a phase transition with features similar to those found in physical systems. The behaviour is analogous to that of an isotropic ferromagnet (in mean field theory), with α playing the role of temperature and the noise that of an external magnetic field: As α approaches α_c from above in zero noise, the relaxation time diverges. At any α below α_c, the relaxation time diverges in the low–noise limit like the reciprocal of the noise variance (in just the way that the relaxation time of the magnetization of the isotropic ferromagnet diverges in the zero–field limit).

We have also considered input patterns that are biased, letting the probability that $\xi_i^\mu = \pm 1$ be $(1 \pm a)/2$ so the average of the ξ's is a. In this case the averaged response function in (11) is no longer diagonal. The diagram technique used in the previous section must then be generalized. Details are given in ref. [13]. The interesting thing that comes out of the analysis is that there are now two eigenmodes of G_{ij} and, correspondingly, two relaxation times. The shorter if these can be interpreted as the time it takes the system to find a "prototype solution" before finding the finer details.

2.2 generalization

It is very important in practical situations to know how well a neural network will generalize from the examples it is trained on to the entire set of possible inputs. This problem is the focus of a lot of recent and current work[15,16,17,18,19,20,21]. All this work, however, deals with the asymptotic state of the network after training. Here we study the evolution of the generalization ability in time under training in our simple linear model, without noise or constraints on the size of the weights. Despite its simplicity, this model exhibits nontrivial behaviour: a dynamical phase transition at a critical number of training examples, with power-law decay right at the transition point and critical slowing down as one approaches it from either side.

Our simple linear neuron has an output $V = N^{-\frac{1}{2}} \sum_i w_i \xi_i$, where ξ_i is the ith input. It learns to imitate a teacher whose weights are u_i by training on p examples of input-output pairs $(\xi_i^\mu, \zeta^\mu = N^{-\frac{1}{2}} \sum_i u_i \xi_i^\mu)$ generated by the teacher. The learning equation

$$\dot{w}_i = \frac{1}{\sqrt{N}} \sum_{\mu=1}^{p} (\zeta^\mu - \frac{1}{\sqrt{N}} \sum_j w_j \xi_j^\mu) \xi_i^\mu \tag{28}$$

then takes the form

$$\dot{w}_i = \frac{1}{N} \sum_{\mu j} (u_j - w_j) \xi_j^\mu \xi_i^\mu = \sum_j A_{ij} (u_j - w_j). \tag{29}$$

We let the example inputs ξ_i^μ take on values ± 1, randomly and independently. For a large number of examples ($p = O(N) \gg 1$), the resulting generalization ability will be independent of just which p of the 2^N possible binary input patterns we choose. All our results will then depend only on the fact that we can calculate the spectrum of the matrix A_{ij} (10).

To measure the generalization ability, we test whether the output of our perceptron with weights w_i agrees with that of the teacher with weights u_i on all possible binary inputs. Our objective function is just the square of the error, averaged over all these inputs:

$$F = \frac{1}{N 2^N} \sum_{\{\sigma\}} \left(\sum_i (u_i - w_i) \sigma_i \right)^2 = \frac{1}{N} \sum_i (u_i - w_i)^2. \tag{30}$$

That is, F is just proportional to the square of the difference between the teacher and pupil weight vectors. The normalization factor N^{-1} is included because we will find it convenient later to fix the mean square magnitude of the u_i's at 1; then F will vary between 1 (*tabula rasa*) and 0 (perfect generalization). During learning, w_i depends on time, so F is a function of t. The complementary quantity $1 - F(t)$ could be called the generalization ability.

Let us define the weight difference vector $\mathbf{v} = \mathbf{u} - \mathbf{w}$. We then write the equation of motion (29) in the basis $\{\alpha\}$ of eigenvectors of A:

$$\dot{v}_\alpha = -A_\alpha v_\alpha \tag{31}$$

where A_α are the eigenvalues of A. This has the solution

$$v_\alpha(t) = v_\alpha(0)\exp(-A_\alpha t) = u_\alpha \exp(-A_\alpha t), \qquad (32)$$

Thus we find

$$F(t) = \frac{1}{N}\sum_{i\alpha\beta}\langle i|\alpha\rangle u_\alpha e^{-(A_\alpha + A_\beta)t}u_\beta\langle\beta|i\rangle$$

$$= \frac{1}{N}\sum_\alpha u_\alpha^2 \exp(-2A_\alpha t). \qquad (33)$$

As mentioned above, we normalize the teacher weight vector **u** to length $\sqrt{N}$. Averaging the result (33) over all directions, we then write our result in the form

$$F(t) = \int d\epsilon\rho(\epsilon)e^{-2\epsilon t}. \qquad (34)$$

where $\rho(\epsilon)$ is the density of eigenvalues of A. (Instead of keeping a fixed training set and averaging over directions of **u**. we could equally well have chosen a fixed **u**, say $(1,1,1,\ldots)$ and averaged over all possible training sets of size p, with the same result.)

For large N, the eigenvalue density can be obtained simply from the imaginary part of the Green's function (18):

$$\rho(\epsilon) = \frac{1}{\pi}\operatorname{Im}G(-\epsilon)$$

$$= \frac{1}{2\pi\epsilon}\sqrt{(\epsilon_+ - \epsilon)(\epsilon - \epsilon_-)} + (1-\alpha)\theta(1-\alpha)\delta(\epsilon), \qquad (35)$$

where

$$\epsilon_\pm = (1\pm\sqrt{\alpha})^2 \qquad (36)$$

and $\theta(\ )$ is the unit step function. The delta-function term appears for $\alpha < 1$ because no learning takes place in the subspace orthogonal to that spanned by the training patterns.

The results at infinite time are immediately evident. For $\alpha < 1$ there is a nonzero limit

$$F(\infty) = 1 - \alpha, \qquad (37)$$

while for $\alpha \geq 1$ $F(\infty)$ vanishes, indicating perfect generalization. While on the one hand it may seem remarkable that perfect generalization can be obtained from a training set which forms an infinitesimal fraction of the entire set of possible examples, the meaning of the result is just that N points are sufficient to determine an $N-1$-dimensional hyperplane in N dimensions. It is interesting to note that the necessary number of examples to produce correct generalization in a perceptron with a threshold output unit instead of a linear one is also of order N[18].

Figure 4 shows $F(t)$ as obtained numerically from (34) and (35). The qualitative form of the approach to $F(\infty)$ can be obtained analytically by inspection. For $\alpha \neq 1$, the asymptotic approach is governed by the smallest nonzero eigenvalue ϵ_-. Thus we have critical slowing down, with a divergent relaxation time

$$\tau = \frac{1}{\epsilon_-} \sim \frac{1}{|\alpha - 1|^2} \qquad (38)$$

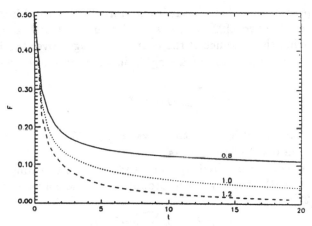

Figure 4: The function $F(t)$ for several values of α.

as the transition at $\alpha = 1$ is approached. Right at the critical point, the eigenvalue density diverges for small ϵ like $\epsilon^{-\frac{1}{2}}$, which leads to the power law

$$F(t) \propto t^{-\frac{1}{2}} \qquad (39)$$

at long times. Thus, while exactly N examples are sufficient to produce perfect generalization, the approach to this desirable state is rather slow. A little bit above $\alpha = 1$, $F(t)$ will also follow this power law for times $t \ll \tau$, going over to (slow) exponential decay at very long times ($t \gg \tau$). By increasing the training set size well above N (say, to $\frac{3}{2}N$), one can achieve exponentially fast generalization. Below $\alpha = 1$, where perfect generalization is never achieved, there is at least the consolation that the approach to the generalization level the network does reach is exponential (though with the same problem of a long relaxation time just below the transition as just above it).

This model is of course much simpler than most real-life training problems. However, it does allow us to examine in detail the dynamical phase transition separating perfect from imperfect generalization. We conjecture that such transitions may be a general feature of the generalization problem in networks.

3 Unsupervised Learning

As mentioned earlier, there is another learning paradigm, called *unsupervised* learning, in which one does not know what the correct output should be. The dynamics (the form of the equation of motion of the w_{ij}'s) may be chosen, for example, to classify input patterns into several classes, where the appropriate definition of the classes is not known *a priori* (clustering algorithms). Although not always described explicitly in terms of minimizing a cost function like (3) or (5), most supervised learning can generally be formulated as such an optimization process.

In the example we shall study here, what is maximized is the total information passed through a simple layered network[22]. Since we will again deal with a linear net, this is equivalent to maximizing the variance of the output (averaged over the input patterns).

We focus first on the simple case of a single, linear output unit, which produces a signal

$$V = \sum_j w_j \xi_j = \mathbf{w} \cdot \xi \tag{40}$$

in response to an N-component input pattern (vector) ξ. We suppose that we know the correlation matrix $C_{ij} \equiv \langle \xi_i \xi_j \rangle$ of the inputs.

The learning dynamics are inspired by Hebb's idea about learning in real brains — that the connection strength should grow in proportion to the correlation between input and output signals [23]:

$$\dot{w}_j = \eta V \xi_j = \eta \sum_{j'} w_{j'} \xi_{j'} \xi_j. \tag{41}$$

Averaging over the input distribution, then, we get

$$\dot{w}_j = \eta \sum_{j'} C_{jj'} w_{j'}. \tag{42}$$

Although no optimization or cost function has been put explicitly into the dynamics here, one can show that this "Hebb dynamics" actually does maximize the output variance.

In ths form the dynamics have an inconvenient instability, which we can see by writing (41) in the basis where the matrix C is diagonal: Letting α label the eigenfunctions of C, we have

$$\dot{w}_\alpha = C_\alpha w_\alpha. \tag{43}$$

Since the eigenvalues of a correlation matrix are positive, we see that the components of w grow exponentially. We want to modify the learning equation of motion so as to limit this growth and restore stability.

A nice way to do this was suggested by Oja[24], who proposed adding a stabilizing term $-\eta S^2 w_j$ to (41), so that the equation of motion now reads

$$\dot{w}_j = \eta V(\xi_j - V w_j). \tag{44}$$

That is, the "back-propagated" output is subtracted from the input in the second factor in the simple Hebb equation (41). Substituting (40) and averaging as in (42) gives ($\eta = 1$ from now on)

$$\dot{w}_j = \sum_{j'} C_{jj'} w_{j'} - \sum_{j'j''} w_{j'} C_{j'j''} w_{j''} w_j. \tag{45}$$

In the basis where C is diagonal, this reads

$$\dot{w}_\alpha = C_\alpha w_\alpha - (\sum_\beta w_\beta^2 C_\beta) w_\alpha. \tag{46}$$

We can get some insight into the physics of this problem by the change of variable

$$\phi_\alpha = \sqrt{C_\alpha} w_\alpha, \tag{47}$$

in terms of which the learning equation becomes

$$\dot{\phi}_\alpha = C_\alpha \phi_\alpha - (\sum_\beta \phi_\beta^2)\phi_\alpha. \tag{48}$$

This is just the Landau-Ginzburg equation for an N-component anisotropic ferromagnet with exchange strength C_α in direction α. We therefore know its equilibrium solution — It just orders in the easiest direction (the one with the largest C_α). Thus we have

$$\phi_0 = \sqrt{C_0}, \tag{49}$$

and all other $\phi_\alpha = 0$. (We order the eigenvectors so that $\alpha = 0$ has the largest eigenvalue, $\alpha = 1$ the next largest, and so on.) In terms of the w variables, the only nonvanishing w_α is

$$w_0 = 1. \tag{50}$$

Going back to the original basis, we see that the w_j produced by the learning process acts as a filter which measures the overlap between the input pattern and the eigenvector of C with largest eigenvalue (the so-called (first) principal component. Thus the Hebb learning algorithm (in this version, at least) functions as a "principal component extractor".

We can also use insight from physics in the case where we add noise to the learning equation in the way we did in (6) for the supervised learning case above. Then we can make a simple Landau-Peierls fluctuation analysis like the one one does to discover that two dimensions is the lower critical dimension of a classical Heisenber ferromagnet. The result is that if the density of eigenvalues C_α is nonzero in the limit where $C_\alpha \to C_0$, this model is also at its "lower critical dimensionality", i.e. it cannot quite order in the presence of noise. The resulting connections will not quite act as a stable principal component extractor, but will always have (flucutating) mixtures of other eigenfunctions of C.

There are several approaches to principal component projection or extraction with more than one output unit (say M of them) within the general Hebb learning paradigm. One way [25] is to start with one output unit which extracts the first principal component as above. The next output unit then does the same thing except that its output is orthogonalized to the first one, and so on. The result is obviously that the principal components are extracted one by one. A closely related approach is to introduce *anti-hebbian* learning (i.e with the opposite sign) for connection weights between output units [26,27]. Here we will follow Oja[28], who suggested modifying (44) to read

$$\dot{w}_{ij} = V_i \left(\xi_j - \sum_{i'} V_{i'} w_{i'j} \right). \tag{51}$$

We will prove that (in the absence of noise) the resulting steady-state weights make an orthogonal projection onto the subspace of the input space spanned by the eigenvfunctions of C with the M largest eigenvalues[29].

To find the fixed points we can do as we did above when deriving (45). Assuming that $\dot{w}_{ij} = 0$ and averaging over the input patterns yields

$$0 = (1/\eta)\dot{w}_{ij} = C_{jj'}w_{ij'} - w_{i'j}(w_{ij'}C_{j'j''}w_{i'j''}) \tag{52}$$

where summation on all repeated indices is implicit. The subspace spanned by the weight vectors $\mathbf{w}_i$ is called V_0 and the orthogonal subspace $V_0^\perp$. We now show that V_0 is invariant under C by multiplying (52) by a vector $\mathbf{a} \in V_0^\perp$:

$$0 = a_j[C_{jj'}w_{ij'} - w_{i'j}(w_{ij'}C_{j'j''}w_{i'j''})] = a_jC_{jj'}w_{ij'} \qquad (53)$$

so $C_{jj'}w_{ij'} \in V_0$ which implies that V_0 is spanned by eigenvectors of C.

Multiplying (52) by one of the weight vectors instead gives

$$0 = w_{i''j}[C_{jj'}w_{ij'} - w_{i'j}(w_{ij'}C_{j'j''}w_{i'j''})] = [\delta_{i'i''} - w_{i''j}w_{i'j}]w_{ij'}C_{j'j''}w_{i'j''}. \qquad (54)$$

For the expression in the brackets to be zero the weight vectors have to be *orthonormal*.

This proves that the fixed points of the learning procedure gives orthonormal weights that project onto a subspace spanned by M eigenvectors of C.

We now prove that only the subspace spanned by the M largest principal components is stable. Assume that $\mathbf{w}_i$ is a fixed point and add a small perturbation $\mathbf{v}_i$, as we did in the previous section. Then to first order in the perturbation the weight changes are

$$\frac{1}{\eta}\langle\dot{v}_{ij}\rangle = C_{jj'}v_{ij'} - C_{j'j''}[v_{ij}w_{i'j''}w_{i'j} + w_{ij'}v_{i'j''}w_{i'j} + w_{ij'}w_{i'j''}v_{i'j}]. \qquad (55)$$

Parts of these vectors in input space ($\dot{\mathbf{v}}_i$) lie in V_0 and the other parts in $V_0^\perp$. To see if V_0 is stable we need to show that the parts in $V_0^\perp$ are decreased by the learning. The vectors are split up into two parts:

$$\mathbf{v}_i = \mathbf{v}_i^0 + \mathbf{v}_i^\perp. \qquad (56)$$

Because V_0 is closed under C it is easy to find the part of (55) in $V_0^\perp$,

$$\frac{1}{\eta}\langle\dot{v}_{ij}^\perp\rangle = C_{jj'}v_{ij'}^\perp - w_{ij}C_{j'j''}w_{i'j''}v_{i'j}^\perp \qquad (57)$$

where the sum is still over repeated indices. We transform this to the basis where C is diagonal and number the eigenvectors $\mathbf{e}_i$ such that $V_0 = span\{e_1, \ldots, e_M\}$ and the rest spans $V_0^\perp$. The last equation then becomes

$$\frac{1}{\eta}\langle\dot{v}_{i\alpha}^\perp\rangle = [\lambda^\alpha\delta_{ii'} - w_{ij'}C_{j'j''}w_{i'j''}]v_{i'\alpha}^\perp. \qquad (58)$$

The last matrix in the braces $\tilde{C}_{ii'} \equiv w_{ij'}C_{j'j''}w_{i'j''}$ is just C in the basis of the weight vectors. Since the weight vectors are orthonormal $\tilde{C}$ will have the eigenvalues $\lambda^1 \ldots \lambda^M$. The matrix in the braces $(\lambda^\alpha\delta_{ii'} - \tilde{C}_{ii'})$ will then have eigenvalues $\lambda^\alpha - \lambda^\beta$ with $\beta \leq M$. Note that $v_{i\alpha}^\perp = 0$ for all $\alpha \leq M$ and for $\alpha > M$ the eigenvalues will be negative if and only if

$$\lambda^\alpha < \lambda^\beta \text{ for } \alpha > M \text{ and } \beta \leq M. \qquad (59)$$

When the eigenvalues are negative the perturbation will die out in $V_0^\perp$, so V_0 is stable.

Oja found that in V_0 rotations that keeps the weights orthonormal are allowed, so we do not expect all perturbations to die out unless there are some special directions of the weights that are preferred. To study what is going on in V_0 we look at the part of (55) that stays in V_0:

$$\frac{1}{\eta}\langle \dot{v}_{ij}^0\rangle = C_{jj'}v_{ij'}^0 - C_{j'j''}[v_{ij}^0 w_{i'j''}w_{i'j} + w_{ij}v_{i'j''}^0 w_{i'j} + w_{ij}w_{i'j''}v_{i'j}^0]. \tag{60}$$

In the basis of the weight vectors with $\tilde{v}_{ik} \equiv \sum_j w_{kj}v_{ij}^0$ this can be written as

$$
\begin{aligned}
\frac{1}{\eta}\langle \dot{\tilde{v}}_{ik}\rangle &= \tilde{C}_{kk'}\tilde{v}_{ik'} - \tilde{C}_{ii'}\tilde{v}_{i'k} - \tilde{C}_{kk'}\tilde{v}_{ik'} - \tilde{C}_{ik'}\tilde{v}_{kk'} \\
&= -(\tilde{C}_{ii'}\tilde{v}_{i'k} + \tilde{C}_{ik'}\tilde{v}_{kk'})
\end{aligned} \tag{61}
$$

$\tilde{C}_{ik}$ and $\tilde{v}_{ik}$ are now considered as $M \times M$ matrices in the subspace V_0. We transform into the basis where $\tilde{C}$ is diagonal using an orthogonal transformation S_{ai}, and $\tilde{v}_{ab} \equiv \sum_{i,k} S_{ai}\tilde{v}_{ik}S_{bk}$:

$$\frac{1}{\eta}\langle \dot{\tilde{v}}_{ab}\rangle = -\lambda^a(\tilde{v}_{ab} + \tilde{v}_{ba}). \tag{62}$$

Thus for every pair (a,b) there are two eigenmodes: the symmetric combination

$$u_{ab}^s \equiv (\tilde{v}_{ab} + \tilde{v}_{ba}) \tag{63}$$

which relaxes according to

$$\dot{u}_{ab}^s = (\dot{v}_{ab} + \dot{v}_{ba}) = -(\lambda^a + \lambda^b)u_{ab}^s, \tag{64}$$

and the asymmetric combination

$$u_{ab}^a \equiv \left(\frac{1}{\lambda^a}\tilde{v}_{ab} - \frac{1}{\lambda^b}\tilde{v}_{ba}\right) \tag{65}$$

which does *not* relax, because $\dot{u}_{ka}^a = 0$.

Hence an arbitrary infinitesimal pertubation will relax to an antisymmetric combination $\tilde{v}_{ab} = -\tilde{v}_{ba}$, that is, to a rotation in the a–b plane.

We have thus proved that for a network of M linear units and the learning rule (51) the M weight vectors converge to an orthonormal basis for the subspace spanned by the first M principal components of the input correlation matrix.

There are also other architectures for achieving principal component projection or extraction within the general Hebb learning paradigm [25,26,27] which generally involve couplings between output units.

4   Final comments

The models we have analyzed are the very simplest ones one could consider. They are much simpler than the networks actually used in most real-life applications. Their role in the broader theoretical perspective is the same one simplified models play in statistical physics: they give us insight into the kinds of properties (e.g. phase transitions) large interacting systems can have. Of course, more work, both numerical and analytic, will be necessary to study how such properties carry over to more complex networks, but at least we now have an idea of what to look for.

acknowledgement

I would like to thank my co-workers Anders Krogh and Ingi Thorbergsson for their extensive contribution to the work which is reviewed here.

References

[1] Amit, D. J., *Modelling Brain Function*. Cambridge University Press (1989)

[2] Gardner, E., *Europhys. Lett.* **4** 481-485 (1987); *J. Phys. A* **21** 257-270 (1988); Gardner, E., and Derrida, B., *J. Phys. A* **21** 271-284 (1988)

[3] see *J. Phys. A* **22**, #12 (1989)

[4] Rosenblatt, F., *Principles of Neurodynamics*, New York: Spartan (1962)

[5] Minsky, M., and Papert, S. *Perceptrons*, Cambridge: MIT Press (1969)

[6] Cybenko, G., *Math. of Controls, Signals, and Systems* **2** 303-314 (1989)

[7] Sejnowski, T. J., and Rosenberg, C. R., *Complex Systems* **1** 145-168 (1987)

[8] Gorman, P., and Sejnowski, T. J., *Neural Networks* **1** 75-90 (1988)

[9] Tesauro, G., and Sejnowski, T. J., in *Neural Information Processing Systems*, D. Anderson, ed., pp 452-456; New York, American Institute of Physics (1988)

[10] Lehky, S., and Sejnowski, T. J. *Nature* **333** 452-454 (1988)

[11] Qian, N. and Sejnowski, T., *J. Mol. Biol.* **202** 865-884 (1988)

[12] Bohr, H., et al, *FEBS Letters* **241** 223-228 (1988)

[13] Hertz, J. A., Thorbergsson, G. I., and Krogh, A., *J. Phys.A* **22** 2133-2150 (1989)

[14] M. Opper, *Europhys. Lett.* **8** 389 (1989)

[15] Gardner and Derrida, *J. Phys. A* **22** 1983-1994 (1989)

[16] Schwartz, D. B., Samalam, V. K., Solla, S. A., and Denker, J. S., *Neural Computation* **2** (1990)

[17] Tishby, N., Levin, E., and Solla, S. A., *Proceedings IEEE Int. Conf. on Neural Networks* II-403-410 (1989)

[18] Baum, E. B., and Haussler, D., *Neural Computation* **1** 151-160 (1989)

[19] Abu-Mostafa, Y. A., *Neural Computation* **1** 312-317 (1989)

[20] Györgyi, G., and N Tishby, *Proc. STATPHYS-17 Workshop on Neural Networks and Spin Glasses* W. K. Theumann and R. Köberle, eds. Singapore: World Scientific (to be published)

[21] Hansel, D., and Sompolinsky, H., *Europhys. Lett.* **11** 687-692 (1990)

[22] Linsker, R., *Proc. Nat Acad. Sci. USA* **83** 7508-7512, 8390-8394, 8779-8793 (1986)

[23] Hebb, D. O., *The Organization of Behavior*, New York: Wiley (1949)

[24] Oja, E., *J. Math. Biol.* **15** 267-273 (1982)

[25] Sanger, T. D., MIT preprint (1989)

[26] Földiák, P., *Proceedings IEEE Int. Conf. on Neural Networks* I-401-405 (1989)

[27] Rubner, J., and Tavan, P., *Europhys. Lett.* **10** 693-698 (1989)

[28] Oja, E., *Int. J. Neural Systems* **1** 61-68 (1989)

[29] Krogh, A., and Hertz, J. A., pp. 183-186 in *Parallel Processing in Neural Systems and Computers*, R. Eckmiller, G. Hartmanna nd G. Hauske, eds. Amsterdam: North-Holland (1989)

Learning and Retrieving Marked Patterns

Stam NICOLIS

I. N. F. N. Sezione di Roma, Italy[1]

ABSTRACT

An extension of the Hopfield model is studied, whereby a certain number of 'privileged' patterns are 'marked', during the learning process. The replica trick is used in studying the retrieval properties of such networks. The relevance of the extension to other biological processes is also discussed.

---

[1]Postal Address: piazzale Aldo Moro 2, Roma 00185, Italy.

1 Introduction

In what follows, we shall try to understand how a large number of 'simple' processing units might be capable of 'complex' behavior; namely, of *learning* a certain number of *patterns*; also, (equally importantly), of *retrieving* such information.

To this end, it is necessary to give meaning to the words *learning, retrieval* and so on, on the basis of a concrete model. The one we shall be concerned with was introduced by Hopfield [1] as a simplified model of an associative memory (a recent review of such models may be found in ref. [2,3]). It may also be relevant to the understanding of other biological processes.

The simple units of the Hopfield model are called, quite evocatively, *neurons*, each of which may be in one of two states, *active* or *passive*. Each neuron is connected to the others via links called *synapses*; in this way we have a network. Through the synapses, it is possible for one neuron to communicate its state to another neuron. This communication, and the way each neuron acts upon receiving it, makes the behavior of the network, as a whole, quite non-trivial.

A *state* of the network may be thought of as a large vector, any given component of which represents the state of a certain neuron. Within this framework, *learning* is defined as finding a set of values for the synapses, such that a given state of the network will be compatible with the communication rule between the neurons; *retrieval* is defined as the act of recognizing that an 'externally' presented state is (or isn't, as the case may be) an allowed state of the network (given the synapses).

The above description may ring a familiar bell through the identification of the two states of a *neuron* with the two allowed values of an Ising spin variable and that of the *synapse* between neuron i and j with the exchange coupling J_{ij} between spin i and spin j; finally, the communication rule corresponds to some spin dynamics (*e.g.* single spin flip, Kawasaki, etc.)

In such familiar terms, learning means, given a spin configuration, finding the values of the couplings such that it be a ground state. Retrieval of a spin configuration means, starting from some configuration, ending up 'near' it, under the dynamics. This incorporates a desired 'error-correcting' property, crucial to the relevance of associativity.

It has been found that the following choice of synaptic couplings allows one to store p N-component patterns (where N denotes the number of neurons) ξ, whose entries take the two values 1 and -1 with equal probability

$$J_{ij} = \frac{1}{N} \sum_{\mu=1}^{p} \xi_i^\mu \xi_j^\mu; \tag{1}$$

this is the so-called Hebb rule [1,4]. In terms of an iterative learning procedure this is translated to mean that the effect of learning pattern μ upon the synapse between neuron i and neuron j is

$$\delta^\mu J_{ij} = \frac{1}{N} \xi_i^\mu \xi_j^\mu;$$

if the neurons are in like states the coupling is reinforced; if in unlike it is weakened. Many other learning schemes have since been developed; for a review see ref. [5,6,7].

The retrieval properties of a fully connected network, with hebbian synapses have been studied using the methods of equilibrium statistical mechanics in ref. [8]; it was found that, in the limit $N \to \infty$, retrieval of a pattern was possible, with less than 1.5% error, as long as $p \leq 0.138N$. Beyond this limit, no retrieval is possible; i.e. the final configuration is 'far' from the initial one. This is the 'blackout' catastrophe.

Such a collapse is quite undesirable and modifications of the original Hebb rule, that avoid total blackout, have been the subject of many studies [1,9,10]; the conclusion is that it is possible to implement a 'forgetting' mechanism of old patterns, so that the network may retrieve the most recently learned patterns.

A related modification has been studied in ref. [11]. They 'marked' a certain, finite, number of patterns, against an infinite (as αN) number of 'unmarked' patterns and studied the retrieval properties of such a network. What they found was that it was possible to retrieve the 'marked' patterns, even when the 'unmarked' exceed $0.138N$; furthermore, if the 'unmarked' are less than $0.138N$ both 'marked' and 'unmarked' may be retrieved and the retrieval quality (*i.e.* error at retrieval) of the 'unmarked' patterns is the same as in the normal hebbian régime. Blackout is not avoided, however, only postponed.

Patterns may be 'marked', through the introduction of different weights in the Hebb rule, *viz.*

$$J_{ij} = \frac{1}{N} \sum_{\mu=1}^{p_1} \xi_i^\mu \xi_j^\mu + \frac{\gamma}{N} \sum_{\nu=1}^{p_2} \varsigma_i^\nu \varsigma_j^\nu \quad 0 < \gamma \le 1 \tag{2}$$

In ref. [11] the number of 'marked' patterns p_1 is finite, as $N \to \infty$, while $p_2 = \alpha_2 N$. It should be noted, at this point, that (2) belongs to a family studied in ref. [10]; in their notation $\Delta(\mu/N) = 1$ for $p \le p_1$ and γ for $p_1 < p \le p_1 + p_2$, *i.e.* a step function. Its novelty lies in the explicit introduction (within this context) of more than one *classes* of patterns, thereby suggesting possible future applications to categorization. More concretely, the analysis of ref. [11] revealed that the transition, from the spin glass to the retrieval phase, in the α_2 *vs.* γ plane, is second order for $\gamma < 1/3$, first order for $\gamma \ge 1/3$. The specific problem to be solved, in what follows, is what happens when $p_1 = \alpha_1 N$. A brief account may be found in ref. [13]. Marking a *finite* number of patterns only, each with a different weight, has been considered in ref. [12].

A word on the physical meaning of the modification is, perhaps, in order. On the one hand, one may claim that it incorporates a 'recency' effect, in that recently learned patterns correspond to the ξ's, while earlier ones correspond to the ς's.

On the other hand, in the context of image processing, one might take the ξ's as foreground and the ς's as background features. More on other applications in the final section.

2 Statistical Mechanics of the spin system

Since the synaptic couplings (2) are symmetric, the properties of the corresponding spin system are determined by an Ising Hamiltonian

$$\mathcal{H} = - \sum_{(i,j)} J_{ij} S_i S_j \tag{3}$$

The crucial quantity, of course, is the free energy, averaged over the probability distribution of the memorized patterns (the ξ's and the ς's are chosen from the same distribution)

$$\rho(\xi_i^\mu) = \frac{1}{2} \left(\delta(\xi_i^\mu + 1) + \delta(\xi_i^\mu - 1) \right) \tag{4}$$

To compute it, we shall use the replica trick, *viz.*

$$
\begin{aligned}
Z &= \mathrm{Tr}\{S\} \exp -\beta \mathcal{H} \\
f &= \left\langle\!\left\langle -\frac{1}{\beta N} \ln Z \right\rangle\!\right\rangle \\
&= \lim_{n \to 0} \frac{\langle\!\langle Z^n \rangle\!\rangle - 1}{\beta n N}
\end{aligned}
\tag{5}
$$

where the double angular brackets denote the average over the probability distribution (4) of both ξ's and ς's.

This computation is, by now, standard (see refs. [2,8,10]). What it amounts to is the interchange of the limits $n \to 0$ and $N \to \infty$ from that implied by (5); the second is carried out first, by calculating f as the saddle point value of an integral.

One ends up with

$$\hat{m}_a^\mu = \left\langle\left\langle \tfrac{1}{N}\sum_{i=1}^N \xi_i^\mu S_{i,a} \right\rangle\right\rangle$$

$$\hat{n}_a^\nu = \left\langle\left\langle \tfrac{1}{N}\sum_{i=1}^N \varsigma_i^\nu S_{i,a} \right\rangle\right\rangle$$

$$q^{ab} = \tfrac{1}{N}\sum_{i=1}^N S_i^a S_i^b \tag{6}$$

$$\mathcal{F} = \tfrac{\alpha_1}{2\beta n}\mathrm{tr}\ln(1-\beta q) + \tfrac{\alpha_2}{2\beta n}\mathrm{tr}\ln(1-\beta\gamma q)+$$

$$\tfrac{1}{2}(\alpha_1 + \alpha_2\gamma) + \tfrac{\beta}{n}\sum_{a<b} r_{ab}q_{ab} + \tfrac{1}{2n}\sum \hat{m}_a^2 + \tfrac{\gamma}{2n}\sum \hat{n}_a^2 - \tfrac{1}{\beta n}\langle\langle \ln Z \rangle\rangle$$

where a, b run from 1 to n, μ runs from 1 to $p_1 = \alpha_1 N$ and ν from 1 to $p_2 = \alpha_2 N$. Z is the one-spin partition function in replica space

$$Z = \mathrm{Tr}\{S\}\exp\left[\sum_{a<b}\beta^2 r_{ab}S^a S^b + \beta\sum_a (\xi\cdot\hat{m}_a + \gamma\varsigma\cdot\hat{n}_a)S^a\right] \tag{7}$$

and the r_{ab} are conjugate to the q_{ab} through a delta function.

The saddle point equations are

$$\frac{\partial \mathcal{F}}{\partial \hat{m}_a^\mu} = 0, \quad \frac{\partial \mathcal{F}}{\partial \hat{n}_a^\nu} = 0, \quad \frac{\partial \mathcal{F}}{\partial q^{ab}} = 0, \quad \frac{\partial \mathcal{F}}{\partial r^{ab}} = 0, \tag{8}$$

leading to values for the parameters $\hat{m}_{SP}$, $\hat{n}_{SP}$, q_{SP}^{ab}, r_{SP}^{ab}. The free energy, then, corresponds to

$$f = \mathcal{F}(\hat{m}_{SP}, \hat{n}_{SP}, q_{SP}^{ab}, r_{SP}^{ab}) \tag{9}$$

In the present study we shall find the zero-temperature saddle-point of $\mathcal{F}$ assuming *replica symmetry*, viz.

$$\hat{m}_a^\mu = \hat{m}^\mu; \ \hat{n}_a^\nu = \hat{n}^\nu; \ q^{ab} = q \ (a\neq b); \ r_{ab} = r \ (a\neq b). \tag{10}$$

It is convenient to define $C = \lim_{\beta\to\infty}\beta(1-q)$. If the initial state is a 'marked' pattern of the type $\hat{m}^\mu = \hat{m}\delta^{\mu,1}$, one ends up with the following saddle-point equations

$$\hat{m} = \mathrm{erf}\left(\tfrac{\hat{m}}{\sqrt{2r}}\right)$$

$$C = \sqrt{\tfrac{2}{\pi r}}e^{-\hat{m}^2/2r} \tag{11}$$

$$r = \tfrac{\alpha_1}{(1-C)^2} + \tfrac{\alpha_2\gamma^2}{(1-\gamma C)^2};$$

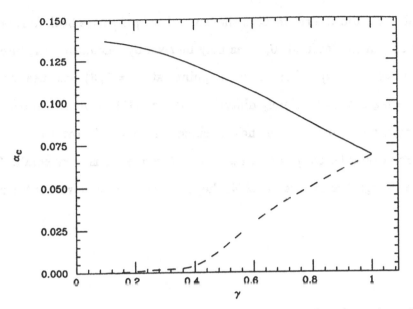

Figure 1: Critical capacity for retrieval of marked (solid line) and unmarked (dashed line) patterns *vs.* γ.

on the other hand, if the initial state is an 'unmarked' pattern of the type $\hat{n}^\nu = \hat{n}\delta^{\nu,1}$, one ends up with

$$\hat{n} = \text{erf}\left(\frac{\gamma\hat{n}}{\sqrt{2}r}\right)$$

$$C = \sqrt{\frac{2}{\pi r}}e^{-\gamma\hat{n}^2/2r} \tag{12}$$

$$r = \frac{\alpha_1}{(1-C)^2} + \frac{\alpha_2\gamma^2}{(1-\gamma C)^2};$$

The above equations lead to one relation between the variables α_1, α_2 and γ, thus defining a critical surface for each case. This is a fairly complicated object to study, so we shall impose an additional condition, namely $\alpha_1 = \alpha_2 = \alpha$, thereby reducing the surface to a line.

Numerical solutions of equations (11, 12), under this constraint, are displayed in Fig, 1, where the critical value of α, α_c is shown as a function of the parameter γ (solid line from (11), dashed from (12)). For a fixed value of γ there correspond two critical values, α_m and α_u. For $\alpha \leq \alpha_u$, the network is capable of retrieving the 'marked' and the 'unmarked' pattern (this means that equations (11, 12) admit solutions $\hat{m} \neq 0$, $\hat{n} \neq 0$); for $\alpha_u < \alpha \leq \alpha_m$, it may retrieve only the 'marked' pattern (i.e. $\hat{m} \neq 0$, $\hat{n} = 0$); and above α_m it cannot retrieve any pattern ($\hat{m} = 0$, $\hat{n} = 0$): it blacks out, entering a 'spin glass' phase.

The transitions from the retrieval to the 'spin glass' phase are both first order, for all values of γ in the interval $(0,1]$ (as may be seen by calculating the internal energy as a function of α). The tricritical point (at $\gamma = 1/3$) has disappeared and the $\alpha - \gamma$ diagram is completely different from ref. [11]. The critical capacity values are significantly lower; nevertheless, since $2\alpha_c > 0.138$, total blackout is postponed. For example, at $\gamma = 0.1$, $\alpha_m = 0.137$, while α_u is very small. This means that, although the load is ≈ 0.274, the network may retrieve the 'marked' half.

Retrieval quality is measured by the overlap of the fixed point pattern with the initial state (i.e. $\hat{m}$ and $\hat{n}$). This is displayed in Fig. 2 as a function of γ for the 'marked' (solid line) and 'unmarked' (dashed line) patterns at criticality: $\alpha = \alpha_m$ in the first case, $\alpha = \alpha_u$ in the second. The overlap for the 'marked' patterns is essentially constant ($\hat{m} \approx 0.97$), while that for the 'unmarked' patterns starts out at 0.97, but very soon goes to 1 as the critical capacity goes to zero with decreasing γ; the network may not recall many 'unmarked' patterns, but those few are recalled perfectly. This is well-known [8]: the number of errors goes to zero exponentially fast as $\alpha \to 0$.

Finally, it is useful to compare the analytical results with direct numerical simulations. While for a finite number of 'marked' patterns, this wouldn't be very meaningful (due to the interplay of essential to nonessential finite size effects), in the present case no such problems arise. An example is displayed in Fig. 2 for $N = 900$ neurons. Each point represents the average over 50 initial conditions, each being a pattern. 'Normal' finite size effects should account for the discrepancies (see [8] for a discussion).

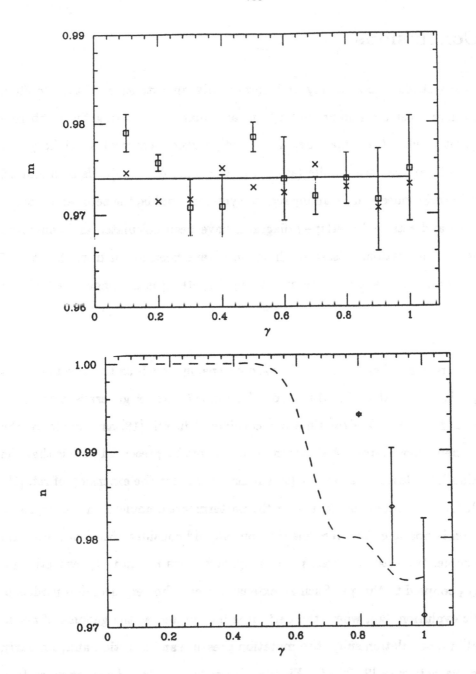

Figure 2: Output overlap near criticality for (a) marked (solid line, squares) and (b) unmarked (dashed line, diamonds) patterns. The data points were obtained from numerical simulations of $N = 900$ neurons, averaged over 50 patterns as initial conditions. Crosses are numerical solutions of the saddle point equations.

3   Conclusions

We have studied analytically and numerically an extension of the Hopfield model, where a certain number of patterns are 'marked', entering the Hebb rule with weight 1, while the remaining, 'unmarked', patterns enter with weight $\gamma < 1$. Both sets of patterns are infinite (as αN, in the limit $N \rightarrow \infty$). Using standard mean-field techniques and assuming replica symmetry holds, the zero temperature capacity–γ and retrieval overlap–γ diagrams have been calculated and compared to numerical simulations. Essential differences have been found from the case of a finite number of 'marked' patterns, namely forgetting is no longer gradual but always abrupt (the retrieval phase to spin glass phase transition is first order rather than second order).

There are a number of straightforward extensions, such as the effects of persistent, noisy external stimuli [14,15,16], dilution [17], etc. A generalization to γ's that depend on space and/or time was considered in ref. [18] as a model of the action of neuromodulators. A shortcoming of the model presented here is that the distinction into classes is arbitrary (this remains true for the extension of ref. [18] as well). The reason, of course, is that the patterns are completely random, so no distinction is possible. To overcome this, one should consider biased patterns [19] (for example, two sets of patterns with magnetizations m_1 and m_2 respectively), with $m_{1,2}$ coupled to the γ's. Such an extension would be very useful in models of prebiotic evolution [20], where the pattern is interpreted as the genome of an individual in a population subject to mutation pressure and selection and, similarly, in adaptive networks [21,22,23]. Finally, one may use other learning rules (e.g. Kepler-Abbott [5] optimized schemes); their possible relevance, especially within these latter models, merits some further investigation.

ACKNOWLEDGEMENTS Stimulating discussions with M. A. Virasoro and K. Y. M. Wong are gratefully acknowledged. I would also like to thank Ton Coolen for a reprint of ref. [18].

Bibliography

[1] J. J. Hopfield, *Proc. Nat. Ac. Sci.(USA)* **79** 2554 (1982).

[2] D. J. Amit *Modeling Brain Function*, Cambridge University Press (1989).

[3] E. Gardner Memorial Volume, *J. Phys.* **A22** #12 (1989).

[4] D. O. Hebb *The Organization of Behavior*, Wiley, New York (1949).

[5] L. F. Abbott "Learning in Neural Network Memories", Brandeis University preprint BRX-274 (1988). T. B. Kepler and L. F, Abbott, *J. Phys.* **A22** 2031 (1989) L. F. Abbott and T. B. Kepler, Brandeis University preprint BRX-255 (1988).

[6] J. A. Hertz these Proceedings.

[7] W. Kinzel these Proceedings.

[8] D. J. Amit, H. Gutfreund and H. Sompolinsky, *Annals of Physics (N. Y.)* **173** 30 (1987).

[9] G. Parisi, *J. Phys.* **A19** L617 (1986).

[10] J. P. Nadal, G. Toulouse, J. P. Changeux and S. Dehaene, *Europhys. Lett.* **1** 535 (1986);
M. Mézard, J. P. Nadal and G. Toulouse, *J. Phys. (France)* **47** 1457 (1986).

[11] J. F. Fontanari and R. Köberle, *J. Phys.* **A21** L253 (1988).

[12] L. Viana, *J. Phys. (France)* **49** 167 (1988) and these Proceedings.

[13] S. Nicolis, "Retrieval Properties of Neural Networks with Infinitely many Marked Patterns", to appear in *Europhys. Lett.*

[14] A. Engel, H. Englisch and A. Schütte, *Europhys. Lett.* **8** 393 (1989).

[15] D. J. Amit, G. Parisi and S. Nicolis, *Network* **1** 75 (1990).

[16] A. Rau and D. Sherrington, *Europhys. Lett.* **11** 499 (1990).

[17] H. Sompolinsky, *Phys. Rev.* **A34** 2571 (1986); see also *Heidelberg Colloquium on Glassy Dynamics*, Springer-Verlag (1987). Strongly asymmetric dilution was studied by B. Derrida, E. Gardner and A. Zippelius, *Europhys. Lett.* **4** 167 (1987).

[18] A. C. C. Coolen and A. J. Noest, *J. Phys.* **A23** 575 (1990).

[19] D. J. Amit, H. Gutfreund and H. Sompolinsky, *Phys. Rev.* **A35** 2293 (1987).

[20] C. Amitrano, L. Peliti and M. Saber, *C.R. Acad.Sci. Paris, série III* **307** 803 (1988); "Population Dynamics in a Spin Glass Model of Chemical Evolution", *J. Mol. Evol. in press.*

[21] S. Kauffman in *Studies of Complexity*, Proc. of the Santa Fe Institute, ed. D. L. Stein, Allison-Wesley (1989).

[22] K. Y. M. Wong, these Proceedings.

[23] E. Hutchins, "How to invent a lexicon: the development of shared symbols in an interaction", University of California at San Diego, preprint (1990).

LEARNING ALGORITHM FOR BINARY SYNAPSES

C.J. Pérez Vicente

Centro Nacional de Microelectrónica

Universitat Autònoma de Barcelona, 08193 Bellaterra, Barcelona

Spain

INTRODUCTION

The analytical calculation of the storage capacity of attractor neural networks (ANN) is now a standard procedure after the generalized use of the techniques developed by E. Gardner [1,2]. Usually, by capacity one refers to the maximum number of patterns $\{\xi_i^\mu\}$ which can be stored in the network satisfying the stability condition

$$\Delta_i^\mu = \xi_i^\mu \sum_j J_{ij} \xi_j^\mu \geq K \sqrt{\sum_j J_{ij}^2} \quad \forall\, i = 1 \ldots N, \mu = 1 \ldots p \tag{1}$$

where N is the number of neurons of the network, p is the number of patterns and $\{J_{ij}\}$ are the synapses. The stability parameter K plays an important role since it is related to the basins of attraction of the memories. The larger K the bigger is the retrieval region associated to the patterns but less can verify the stability condition. Therefore, the control of this parameter allows us to store more or less memories according to the degree of associativity desired for the ANN.

If the number of patterns is small is quite intuitive that there will be a certain amount of different possible configurations J_{ij} which satisfy (1), defining in this way a volume of solutions in the space phase of interactions. As the number of patterns increases the volume becomes smaller and for a given number of patterns p it collapses in a point. This situation defines the storage capacity α_c ($\alpha = p/N$) of the network.

Assuming the spherical constraint for the synapses

$$\sum_j J_{ij}^2 = N \tag{2}$$

Gardner showed that the capacity of an ANN storing uncorrelated patterns is

$$\alpha_c = \int_{-K}^{\infty} \frac{1}{Dt(t+K)^2} \tag{3}$$

where

$$Dt = \frac{1}{\sqrt{2\pi}} \exp(-t^2/2)dt \tag{4}$$

whose maximum value is

$$\alpha_c(K = 0) = 2 \tag{5}$$

Since this result does not give any information about the optimal learning rule it would be important to know some algorithms or iterative methods able to find a solution if it exists. Several alternatives have been proposed in the last years [1,3,4,5]. Perhaps the most known is the Minover algorithm [4] which finds a solution with optimal stability. The weakest feature of this algorithm is its slow convergence to the solution. Recently, Kepler and Abbott [5] have improved this point by proposing an accelerated version which is probably the fastest way of finding a set $\{J_{ij}\}$ verifying all the conditions (1).

However, a deeper analysis of the problem shows that for physical implementations in hardware or VLSI some restrictions should be taken into account. The value of the connections, which in such environment are resistors, achieved after using one of these algorithms previously mentioned are very precise, demanding a level of accuracy so high which cannot be provided by any technology. But this is not the only problem because a neural network should be able to work efficiently in quite different situations simply by changing the values of the weights, which means to built programmable resistors. If their values are analog we have to pay a higher price in area of silicon than if they are discretized, independently of the higher technical complexity needed to implement the system.

Therefore, the most commonly adopted solution is to work with discrete synapses and as a simplification it is interesting the case with only binary couplings, $J_{ij} = \pm 1$. It is clear that the properties of ANN with these new ingredients are necessarily different because now the only solutions in the space phase of connections are those restricted to the corners of the N-dimensional hypercube. Therefore a specific analysis of this new situation must be carry out.

The first theoretical approach aimed at the calculation of the storage capacity of these networks, using the replica method with replica symmetry hypothesis led to a wrong result $\alpha_c = 4/\pi$ because the solution is locally unstable [2]. Going one step forward Krauth and Mezard have shown recently, by using replica symmetry breaking, that the new storage capacity is [6]

$$\alpha_c(K = 0) = 0.83 \tag{6}$$

They also studied extensively the physical interpretation of all the possible results which can be obtained from the replica approach.

This theoretical result has been corroborated in simulations by using different techniques. Krauth and Opper [7] analyzing a small system ($N \leq 25$) found that $\alpha_c \approx 0.82$

by counting the 2^N possible solutions $\{J\}$ and evaluating for each its associated stability K_J. Independently of its own interest one can imagine that this approach is not useful for practical applications in networks of reasonable size. Amaldi and Nicolis [8] applied a sophisticated search procedure called "tabu search" to find solutions for bigger systems. They estimated that for $N \leq 81$ the storage capacity is between 0.6 and 0.9 showing that it is not easy to find precise results due to a strong size effects. Both methods are based on the same strategy: to restrict the search of solutions to the discrete space, i.e. to the corners of the hypercube, evaluating in every step the stability of the chosen vertex and taking as a final result the greatest K.

Recently Kohler et al. have proposed a new approach [9]. Their idea is to minimize the cost function

$$E = \sum_{i, \mu} \left(\Delta_i^\mu - K \sqrt{\sum_j J_{ij}^2} \right)^2 \tag{7}$$

where Δ_i^μ is given by (1). As they are still working on the discrete space they cannot define a gradient descent procedure and therefore other type of techniques such as sequential descent must be applied. In this way the synapses J_{ij} are changed to $-J_{ij}$ in a fixed sequence only if the cost function is lowered by such modification. Using this procedure they have found $\alpha_c \approx 0.4$. The main drawback of this method is the convergence to a local minimum instead of the global one, but as a positive feature I have to point out that it is a real learning procedure, an algorithm not based on a simply counting of corners and stabilities.

Other classical techniques used in optimization problems such as simulated annealing have been also applied to this problem. However, this procedure is very expensive in time which makes it useless for practical applications.

Nevertheless, one may think in a different strategy. Perhaps it is worthwhile to work without any restriction in the phase space of connections looking for solutions satisfying certain requirements and then make a transformation from the continuous space to the corners of the hypercube. Following this idea a straightforward approach is to find an optimal solution verifying all the stability conditions (continuous space) using a Minover algorithm and then make a clipping (discrete space). The storage capacity found using this approach is $\alpha_c \approx 0.3$, much smaller than the theoretical limit. The results are so bad because this procedure does not take into account the features of the problem, since after the clipping we find the closest corner to the real solution which can be far away from the optimal one in the discrete space. The main goal of this article is to show that other strategies can be developed in the continuous space.

THE ALGORITHM

Let's try to built a cost function whose minima are identified with the corners of a N-dimensional hypercube. A first attempt might be to create a spherical-type cost function

$$E = \sum_{ij}(J_{ij}^2 - 1) \tag{8}$$

Actually this choice is not useful for our purposes because there are many ways of minimize E without the specific requirement of having $J_{ij} = \pm 1$ as a solution. However, if instead of (8) we take

$$E = \sum_{ij}(J_{ij}^2 - 1)^2 \tag{9}$$

we ensure that the only points where E takes value 0 are in the corners of the hypercube. In general any function with an even exponent n has this property but in this paper it will be considered only the case n=2. The main advantage of (9) is that it is defined in the whole phase space of J_{ij} since we have put some conditions on the final solution of E but not on the path which leads to it. Therefore, it is possible to apply classical optimization techniques such as gradient descent which cannot be used in the discrete space because they are not defined.

Up to now, nothing has been said about the characteristics of the main problem. In our case they are the set of restrictions (1) which must be verified by all the patterns. Usually this information is introduced directly on the cost function but as an alternative we can introduce it as a set of constraints. In this way we define a non-linear constrained optimization problem with certain differences respect to other conventional studies. Instead of having a cost function where all the relevant information appears in its own definition and all the restrictions (discretized solution) as a constraints (which defines an integer programming problem) we propose a different situation where only the conditions on the solutions appear in the cost function whereas the relevant information is introduced as a constraint.

There are several methods able to deal with such type of problems but perhaps the most known are those called "penalty function methods". This methods transforms the original constrained optimization problem in a sequence of unconstrained minimization processes. The most general case could be stated as follows:

Find the set X which minimizes E(X) subject to

$$C_j(X) \leq 0 \quad j=1, \ldots, n$$

which can be converted into an unconstrained minimization problem by constructing the following function

$$\phi_t(X, R_t) = E(X) + R_t \sum_{j}^{n} H_j[C_j(X)] \tag{10}$$

where R_t is a positive constant called penalty parameter and H_j is some function of the constraints $C_j(X)$. The basic idea is to minimize $\phi_t(X, R_t)$ for a given sequence of R_t ($R_t > R_{t+1}$), such that the final solution converges to that of the original constrained problem.

Two different approachs can be applied. In the first case the minimization process is developed in the feasible region defined by the constraints, so all the unconstrained minima of ϕ_t remain inside. This procedure is called interior penalty method. In the second case, the unconstrained minima of the cost function lie in the infeasible region and converge to the solution from the outside. These are the exterior penalty methods. Only the interior approach will be discussed here. For more information about this methods it is worthwhile to consult [10].

The function H_j is closely related with the chosen penalty method. Although there are several alternatives the most commonly used function for the interior case is

$$H_j = -\frac{1}{C_j(X)} \tag{11}$$

Therefore, the final cost function ϕ becomes

$$\phi_t(X, R_t) = E(X) - R_t \sum_{j}^{n} \frac{1}{C_j(X)} \tag{12}$$

Now, it is easier to understand how does the method work. Initially we need a starting feasible point X_1 satisfying all the constraints $C_j(X_1)$ with strict inequality sign. This is important because if one constraint is satisfied with equality sign the second term in the cost function ϕ_t goes to infinity. In our particular problem it is not difficult to find an initial feasible point because we can use one of these optimal algorithms commented in the introduction, for instance the Kepler-Abbott algorithm. Let's notice that the for given K there is only one feasible region, so all the possible solutions with $\Delta_i^\mu \leq K$ lie inside.

We also take an initial penalty parameter R_1 big enough to be far away from the borders generated by the constraints. However, we have to take into account that a big penalty parameter means a slow minimization process.

Next step is to minimize the cost function $\phi(X, R_1)$ by using an unconstrained minimization process such as gradient descent or a quasi-Newton method. During this process it is not allowed to violate any constraint because the borders act as a barriers ensuring that all the generated points will remain in the feasible domain. As a final result we will have found a certain minimum X_1^*. If this minimum is a solution of the problem ($J_{ij} = \pm 1$) then we stop the process, if not we modify the penalty parameter from R_1 to R_2 following the next rule

$$R_2 = aR_1; \quad a < 1 \tag{13}$$

Let's notice that this procedure change the topology of the cost function ϕ in each step, so a descent direction in the step t may be different of a descent direction in the step t+1. The process continues by minimizing the new cost function $\phi(X, R_2)$ taking as starting point of the process X_1^*. An iterative sequence of this procedure will converge finally to the minimum of the cost function ϕ_t and for R=0 will coincide with the minimum of the objective function E. To be sure that the final solution X_{sol} is the solution of the original constrained problem we have the following theorem:

If the function

$$\phi_t(X, R_t) = E(X) - R_t \sum_j^n \frac{1}{C_j(X)} \tag{14}$$

is minimized for a decreasing sequence of values of R_t, the unconstrained minima X_t^* converge to the optimal solution of the constrained problem as $R_t \leftarrow 0$.

The proof of this theorem can be found in [10]. Another important point is related to the nature of the minimum found by the interior penalty method. The answer is given by the following two theorems:

1.- If E(X) and $C_j(X)$ are convex and at least one of E(X) and $C_j(X)$ is strictly convex, then the function $\phi(X, R_t)$ will be a strictly convex function of X.

2.- Any positive combination of convex functions is convex.

The proof can also be found in [10]. It is clear that if we can proof that both terms in the cost function ϕ_t are globally convex we have solved the optimization problem since in this case any local minimum is a global one and therefore any solution found by the minimization process (if it exists) will coincide with some corner of the hypercube.

Let's consider the second term of the cost function $\phi_t(X, R_t)$.

$$\sum_{i,\mu} \frac{-R}{\lambda - \xi_i^\mu \sum_j J_{ij} \xi_j^\mu} \tag{15}$$

with $\lambda < K$. As we know that the whole unconstrained minimization process is developed in the feasible region (remember that it is not allowed to go outside) the denominator is always negative but the term is positive because of the minus sign in the numerator. Therefore this term is equivalent to the hyperbola $\frac{1}{y}$ for y positive and it is straightforward to show that in such conditions we have a strictly convex function of J_{ij} (the Hessian is positive definite). It is worhtwhile to point out that the normalization condition present in (1) is necessary in the search of the starting feasible point which defines the initial shape of the borders but not in the minimization process of the cost function. We have to take into account that the final solution has fixed norm.

Finally, the first term of the cost function ϕ_t, given by the expression (9), is convex for

$$J_{ij}^2 > \frac{1}{3} \tag{16}$$

and in the most general case

$$E = \sum_{ij}(J_{ij}^2 - 1)^n \quad \forall\, i,j \tag{17}$$

for

$$J_{ij}^2 > \frac{1}{2n-1} \quad \forall\, i,j \tag{18}$$

Although all the corners of the N-dimensional hypercube lie in the convex region it is necessary to point out that some problems may appear very close to the borders because E is not convex in the whole space of interconnections. For instance, in the case of having a solution nearby a border but outside, we need an extremely accurate minimization method to be sensitive to the topology of the problem. If the solution is not in the borders then the algorithm should work quite well.

Therefore, though the algorithm cannot ensure that a solution will always be found (if it exists) because the objective function E is not globally convex I think that it is very promising. A careful analysis of its performance will be presented elsewhere.

CONCLUSION

In this paper has been introduced a new algorithm for discrete synapses. Its advantage respect to previous algorithms is that the search of the solutions is not restricted to the discrete space. Therefore, classical optimization techniques such as gradent descent can be applied. Its good convexity properties makes it quite interesting. The study of this type of algorithms deserve more attention.

ACKNOWLEDGEMENTS

I am grateful to Dra. E. Valderrama and J. Carrabina for very useful discussions.

REFERENCES

[1] E. Gardner, J. Phys. A: Math. Gen. 21, 257 (1988).

[2] E. Gardner and B. Derrida, J. Phys. A: Math. Gen. 21, 271 (1988).

[3] S. Diederich and M. Opper, Phys. Rev. Lett. 58, 949 (1987).

[4] W. Krauth and M. Mezard, J. Phys. A: Math. Gen. 20, L745 (1987).

[5] L. Abbott and T. Kepler, Preprint Brandeis University BRX-TH-255.

[6] W. Krauth and M. Mezard, J. Phys. France 50, 3057 (1989).

[7] W. Krauth and M. Opper, J. Phys. A: Math. Gen. 22, L519 (1989).

[8] E. Amaldi and S. Nicolis, J. Phys. France 50, 2333 (1989).

[9] H. Kohler, S. Diederich, W. Kinzel amd M. Opper, Z. Phys. B 78, 333 (1990).

[10] S.S. Rao in Optimization theory and applications, Wiley Eastern Limited (1984).

STATISTICAL MECHANICS OF THE PERCEPTRON WITH MAXIMAL STABILITY

Wolfgang Kinzel

Institut für Theoretische Physik

Justus–Liebig–Universität, 6300 Giessen

Abstract

The replica method of E. Gardner is used to calculate several properties of the perceptron with maximal stability. In particular results on learning times for biased patterns, the problem of generalization, unsupervised learning and a fast learning algorithm (Adatron) are reported.

1. Introduction

The "perceptron" reacts to a layer of input signals transmitted by synaptic weights [1]. Usually the input signals are binary values $S_i \epsilon \{-1,+1\}$, $i=1,...,N$ and the synaptic weights J_j are real numbers. The output signal is given by

$$S_o = \text{sign}(\underline{J}\ \underline{S}) \qquad (1)$$

with $\underline{J}\ \underline{S} = \sum_j J_j S_j$. Hence the perceptron classifies all of the 2^N many input configurations $\{S_j\}$ into $S_o=+1$ and $S_o=-1$ according to their overlap with the synaptic vector $\underline{J}$. It is a linearly separable Boolean function.

Although the perceptron only represents a rather limited set of Boolean functions, it may be a part of a multilayer network [2–4] which can represent any function between binary configurations.

By synaptic plasticity the perceptron is able to learn specific classifications from a set of examples. Consider a set of αN input/output configurations (=patterns or examples)

$$(\underline{\xi}^{\nu}, \xi_{o}) \qquad \nu = 1, \ldots, \alpha N.$$

The constant α is called the storage capacity. A learning algorithm changes the synaptic weights $\underline{J}$ according to a presented example $(\underline{\xi}^{\nu}, \xi_{o})$. The expamples are classified correctly by Eq.(1) if

$$E^{\nu} = \xi_{o}^{\nu} \, \underline{J} \, \underline{\xi}^{\nu} > 0 \tag{2}$$

Obviously a perceptron maps similar inputs to the same output if the fields E^{ν} are large. In this sense a perceptron generalizes. Hence it may be useful to investigate networks which maximize the "stability"

$$\Delta = \min_{\nu} E^{\nu} / |\underline{J}| \tag{3}$$

with $|\underline{J}|^{2} = \underline{J} \, \underline{J}$.

Recently the late E. Gardner applied methods of statistical mechanics to the perceptron [5]. For random patterns she calculated the maximal possible stability Δ as a function of storage capacity α in the limit of infinite networks $N \to \infty$. The result is given by equation

$$\alpha \int_{-\Delta}^{\infty} Dz(z+\Delta)^{2} = 1$$

with

$$Dz = \frac{1}{\sqrt{2\pi}} e^{-z^{2}/2} \, dz \tag{4}$$

For $\Delta = 0$ one obtains the maximal storage capacity $\alpha_{c} = 2$, which can also be obtained from geometrical arguments [6]. Gardner's calculation shows that there exists a nonzero volume in $\underline{J}$-space which classifies αN random patterns according to Eq.(2). At the border $\Delta(\alpha)$, Eq.(4), this volumen shrinks to zero.

There is also a learning algorithm which finds the perceptron with maximal stability $\Delta(\alpha)$, namely the "minover" algorithm by Krauth and Mezard [7], which runs as follows: The pattern $\underline{\xi}^{\nu}$ with the lowest field E^{ν} is selected and the synapses J_{j} are changed to

$$J_{j} \rightarrow J_{j} + \frac{1}{N} \xi_{o}^{\nu} \xi_{j}^{\nu} \tag{5}$$

This procedure is iterated. If a perceptron exists which classifies all of the examples then the minover algorithm converges to the perceptron with maximal stability.

In this paper I want to report on some extensions of Gardner's method to problems of learning times, generalization, unsupervised learning and perceptron above threshold. I will concentrate on the results and refer to the original literature for technical details.

2. Learning Times

It is useful to write the synaptic weight $\underline{J}$ als a linear combination of the pattern vectors $\xi_0 \xi^\nu$

$$\underline{J} = \frac{1}{N} \sum_\nu x_\nu \, \xi_0^\nu \xi^\nu + \underline{W} \tag{6}$$

where $\underline{W}$ is orthogonal to all of the patterns. Since $\underline{W}$ does not change the fields E^ν and since it increases the norm of the vector J, it decreases the stability Δ, Eq.(3). Therefore the optimal perceptron has $\underline{W}=0$.

The coefficient x_ν is a measure of how strongly the pattern ξ^ν is embedded in the synaptic weight $\underline{J}$. Since the minover algorithm (5) increases x_ν by $\delta x_\nu = 1$ if pattern ξ^ν is learnt, the coefficients x_ν give the number of times the pattern ν is presented during the learning procedure (5). Hence in this case the learning times can be determined just from the properties of the final perceptron $\underline{J}$, without studying the dynamics of the learning rule explicitely.

For the optimal perceptron the distribution of the coefficients x_ν has been calculated by Opper [8]. Fixing the scale of the fields E^ν to $\min_\nu E_\nu = 1$ the vector $\underline{J}$ has two properties

(i) $E^\nu \geq 1$ for all patterns ν

(ii) $\Delta = 1/|\underline{J}|$ is maximal.

Using Eq.(6) these two properties can be expressed in terms of the embedding strengths x_ν:

(i) $\sum_\nu B_{\mu\nu} x_\nu \geq 1$

(ii) $N\Delta^{-2} = \sum_{\nu\mu} x_\mu B_{\mu\nu} x_\nu$ is minimal   (7) where B is the correlation matrix

$$B_{\nu\mu} = \frac{1}{N} \xi_o^\nu \xi_o^\mu \sum_i \xi_i^\nu \xi_i^\mu$$

Therefore, finding the optimal perceptron may be considered as a quadratic optimization problem (ii) with boundary condition (i).

Following Gardner [5] both conditions (7) are expressed in a partition sum

$$Z = \int \prod_\mu \left[dx_\mu \; \Theta \left[\sum_\nu B_{\mu\nu} x_\nu - 1 \right] \right] \exp \left[-\beta \, N \Delta^{-2} \right] \tag{8}$$

which is evaluated by the replica method in the limit of $\beta \to \infty$. In a similar way one obtains the distribution $w(x)$ of coefficients x_ν. For random patterns one finds [8]

$$w(x) = P_o \, \delta(x) + \frac{\Theta(x)}{\sqrt{2\pi\sigma^2}} \exp \left[\frac{(x_o - m)^2}{2\sigma^2} \right] \tag{9a}$$

with

$$P_o = \int\limits_\Delta^\infty Dz \quad ; \quad \alpha_{eff} = \alpha(1 - P_o) \tag{9b}$$

$$m = (1 - \alpha_{eff})^{-1} \quad ; \quad \sigma^2 = \Delta^{-2} (1 - \alpha_{eff})^{-2} \tag{9c}$$

Δ is obtained from Eq.(4).

$w(x)$ has a delta–peak at $x=0$ and a Gaussian distribution for $x>0$ with mean m and width σ. Hence there is a fraction P_o of patterns ξ^ν which is **not** embedded in the synaptic weights $\underline{J}$, i.e. one has $x_\nu = 0$. The minover algorithm (5) (after some transient time) does not learn these patterns, they are learnt from accidential correlations to the other examples. These patterns have $E^\nu > 1$. Only $\alpha_{eff} N$ many patterns are embedded in the vector $\underline{J}$, these patterns are learnt perfectly on to the threshold; for them one has $E^\nu = 1$.

The average learning time is proportional to $T = \frac{1}{N} \sum_\nu x_\nu$ which can be calculated from the distribution $w(x)$. One finds that T diverges as $(2-\alpha)^{-2}$ when the number of examples is increased to the maximal storage capacity $\alpha_c = 2$.

We do not know of any method which before learning finds the patterns which are not embedded ($x_\nu = 0$). If one would know those patterns one could learn the rest (α_{eff} N many) by the linear ADALINE learning rule [10]; this would be a fast learning algorithm.

It is illustrative to derive the distribution $w(x)$ of learning times by a cavity approach [11]: Add a new pattern ξ^ν to a perceptron $\underline{J}$ without learning it. Then the field $u^\nu = \xi^\nu \underline{J}$ is just a random variable which (for several new patterns ν) is Gaussian distributed with zero mean and variance $\underline{J} \, \underline{J} = \Delta^{-2}$. The probability P_0, that this new pattern has to be learnt by Eq.(5) is given by the probability that $u_\circ \geq 1$ which is

$$P_0 = \int_1^\infty \frac{dz}{\sqrt{2\pi}} \, e^{-z^2 \Delta^2 / 2} = \int_\Delta^\infty Dz \tag{10}$$

This argument immediately gives the effective storage capacity α_{eff}, Eq.(9b). Similarly $w(x)$ and Gardner's result, Eq.(4) can be derived.

3. Correlated patterns

Up to now we have considered completely random patterns ξ_0, ξ^ν. However, it is also interesting to study the effect of correlations between the presented examples. The simplest correlation is a bias in the random distribution of ξ_0 and ξ_j:

$$\xi_j = \pm 1 \qquad \text{with probability} \quad \frac{1 \pm m}{2}$$

$$\xi_0 = \pm 1 \qquad \text{with probability} \quad \frac{1 \pm m'}{2} \tag{11}$$

In the following we consider $m = m'$ only, since in this case the results are also applicable to attractor networks [12].

Gardner has already calculated the stability Δ as a function of α and m [5]. It turns out that the maximal storage capacity α_c increases with increasing correlations m and diverges for $m \to 1$.

Using the same methods as in Sec.2 the distribution $w(x)$ of (rescaled) learning steps x_ν has been calculated [13]. Again one finds a fraction P_0 of patterns which are not embedded in the synaptic weight $\underline{J}$. Fig.1 shows P_0 as a function of correlation m for different storage capacities α. With increasing m the effective capacity $\alpha_{eff} = \alpha(1 - P_0)$ decreases, i.e. less patterns are embedded.

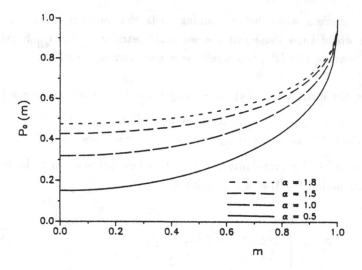

Fig. 1:   Fraction P_o of patterns which are not embedded in the synaptic
vector as a function of bias m. (From Ref.13).

The average learning time $T = \langle x_\nu \rangle$ is shown in Fig. 2 for different values of m. With
increasing correlation α_c increases, but the power law $T \sim (\alpha_c - \alpha)^{-2}$ always holds.

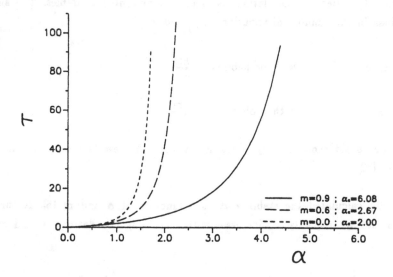

Fig. 2:   Average learning time T as a function of storage capacity α for different
correlations m of the stored patterns. (From Ref. 13).

Note that the learning algorithm becomes faster with increasing correlation m, in agreement with α_{eff} decreasing. However, if one considers T for fixed information content I of the patterns given by [5]

$$I = -\alpha\left[\frac{1+m}{2} \ln \frac{1+m}{2} + \frac{1-m}{2} \ln \frac{1-m}{2}\right] \tag{12}$$

one finds that the average learning time T is larger for correlated patterns.

4. Generalization

In the previous section the output bit ξ_0^ν of the pattern $\underline{\xi}^\nu$ was selected randomly. Therefore for any random input $\underline{\xi}$, which is not learnt, the output variable $\xi_0 =$ sign $(\underline{J}\ \underline{\xi})$ is just a random number. The perceptron cannot generalise random inputs, since the function which gives the examples $(\xi_0, \underline{\xi}^\nu)$ is just random.

The situation is different, if the examples are presented by another nonrandom function which we call the "teacher". In the simplest case the teacher himself is a perceptron with weights $\underline{B}$

$$S_0 = \text{sign}(\underline{B}\ \underline{S}) \tag{13}$$

As before we consider a set of αN random inputs $\underline{\xi}^\nu$ but now the output ξ_0^ν is given by the teacher (13). From these examples $(\xi_0^\nu, \underline{\xi}^\nu)$ another perceptron $\underline{J}$ is constructed by the minover algorithm (5). Note that for each value of the storage capacity α the algorithm converges to the perceptron with maximal stability Δ due to the existence of the teacher $\underline{B}$ (which has zero stability $\Delta=0$).

In contrast to the previous section the output bit ξ_0 is correlated to the input bits, although this correlation is very weak. One finds that $<\xi_0^\nu, \xi_j^\nu>$ is of the order of $1/\sqrt{N}$.

The examples are learnt perfectly: one has $\xi_0^\nu\ \underline{J}\ \underline{\xi}^\nu > 0$ for all patterns ν. However, now also a random input $\underline{S}$ which has not been learnt is classified correctly with a probability $G(\alpha)>0.5$

$$G(\alpha) = \text{prob. that } (\underline{J}\ \underline{S})(\underline{B}\ \underline{S})>0 \tag{14}$$

It is easy to show that $G(\alpha)$ is determined by the angle Θ between the learnt perceptron $\underline{J}$

and the teacher $\underline{B}$; one finds

$$G(\alpha) = 1 - \frac{\Theta}{\pi} = 1 - \frac{1}{\pi} \arccos \frac{\underline{B} \quad \underline{J}}{|\underline{B}| \, |\underline{J}|} \tag{15}$$

The angle Θ has been calculated using the replica method [14]. Compared to Gardner's calculation [5] the input/output correlations introduce an additional Gaussian field in the saddle point equation (4), one finds

$$\alpha \iint\limits_{A} DuDz \, [\Delta - z \sin\Theta - |u|\cos\Theta]^2 = \sin^2\Theta \tag{16}$$

plus the derivative of this equation with respect to $\cos\Theta$. The integral is taken over the two–dimensional domain A given by

$$\Delta - z \sin\Theta - |u|\cos\Theta > 0$$

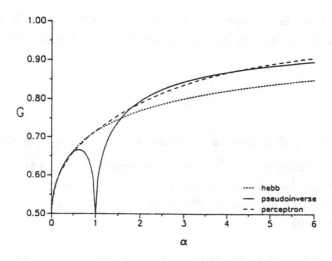

Fig. 3: Generalization probability G as a function of storage capacity α. (From Ref. 14).

These equations are solved numerically and give Θ and by Eq.(12) the generalization probability $G(\alpha)$, which is shown in Fig.3. The result is compared with the corresponding weights $\underline{J}$ obtained from the Hebb rule and the pseudoinverse [15]. The latter vector $\underline{J}$ enforces $\underline{J} \, \underline{\xi}^{\nu} = \Delta \xi_{o}^{\nu}$ for all of the patterns. This works for $\alpha<1$, only. Although the pseudoinverse learns all of the patterns, perfectly, it fails to generalize for $\alpha<1$. This property has been called "overfitting". However, our results show that the optimal perceptron does not overfit; it learns perfectly and generalizes well.

Fig.4 shows the effective storage capacity α_{eff} as a function of the fraction α of learnt examples. As one may expect, due to the correlations only a small number of examples is embedded in the perceptron $\underline{J}$. More than half the examples are not stored in the synapses.

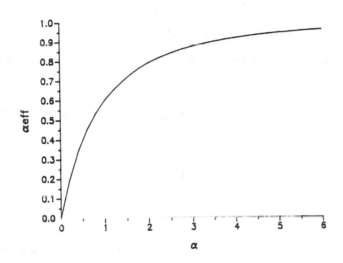

Fig. 4:   Effective storage capacity α_{eff} as a function of real storage capacity α.
(From Ref. 14).

5. Unsupervised learning

In the previous sections the set of examples (ξ_o^ν, ξ^ν) consists of given input/output configuration. However, in particular in connection with multilayer networks [2—4,16] it is interesting to study the case where the network itself is searching for an optimal classification ξ_o^ν.

This motivated us to investigate the following problem: Given a set of αN random input configurations $,\xi^\nu$, what is the maximal possible stability Δ, Eq.(3), when the network can adjust the output bits ξ_o^ν as well as the synaptic weights $\underline{J}$.

Δ has been calculated by extending Gardner's calculation of the phase space of synapses [5]; now the partition function also sums over the additional degrees of freedom $\xi_o^\nu \in \{+1,-1\}$.

Note that zero stability $\Delta=0$ can always be obtained, for any storage capacity α. Namely for any arbitrary vector $\underline{J}$ one can choose

$$\xi_o^\nu = \text{sign}(\underline{J}\ \underline{\xi}^\nu) \tag{17}$$

which gives $\Delta>0$ from the definitions (2) and (3). But the system can do better; by adjusting $\underline{J}$ as well as $\{\xi_o^\nu\}$ to the set of input patterns $\{\underline{\xi}^\nu\}$ it can increase the stability Δ.

Fig.5 shows the result of the (replica-symmetric) calculation which gives [17]

$$2\ \alpha \int_{-\Delta}^{o} Dz(z+\Delta)^2 = 1 \tag{18}$$

For supervised learning (fixed given values of ξ_o^ν) one obtains Gardner's result, Eq.(4); with a maximal storage capacity $\alpha_c=2$. Unsupervised learning gives a correct classification up to $\alpha=\infty$ and the stability increases.

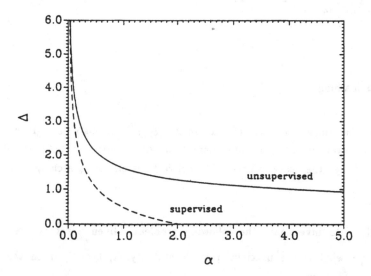

Fig. 5: Stability Δ as a function of storage capacity α for unsupervised and supervised learning. (From Ref. 17).

6. Adatron

From the evaluation of the space of all possible synaptic weights $\underline{J}$ satisfying the classification of the set of examples one knows for which parameters there exists a vector $\underline{J}$ with an maximal stability Δ. As mentioned above, the minover algorithm (5) converges to these synapses $\underline{J}$.

But there is a faster algorithm, called Adatron [18], which is a combination of the Adaline and the perceptron algorithm [11,19]. The idea is: If a pattern ξ^ν is presented the adaption step (5) is weighted such that the field E^ν is moved precisely onto the threshold, say $E^\nu=1$. This is the Adaline rule

$$\underline{J} \rightarrow \underline{J} + \xi_0^\nu(1-E^\nu)\underline{\xi}^\nu \tag{19}$$

However, this algorithm may lead to negative embedding strengths x_ν, Eq.(6), which do not existent in the optimal perceptron (Eq.(19) converges to the pseudoinverse [11,19]). Hence the Adatron algorithm performs step (19) only if the coefficient x_ν remains positive, otherwise is x_ν set to zero. The rule can be directly expressed in the change of x_ν

$$x_\nu \rightarrow x_\nu + \max \{-x_\nu, (1-E^\nu)\} \tag{20}$$

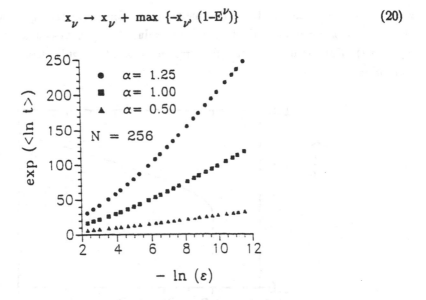

Fig. 6: Average time $\exp<\ell nt>$ after which the difference $\Delta(t)-\Delta$ is smaller than ϵ. (From Ref. 18).

It has been shown [18], using theorems of optimization theory, that this algorithm converges to the perceptron with optimal stability Δ, if a perceptron exists at all. While the minover algorithm converges to Δ with a power law in time t

$$\epsilon = \Delta(t) - \Delta \sim 1/t \tag{21}$$

numerical simulations have shown that for the Adatron (20) the error $\epsilon(t)$ decreases exponentially fast to zero (Fig.6)

$$\epsilon(t) \sim e^{-t/\tau} \tag{22}$$

τ diverges as $(\alpha_c-\alpha)^{-2}$ when the maximal storage capacity $\alpha_c=2$ (for random patterns) is approached.

For $\alpha>\alpha_c$ there does not exist a perceptron which classifies all of the patterns, but the algorithm (20) is still defined. It converges to a synaptic vector $\underline{J}$ which minimizes the function

$$g(\underline{J}) = \sum_\nu (1-E^\nu)^2\ \Theta(1-E^\nu) \tag{23}$$

g has been calculated by the replica method [20]. One finds that g as well as the number of errors jumps discontinuously to a nonzero value if α is increased above the critical value α_c (Fig.7). Note that for $\alpha<\alpha_c$ the gradient descent of $g(\underline{J})$ does not give the optimal perceptron.

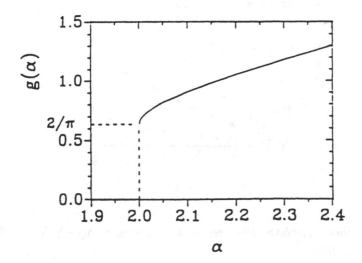

Fig. 7:   Cost function g, Eq.(23), vs. storage capacity α. For $\alpha>2$ there are no errors (g=0). (From Ref. 20).

7. Summary

The phase space calculation of Gardner has been used to calculate properties of the perceptron with optimal stability, which maps similar input configurations to the same output bit. For random patterns with a bias m the distribution of embedding strengths x_ν has been calculated. x_ν is proportional to the number of learning steps the patterns ξ^ν has to presented by the "minover" algorithm of Krauth and Mezard [7]. There is a fraction of patterns which is not embedded at all ($x_\nu = 0$). The average learning time diverges as the maximal storage capacity α_c is approached.

If the examples are presented by a "teacher" which itself is a perceptron, it is interesting to know how well the learnt perceptron can map configurationsorrectly which have not been learnt before. The probability to generalize has been calculated analytically. In contrast to the "pseudoinverse" synaptic weights, the optimal perceptron does not overfit; it learns perfectly and generalizes well.

If the perceptron obtains the freedom to choose the classification bits ξ_o^ν as well as the weights $\underline{J}$, it increases its stability parameter Δ; in particular α_c extends to infinity.

Based on the insight into the optimal perceptron given by the phase space calculation, a fast learning algorithm (Adatron) has been developed. It has been proven that it converges to the optimal perceptron (if it exists), and it minimizes a function $g(\underline{J})$ above the maximal storage capacity $\alpha > \alpha_c$.

Acknowledgement

This report summarizes results of collaborations with M. Opper, J.K. Anlauf, M. Biehl, J. Kleinz, R. Nehl and A. Wendemuth. This work has been supported by the Deutsche Forschungsgemeinschaft, the Stiftung Volkswagenwerk and the Höchstleistungsrechenzentrum Jülich.

References

1. M. Minsky and S. Papert, Perceptrons (MIT press, Cambridge 1988)
2. T. Grossman, R. Meir and E. Domany, Complex Systems 2, 555 (1989)
3. M. Mezard and J.P. Nadal, J. Phys. A 22, 2191 (1989)
4. P. Rujan and M. Marchand, Complex Systems 3, 229 (1989)

7. W. Krauth and M. Mezard, J. Phys. A 20, L 745 (1987)

8. M. Opper, Phys. Rev. A 38, 3824 (1988)

9. M. Mezard, G. Parisi and M.A. Virasoro, Spin Glass Theory and Beyond, World Scientific (Singapore 1987)

10. M. Opper, Europhys. Lett. 8, 389 (1989)

11. W. Kinzel and M. Opper in: Physics of Neural Networks, ed. by E. Domany, J.L. van Hemmen and K. Schulten (Springer, Berlin 1990)

12. D.J. Amit, Modeling Brain Function (University Press, Cambridge 1989)

13. A. Wendemuth, M. Opper, W. Kinzel, unpublished

14. M. Opper, W. Kinzel, J. Kleinz and R. Nehl, J.Phys. A (1990)

15. F. Vallet, Europhys. Lett. 8 747 (1989);
    F. Vallet, J. Cailton, P. Refregier, Europhys. Lett. 9, 315 (1989)

16. R. Meir and E. Domany, Phys. Rev. A 37, 2660 (1988)
    E. Domany, R. Meir and W. Kinzel, Europhys. Lett. 2, 175 (1986)

17. J.K. Anlauf, M. Opper, K. Pöppel, unpublished

18. J.K. Anlauf and M. Biehl, Europhys. Lett 10, 687 (1989)

19. S. Diederich and M. Opper, Phys. Rev. Lett. 58, 949 (1987)

20. J.K. Anlauf and M. Biehl, in: Parallel processing in neural systems and computers, ed. by R. Eckmiller, G. Hartmann and G. Hauske, p. 153, (Elsevier, North Holland 1990); H. Gutfreund, private communication

SIMULATION AND HARDWARE IMPLEMENTATION OF COMPETITIVE LEARNING NEURAL NETWORKS

PRIETO, A.; MARTIN-SMITH, P.; MERELO, J.J.; PELAYO, F.J.;
ORTEGA, J.; FERNANDEZ, F.J.; PINO, B.
Departamento de Electrónica y Tecnología de Computadores
Universidad de Granada. 18071 GRANADA (Spain)

Abstract
One of the main research topics within neuron-like networks is related to learning techniques. Competitive learning has got an special interest among them, because a great network automation is achieved with it, ie, autonomously and without explicit indication of the correct output patterns, the network extracts general features that can be used in order to cluster a set of patterns. In this paper, after giving a brief overview about learning procedures, the most peculiar characteristics of competitive learning are pointed out, and the different ways of implementing neuron-like networks are quoted, describing as an implementation instance our present project of hardware construction of a neural chip to be included in a coprocessor board with competitive learning.

1. INTRODUCTION

The interest in obtaining artificial neural networks is due to the widespread opinion that tasks similar to those performed by the human brain, such as speech [HOP88, KOH88b, LIP89] and image recognition [FUK88], can be carried out with this type of networks whereas normal processors require a huge amount of processing time. Neural networks provide the means for the required processing capabilities and new learning-based programming methods, thus appearing as a clear alternative to a formal manipulation of symbolic expressions (conventional computers).

In this paper we will first present (§2) a brief qualitative description of neuron-like networks learning procedures. The adopted descriptive approach is judged as the best background for appreciating the material in following sections without getting involved in an elaborated mathematical exposition. Section §3 deals specifically with competitive learning, and shows Kohonen network as a particular case of this kind of learning. Finally (§4), the different ways of implementing neuron-like networks are enumerated, and as an instance, our present project is described. It consists on the design and development of a coprocessor board which includes a neuron-like network integrated within an analog chip designed by us.

2. A BRIEF OVERVIEW ON LEARNING PROCEDURES

An artificial neural network is formed by many computational elements (nodes) operating in parallel, linked by weights, that simulate neuron synapses. The task commended to a neuron-like network can be one of the following [RUM86a, ALE89]:

- **Pattern completion and pattern association** . The aim is to complete or recreate an input pattern; that is to say, to get an original pattern from a part of it or a deformed pattern paired with the original one. This is an autoassociation process in which a pattern is associated with itself or with its pair, so that a degraded version of the original pattern can act as a retrieval cue.
- **Classification.** Pattern classes are mapped into distint output codes. There is an a priori fixed and known set of categories into which the stimulus patterns are going to be classified.
- **Regularity detection.** The system must statistically discover the most salient features of the input population to develop its own natural representation.

The simplest neuron model adds n weighted inputs and submits the result to a non-linear function, f. The neuron **total input** or **excitation** can be defined as:

$$X_T = \sum_i w_i x_i + I; \qquad i=1,\ldots,n \tag{1a}$$

where w_i is the weight associated to the x_i input coming from other neuron output, and I is an external or control input. Excitatory and inhibitory connections between neurons can be emulated by assuming positive and negative weights respectively.

The neuron **output** or **state**, y, is given by:

$$y = f(X_T - t) = f(u) \tag{1b}$$

where t is the **threshold** associated to the neuron, $u \equiv X_T - t$ is the **activation potential**, and f is a non-linear function called **activation function**. Frequently I and t are not explicitly expressed, supposing that they are lumped together within the inputs x_i (t with weight w=-1), in such a way that:

$$u = X_T = \sum_i w_i x_i \tag{2}$$

is verified.

A hard-limiting function (f_h), a unit step (f_u), a sigmoid function (f_s), or a pseudolinear function (f_p), can be considered as an activation function[TRE89, LIP87]; however, from a theroretical point of view [ALM89], any continuously differentiable function is also acceptable.

For a binary neuron-like network, i.e. $x_i, y \varepsilon \{0,1\}$, with N neurons interconnected to each others, Hopfield [HOP84] defined a Hamiltonian or energy function E :

$$E = -\frac{1}{2}\sum_i \sum_j w_{ij}\, y_i\, y_j + \sum_i (t_i - I_i)\, y_i \tag{3}$$
$$j, i \in \{1,\ldots,N\} \; ; \quad i \neq j$$

The energy is thus determined by the connectivity matrix **W** (the weight array), the threshold levels t_i, and the external inputs I_i The whole network can be considered as a system whose evolution in the space of the states of the neuron-like elements describes the computation that the neuron set is performing. The final (stable) system state corresponds to an state that minimizes the energy (3) and thus, being the response or solution pattern to the stimulus or input pattern. Any symetric **W** with zero diagonal elements (i.e., $w_{ij} = w_{ji}$; $w_{ii} = 0$) will produce such a flow. If the network is stable when the control inputs I_i are changed it always is led to an steady state of local or global energy minimum.

There are some other more realistic neuron models in which time, τ, is taken as a continuous variable. One of these models states that evolution in time of the "membrane potential" of the jth neuron, u_j, can be described by [VEM88]:

$$\frac{du_j}{d\tau} = -u_j + \sum_i w_{ij} x_i + I_j - t_j \; ; \qquad i=1, 2, \ldots, n \tag{4a}$$

where the output y_i is related to the state u_j through the relation

$$u_j = h(y_j) \tag{4b}$$

In this model a new variable called **firing rate** is considered; it represents the number of spikes traversing the axon in the last $\delta\tau$ (usually 20 ms). If the inputs x_i vary slowly -or are held stationary- $du_j/d\tau \approx 0$, and the equations (4) result

$$y_j = h^{-1}[\sum_i w_{ij} - u_j]; \qquad i = 1, 2, \ldots, n \tag{5}$$

If h^{-1} is identified with f, these equations are analogous to equations (1).

Most of the neuron models used in the literature are variants of the ones above mentioned. They differ basically in the complexity of the activation function, the temporal dependence, etc. Major differences are in neuron-like network architecture and connectivity, which are more application dependent.

TRAINING OR LEARNING

Learning means [HIN89] forming simple associations between representations that are externally specified, or building complex internal representations of their environment. A major goal is to produce networks that correctly generalize to new cases after training on a sufficiently large set of typical cases from some domain. The learning procedures must be able to modify the connection weights in such a way that neuron-like units associate patterns or represent outstanding features of the task domain. Learning methods fall into two broad cathegories: supervised, and unsupervised procedures - within the latter cathegory competitive learning can be included.

Supervised learning

Supervised learning requires the teacher to provide a training pair: an input pattern and the desired output, d, to the network. The network adjusts weights based upon an error function.

The LMS (Least Mean Squares) method uses

$$e = \frac{1}{2}\sum (y_{q,k} - d_{q,k})^2; \qquad q \in O \tag{6}$$

as an error measure, where O is the set of neuron-like units that produce the external outputs, y_{qk} is the actual output of the unit q in the input-output learning case k, and d_{qk} is the corresponding desired output. The error can be minimized by starting with a random set of weights and iterativelly changing them by a value equal to Δw_{ij}.

Hebb procedure [HEB49] consists of an error minimization, starting with an arbitrary set of weights and systematically changing them by a value equal to

$$\Delta w_{ij} = g y_i y_j \tag{7}$$

where g is a constant. This procedure is a reasonable one, as it implies that the connection reinforcement between elements of the network happens only when both pre-synaptic and post-synaptic units are active simultaneously.

Grossberg [GRO76] proposed a modification to this scheme in such a way that

$$\Delta w_{ij} = kf(y_j - w_{ij}) \tag{8}$$

where f is the activation function.

In a feedforward network with one layer of neuron-like elements that have continuous-valued outputs, Widrow and Hoff's [WID60] LMS algorithm or the perceptron learning algorithm [ROS58] can be used. These error correction algorithms require the weight from the i-th input unit to j-th output unit to change by

$$\Delta w_{ij} = k(d_{q,k} - y_{j,k}) y_i \tag{9}$$

The patterns that can be correctly classified with only one layer of neuron-like elements are limited to those that are geometrically equivalent to regions of a vector space bounded by a plane [MIN69; SEJ89]. Adding a single intermediate layer of hidden units between the input and output layers is enough to perform any desired transformation [PAL79]. **Back propagation** is an often used procedure in feedforwared neuron-like networks with hidden units (to which there is no access from outside the network); this procedure uses real-valued deterministic units and consists on actualizing weights once output error is obtained, starting at the output layer and work in a backward sweep [RUM86; PAR86; LeC85].

There are also stochastic learning procedures. In them, each neuron updates its state according to an stochastic decision. One example of such procedure is the Boltzmann machine [HIN84], in which the possible states of each unit are equal to 0 or 1. The probability for a unit j to adopt the state 1 is

$$p_i = \frac{1}{1 - e^{-u_i / T}} \tag{10}$$

where u_i is the activation function of the i unit (expression (2)) and T is a parameter called temperature because of its similarity with Boltzmann's distribution function. Analogously to a high-temperature atom system, the network quickly approaches equilibrium but low energy states are not much more probable than high energy states. At low temperature the network approaches equilibrium more slowly, but reaches low energy states [KIR83, SEJ89].

Unsupervised learning

It consists of a building of internal models that capture regularities in their input vectors without receiving any additional information.

In **unsupervised Hebbian learning**, a unit can develop selectivity to certain kinds of features in its set of input vectors by using a simple weight modification procedure that depends on the correlation between the activity of the unit and the activity on each of its input lines [HIN89, COO79, BIE82].

In **competitive learning**, the network must detect the outstanding pattern regularities, i.e., general features which can be used in order to cluster a set of patterns. It is a procedure that divides a set of input vectors into a number of disjoint clusters in such a way that the input vectors within each cluster are all similar to each other. It is called "competitive" because there is a set of hidden units wich "compete" with each other to become active.

Several researchers [GRO88, VEM88, TRE89] claim that supervised learning is biologically not plausible; there is no evidence that synapses can be used in this form. Competitive learning seems, to a certain point, analogous to natural learning [GRO88, VEM88, TRE89].

3. COMPETITIVE LEARNING

In the first place, supported by Rumelhart and Zipser work [RUM86a], the ideas in which competitive learning is based are briefly presented. Then we will comment some specific neuron-like networks that in some way include the competitive learning procedures. We will devote some more attention to Kohonen's netwok [KOH82, KOH84, LIP87] because of two important reasons, first because it illustrates well the way in which basic competitive learning ideas are used, and second because it will serve us to discuss some general problems related with the development of neuron-like networks that use that kind of learning.

A neuron-like network with competitive learning is shown in Fig.1. It consists of a hierarchized layer set where each unit is connected (with excitatory connection) to upper layer units. In the most general case all units between two consecutive layers are connected to every other. The first layer units are nothing else than inputs to second layer units, through which stimuli patterns are received. In each layer, excluding the first one, units are grouped into disjoint subsets or **clusters** where all units are connected to each other through inhibitory connections (**inhibitory clusters**).

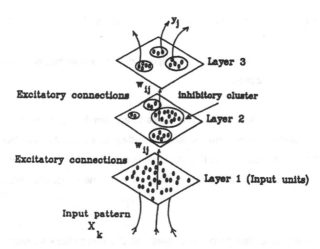

Figure 1: *General scheme of a neuron-like network with competitive learning.*

From now on, we will consider the general case in which every inhibitory cluster element is connected to all the previous layer outputs. Definitions and rules upon which competitive learning is based can be resumed as follows:

Rule 1)

An unit Uj is said to win the "competition" inside its cluster when an stimulus $X_k=(x_{1k},x_{2k},....,x_{nk})$ is applied to its inputs, if the total contribution $X_{Tj}(k)$ of its inputs is such that:

$$x_{Tj}(k) > x_{Ti}(k) \qquad \forall i \neq j \qquad\qquad (11)$$

where

a)   indices i and j correspond to units U_i and U_j from the same inhibitory cluster,

b) $X_{Tj}(k)$ function is an scalar that depends on vectors W_j and X_k. Usually $X_{Tj}(k)$ is evaluated as expression (2) states, more explicitly:

$$X_{Tj}(k) = \sum_i w_{ij} \, x_{1k}$$

Please do note that this rule implies that in an inhibitory cluster only one unit U_j wins at every moment.

Rule 2)

The output y_j from an unit U_j depends, generally, on the total contribution of its inputs and on the fact that it wins or not in its inhibitory cluster. These outputs y_j are defined in such a way that they are compatible with upper layer inputs. To go to the point, along our exposition we will assume that:

$$\begin{aligned} y_j &= 1 \qquad & if\ U_j\ wins \\ y_j &= 0 \qquad & otherwise \end{aligned} \qquad\qquad (12)$$

Please note that rules 1 and 2 imply that in an inhibitory cluster only one unit Uj at every moment can be active (y_j=1).

Rule 3)

The total contribution of inputs X_{Tj} for an unit U_j must be bounded.

Rule 4)

The units Uj in a cluster must be allowed to compete in such a way that each unit can win in some subset of the input patterns applied.

Rule 5)

It is said that an unit U_j in an inhibitory cluster learns to an stimulus X_j, when that unit wins in such a cluster. In that case, its weights, wij, are modified according to an specific learning rule in such a way that the weights of the most active input line are incremented by an small amount and the least active are decreased. The specific way in which this operation is performed depends on the particular learning rule.

An useful geometrical image that helps to understand the competitive learning procedure can be grasped with the following simile.

Let us suppose that input patterns and weights of each unit U_j are represented as normalized vectors X_k y W_j. Such vectors can be represented as points in an n-dimensional hypersphere. In Fig.2 the patterns, X_k, and the weight vectors, W_j for each unit U_j, are respectively marked with "*" and "o_j". Originally W_j marks for each unit are randomly distributed in the hypersphere.

The learning procedure consists of an iterative exposition of an input pattern population X_j. Every time an input X_j is presented, only one unit U_j from its inhibitory cluster wins competition. Explicitly, the winner will be the one whose mark W_j is closer to the presented X_k mark. Only that unit will displace its mark W_j in the address of pattern X_k. If marks X_k statistically show some sort of "natural" agrupation, competitive learning will lead to a situation in which marks Wj of each inhibitory cluster units are approximately centered in those natural assemblies. This learning procedure will thus "discover" some statistically important characteristics of the X_k inputs population.

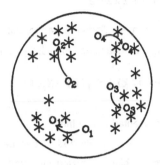

Figure 2: A geometric interpretation of competitive learning [HIN89]

The more structured input patterns are, the sooner an equilibrium situation will be reached, in which U_j units will thus perform an stable classification of input patterns, according to some statistical properties intrinsic to the population of patterns. On the other hand, if the input patterns are little structured the classifications generated by Uj units will be shakier. Rumelhart and Zipser study learning procedure in those situations. We comment here some of their results.

The competitive learning scheme used by Rumelhart and Zipser [RUM86] is the following:

a) Input patterns X_k are binary: $x_{ik}\varepsilon\{0,1\}$

b) Each unit in a cluster receives the same inputs

c) The weight in each unit U_j are such that:

$$\sum_i w_{ij} = 1 \qquad (13)$$

where wij takes real values

d) Learning rule is the following one

$$\Delta w_{ij} = 0 \qquad \qquad if\ U_j\ loses$$
$$\Delta w_{ij} = g\ \frac{x_{ik}}{n_k} - gw_{ij} \quad if\ U_j\ wins\ on\ stimulus\ x_k \qquad (14)$$

where x_{ik} is the stimulus i-th bit X_k and $n_k=\Sigma xik$

e) $X_{Tj}(k)$ is evaluated as expression (2) states.

Formally, for a neuron-like network that uses the previously shown competitive learning procedure we call network **equilibrium states** to those situations in which the average inflow of weights to a particular line, i, is equal to the average outflow. If stimulus patterns are of the same size ($n_k=n=cte$, for every k), it can be proved that the weights, wij, values in the equilibrium states are proportional to the conditional probability that line i is active ($x_i=1$) given that unit uj wins ($y_j=1$)

$$w_{ij} \propto P(line_i\text{=}1 \mid U_j \ wins)/n \tag{15}$$

However, the probability that unit U_j responds to a stimulus X_k, $P(j,k)$, is submitted to the constraint:

$$
\begin{aligned}
P(i,k) &= 1 \quad && if \ X_{Tj}(k)>X_{Ti}(k) \quad \forall i \neq j \\
P(i,k) &= 0 \quad && otherwise
\end{aligned}
\tag{16}
$$

This means that usually there can exist many different equilibrium states for the network. Some equilibrium states can be more stable than others. A quantity that measures the stability degree of equilibrium state is:

$$T = \sum_k P(k) \sum_{j,i} P(j,k) \ [X_{Tj}(k) - X_{Ti}(k)] \tag{17}$$

where P(k) is the probability that stimulus X_k is presented in any trial.

The network evolves towards states in which T is maximized. It can be proved that this quantity maximizes overlap among patterns within a group while minimizing the overlap between groups. In the geometrical model previously used this would correspond to a vector displacement to compact stimulus X_k groupings, which are far away from each other.

Rumelhart and Zipser have developed simple and clever experiments, as well as formal analyses in which some learning general trends are illustrated with cases in which pattern populations are little structured. For these cases the most stable equilibrium states for the network are those in which the following goals are met:

a) Bias to a minimization of the number of cases in which unit U_i responds to a pattern and does not respond to another that overlaps with the previous one. That is, it minimizes the number of "border" patterns for which unit U_i responds.

b) There is a tendency to "allot" stimuli patterns among cluster units as a function of the total sum of each unit weights.

There exist some other situations in which the stimulus population can be such that some lines never become active. This could randomly lead to a situation in which an unit in an inhibitory cluster never gets to win in the cluster, and so the competitive learning does not play its role. Rumelhart and Zipser solve this situation by slightly modifying expression (18) in the following way:

$$
\begin{aligned}
\Delta w_{ij} &= g_l \frac{x_{ik}}{n_k} - g_l w_{ij} \quad && if \ unit \ U_j \ loses \ on \ stimulus \ X_k \\
\Delta w_{ij} &= g_w \frac{x_{ik}}{n_k} - g_w w_{ij} \quad && if \ unit \ U_j \ wins \ on \ stimulus \ X_k
\end{aligned}
\tag{18}
$$

where $g_l \ll g_w$

This means that losing units get slowly displaced towards where stimuli are produced. Those units will reach the point of "capturing" patterns X_k in such a way that competitive learning process gets established.

It must be enphasized that the basic ideas in which competitive learning is based are not something that arise isolatedly, but it rather constitutes a set of ideas totally or partially used by researchers on neural network learning procedures [MAL73,FUK75, GRO76, KOH82, AMA83, etc.]. From this point of view, we can then consider competitive learning procedures as an available resource for design of specific neuron-like networks.

One of the important properties of competitive learning is, as we have said before, that input patterns get self-classified based on some statistical characteristics of the input stimuli population. This means that in a complex scheme those characteristics may not coincide with the classification criterium looked for; so possible solutions should be researched. For instance, the task commended to Kohonen's network is to clasify preprocessed speech signals [KOH88] (spectral components of speech acoustic signal) into phonemes. Kohonen utilizes a proper acoustic signal preprocessing in order to prefix some "natural" assemblies (using a simple euclidean metric) around the phonemes. The competitive learning procedure of Kohonen's network, that will be treated with greater detail in the following section, will then "capture" those natural assemblies in order to classify them, thus achieving spatial reduction of these assemblies.

Another important property that competitive learning procedures have got is that the number of units in a inhibitory cluster is limited to some fixed value. This value is equal to the number of cathegories into which input patterns must fall. This implies an advantage in the sense that stimulus patterns with small differences will fall into the same cathegory, avoiding the fact that a somewhat deformed or noisy pattern is misclasified by the network.

Another instance of net that uses competitive learning is *Fukushima*'s network [FUK82]. This network is composed of two kinds of layers, one with units that learn and another one with fixed weights (pre-wired). The network has got mixed layers of one type and the other. The network is able to recognize simple, slightly deformed shapes or layed in different positions. The generalization regarding pattern traslation is not actually achieved through competitive learning mechanism, but through the layer of pre-wired units [HIN89].

Kohonen's network

Kohonen's network learning is a variant of previously explained competitive learning. Such network thus holds the basic characteristics of competitive learning with real values.

Kohonen's network consists of a neuron-like network, that is composed of two layers as it is shown in Fig.3. The first layer is composed of inputs for the second layer units. All the second layer units form <u>an unique inhibitory cluster</u> and have common and analog inputs.

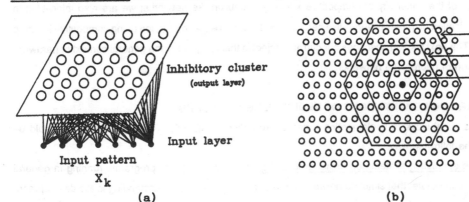

(a)   (b)

Figure 3: Kohonen's network topology

The learning rule (Rule 5) is the following

$$\Delta \overline{W}_{ij} = g(t) x_{ik}(t) - g(t) w_{ij}(t) \qquad (19)$$

This expression is similar to (14), where $x_{ik}(t)$ is of analog type and $g(t)$ is not a constant. Parameter t counts the number of stimulus vectors X_k presented during training. If vectors X_k are supplied at constant time intervals, then parameter t measures the time spent since the network started to be trained until the presentation of a specific input vector $X_k(t)$.

In Kohonen's algorithm total contribution X_{Tj} is not computed (c.f. rule 1) in order to ascertain the winning unit of the inhibitory cluster. Kohonen uses an euclidean metric and computes the winning unit in the following way: given an input stimulus $X_k(t)$, the inhibitory cluster unit U_j chosen as winner is the one whose weight vector $W_j(t)$ comes closer to vector $X_k(t)$. That is, unit U_j wins if the quantity d_j:

$$d_j = \sum_i (x_{ik} - w_{ij}); \qquad i=1,\ldots,n \qquad (20)$$

is minimum among all cluster units.

The essential variation in Kohonen's network with respect to competitive learning is in the fact that not only the winner unit learns at every moment, but also a neighborhood $NE_j(t)$ is defined around the winner unit (ver Fig.3b) to which learning rule is (19) applied. Functions $g(t)$ and $NE_j(t)$ are empirically chosen, and decrease more or less monotonically with time [KOH84].

We are using the general scheme of competitive learning not only for specific neuron-like networks, but also as a resource to explore some statistical properties in pattern populations. These properties will allow to define parameters that can serve as inputs to a competitive learning network, in such a way that the self-organization this one performs classifies input patterns in a desirable way. To be concrete, we try to determine, by using competitive learning, complementary or alternative parameters to conventional accoustic speech preprocessing (FFT transform, LPC coefficientes). These parameters would serve as an input to a network similar to Kohonen's network, in such a way that this one properly clasifies Spanish speech phonemes. We have developed software implementations of a Kohonen type network for SUN and PC-AT systems, and we are exploring some algorithms and hardware-preprocessing optimizations.

Because of the generality of competitive learning in neuron-like networks; we are also interested in hardware realization of general neuron-like networks that ease up experimentation using competitive learning procedures. The next section describes our current project in this field, within the general context of hardware implementation of neuron-like networks.

4. HARDWARE IMPLEMENTATIONS OF NEURON-LIKE NETWORKS

There exist several ways of implementing a neural-like network [HEC89; TRE89; GOS89], that could be grouped in the following way:

- Software simulation systems on digital processors. They are computer programs, running in general purpose computers, that simulate network behaviour. In this field most outstanding is the development of neurocomputing languages for expresing neural networks configurations [HEC89].
- General-purpose neurocomputers or simulation hardware systems. They are usually digital architectures, specifically devoted to simulate neuron-like networks. They usually consist of parallel processor arrays,

floating-point or signal-processing accelerators boards provided with a large memory. Designers usually implement these systems by means of standard chips [TRE89], obtaining physical structures in which software simulation algorithms can be run efficiently. These programmable machines can support a wide spectrum of neural network models, providing a framework analogous to traditional computers. Realizations in this field can be found in the following papers [HEC89; AKE89; GAR89; KUN90].

■ Special-purpose neurocomputers and neural-chips. The above-said systems can check by means of simulation theoretical developments or partial functions of a neuron-like network, they can not get a high degree of parallelism. In this approach, designers directly implement a specific neural network model in hardware. Here we would include VLSI realisations of neuron-like networks [MAC88, MAC89, MAH89,VER89a, VER89b]. Each neuron and synapse takes a dedicated piece of harware. The final objective is the functional integration of a whole system on a silicon chip [MEA88, GOS89]. Integrated networks can have programmable weights [KUB90, BOR90, PEL90a, PEL90b] or fixed weights [GOS89, MAR89], including [ALS88] or not learning specific circuits. With these neural-chips we intend a functional and structural emulation of natural neural networks, setting off from its simplified models.

In what follows we will briefly describe our actual project of hardware implementation of a neuron-like network with competitive learning. The system is in a board to be connected to a PC-AT. This computer will pose as a peripheral and output interface for the neuron-like network (data input and result storage for further study). Our main targets are:

■ Simplify computation, looking for hardware emulation of the used neuron model.

■ Achieve true parallel and asynchronous functioning.

■ Ease the evaluation of various learning algorithms, with modifiable learning law.

The board will include a neural-chip devised by us, as well as some programmable circuits that implement the modification of weights during learning phase.

The devised chip includes a neural network that implements an inhibitory cluster. A scheme of the integrated network architecture and the layout of a VLSI prototype are shown in Fig.4. The circuit has got the following characteristics:

■ The synaptic weights (w_{ij}) are externally alterable asynchronously with network operation.

■ The weights are supplied to the chip as binary patterns ($z_1,..,z_m$), which are stored in a external conventional binary memory. These patterns are converted to analog values into the chip.

■ Each neuron-like element in the network computes X_T (see expression (2)) as an analog sum of currents. For each input pattern ($x_1,x_2,....,x_n$) only the neuron with greatest X_T is activated. In order to implement this operation, maximum X_T is substracted from all neurons in the network. The output y_j is given by:

$$y_j = 1 \quad iff \quad x_{Tj} - \max\{x_{Tj}\} + t > 0; \quad j = 1,..,k \qquad (21)$$

where t is a offset value.

- Because of the fully-parallel operation of neurons and the analog computation of weighted-sum and maximum, the output evaluation times are in the range of nanoseconds.
- A high integration density is achieved: about 200 synaptic connections per square milimeter.   The Fig.5 shows a symplified scheme of the CMOS implementation of two neuron-like elements included into the network.

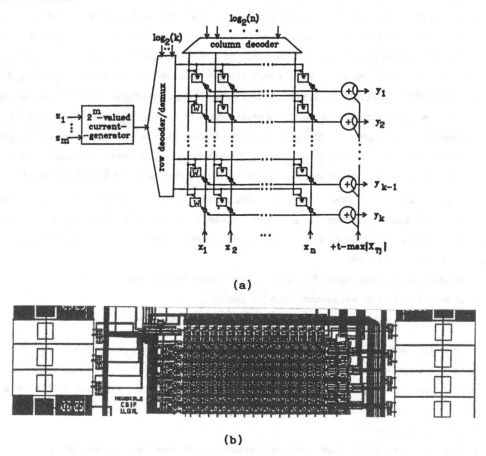

(a)

(b)

Figure 4: *Hardware implementation of an inhibitory cluster: (a) Architecture of the neural network included in the chip. A row-and-column scheme is used to select each synaptic connection for weight up-dating. The weights are temporarily memorized at synaptic connections. The maximum X_{Ti} is obtained simultaneously with the network functioning by a circuitry not included in the figure; (b) VLSI implementation of a network (with de previous architecture) including 8 neuron-like elements with 16 inputs, designed with the rules of the 2-μm CMOS process of ES2.*

We have decided not to include into the neural-chip the complete implementation of the learning procedure, as we think that it imply the loss of generality that would mean the a priori choice of an specific algorithm.

The above said network has got binary inputs; however, we are designing an alternative neuron model in which inputs are analog in order to hardware-implement a Kohonen neuron-like network to be applied to Spanish speech recognition.

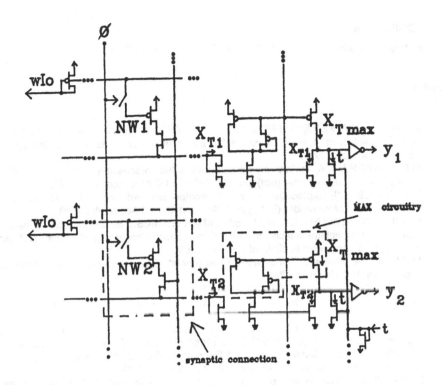

Figure 5: *Scheme of the CMOS implementation of two neuron-like elements in the inhibitory cluster. The weights are temporarily stored as analog voltages at gate nodes NWi. The ϕ signal is produced by an output of the column decoder (see Fig.4). The w.l$_o$ current comes from the current generator (through the row decoder). The currents representing $X_{Ti}+t$ and X_{Tmax} are compared at inputs nodes of the output inverters.*

5. CONCLUSION

Competitive learning is one of the techniques that presents a broader interest within neural-like networks, as it allows to extract the most interesting features without an a priori knowledge of them.

In this paper we show, within the general frame of training methods, the most outstanding characteristics of competitive learning.

Our current research in neural networks with competitive learning for real-time Spanish speech recognition, is oriented towards:

- Network development with the goal of extracting features from input signals (preprocessing).
- Software implementation of classifier networks.
- Analog VLSI circuits development for speech preprocessing.
- Neural-chip and hardware system design with competitive learning. In the current paper we have presented a prototype of a VLSI circuit that emulates an inhibitory cluster.

ACKNOWLEDGEMENTS

This work has been supported by Project MIC 89-04-15, financed by CICYT-SPAIN.

REFERENCES

[AKE89]   AKERS,L.A.; WALKER,M.R.; FERRY,D.K.; GRONDIN,R.O.: "Limited interconnectivity in synthetic neural systems". In: "Neural Computers" R.Eckmiller & C.v.d. Malsburg (Eds); pp. 407-416. Springer-Verlag. 1989

[ALE89]   ALEKSANDER,I.: "A review of parallel distributed processing". In: "Neural Computing Architectures", I. Aleksander (Edt), pp. 329-380. Nort Oxford Academic. 1989

[ALM89]   ALMEIDA,L.B.: "Backpropagation in non-feedforward networks". In: "Neural Computing Architectures", I. Aleksander (Edt), pp. 74-91. Nort Oxford Academic. 1989

[ALS88]   ALSPECTOR,J.; ALLEN,R.B.; HU,V.; SATYANARAYANA,S.: "Stochastic learning networks and their electronic implementation". In: "Neural Information Processing Systems", D.Z. Anderson (Edt), pp.9-21. American Institute of Physics. 1988

[AMA83]   AMARI,S.: "Field theory of self-organizing beural nets". IEEE Transactions on Systems, Man and Cybernetics. Vol.13, pp.741-748. 1983

[BIE82]   BIENENSTOCK,E.L.; COOPER,L.N.; MUNRO,P.W.: "Theory for the development of neuron selectivity: orientation specifity and binocular interaction in visual cortex". Journal of Neuroscience; Vol.2, pp. 32-48. 1982

[BOR90]   BORGSTROM,T.;ISMAIL,M.;BIBYK,S.: Programmable current-mode neural network for implementation in analogue MOS VLSI. IEE Proc. vol 137, Pt.G, no.2, April 1990. pp. 175-184.

[COO79]   COOPER,L.N.; LIBERMAN,F.; OJA,E.: "A theory for the acquisition and loss of neuron specificity in visual cortex". Biological Cybernetics; Vol.33, pp.9-28. 1979

[FUK75]   FUKUSHIMA K.: "Cognitron: A self-organizing multilayer neural network", Biological Cybernetics, Vol.20, pp.121-136, 1975.

[FUK82]   FUKUSHIMA K.: "Neocognitron: A new algorithm for pattern recognition tolerant of deformations and shifts in position". Pattern Recognition; Vol.15, pp.455-469. 1982

[FUK88a]   FUKUSHIMA K.: "Neocognitron: A Hierarchical Neural Network Capable of Visual Pattern Recognition." Neural Networks.; Vol.1, No.2; 1988

[FUK88b]   FUKUSHIMA K.: "A Neural Network for Visual Patern Recognition" IEEE Computer; Vol.21, N°3, pp.65-75; 1988

[GAR89]   GARTH,S.: "A dedicated computer for simulation of large systems of neural nets". In: "Neural Computers" R.Eckmiller & C.v.d. Malsburg (Eds); pp. 435-444. Springer-Verlag. 1989.

[GOS8]   GOSER,K.;HILLERINGMANN,U;RUECKERT,U.;SCHUMACHER,K.: VLSI technologies for artificial neural networks. IEEE MICRO; Vol.9, No.6, pp 28-44. 1989

[GRA88]   GRAF H. P.; JACKEL L. D.; HUBBARD W. E. "VLSI Implementation of a Neural Network Model" IEEE Computer; Vol. 21, N°3, pp.41-49; 1988

[GRO76]   GROSSBERG S.: "Adaptive Pattern Classification and Universal Recoding, I: Parallel Development and Coding of Neural Feature Detectors." Biological Cybernetics.; Vol.23, pp.121-134; 1976.

[GRO88]   GROSSBERG S.: "neural Networks research: From a personal perspective". Electronic Engineering Times; pp.A12-A44; March 7. 1988.

[HEB49]   HEBB,D.O.: "Organization of behavior". John Wiley. 1949

[HEC88]   HECHT-NIELSEN R.: "Neurocomputing: picking the human brain." IEEE Spectrum.; pp.36-41; 1988

[HEC89]   HECHT-NIELSEN R.: "Neurocomputer applications". In: "Neural Computers" R.Eckmiller & C.v.d. Malsburg (Eds); Springer-Verlag. 1989

[HIN84]   HINTON,G.E.; SEJNOWSKI,T.; ACKLEY,D.: "Boltzmann Machines: Constraint Satisfaction Networks that Learn". Tech. Rep. CMU CS 84 119, Carnegie-Mellon Univ. 1984.

[HIN89]   HINTON,G.E.: "Connectionist Learning Procedures"; Artificial Intelligence. Vol.40, No.1-3, pp.185-234. 1989

[HOP84]   HOPFIELD, J.J.: "Neurons with Graded Response Have Collective Computational Properties Like Those of Two State Neurons." Proc. Academy of Science USA.; Vol.81, pp.3088-3092; 1984.

[HOP88]   HOPFIELD, J.J.: "Artificial Neural Networks"; IEEE Circuits and Devices Magazine; pp.3-10; Sept. 1988

[KIR83]   KIRKPATRICK,S.; GELATT,C.D.;VECCHI,M.P.: "Optimization by simulated annealing"; Science. Vol.220, pp.671-680. 1983

[KOH82]   KOHONEN,T.: "Clustering, taxonomy, and topological maps of patterns". In: Proceedings of the 6th Int. Conf. on Pattern Recognition. IEEE Computer Society Press. 1982.

[KOH84]   KOHONEN,T; MÄKISARA,K.; SARAMÄKI,T.: "Phonotopic maps -insightfull representation of phonological features for speech recognition"; Proceedings of IEEE 6th Int. Conf. on Pattern Recognition. Montreal (Canada). pp.182-185. 1984

[KOH88a]   KOHONEN, T.: "Self-Organization and Associative Memory"; Springer-Verlag; 1st Edt. 1984; 2nd Edt. 1988

[KOH88b]   KOHONEN T. "The 'Neural' Phonetic Typewriter" IEEE Computer; Vol.21, N°3, pp.11-22; 1988

[KUB90]   KUB,F.;MOON,K.;MACK,I.;LONG,F.: Programmable analog vector-matrix multipliers. IEEE Journ. of Solid-State circuits, vol 25, no.1, February 1990. pp 207-214.

[KUN90]   KUNG,S.Y.; HWANG,J.N.: "VLSI array processors for neural network simulation". Journal of Neural Network Computing. 1990.

[LeC85]   LE CUN,Y.: "A learning procedure for asymmetric networks". Proceedings of Cognitiva; Vol.85, pp.599-604. Paris. 1985

[LIP87]   LIPPMANN R.P.: "An Introduction to Computing with Neural Nets." IEEE Acoustics, Speech, and Signal Processing.; Vol.4, pp.4-22; 1987.

[LIP89]   LIPPMANN,R.P.: "Review of Neural Networks for Speech Recognition". Neural Computation, Vol.1; No.1; pp.1-38. 1989.

[MAC88]   MACKIE,S.; GRAF,H.P.; SCHWARTZ,D.B., DENKER,J.S.: "Microelectronic implementations of connectionist Neural Networks". In: "Neural Information Processing Systems", D.Z. Anderson (Edt); American Institute of Physics. 1988

[MAC89]   MACKIE,S.; GRAF,H.P.; SCHWARTZ,D.B.: "Implementation of Neural Networks Models in Silicon". In: "Neural Computers" R.Eckmiller & C.v.d. Malsburg (Eds); Springer-Verlag. 1989

[MAH89]   MAHER,M.A.; DEWEERTH,S.; MAHOWALD,M.A.; MEAD,C.A.: "Implementing Neural Architectures Using Analog VLSI Circuits". IEEE Transactions on Circuits ans Systems; Vol.36, No.5, pp.643-652. 1989

[MAL73]   MALSBURG, C. von der: "Self-organization of orientation sensitive cells in striate cortex". Kybernetik; Vol.14, pp.85-100. 1973

[MAR89]   MARTIN-SMITH,P.; PRIETO,A.; PELAYO,F.J; ORTEGA,J.; LLORIS,A.: "CMOS Design of a Neural Network Model". EUROCAST'89 International Workshop on Computer Aided Systems Theory. Canary Islans; pp.119-124; March 1989.

[MEA88]   MEAD,C.A.: "Analog and neural VLSI Systems". Addison-Wesley. 1988.

[MIN69]   MINSKY,M.; PAPERT,S.: "Perceptrons"; MIT Press. 1969

[PAL79]   PALM,G.: "On representation and approximation of nonlinear systems". Part II: "Discrete time". Biological Cybernetics; Vol.34, pp.49-52; 1979.

[PAR86]   PARKER,D.B.: "A comparison of algorithms for neural-like cells".In: "Neural Information Processing Systems", D.Z. Anderson (Edt); Americam Institute of Physics; pp.327-332. 1986

[PEL90a]   PELAYO, F; PRIETO, A; PINO, B; ORTEGA, J; MARTIN-SMITH, P.: "A New Computational Element for Neural Networks"; 4th Int. Symp. on Knowledge Engineering; Barcelona, May 1990.

[PEL90b]   PELAYO, F; PRIETO, A; PINO, B; ORTEGA, J; MARTIN-SMITH, P.: "Hardware Implementation of a Neuron Model"; 8th Int Symp Applied Informatics, IASTED, Innsbruck, Austria, Acta Press; February 1990.

[RUM86a]   RUMELHART D. E.; ZIPSER,D.: "Feature discovery by competitive learning". In: D.E. "Parallel Distributed Processing" Rumelhart & J.L. McClelland (Edts) MIT Press; Vol.I, 151-193. 1986

[RUM86b]   RUMELHART D. E.; HINTON,G.E.; WILLIANS,R.J.: "Learning internal representations by error propagation"; in: Rumelhart,D.E. & McCLELLAND J. L. "Parallel Distributed Processing" MIT Press; Vol.I. 1986

[SEJ89]   SEJNOWSKI,T.J.: "Natural Network Learning Algorithms". In: "Neural Computers" R.Eckmiller & C.v.d. Malsburg (Eds); Springer-Verlag. 1989

[TRE89]   TRELEAVEN,P.; PACHECO,M.; VELLASCO,M.: "VLSI Architectures for neural networks"; IEEE Micro; Vol.9, No.6, pp.8-27. 1989.

[VER89a]   VERLEYSEN,M.; SIRLETTI,B.; VANDEMEULEBROECKE,A.M.; JESPERS,P.G.A.: "Neural Networks for high-storage Content-Addressable Memory: VLSI circuit and learning algorithm"; IEEE Journal of Solid-State Circuits; Vol.24, No.3, pp.562-569. 1989

[VER89b]   VERLEYSEN,M.; JESPERS,P.G.A.: "An analog VLSI implementation of Hopfield's neural network"; IEEE Micro; Vol.9, No.6, pp.46-55. 1989

[VEM88]   VEMURI,V.: "Artificial Neural Networks: An introduction". In: "Artificial Neural Networks: Theoretical conceps" V.Vemuri (Edt). pp. 1-12; IEEE Computer Society Press. 1988

[WID60]   WIDROW,G.; HOFF,M.E.: "Adaptive switching circuits" Institute of Radio Engineers Western Electronic Show and Convention, Convention Record 4, pp.96-194. 1960.

LEARNING IN MULTILAYER NETWORKS:
A GEOMETRIC COMPUTATIONAL APPROACH

Pál Ruján

KOSY AG, Fachbereich 8 Physik, Oldenburg Universität
Postfach 2503, D-2900 Oldenburg, Federal Republic of Germany

1 PERCEPTRONS AND FEEDFORWARD NETWORKS

Artificial neural networks are simple models attempting to capture some essential features of the central nervous system. The arguments on exactly what are these essential features and to what extent they are missing from the fashionable theoretical models will certainly continue to generate heated debates in the foreseeable future. Here we will discuss several properties important for possible *technological* applications, namely learning and generalization abilities of multilayer feedforward networks. Our approach is based on a geometric picture of how feedforward networks process information. This forms the basis for constructing 'good' networks. Several network growing algorithms are discussed.

Until recently, the attention of physicists has focused on simple Perceptrons like models, which can be treated by methods borrowed from the statical mechanics of random systems. These methods are best applicable when the set of examples consists of uncorrelated bit strings. This is the average 'worst case' for such networks, since a random boolean is by definition an unpredictable function. On the other hand, the overwhelming majority of boolean functions are random ones and these results provide strict upper bounds on the information capacity, for example. Our own experience shows that a geometric view is a helpful alternative for understanding neural networks[1,2], as suggested long ago by Pellionisz and Llinàs[3] in a different context. This geometric approach is strongly linked to combinatorial optimization theory [4] and shares many of its concepts and difficulties.

Our basic processing unit is a McCullogh-Pitts [6] binary unit (or linear threshold gate) which is activated only when the weighed sum of its inputs

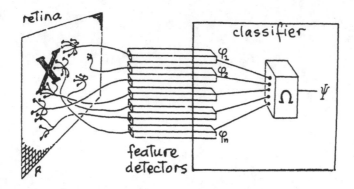

Figure 1: *The Perceptron (from Minsky and Papert). The preprocessing stages of encoding and noise reduction are represented here by 'feature detectors' and are not discussed here.*

is larger than some threshold:

$$\sigma = \mathrm{sgn}\left(\sum_i w_i s_i - \Theta\right) \tag{1}$$

where the vector $\vec{w} = (w_1, \ldots w_N)$ is a vector of weights representing the strengths of N impinging synapses from N other neurons with activations $(s_1, \ldots, s_N)$, $s_i = \pm 1$. Θ is a threshold value. The output of the processing unit is $\sigma = 1$ if the 'neuron' fires a high frequency train of spikes or $\sigma = -1$ if it remains inactivated. From a geometric point of view Eq. (1) says that every elementary processor of our network corresponds to a N-dimensional hyperplane

$$\vec{w}\vec{s} = \Theta \tag{2}$$

and that the points of the N-dimensional configuration space are partitioned in two classes, which are called black for $\sigma = 1$ and white for $\sigma = -1$. In what follows the terms 'hyperplane' and 'unit' are used interchangeably, associating the vector $\vec{w}$ and the threshold Θ with unit σ. In many models the activation values σ, $s_i \in (0, 1)$ are continuous variables and the step-like activation function (1) is a sigmoid. Physically, such situations correspond to a probabilistic updating scheme and the output variable σ is then essentially a time or ensemble average activation [5]. The geometric picture remains valid, provided that one works with slabs of finite width instead of geometrical hyperplanes.

If the input units are the output of some *feature detectors*, $s_i = \pm 1$, the resulting one-processor network is a simple classifier better known under the name of *Perceptron* [8] see Fig. 1. Note that the Hopfield model, for example, is an ensemble of N Perceptrons, one for each unit. As discussed by Minsky and Papert [7], Perceptrons can represent only the class of 'linearly separable' boolean functions (LSF) and many practical tasks fall outside this class. Adding one or more layers of so-called hidden units but without allowing feedback connections leads to a multilayer feedforward architecture. Such devices can implement arbitrary boolean functions at the price of using most of the times an exponentially large number of hidden units.

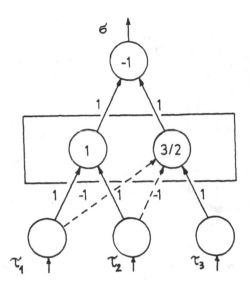

Figure 2: *A feedforward network with three input units, two hidden units and one output unit. The numbers inside the circles represent threshold values and those on connections connection strengths, respectively.*

Only connections between nearest neighbour layers will be allowed. Fig. 2 shows an example of a simple feedforward network.

Unlike attractor networks, which relax to their low energy states by a Monte Carlo-like dynamics, feedforward networks run only for a number of steps limited by the number of layers, very much like a fast arithmetic unit or a Programmable Array Logic chip. This is an attractive feature from a technological point of view and makes such networks a viable model of parallel computation.

Assume now that a (predictable) source is generating a sequence (set) of digitized examples and that a 'teacher' provides information on the class to which each particular example belongs. In our case every example (object) is represented by an input bit string and may belong to one of two classes. Following some *learning algorithm* the network (machine) adjusts its internal parameters (connections and thresholds) so as to commit a minimal error on the presented training set. This task is called *supervised learning* or learning from examples. When the example set is loaded (learnt) into an *appropriate architecture* the training process leads usually to a good representation of the example generating function. This means that further examples from the same source are with high probability correctly classified, provided the function is predictable at all. In contrast, a look-up (or hash) table does not provide any information on new, unknown examples.

2 GEOMETRIC REPRESENTATION

A feedforward network can be represented as follows:

$$\sigma = f(s_1, \ldots, \sigma_N; \{w_{i,j}\}, \{\Theta_i\}) \tag{3}$$

where the dependence on the adjustable connection weights $\{w_{i,j}\}$ and the set of thresholds $\{\Theta_i\}$ has been explicitly displayed. The boolean function ($\sigma = \pm 1$) satisfies a set of input-output (IO) examples $\vec{\xi}^{(\nu)} = \left(\xi_1^{(\nu)}, \ldots, \xi_N^{(\nu)} | \sigma^{(\nu)}\right)$ without any error. Note that the last element of the

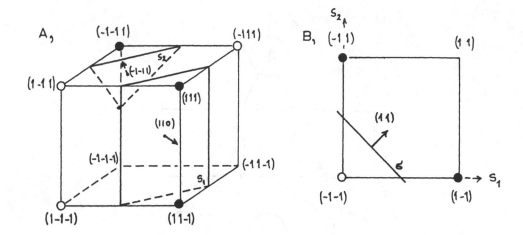

Figure 3: *The hypercube representation of the network shown in Fig. 2. a)
The input configuration cube b) The hidden configuration square. Note the
one-to-one correspondence between the network form and the components
of the normal vectors of the planes shown here.*

vector $\vec{\xi}^{(\nu)}$ is the desired output value and $\xi_i^{(\nu)} = \pm 1$. The *configuration
input space* is defined as the set of vertices in a N-dimensional unit hypercube. All possible input configurations are corners of this hypercube.
Only some of these corners correspond to given examples: we will colour
them according to the desired output value. For the feedforward network
shown in Fig. 2 this is shown in Fig. 3a. Any such function can be rotated,
reflected, etc. according to the symmetries of the N-dimensional unit hypercube. Other, gauge type symmetries are related to changing signs of
connections, thresholds and of some input units but leaving σ invariant. A
good network function (3) should capture all these symmetries.

Consider once again the feedforward architecture shown in Fig. 2 : it
has three input units $N = 3$ and two hidden units, $N_h = 2$. Depending
on the connections between the input and the middle layer and on the
threshold values for the hidden units each input string will induce a given
configuration of hidden units. An N_h dimensional unit hypercube represents now the *hidden configuration space*. The set of coloured (marked)
corners of this cube is the image of the original problem on the hidden
layer. The relation between the configuration cube representations and the
feedforward architecture of Fig. 2 is explicitly shown in Fig. 3. Every set
of connections to a hidden unit corresponds to a hyperplane partitioning
the input configuration hypercube in two. The connections bewteen the
output and the hidden units are represented by a single hyperplane (here a
line) lying in the *hidden* unit hypercube. Each layer of a multilayer network
forms its own image (or internal representation) of the original function.

A convex combination of M vectors $\{\vec{\xi}^{(\nu)}\}_{\nu=1}^M$ is defined as

$$S = \{\vec{x} \in S \mid \vec{x} = \sum_{\nu=1}^M a_\nu \vec{\xi}^{(\nu)} \ ; \ \sum_\nu a_\nu = 1, \ a_\nu \geq 0\} \tag{4}$$

An *extremal* point of a convex combination cannot be expressed as a linear
combination of other points in the set. Given a set of points on the configuration hypercube one can construct the convex hull of this set as a convex
combination of the form (4). The convex hull forms an N-dimensional con-

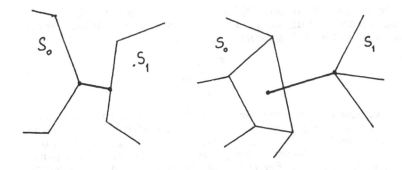

Figure 4: *A schematic representation of a minimal connector in 2 and 3 dimensions.*

vex polytope. The vertices, edges,.., faces of this polytope have dimensions $0, 1, \ldots, n$ and are themselves convex polytopes of lower dimensionality. The *facets* are faces of $n = N$ dimensionality.

The intersection of two convex polytopes is also a convex polytope. The minimal distance between two nonintersecting convex polytopes is called the minimal connector. A schematic view of a minimal connector is shown in Fig. 4.

Consider now a vector $\vec{w}$ in the configuration space. The *map* associated with this direction is the set $\{h^{(\nu)}\}$,

$$h_\nu = \vec{w}\vec{\xi}^{(\nu)} \qquad (5)$$

and consists of the set of projections (internal fields) of the marked unit cube corners on this direction.

Every learning algorithm for multilayer networks incorporates the following two subtasks:

- Problem 1. Is the problem linearly separable or not? If the problem is linearly separable no (further) hidden units are necessary and one can go to

- Problem 2. Find a hyperplane correctly classifying the set of examples of a linearly separable problem.

Both problems are soluble in a polynomial number of steps (as a function of the number of examples and dimension N). The link to the geometrical picture presented above is very simple. First, construct the convex hull S_1 corresponding to the M_1 black points and the polytope S_0 corresponding to the M_0 white points ($M = M_0 + M_1$):

$$S_0 = \{\vec{X}_0 \in S_0 \mid \vec{x} = \sum_{\nu=1}^{M_0} a_\nu \vec{\xi}^{(\nu)} \; ; \; \sum_\nu a_\nu = 1, \; a_\nu \geq 0\} \qquad (6)$$

$$S_1 = \{\vec{X}_1 \in S_1 \mid \vec{x} = \sum_{\nu=1}^{M_1} b_\nu \vec{\xi}^{(\nu)} \; ; \; \sum_\nu b_\nu = 1, \; b_\nu \geq 0\} \qquad (7)$$

Here I_1 and I_0 is the set of indices corresponding to black and white points, respectively. If the intersection of S_0 and S_1 $P = S_0 \cap S_1 = \varnothing$ the problem is linearly separable, otherwise it is not. Actually, if any two faces of the convex hulls S_0 and S_1 are not disjoint the problem is not linearly separable. In terms of maps the same question can be formulated as following: is there a direction $\vec{w}$ in the N dimensional space such that

$$
\begin{aligned}
h_\nu &\geq \alpha \; \forall \, \nu \in I_1 \\
h_\nu &\leq \beta \; \forall \, \nu \in I_0 \\
\alpha &> \beta
\end{aligned}
\tag{8}
$$

Assume now that we have found an architecture which reproduces exactly the whole training set. What are the geometric properties of such a solution? The hidden units are partitioning (or tiling) the space of the input configuration cube with hyperplanes, each hyperplane corresponding to the synaptic strengths afferent to a hidden unit. The first observation is that an exact representation of the example function is possible only if in every *box*, part of space bounded by hyperplanes and by the facets of the unit hypercube, one finds only corners of the same colour. In fact, when considering the image of the function on the first layer for every input cube box there will be only one representative corner of the *first layer* configuration cube. In other words: every marked corner within a given box is mapped into a single configuration of the hidden units. This is evident since the activation patterns on the first hidden layer can change only when one moves across the boundary of a box - and then only one of the coordinates will change. Thus the number of distinct activation patterns for the hidden units equals the number of non empty boxes.

The following construction will always reproduce correctly all IO examples. Separate every black corner by a hyperplane whose normal vector $\vec{w} = \vec{\xi}_1^{(\nu)}$, $\nu \in I_1$ and the threshold is such that the plane 'cuts out' (separates) only the point in question. This kind of solution is of 'grandmother' type, because every black input pattern corresponds now to an uniquely specified hidden configuration, very much like grandmother neurons (if they exist) should be activated only on recognition of your grandmother. One can show (see the regular partition theorem below) that this construction leads to a linearly separable hidden configuration space, thus only one layer of hidden units suffices. Grandmother type solutions are extremely localized representations without any generalization ability: they are the neural network counterparts of usual storage devices. Many theories of neural information processing are based explicitly or implicitly on this type of knowledge representation (including Grossberg's ART networks) and - except perhaps for existence proofs - should be avoided. Since physicists have a lot of experience with cooperative phenomena we perhaps do not need here a long discussion on the advantages of distributed representations[9].

The geometric picture suggests that the right thing to do is try packing as many marked corners in a box as possible. In this case the number of distinct hidden configurations will be a small fraction of the original number of examples, implying a strong contraction of the 'excitable' configuration space from one layer to the next. This contraction is quite natural, since in output the configuration space will consist of only one output variable. This heuristic principle of maximal contraction is a good guide in constructing algorithms. If one can prove that in a concrete procedure the image of the original function is contracting from layer to layer, then this algorithm will eventually find an architecture representing correctly all presented examples. Such a strategy has been used by Mézard and Nadal [10] to prove that their tiling algorithm is converging.

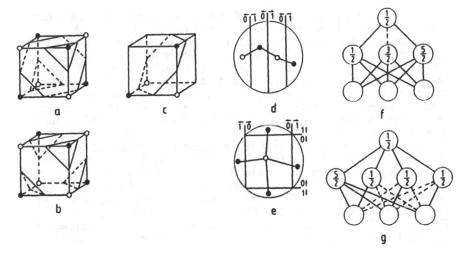

Figure 5: *Constructing two networks solving the parity-3 problem (see text). Full (broken) lines have connection strengths +1 (-1).*

In conclusion: a good architecture is obtained by a partition of the unit hypercube with hyperplanes so that the formed boxes contain as many corners of the same colour as possible. The next question then arises naturally:

- *Problem 9.* Under what conditions is the configuration space of the last hidden layer linearly separable?

This question is again related to linear separability. The hidden unit configurations, however, have some special properties leading to tight sufficient conditions so that the image of a function on a hidden layer is linearly separable. We give here two such conditions. One is restricting the way the space is partitioned in boxes, namely requiring that the hyperplanes corresponding to the hidden unit connections be the boundary of exactly two boxes, one containing black and one containing white corners. We called such tilings regular partitions[1,2]. Two such cases are shown in Fig. 5a-b for the parity-3 problem (Fig.4b shows the grandmother solution). Figs. 5d and 5e are schematic representations of these regular partitionings. The center points of each box represent a whole cluster of corners and are connected into a bipartite tree. Figs. 5f, 5g display the same solutions in the more conventional network form.

The schematic section of the configuration space (Figs. 5f, 5g) shows the main idea of the proof: construct the dual graph of the tiled configuration space. The vertices of this graph are representative internal points of the non empty boxes and are coloured accordingly, similar to Fig. 3b. These points are connected by straight edges. By the definition of regular partitions, every edge of the graph is intersecting only one hyperplane boundary. From the conditions of the tiling it is easy to see that the dual graph must be a bipartite tree and that every edge points on a different direction. A separating plane can be constrained to pass through the median point of each edge by solving a sparse set of linear equations. It is easy to show that all black (white) vertices of the graph will fall (at the same distance) in one side of this plane. In this respect the regular partitions are basically generalizations of the Adaline method at the hidden layer level.

The second formulation is related to a sequential learning algorithm which requires only that the hidden units are set successively and at every step only a box containing corners of the same colour is separated from the input configuration space (sequential learning procedure [11]). A proof similar to the stratification method of Minsky and Papert [7] shows that the resulting tiling is linearly separable and is given in Appendix A.

These two sufficient conditions ensure that a wide class of feedforward networks with one single hidden layer can prfectly represent an arbitrary set of IO examples. A simple criterion for 'good' architectures in terms of random boolean functions, constituting the *average worst case* situation, is the following. As explained above, the main obstacle to linear separability is the presence of a non void intersection polytope, $P = S_0 \cap S_1$. Every new hyperplane (hidden unit) cuts a chunk out of P and a sequential learning procedure, for example, ends when P has been 'eaten' up. It is obvious that the facets of P are parallel to either a facet of S_0 or a facet of S_1 and can be identified by solving a linear programming problem. Looking at the map in the direction normal to the facet, one can define two parallel planes cutting off at least the N linearly independent corners defining the facet. On the other hand, it is obvious that the number of example patterns per hidden unit cannot be larger than $2N$, the maximal storage capacity of the Perceptron. Thus the number of hidden units N_h needed to represent a random boolean function in a one internal layer network is bounded by

$$\frac{M}{2N} < N_h < \frac{2M}{N} \tag{9}$$

where M is the number of examples and N the number of input units. For more hidden layers, including tree like structures, N_h should be smaller. This particularly simple result shows that in average $2M/N$ hidden units are enough to ensure a solution for a typical set of examples. Although this bound does not tell us anything about classes of functions with particular symmetries, it suggests that studying the facet structure of the convex polytope P might help in finding bounds on the sufficient number of hidden units. Thus, it seems plausible that the structure of P facets is also determining to what extent is a function learnable in Valiant's sense [13].

3 LEARNING ALGORITHMS AND THEIR COMPLEXITY

Learning is the search in the space of connections and thresholds (fixed architecture) or in more general, the search in the space of all possible architectures as to reproduce without errors a given set of IO examples. The problem of noisy inputs is not discussed here, nor are learning algorithms leading to a minimal error probability for further examples sampled from the same source. Before discussing algorithms for multilayer networks we shortly summarize the methods known for the Perceptron (for more details see [12]), in particular, the Fisher discriminant method[14], the least mean square error (Adaline) method [15] and the optimal Perceptron method[16,17].

The Fisher discriminant method

This is a standard method in pattern recognition and is based in the simple statistical picture shown schematically in Fig. 6. Assume that the classes of white and black points follow a distribution similar to the normal distribution. Consider a unit vector $\vec{w}$ and the projection of the center of mass and of the variance of these distributions into direction $\vec{w}$ ($\alpha = 0, 1$):

$$\mu_\alpha = \frac{1}{M_\alpha} \sum_{\nu \in I_\alpha} \vec{w} \vec{\xi}^{(\nu)}$$

$$\sigma_\alpha = \frac{1}{M_\alpha} \sum_{\nu \in I_\alpha} (\vec{w} \vec{\xi}^{(\nu)} - M_\alpha \mu_\alpha)^2 \tag{10}$$

The Fisher discriminant method is based on the idea that a good classifying hyperplane should maximize the distance between the average position of the two distributions and minimizes the sum of the two variances.

$$\max_{\{\vec{w}\}} F = \frac{(\mu_1 - \mu_0)^2}{\sigma_1^2 + \sigma_0^2} \tag{11}$$

Note that the direction pointing from the center of mass of the distribution of white points to the center of mass of the black points,

$$\vec{w} = \frac{1}{M_1} \sum_{\nu \in I_1} \vec{\xi}^{(\nu)} - \frac{1}{M_0} \sum_{\nu \in I_0} \vec{\xi}^{(\nu)} \tag{12}$$

corresponds to the Hopfield learning rule and maximizes the distance between the two centers of mass.

The linear discriminant method implies the inversion of an $N \times N$ matrix or can be iteratively obtained by a gradient descent method.

The least mean square error (Adaline) method

The Adaline method[15] is a similar method, based on the observation that Eq. (9) can be rewritten as

$$\sigma h^\nu - \sigma \alpha \geq 0 \text{ for } \nu \in I_1, \ \sigma = 1 \tag{13}$$

$$-\sigma h^\nu - \sigma \beta \geq 0 \text{ for } \nu \in I_0, \ \sigma = -1 \tag{14}$$

Since this set of inequalities is homogeneous, the scale can be fixed by one of the two constants α, β or by their difference. Using $\alpha = \theta + 1$, $\beta = \theta - 1$ the squared error cost function is

$$E_{Adaline} = \sum_{\nu in I_1} (h_\nu - \theta - 1)^2 + \sum_{\nu in I_0} (\theta + 1 - h_\nu)^2 \tag{15}$$

The Adaline method is basically constructing a hyperplane passing through M linearly independent points, thus the storage capacity—the number of examples which can be exactly stored—is N. Recalling Eq.(7) one obtains that in this case the direction of the plane $\vec{w}$ is a linear combination of only M linearly independent set of examples:

$$\vec{w} = \sum_{\mu=1}^{M} x_\mu \xi^{(\mu)} \tag{16}$$

When the set of examples is larger than N, Adaline makes a least mean squared fit, so as to minimize the average squared distance of black (white) points from the corresponding planes. As shown in Fig. 6 for a special (schematic) distribution of example configurations, this separating plane is not particularly well suited for generalizations, a fact supported by recent analytic calculations[18]. The minimization implies the calculation of the generalized inverse of a $M \times N$ matrix or can be obtained by the fa-

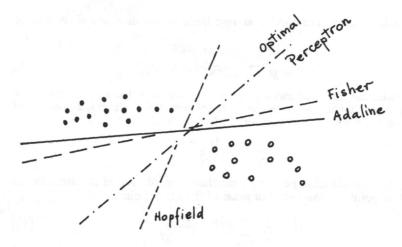

Figure 6: *A schematic representation of different methods for solving linearly separable problems. a) Hopfield-rule, b) Fisher determinant, c) Adaline, d) optimal Perceptron.*

mous Widrow-Hoff Adaline iterative method (a special variant of gradient descent).

Let us remark that the output units in networks trained with Adaline ($M \leq N$) are 'transparent', in the sense that they need be only linear pocessing units. When $\theta = 0$ is a solution and the error function vanishes the output map has only two points, ± 1, and thresholding is not necessary. The same is true for the output units obtained by constructing regular partitions. When the output is further processed by threshold units, transparent units can be deleted from the network after appropriate scaling and redirection of the connection weights.

The optimal Perceptron

Originally, the optimal Perceptron method has been introduced by P. Lambert [16] in a form which slightly differs from the definition used nowadays in the physics literature. Consider once again the geometric picture of the convex hull of black and white examples and assume for a moment that their intersection is void, so the problem is linearly separable. The 'best' separating plane is the one which is most robust to changes in connection strengths and thresholds, corresponding to rotations and translations, respectively. Treating rotations in the N dimensional hyercube is ackward. Translations are easier to treat: one looks for two parallel planes Eq. (9) such that the gap $(\alpha - \beta)/\sqrt{\vec{w}\vec{w}}$ is maximal

$$\max_{\{\vec{w}\}} G = \frac{\alpha - \beta}{\gamma}, \text{ where } \gamma^2 = \vec{w}\vec{w} \tag{17}$$

This is the *primal* quadratic programming problem, subject to the linear set of inequalities (9). In the geometric language one can alternatively ask for the minimal connector, defined as the minimal distance between any point of S_1 from any point of S_0:

$$\min_{\{a_\nu, b_\nu\}} L^2 = (\vec{X}_1 - \vec{X}_0)^2 \tag{18}$$

where the vectors $\vec{X}_i$ are defined in (7) and the coefficients b_ν and a_ν are convex combination coefficients. This equation is called the *dual* quadratic programming problem and one can show [33] that at optimality $G = L$. Obviously, if the set of examples is not linearly separable then G is not defined but $L = 0$. This allows us to use the dual formulation of the problem to decide (in a polynomial number of steps) whether a set of examples (or its image) is linearly separable or not.

The optimal Perceptron used by physicists is slightly different. It is defined by maximizing the gap between the origin and the union $S_1 \cup S_0$ - that is the gap between the origin and the set S_i ($i = 0, 1$) closest to the origin. For large systems and random examples the two definitions agree in average. We prefer to work with Lambert's definition because it leads to a possibly larger gap and works for any distribution of examples.

Quadratic programming algorithms are basic tools of nonlinear optimization and fast implementations are commonly available. One example is the routine QPROG of IMSL version 10.0 implemented by Powell. Two iterative methods have been also developed recently, the MinOver algorithm [17] and the AdaTron algorithm [19]. Both can be generalized to solve the Eqs. (17,18). In our variable step implementation the MinOver is numerically more stable and only slightly slower than the AdaTron (see Appendix B for details).

The optimal Perceptron plays an important role in our discussion for several reasons. First, it makes clear that from all presented examples only the so-called *active* set of N linearly independent edges (difference vectors) is instrumental in determining the optimal hyperplane. This fact is also emphasized by the so-called active set algorithms of quadratic optimization. This allows to define an optimal set of examples as the minimal IO set from which the optimal Perceptron can be uniquely reconstructed. Obviously, the optimal IO set contains only the active extremal points. Hence, any linearly separable function can be stored in $(N+1) \times (N+1)$ bits, independently of the precision of the $N + 1$ real components of $(\vec{w}; \Theta)$. Secondly, the generalization properties of the optimal Perceptron have been calculated analytically[18] and they are superior to any other known learning method (the Fisher discriminant has not been considered yet). This hints at the following intuitive situation: the more robust against rotations and translations a hyperplane is, the better its generalization abilities. This simple intuitive principle is very helpful when dealing with more complex networks, where analytical calculations are difficult to perform.

Multilayer feedforward networks

For fixed architecture feedforward network the training process uses some gradient descent method, for example back-propagation [20], to minimize an error function. Unfortunately, this learning process is quite slow and even if simulated annealing is used the running time increases in a prohibitive way with the network size. This is explained theoretically by the fact that, as shown by Judd[21], the simpler problem of deciding whether a set of examples can be represented by a given feedforward network with hidden units or not is $\mathcal{NP}$-complete. Later Blum and Rivest [22] have mapped the learning problem for a network with only two hidden units into the set splitting problem, which is known to be $\mathcal{NP}$-complete. This means that even for such a simple architecture there are sets of examples which are very difficult to learn. $\mathcal{NP}$-complete problems are characterized by an exponential number of metastable states, leading usually to a slow and erratic dynamics. One should, however, keep in mind that these negative results apply to the worst case of general algorithms: for special cases one might develop fast special algorithms.

According to Valiant's definition[13] a set of examples (or boolean functions) is learnable only if there is a polynomial time algorithm which can approximate the function from a polynomial set of examples within a *controllable* error, that is when the error can be made arbitrarily small. A necessary condition for learnability in feedforward networks is thus that the minimal unit representation of the function has a number of hidden units which is a power of the number of the input units and of the required accuracy. Otherwise the polynomial set of examples cannot 'cover' densely enough the function. Such architectures are possible only if the set of examples has enough symmetries and correlations. Assume that one tries to model these correlations by an energy functional including some interactions between the different corners of the input hypercube (the IO strings). In other words, define an energy for each boolean function which depends, for instance, on the Hamming distance of the different input strings. Intuitively, learnable functions are such that in the in N dimensional input space black and white points form clusters separated by planar interfaces. This restricts the class of interactions to Ising type models with smooth (not rough) planar interfaces.

By solving *Problems 1—3* we have now the tools for a class of algorithms *constructing* an appropriate network from the set of examples. The main idea is to add sequentially new hidden units until the problem is linearly separable by the output unit. This type of algorithms are called *growing algorithms* and so far two main approaches have been used. They differ mainly in the way a new hidden unit is set up given the yet unclassified IO set. The first method is to partition in two parts the example set by a hyperplane which optimizes some appropriately defined cost function. This strategy has been used by Mźard and Nadal [10] in the tiling algorithm and by Nadal [23] and Marchand and Golea [24] for constructing tree-like architectures.

The second approach is using the theorems of the previous Section by separating in every step a box containing as many points of the same colors as possible. This separation step is repeated on the remaining example set until the linear separability test is positive. Then one uses either the solution suggested by the theorems or the (optimal) Perceptron method to set up the output unit.

As long as the boxes contain only points of the same colour, any construction scheme will lead to a good representation of the function, even when several layers are constructed in this way. The number of network architectures implementing a large set of examples is not countable (like the number of functions defined on integer points). Which networks are good and which ones present no interest? Wilhelm Occam's razor says that among different hypotheses explaining the same facts the simpler ones should be preferred. The hint that networks of minimal size have the best generalization properties is also implied by some mathematical work[25]. This should not be surprising since minimal networks must fully incorporate all the correlations and the symmetries of the input set. In our heuristic algorithms we try thus to minimize the number of hidden units by using the maximal contraction principle. It is important to remark that finding the minimal network representing a set of examples must be at least as difficult as deciding whether a given architecture can or cannot implement without error a set of examples. Thus the minimal architecture problem is also $\mathcal{NP}$-hard and no polynomial time general algorithm can decide whether an architecture is minimal or not (unless $\mathcal{P} = \mathcal{NP}$).

The sequential learning algorithm [11] has found the known minimal architectures for the parity problem, the mirror symmetry problem and very good solutions (closed to the optimal bound) for random boolean functions. A similar algorithm has been recently applied to the handwritten numeral recognition problem with very impressive results [26].

The main problem remains to put these ideas in a mathematically more rigorous form. Steps in the direction of optimal predicting strategies have already been made recently [27] and, interesting enough, are also based on a combinatorial geometrical description similar to the one used here.

In conclusion: a very nice feature of the growing network algorithms is that they use only the information contained in the set of examples but do not need *task-dependent* fine tuning of many parameters, like learning rates, acceleration terms, starting conditions, noise levels in simulated annealing, which is the case for back-propagation, competitive learning or Boltzmann machines [9]. In addition, we have been able to give a simple example where the combination of the sequential learning procedure linked to a facet indentification routine leads to a procedure to construct in worst case only 4 times more hidden units than the best possible case for random IO examples. The random IO problem can be considered as a test characterizing the average worst case for any network growing algorithm. According to this test the Perceptron based sequential learning algorithm of Marchand, Golea and Rujan [11] is performing near optimality and is consistently better than the tiling algorithm, the regular partition algorithm or the recently suggested tree generating algorithms.

4 DATA REPRESENTATION AND OPTIMAL ARCHITECTURES

Have we thus solved the boolean function representation problem? Not really: more questions have been actually raised than answered. In what follows some of these questions will be discussed, other ones will be only stated and left at the Reader's discretion.

What is the optimal architecture if several technological or biological constraints have to be taken into account? Consider the mirror-symmetry problem: output a 1 only if the input string is mirror symmetric about its center. This problem can be solved with two hidden units obtained with the help of the stratification method [7] (Fig.7a). However, this implies very oblique planes, with some connection strength ratios growing exponentially with size. Another realization, which follows the disjunctive form of this function and involves N hidden units of unit strength seems more useful from a practical point of view (Fig. 7b). Yet another realization, which uses exact identities to eliminate a hidden units by allowing for direct connections between the input and output units is shown in Fig. 7c. This is the minimal architecture, since the problem is not linearly separable. Which architecture is preferable? This obviously depends on the price of connections, processing units, etc. and cannot be answered in general.

How does the complexity of the network change with the degree of parallelism (number of hidden layers)? Fig. 8a-c shows three possible

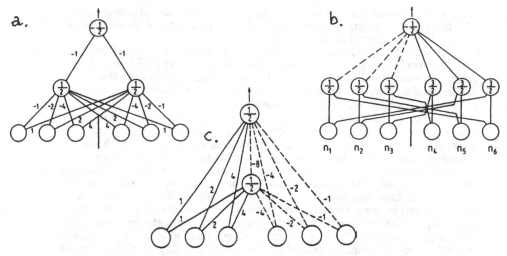

Figure 7: *Three networks solving the mirror symmetry problem. a) Stratification solution, b) Disjunction of regular partitions, c) Minimal solution with one oblique hyperplane.*

realizations of the parity-N (here $N = 7$) problem: the fastest architecture has N hidden units ($N/2$ if input-output connections are allowed), a tree-like 'canonical' realization of the parity function [28] with an $\ln N$ number of layers and hidden units and the two processor network with feedback implementing directly the recursion form of the parity function and running in $2N$ steps. The execution time can be a very important factor in the choice of the network architecture, as every VLSI designer knows. More speed means usually more transistors and here more hidden units. In biological systems one might expect the same effect when and where fast reflexes are crucial.

What classes of simple recursive functions are learnable by feedforward architectures? An interesting question is whether simple arithmetic operations, like addition or multiplication are learnable functions in Valiant's sense. Our numerical results indicate that indeed addition may require only a polynomial number of hidden units. For boolean functions symmetric under the permutation of all input variables the same results apply as above for the parity function. A systematic analysis along these lines seems most desirable.

What happens if we allow the elementary processors to be more complex? Assume for example, that the input and output units can have more colours than two. The basic strategy of the growing algorithms can be easily adapted to such cases provided the hidden units remain hyperplanes, (binary units). Several combinations of binary units have also simple geometric representations. Two parallel planes, for instance, cut the space into three regions and will correspond to a three state unit. Obviously, these cases are related to the data representation problem.

What is the fastest learning algorithm leading to a still acceptably performing network? Can one give some bounds on the expected probability of error for new examples? Can the above learning algorithms be implemented locally and iteratively? How to generalize our strategies when the problem to represent is a probability distribution and the set of examples contains also contradictory examples? These question pertain to a statistical theory of such networks, which is still in its infancy[29].

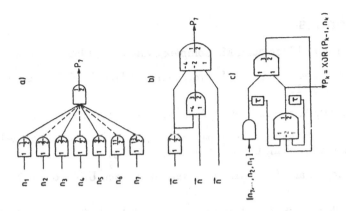

Figure 8: *Three architectures for the parity problem a) The regular partition. The number of processing units is linear in N, running time is minimal (2). b) The 'canonical' form with a logarithmic number of hidden units. The number of processing units and of hidden layers is $\log N$. Running time is also of the same order. c) The feedback network corresponding to a sequential program: $f_{parity}^N = \mathrm{XOR}\left(f_{parity}^{N-1}, n_N\right)$. It requires only two delays and three processing units but execution time is $T \sim 2N$.*

And above all, there hovers the problem which we avoided from the beginning: how are the feature detectors created and to what extent is their data representation a function of the classification process? This question goes deep into the problem of what the process of perception actually is. In general, the number of input units can be increased by a sparse representation of the data. An extreme example is an analogue representation: for example one can use (1000) to represent 0, (0100) to represent 1, etc., instead of the only 2 bits needed to store 4 values in digital code. More sophisticated external coding techniques[30] can be also used to enhance the number of input units in such a way as to increase the probability that the function is linearly separable. A layer of hidden units with random couplings might play a similar role[31]. The question is then: when is the addition of input units helping to minimize the total number of processors ? What is the impact on learning times, since it is always more expensive to search a large input configuration space than a smaller one. At some point these question should be faced in a global way: from the choice of 'sensorial' detectors up to the form of the classifier system and/or the autoassociative features, including their mutual interactions. Nature has solved this problem in its own inimitable way. With the introduction of adaptable Random Markov Field methods in the early stages of feature detection [32] and a better understanding of the geometry of learning we are now closing up on a statistical mechanical information processing machine with exciting abilities.

References

[1] Ruján P. and Marchand M. *Complex Systems* **3** (1989) 229-242

[2] Ruján P. and Marchand M. *Proceedings of IJCNN 1989, Washington* Vol. II 105-110

[3] Pellionisz, A. and LLinàs, R. *Neuroscience* **5** (1979) 1125-1136

[4] Grötschel A. and Padberg, M. *Polyhedral Theory* in The Traveling Salesman Problem, Lenstra *et al* Eds, J. Wiley, 1986

[5] Burnod, Y. and Korn, H. *Proc. Natl. Acad. Sci. USA* **86** (1989) 352-356

[6] McCullogh W. S. and Pitts W. (1943) *Bull. Math. Biophys.* **5** 115-133

[7] M. L. Minsky, S. Papert: *Perceptrons: An Introduction to Computational Geometry* Cambridge Ma., MIT Press, 1969 and 1988

[8] Rosenblatt, F. *Psychoanalytic Review* **65** (1958) 386

[9] Rumelhart,D. E. and McClelland, J. L. (Eds) *Parallel Distributed Processing* Vol. 1-2 , Cambridge Ma., Bradford Books, MIT Press, 1986

[10] Mézard M. and Nadal J. P. *J. Phys.* A **22** (1989) 2191-2203

[11] Marchand, M., Golea, M. and P. Ruján *Europhys. Lett.* **11** (1990) 487-492

[12] Kinzel, W. and Opper, M. in *Physics of Neural Networks*, van Hemmen, Domany and Schulten (Eds) Spinger Verlag, 1989

[13] Valiant L. G. *Comm. of the ACM* **27** pp. 1134-1142 (1984)

[14] Fisher R. A. *Contributions to Mathematical Statistics* 32.179-32.188 John Wiley, New York, 1950

[15] Widrow, B. *Self-Organizing Systems 1962* Yovits et al (Eds) Spartan Books, Washington, D.C., 435-461 John Wiley, New York, 1950

[16] Lambert, P. F. *Methodologies of Pattern Recognition* Watanabe M. S. (Ed.) 359-381 Academic Press, New York, 1969

[17] Krauth, W. and Mézard M. *J. Phys.* A **20** (1987) L745

[18] Opper M., Kinzel, W, Kleinz J. and Nehl, R. On the ability of the optimal Perceptron to generalize preprint Universität Gißen, February 1990

[19] Anlauf J. K. and Biehl M. *Europhys. Lett.* **10** (1990) 687-692

[20] D. E. Rumelhart, G. E. Hinton, R. J. Williams: *Nature* **323**, 533-536 (1986)

[21] Judd S. *Proc. IEEE First Conference on Neural Networks San Diego 1987* Vol. II pp. 685-692 (IEEE Cat. No. 87TH0191-7)

[22] Blum A. and Rivest R. *Proc. of the 1988 Workshop on Computational Learning Theory* pp. 9-18 Hussler D. and Pitt L. (Eds.) Morgan Kaufman, San Mateo, Ca.

[23] Nadal J-P. *Int. J. of Neural Sys.* **1** (1989) 55

[24] Golea M. and Marchand M. A growth algorithm for neural decision trees, submitted to *Europhys. Lett.*

[25] Blumer A., Ehrenfeucht, A., Haussler, D. and Warmuth M. K. *Inf. Proc. Lett.* **24** (1987) 377-380

[26] Knerr, S., Personnaz, L. and Dreyfus, G. to appear in *Neurocomputing: Algorithms, Architectures and Applications* NATO ASI Series, Springer Verlag 1990

[27] Haussler, D., Littlestone, N. and Warmuth M. K. Predicting $\{0,1\}$-Functions on Randomly Drawn Points to be published

[28] Lewis II P. M. and Coates C. L. *Threshold Logic* John Wiley & Sons, New York, 1967

[29] Levin E., Tishby N. and Solla S. A. to appear in *Proc. of the 1989 Workshop in Computational Learning Theory* Györgyi G. and Tishby N. to be published

[30] Saad D. and Marom E. Capacity expansion of neural network models using external coding, IJCNN 1989, Washington, D.C.

[31] Gallant, S. L. *Neural Networks.* **3** (1990) 191-201

[32] Poggio, T. A Parallel Vision Machine that Learns, in *Models of Brain Function*, Cotterill, R. Ed. Cambridge University Press, 1989

[33] Dorn W. S. *Quarterly of Applied Mathematics* **18** (1960) 155-162

APPENDIX A: SEQUENTIAL LEARNING THEOREM

A sequential learning procedure is defined as following[11]. Assume one has some method to find hyperplanes (here hidden units) such that on one side of the hyperplane there are only points of the same colour σ. By adding such a plane to the set of hidden units we say that we have cut the later set of points (they are removed from the example list). Choose the sign of the hidden unit (this can be achieved by negating all connections and the threshold if necessary) so as to match that of σ. This is not really necessary but will simplify our discussion. Assume now that after H such steps the remaining set of IO examples is found to be linearly separable. This means that we have created H hyperplanes (hidden units) which partition the unit input hypercube in at least $H+1$ boxes. Each hidden unit separates a subset Ω_i of M_i points from the input IO set and assumes the value $s_i = \sigma$ when a member of the subset is processed. The subset Ω_{H+1} is defined to be the second set of points obtained by linear separation in the last step of the algorithm and the colour of its output is $s_{H+1} = -s_H$. We give now a recursive algorithm to calculate the connection strengths $u_1, \ldots, u_H$ of the hidden units to the output unit σ and the threshold Θ.

When presenting the whole IO set to the input one can easily see that the form of the activations $\vec{\zeta} = (\zeta_1, \ldots, \zeta_H)$ on the hidden layer will have the following form

$$
\begin{aligned}
\vec{\zeta}^{(1)} &= (s_1, *, *, \ldots, *) & \text{if } \vec{\xi} \in \Omega_1 \\
\vec{\zeta}^{(2)} &= (-s_1, s_2, *, \ldots, *) & \text{if } \vec{\xi} \in \Omega_2 \\
&\vdots \\
\vec{\zeta}^{(k)} &= (-s_1, -s_2, \ldots, s_k, *, \ldots, *) & \text{if } \vec{\xi} \in \Omega_k \\
&\vdots \\
\vec{\zeta}^{(H)} &= (-s_1, -s_2, \ldots, -s_k, \ldots, s_H) & \text{if } \vec{\xi} \in \Omega_H \\
\vec{\zeta}^{(H+1)} &= (-s_1, -s_2, \ldots, -s_k, \ldots, -s_H) & \text{if } \vec{\xi} \in \Omega_{H+1}
\end{aligned}
$$

where $\vec{\xi}$ are the input examples and $\vec{\zeta}$ their image on the first layer. The wild card $*$ signals that the corresponding component can have either a $+1$ or a -1 value.

Choose the threshold in the following form:

$$
\Theta = -\sum_{i=1}^{H+1} u_i s_i \tag{19}
$$

where $u_{H+1} \equiv 1$ by definition. In order to prove that the space of hidden unit activations is linearly separable one has to satisfy

$$
\sigma^{(i)} = s_i = \text{sgn}\left(\sum_{j=1}^{H} u_j \zeta_j^{(i)} - \Theta\right) \tag{20}
$$

for all classes of inputs $\in \Omega_i$, $i = 1, 2, \ldots, H$. These equations are equivalent to the set of inequalities

$$s_i \sum_{j=1}^{H} u_j \zeta_j^{(i)} - s_i \Theta \geq 0 \tag{21}$$

Start with class Ω_{H+1}. Using the general form of $\vec{\zeta}^{(H+1)}$ and the definition of the threshold Θ one can see that the choice $u_{H+1} = 1$ satisfies this inequality. Now repeat the procedure for the class Ω_H. A simple calculations give an inequality involving only u_H (as a function of u_{H+1}). Continue this procedure. In step k one recovers a set of M_{H+1-k} inequalities for the component u_{H+1-k}. Choosing as solution the most stringent inequality, the procedure is continued until the whole vector $\vec{u}$ is obtained. A solution always exists: the choice

$$u_k = 2^{k-H-1} \tag{22}$$

provides a sufficient condition due to the property

$$2^N > \sum_{i=1}^{N-1} s_i 2^i, \; s_i = \pm 1 \tag{23}$$

Whenever the actual condition for u_k has the form $u_k > -1$ the choice $u_k = 0$ will imply that the kth hidden unit is not connected to the output unit and will result in a simplification of the network. For regular partitions one can show that this choice is the algebraic correspondent to the 'sleep' phase[1,2].

A better solution than the one obtained by stratification can be obtained by using the optimal Perceptron learning algorithm discussed in Appendix B. However, the 'stratification' solution is a constructive one and thus runs much faster.

APPENDIX B: A VARIABLE-STEP MINOVER ALGORITHM

In this Appendix we discuss a variant of the MinOver algorithm [17] for the optimal Perceptron defined in Section 3. Recall the definition of the direction $\vec{w}$ connecting a point from S_1 to a point from S_0:

$$\vec{w} = \vec{X}_1 - \vec{X}_0$$
$$\vec{X}_1 = \frac{\sum_{\nu \in I_1} b_\nu \vec{\xi}^{(\nu)}}{\sum_{\nu \in I_1} b_\nu}$$
$$\vec{X}_0 = \frac{\sum_{\nu \in I_0} a_\nu \vec{\xi}^{(\nu)}}{\sum_{\nu \in I_0} a_\nu}$$

where the coefficients a_ν and b_ν are nonnegative. The map corresponding to the direction $\vec{w}$ is then

$$h^{(\nu)} = \vec{w}\vec{\xi}^{(\nu)} = \frac{\sum_{\mu \in I_1} b_\mu Q_{\mu,\nu}}{\sum_{\mu \in I_1} b_\mu} - \frac{\sum_{\mu \in I_0} a_\mu Q_{\mu,\nu}}{\sum_{\mu \in I_0} a_\mu} \tag{24}$$

where $Q_{\nu,\mu}$ is the correlation (overlap) matrix

$$Q_{\nu,\mu} = \vec{\xi}^{(\nu)} \vec{\xi}^{(\mu)} \tag{25}$$

The length of the $\vec{w}$ is given by

$$L^2 = \vec{w}\vec{w} = \frac{\sum_{\nu \in I_1} b_\nu h^{(\nu)}}{\sum_{\nu \in I_1} b_\nu} - \frac{\sum_{\nu \in I_0} a_\nu h^{(\nu)}}{\sum_{\nu \in I_0} a_\nu} \tag{26}$$

It is this length that we want to minimize. In our definition of the map there are two parts, one contains the projections of positive examples, while the other consists of the projections of the negative example points on the direction $\vec{w}$. In the MinOver algorithm one shall update only the coefficients of the black point corresponding to the minimal field $h_1^{(min)}$ of all positive examples and of the white point corresponding to the maximal field $h_0^{(max)}$ among all negative examples. In other words, we push the white points on the negative direction from below and the black points on the positive direction from the top. Therefore, after finding the two indices (min) and (max) we update the vectors of coefficients a_ν and b_ν as following:

$$
\begin{aligned}
a'_\nu &= a_\nu + \eta \delta_{min,\nu}, \quad \nu \in I_0 \\
b'_\nu &= b_\nu + \eta \delta_{max,\nu}, \quad \nu \in I_1
\end{aligned}
\tag{27}
$$

Using the definitions above we first calculate in second order the new values of the map:

$$(h^{(\nu)})' = h^{(\nu)} + y(1-y)[Q_{min,\nu} - Q_{max,\nu} - h^{(\nu)}] \tag{28}$$

where we assumed that initially $\sum_{\nu \in I_0} a_\nu^0 = \sum_{\nu \in I_1} b_\nu^0 = S^0$ and according to our updating rule the norm factor of $a'_\nu s$ and $b'_\nu s$ remains equal during the iteration. $y = \eta/S$, S being th value of the normalization factor before updating. Iterating this procedure but now for L^2 one obtains after some simple algebra

$$L'^2 = L^2 + 2y(h^{(min)} - h^{(max)} - L^2) + y^2 L^2 \tag{29}$$

Since the gap is defined as

$$G = \frac{h^{(min)} - h^{(max)}}{\sqrt{\vec{w}\vec{w}}} \Theta(h^{(min)} - h^{(max)}) \tag{30}$$

where $\Theta(x)$ is the step function, the previous equation has the form

$$L'^2 = L^2 - 2yL(L - G) + y^2 L^2 \tag{31}$$

From the duality theorem of quadratic programming[33] $G = L$ only at optimality, which proves that for small enough η this procedure converges to the optimal solution. The scale of the optimal step size can be also calculated and reads

$$\eta \sim S(L - G) \tag{32}$$

which is typical for a quadratic problem.

STORAGE CAPACITY OF DILUTED NEURAL NETWORKS

M. Bouten

Limburgs Universitair Centrum, Universitaire Campus,
3610 Diepenbeek

1. Introduction

The phase space of interactions, introduced by Elizabeth Gardner [1], has rapidly become a familiar concept and a powerful tool for physicists engaged in research on neural networks. Gardner introduced the concept originally while investigating the maximum possible storage capacity of neural nets. Since then, many different people have extended its use to other problems like the determination of probability distributions of several quantities of interest. A nice collection of papers that make use of the space of interactions can be found in the special issue of the Journal of Physics [2] dedicated to the memory of Elizabeth Gardner. In the present paper, I will describe a further application of this fruitful approach to the study of the storage capacity of diluted networks.

I will distinguish three different types of diluted networks, depending on the time order of the learning and the dilution process. I will start by considering the case where the learning process is carried out first in the fully connected network. (Dilution after learning). When all the memories have been safely stored in the synapses of the network, a certain fraction of these synapses are cut at random. This could, for example, be the result of an accident or of an ageing process. The question to be studied will be : how well are the learnt memories preserved in the remaining synapses ? Or put in different words : are the memories robust to random cutting of connections ?

In the second type (dilution before learning), the time order of the learning and the dilution processes is reversed. Here the fully connected network is first diluted by cutting a certain fraction of the synapses. The cutting may be done in a random or in a systematic way. One may, for example, consider the case of a diluted network in which each neuron remains connected only to its neighbours. The important point here is that the dilution process must be carried out in a way which is independent of the patterns that must be stored. The problem to be studied is the determination of the maximum number of patterns that can be stored in the remaining synapses.

The third type of diluted network will be called : dilution during learning. In contrast to the previous two cases where the learning and the dilution process are totally independent of each other, both processes will become closely interrelated in the third type and must therefore be carried out at the same time. The sole constraint of the dilution is now the total number of broken connections. The position of the broken bounds as well as the strength of the remaining synapses have to be determined in order to maximize the number of stored patterns. The problem I will consider is again the determination of the storage capacity of this third type of diluted network.

In this paper, I will restrict myself to a general description of the calculations and a discussion of the results. The technical details of the calculation can be found in a forthcoming paper [3].

2. Dilution AFTER learning

1. First phase : learning

Consider a fully connected network with N neurons i=1...N and denote p random patterns by $\{\xi_i^\mu\}$, i=1...N, μ=1...p. Storing these patterns as fixed-point attractors requires the determination of couplings coefficients J_{ij} which satisfy the conditions :

$$\sum_{j=1}^{N} J_{ij}^2 = N$$

$$\Delta_\mu = \frac{1}{\sqrt{N}} \sum_{j=1}^{N} J_{ij} \xi_i^\mu \xi_j^\mu > K \qquad (\mu=1...p)$$

(1)

These conditions must be satisfied for all values of i=1...N. Since they fall apart in separate groups, I will concentrate on a single value i. The first condition is the normalization condition. It ensures that J_{ij} is of order one on average. The second condition, the p inequalities $\Delta_\mu > K$ called the stability conditions, express the fact that the patterns $\{\xi_i^\mu\}$ are fixed-point attractors of the neural network dynamics. To be more exact, it would be sufficient to impose these inequalities with K=0 but a larger value of K is usually chosen to increase the size of the basins of attraction [4].

The storage capacity α_c is defined by

$$\alpha_c = \frac{p_{max}}{N}$$

(2)

where p_{max} is the maximum number of random patterns for which the conditions (1) have at least one solution. In the limit $N \to \infty$, Gardner obtained the simple result

$$\frac{1}{\alpha_c\,(K)} = \int\limits_{-K}^{+\infty} Dz(z+K)^2 \tag{3}$$

where Dz is the gaussian measure

$$Dz = \exp[-\frac{z^2}{2}]\,\frac{dz}{\sqrt{2\pi}}\;. \tag{4}$$

When $\alpha = \frac{p}{N} < \alpha_c(K)$, there exist many solutions J_{ij} which satisfy all conditions (1). When $\alpha > \alpha_c(K)$, no solution of (1) exists.

2. Second phase : dilution.

We now choose randomly $(1-f)N$ integers j in the set $(1...N)$ and put the corresponding $J_{ij} = 0$. The parameter f indicates the fraction of non-zero synapses afferent to neuron i. It will be used to characterize the degree of dilution : f=1 represents the fully connected network (no dilution) while f=0 stands for the disconnected network (extreme dilution).

After the dilution process is carried out, there remain only fN non-zero terms in the sums in (1) and many of the stability conditions will become violated. The question now is : do the learnt patterns still remain stored in the network as fixed-point attractors or have they been destroyed during the dilution process ?

The conditions for $\{\xi_i^\mu\}$ to be a fixed point in the diluted network are

$$\gamma_\mu = \frac{1}{\sqrt{N}} \sum_{j=1}^{N}{}' J_{ij}\, \xi_i^\mu\, \xi_j^\mu > 0 \tag{5}$$

where the prime signifies that the synapses j, cut during the dilution process, are left out of the sum. The above question can thus be reformulated as a mathematical problem: what is the probability that γ_μ is positive when we know that Δ_μ is larger than K ?

The probability distribution for γ_μ is given by

$$P(\gamma) = \overline{\delta(\gamma\text{-}\gamma_\mu)} \tag{6}$$

where the bar means an average over all synaptic matrices which satisfy the stability conditions (1) as well as over all choices of the random patterns $\{\xi_i^\mu\}$. The distribution $P(\gamma)$ depends on the number of stored patterns α, on the degree of dilution (i.e. on f) and on the parameter K which is used in the stability conditions (1) during the learning process.

The distribution P(γ) was first calculated by Kepler and Abbott [5] for the fully connected network f=1. Their result has a particularly appealing form near saturation α → $\alpha_c(K)$:

$$P_K(\gamma) = \delta(\gamma\text{-}K)H(\text{-}K) + \theta(\gamma\text{-}K) \frac{1}{\sqrt{2\pi}} \exp[-\frac{\gamma^2}{2}] \tag{7}$$

It is the sum of a delta function at γ=K and the tail of a standard gauss distribution for γ>K. The function H(x) is related to the error function by

$$H(x) = \frac{1}{2}[1 - Erf\frac{x}{\sqrt{2}}] \tag{8}$$

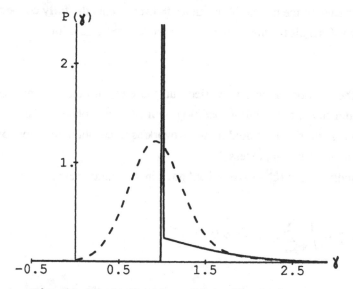

Fig 1: Distribution P(γ) for K=1.
Solid line for f=1. Dashed line for f=0.9.

The calculation of P(γ) becomes more complicated in de diluted case f<1. Using the techniques of Gardner, we obtain the following expression, valid near saturation :

$$P_{K,f}(\gamma) = \frac{1}{\sqrt{2\pi f(1\text{-}f)}} \exp[-\frac{(\gamma\text{-}fK)^2}{2f(1\text{-}f)}] H(\text{-}K) + H(\frac{K\text{-}\gamma}{\sqrt{1\text{-}f}}) \frac{1}{\sqrt{2\pi f}} \exp[-\frac{\gamma^2}{2f}] \tag{9}$$

The delta function in (7) spreads open as a gauss function with variance f(1-f) while the step function is replaced by the smooth function (8). This is illustrated in figure 1 which shows the comparison of the distribution (7) and the distribution (9) for f = 0.9 in the case K=1.

The distribution $P(\gamma)$ can now be used to study the robustness of the memories to random cutting of synapses. Consider first the case K=0 for which the maximum storage capacity $\alpha_c = 2$ can be attained. Figure 2 shows the distribution (9) for two degrees of dilution : f = 0.99 or 1% dilution and f = 0.9 or 10% dilution. It is seen immediately that, in both cases, a substantial part of the distribution is located at negative values of γ. Calculating the integral of $P(\gamma)$ over the negative γ axis yields 0.266 and 0.30 respectively. These large values are easily understood from the Kepler-Abbott distribution (7) because, when K=0, the delta function is located at the origin $\gamma = 0$ and it carries 50% of the total probability. (Since H(0) = 0.5 as is seen from (8)). When the delta function spreads open, we immediately get 25% negative γ_μ's from the first term in $P(\gamma)$. The extra few per cent come from the spreading of the gaussian tail. (For K=0, this "tail" is one half of the gauss function.) The conclusion is clear. For a saturated network with K=0, the robustness of the memories is nil. This result is of course not surprising because for $\alpha = 2$ the storage possibilities of the network are stretched to their ultimate limit.

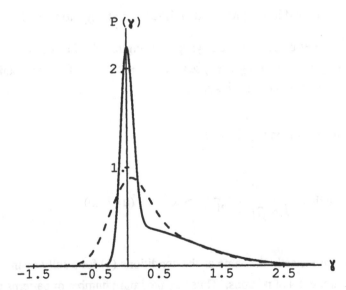

Fig 2: Distribution P(γ) for K=0.
Solid line for f=0.99. Dashed line for f=0.9.

The robustness to random dilution can be improved either by staying away from the saturation limit or by using a large value of K during the learning process. The first possibility gives only a small improvement. If we keep K=0 and store $\frac{N}{2}$ patterns i.e. only one quarter of the maximum possible, random cutting of 1% synapses (f = 0.99) still yields 4% negative γ's. The same number $\frac{N}{2}$ of patterns can also be stored using a larger value of K, the largest possible value K=1.033 being determined by Gardner's formula (3). Using K = 1.033 during the learning process and cutting 1% of the synapses yields only a fraction 10^{-25} negative γ's. Cutting 10% of the synapses still yields less than 0.1% negative γ's.(see figure 1). This illustrates the vast superiority of storing any given set of patterns with the largest possible value of the stability bound K.

3. Dilution BEFORE learning

1. First phase : dilution
We now start by cutting the fraction (1-f) of all the synapses entering into every neuron. The cutting may be done at random or in a systematic way. But once the cutting has been done, it is final. The eliminated bonds cannot be replaced later on.

2. Second phase : learning
After the dilution process if finished, we attempt to store the p patterns $\{\xi_i^\mu\}$ in the remaining synapses by altering their synaptic strengths. Again focussing on a single neuron i and renumbering its remaining synapses from j = 1 to j = fN, we obtain the following conditions for the coupling coefficients J_{ij} :

$$\text{Normalization} : \sum_{j=1}^{fN} J_{ij}^2 = fN$$

(10)

$$\text{Stability} : \frac{1}{\sqrt{fN}} \sum_{j=1}^{fN} J_{ij} \xi_i^\mu \xi_j^\mu > K \qquad (\mu=1...p)$$

These conditions are exactly the same as the conditions (1) for a fully connected network but with a smaller number fN of neurons. Thus the maximum number of patterns p_{max} for which a solution of (10) exists is given by Gardner's equation (3)

$$\frac{p_{max}}{fN} = \alpha_c (K)$$

(11)

The storage capacity per neuron in the diluted network is then

$$\alpha_c(K,f) = \frac{p_{max}}{N} = f\,\alpha_c(K) \tag{12}$$

This storage capacity is a linear function of the fraction f of bounds. The dependence on K is the same for all values of f and is given by Gardner's formula (3).

4. Dilution DURING learning

In the previous two cases, the dilution process and the learning process were completely separated, not only in time but also in procedure. The dilution process was carried out without ever looking at the patterns stored in the network.

In the case which we consider now, the opposite is true. The choice of broken bonds will be made while keeping a close eye on the patterns that must be stored. Whereas the number of cut synapses (the degree of dilution) is prescribed, their position in the network is completely free. The broken bonds can be shifted around until their optimal positions are found to maximize the total number of patterns stored. This type of dilution therefore can aptly be called "annealed" dilution, in contrast to the previous cases where the cutting was definitive ("quenched" dilution).

The calculation of the storage capacity for a diluted network in the annealed case follows closely the calculation of Gardner [1] for the fully connected network. To express that $(1-f)N$ coupling coefficients J_{ij} must vanish for every value of i, it is convenient to write J_{ij} as the product of two factors

$$J_{ij} = c_{ij}\,T_{ij} \tag{13}$$

where $c_{ij} = 0$ or 1. The new coefficients c_{ij} and T_{ij} must then satisfy the conditions :

$$\text{Dilution}: \sum_{j=1}^{N} c_{ij} = fN$$

$$\text{Normalization}: \sum_{j=1}^{N} J_{ij}^2 = \sum_{j=1}^{N} c_{ij}\,T_{ij}^2 = fN \tag{14}$$

$$\text{Stability}: \frac{1}{\sqrt{fN}} \sum_{j=1}^{N} J_{ij}\,\xi_i^\mu\,\xi_j^\mu > K \qquad (\mu = 1...p)$$

We now calculate the fractional volume in the space of interactions where all conditions (14) are satisfied. This leads to four saddle-point equations for four parameters q F E ψ. The first three parameters are the same as Gardner's. The extra parameter ψ appears as a consequence of the dilution condition. The storage capacity α_c (f,K) is obtained from the saddle-point equations in the limit $q \to 1$. Its dependence on f is given by the parametric equations.

$$f = 1 - \text{Erf } u$$

(15)

$$\alpha_c(f,K) = \alpha_c(K) \left[1 - \text{Erf } u + \frac{2}{\sqrt{\pi}} u \exp(-u^2) \right]$$

where the parameter u can take on all positive values. When $u \to 0$, we obtain f=1 and recover Gardner's expression $\alpha_c(K)$. Note that the dependence on K is the same for all values of f.

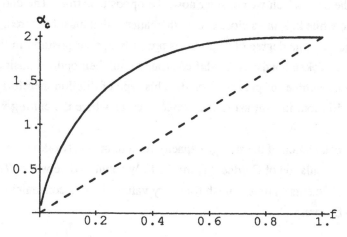

Fig 3:Storage capacity as function of f.
Annealed case:solid line.Quenched case:dashed line.

The storage capacity $\alpha_c(f,K)$ is shown in figure 3 as function of f for the case K=0 (solid line). At f=1 where $\alpha_c = 2$, the curve $\alpha_c(f)$ has a horizontal tangent. A small dilution will thus reduce the storage capacity by a very small amount only. For f = 0.95 for example, we obtain $\alpha_c = 1.9999$. For smaller values of f, the curve $\alpha_c(f)$ stays well above the linearly decreasing dashed line which represents the storage capacity (12) for the quenched case. As an example , for f = 0.5 when 50% of all bonds have been severed the value of α_c is still 1.86, a decrease of only 7%. When f becomes very small and tends to zero, $\alpha_c(f)$ also tends to zero but very slowly :

$$\alpha_c \to -4f \ln f \qquad \text{as } f \to 0$$

(16)

This shows that $\alpha_c(f)$ has the vertical axis as tangent at f=0.

5. Distribution of the synaptic strengths in diluted networks

The difference between diluted networks of the first two types (quenched) and those of the third type (annealed) is highlighted by the different distributions for the coupling coefficients J_{ij}. The distribution $P(J)$ is given by

$$P(J) = \overline{\delta(J - J_{ij})} \tag{17}$$

where the average is again over all synaptic matrices which satisfy the stability conditions and over all choices of random patterns. For a diluted network in which the fraction $(1-f)$ of all synapses is cut, $P(J)$ has the obvious form

$$P(J) = (1-f)\,\delta(J) + P_r(J) \tag{18}$$

The first term represents the vanishing coefficients imposed by the dilution constraint. The second term $P_r(J)$, normalized to f, describes the distribution of the remaining synaptic strengths.

In our first case, dilution after learning, we start with a fully connected network in which the coupling coefficients satisfy the stability conditions (1). The distribution $P(J)$ is easily calculated for this case, yielding a gaussian with mean zero and variance one

$$P(J) = \frac{1}{\sqrt{2\pi}}\exp\left[-\frac{J^2}{2}\right] \tag{19}$$

This result shows that many coupling coefficients in the fully connected network have values close to zero. This reflects the large degree of frustration in the network due to the many conflicting requirements imposed on the coupling coefficients by the different stored patterns. Using a random procedure for cutting synapses will pick out the values of J in proportion to their occurrence in the network, thereby preserving the gaussian form for the distribution of the non vanishing coefficients. When the fraction $(1-f)$ of all synapses has been cut, the distribution $P(J)$ will thus be

$$P(J) = (1-f)\,\delta(J) + \frac{f}{\sqrt{2\pi}}\exp\left[-\frac{J^2}{2}\right] \tag{20}$$

It is shown in figure 4a (solid line) for f=0.50. The distribution (19) for the fully connected network (dashed line) is shown for comparison.

In our second case, dilution before learning, the fN non vanishing coupling coefficients satisfy the conditions (10). These conditions are the same as those for a fully connected network with fN neurons. It follows that $P_r(J)$ is a gaussian with mean zero and variance one but normalized to f. Thus we obtain the same distribution (20) in both cases of quenched dilution. Many of the non-zero coupling coefficients are small in absolute value indicating that the large frustration of fully connected networks is also present in the diluted networks of the first two types.

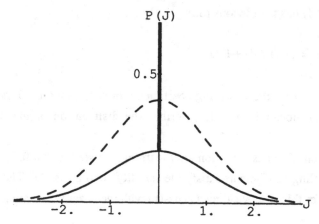

Fig 4a:Distribution P(J) for quenched dilution. Solid line:f=0.5.Dashed line:f=1.

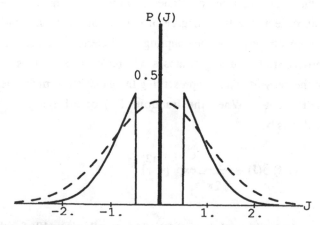

Fig 4b:Distribution P(J) for annealed dilution. Solid line:f=0.5.Dashed line:f=1.

The calculation of $P_r(J)$ is more complex in diluted networks of the third type due to the more complicated conditions (14). The resulting expression for $P_r(J)$ also is more complicated but it simplifies considerably near the saturation limit $\alpha \rightarrow \alpha_c(f,K)$:

$$P(J) = (1-f) \; \delta(J) + \frac{1}{\sqrt{2\pi} \; b} \; \exp\left[-\frac{J^2}{2b^2}\right] \theta(|\,J\,| - J_c) \tag{21}$$

The distribution $P_r(J)$ of the non vanishing coupling coefficients is a gaussian from which the middle section has been removed. The width of the gaussian, determined by b, as well as the width of the gap $2J_c$ depend only on the degree of dilution. The expression for these parameters is given in [3]. The gap grows gradually with increasing dilution to its maximal value 1. The value of b decreases concurrently in order to satisfy the normalisation condition (14). The distribution (21) is shown in figure 4b (solid line) for $f = 0.50$. No small coupling coefficients are present among the non vanishing coefficients. This indicates that the frustration of the fully connected network has been lifted to a large degree by a judicious choice of the position of the broken bonds. On both sides of the gap, the value of P(J) is of the same magnitude as in the fully connected network.(dashed line) This suggests that annealed dilution preserves the large coupling coefficients of the fully connected network.

6. Conclusion

The main conclusions can be summarized as follows.

1. Memories stored in a neural network are robust to random cutting of synapses only if a large value has been used for the stability bound K during the learning process. A learning rule for implementing this objective could be the minimum overlap algorithm of Krauth and Mézard [6].

2. The problem of storing patterns in a previously diluted network is equivalent to the same problem for a fully connected network of smaller size.

3. When dilution and learning are performed together in a correlated way, it becomes possible to store many more patterns than when both processes are carried out independently. The increased storage capacity is achieved by locating the broken bonds at those positions in the network where very small coupling coefficients would have been required in the optimized fully connected network.

4. The distribution P(J) for the annealed case suggests the following learning rule. Starting with a fully connected network, use the minimum overlap algorithm to determine the coupling coefficients for the largest possible value of K. Afterwards, put selectively all small coupling coefficients equal to zero in the order of their increasing absolute value until the desired degree of dilution is obtained. All other coupling coefficients remain unaltered.

This learning rule is similar to the procedure of Sompolinsky for obtaining his three-state synapses model [7].

The results presented in this paper have been obtained in collaboration with A. Komoda and R. Serneels at Diepenbeek and with A. Engel from Berlin. The work has been financially supported by the Belgian Programme on Inter-University Attraction Poles (IUAP) and by the Inter-University Institute for Nuclear Sciences.

References :
[1] E. Gardner, J. Phys. A : Math. Gen. 21 (1988) 257
[2] J. Phys. A : Math. Gen. 22 (1989) number 12
[3] M. Bouten, A. Engel, A. Komoda and R. Serneels, Quenched versus Annealed Dilution in Neural Networks, submitted to J. Phys. A
[4] B.M. Forrest, J. Phys. A : Math. Gen. 21 (1988) 245
[5] T.B. Kepler and L.F. Abbott, J. Physique 49 (1988) 1657
[6] W. Krauth and M. Mézard, J. Phys. A : Math. Gen. 20 (1987) 745
[7] H. Sompolinsky, in Heidelberg Colloq. on Glassy Dynamics and Optimization, eds. Van Hemmen J.L. and Morgenstern I, Springer (1987)

Dynamics and Storage Capacity of Neural Networks with Sign-Constrained Weights

C. Campbell
Department of Engineering Mathematics, Bristol University
University Walk, Bristol BS8 1TR, U.K.

and

K.Y.M. Wong
Department of Theoretical Physics, Oxford University
1 Koble Road, Oxford OX1 3NP, U.K.

Abstract. This paper reviews recent work on neural network models with sign constrained weights. The review will cover learning rules for neural networks with sign constrained weights, the optimal storage capacity of these models and the effect on the dynamics of retrieval, non-retrieval and uniform attractors.

1. INTRODUCTION

In this paper we will consider neural network models in which the signs of weights are prescribed *a priori*. One reason for considering sign constrained weights arises from neurobiology. In biological neural networks there are two types of neurons, a population with excitatory efferent synapses and a second population with inhibitory efferent synapses. Thus in general a particular neuron will send excitatory or inhibitory potentials to other neurons but not a mixture of the two. This property of biological networks was originally observed by Dale [1] and can be imitated in neural network models by constraining all weights emanating from a given neuron so that they all have the same sign.

Neural networks satisfying Dale's hypothesis are only a subcategory of neural network models with sign constrained weights. In general we could enforce any mixture of signs for the weights emanating from each neuron. The results established in this paper will also apply to this more general case.

Apart from a neurobiological motivation the enforcement of weight-sign constraints can also give rise to an exploitable cognitive feature - namely an ability to distinguish between the recognition or non-recognition of an input pattern [2]. If there is an imbalance between the number of excitatory and inhibitory weights, then it is possible to have uniform attractors with all neurons firing or quiescent simultaneously. Furthermore if this weight-sign imbalance is sufficiently large, only retrieval and uniform attractors are stable. The spurious attractors present in neural networks with unconstrained weights are destabilised in this case. Thus an initial input state will flow into a retrieval attractor if it has a sufficient overlap with a stored pattern, or it will flow into a uniform attractor if the initial overlap with all the stored patterns is too small. The uniform states can therefore be regarded as the attractors for unrecognised inputs, and due to their different mean activity they can be readily distinguished from retrieval states.

In this paper we will consider (1) learning rules for storing information in networks with sign-constrained weights (Section 2); (2) the optimal storage capacity of these models (Section 3) and (3) the dynamics and stability conditions for the spurious, uniform and retrieval attractors [3] (Sections 4 and 5).

2. LEARNING RULES FOR NEURAL NETWORKS WITH SIGN CONSTRAINED WEIGHTS

2.1 Introduction. We will consider neural networks with weights J_{ij} connecting neuron j to i$\neq$j (where $i, j = 1, \ldots, N$) and for which J_{ij} and J_{ji} are independent variables. Sign constraints on weights can then be enforced by introducing a matrix g_{ij} with components ± 1 and constraining the weights to satisfy:

$$J_{ij}g_{ij} \geq 0 \tag{1}$$

In the special case $g_{ij} = g_j$ the weight constraints enforce the same sign for all weights leading away from each neuron j. In this section we will consider learning rules for storing information in the weights J_{ij} subject to the constraint (1). First we consider a rule which is guarranteed to converge on a suitable weight set solution in simple perceptron networks [4] provided a solution exists. Following this we consider a gradient descent algorithm for networks with sign constrained weights. Though convergence is not guarranteed for gradient descent algorithms they do have a generalisation to multi-layered networks [5] circumventing well known limitations of single-layered networks [4].

2.2 A Perceptron Rule. Amit, Campbell and Wong [6] have proposed a perceptron learning rule capable of storing patterns in a neural network with sign constrained weights. The p pattern vectors are specified by $\xi_j^\mu = \pm 1$, where $\mu = 1, \ldots, p$. The learning rule is a procedure for encoding these patterns in the weights, J_{ij}, so that on presentation of a particular pattern vector, the same pattern vector is output:

$$\xi_i^\mu = \text{sign}(\sum_j J_{ij}\xi_j^\mu) \quad \text{or equivalently}: \quad \xi_i^\mu \sum_j J_{ij}\xi_j^\mu > 0 \tag{2}$$

Satisfaction of (2) guarantees correct alignment of the local fields though they may be only weakly aligned (i.e. basins of attractions may be narrow). Strong alignment can be enforced by storing the patterns in a set of weights with the local fields bounded from below:

$$\xi_i^\mu \sum_j J_{ij}\xi_j^\mu > \kappa|J|; \qquad |J|^2 = \sum_j J_{ij}^2 \tag{3}$$

For a given stability, κ, pattern vectors can be stored in a network with sign constrained weights by using the weight adjustment procedure:

$$J_{ij}^{(k+1)} = J_{ij}^{(k)} + \Delta J_{ij}^{(k)}$$

$$\Delta J_{ij}^{(k)} = \xi_i^\mu \xi_j^\mu \Theta(\kappa - \gamma_i^\mu)\Theta[J_{ij}^{(k)}(J_{ij}^{(k)} + \xi_i^\mu\xi_j^\mu)] \qquad \gamma_i^\mu = \frac{1}{|J|}\xi_i^\mu \sum_{j=1}^N J_{ij}^{(k)}\xi_j^\mu \tag{4}$$

This algorithm ensures that the weight-signs are preserved at each step (i.e. $J_{ij}^{(0)}g_{ij} \geq 0$ and later adjustments are allowed if $J_{ij}^{(k+1)}g_{ij} \geq 0$), and is guaranteed to converge to a suitable weight matrix if such a solution exists [6].

2.3 Gradient Descent algorithms. For unconstrained networks with continuously valued neuronal states, a common technique for constructing the weights J_{ij} involves the iterative reduction of the total error function by gradient descent in weight space:

$$\Delta J_{ij} = -\eta \frac{\delta E}{\delta J_{ij}}; \quad E = \frac{1}{2}\sum_{i,\mu}(T_i^\mu - O_i^\mu)^2 \tag{5}$$

T_i^μ is the target output for pattern μ and O_i^μ is the output of the updating function, for example:

$$O_i^\mu = f(h_i^\mu); \quad f(x) = (1 - exp(-\beta x))/(1 + exp(-\beta x)) \tag{6}$$

where $h_i^\mu = \sum_k J_{ik}I_k^\mu$, and I_k^μ are the input patterns. $f(x)$ is a differentiable sigmoidal updating function but thresholding binary-valued neurons can be used once training has been completed. β is a parameter controlling the slope of the sigmoid function. (Thus $f(x) \to \text{sign}(x)$ as $\beta \to \infty$.)

The best method to incorporate sign constraints on the weights (1) involves a simple alteration of the usual gradient descent algorithm. Using the updating function (6) the weight adjustments are calculated:

$$J_{ij}^{t+1} = J_{ij}^t + \Delta J_{ij}; \quad \Delta J_{ij} = \frac{\eta}{2}\beta\sum_\mu (T_i^\mu - O_i^\mu)(1 - (O_i^\mu)^2)I_j^\mu \tag{7}$$

where t is the iteration index. If J_{ij}^{t+1} does not violate the constraint condition (1) then the weight adjustment is made while if the adjustment would violate the constraint the corresponding weight J_{ij}^{t+1} is set to zero. This procedure is iteratively repeated until the error function falls below a prescribed limit. Numerical simulations up to a storage capacity of 0.5 stored patterns per neuron suggest that this procedure will generally converge on a suitable weight set solution. The above procedure for constrained neural networks can also be generalised to multi-layered networks [7].

3. THE OPTIMAL STORAGE CAPACITY OF NEURAL NETWORKS WITH SIGN CONSTRAINED WEIGHTS

In general the perceptron rule outlined above will succesfully converge on a suitable weight set solution, provided the optimal storage capacity of the network is not exceeded. In this section we will calculate the optimal storage capacity using a method pioneered by Gardner [9], which considers the fractional volume in weight space for which the pattern vectors are correctly embedded (with error free retrieval) . The optimal storage capacity is determined by the vanishing of this fractional volume.

If all the patterns are correctly embedded, and the weights also correctly satisfy the sign constraint conditions then:

$$\prod_{\mu,i}\Theta(\frac{\xi_i^\mu}{\sqrt{N}}\sum_j J_{ij}\xi_j^\mu) = 1; \quad \prod_{i,j}\Theta(J_{ij}g_{ij}) = 1 \tag{8}$$

To avoid ambiguity we also impose the normalisation [9]:

$$\sum_j J_{ij}^2 = N \tag{9}$$

In a perceptron network the set of weights feeding each node are independent, thus we shall only focus on the calculation for a single output node i (the label i will be implicit henceforth). The fractional volume of weight space satisfying (8-9) is then given by:

$$V = Z^{-1}\int_{-\infty}^{\infty}\prod_j dJ_j\delta(\sum_j J_j^2 - N)\Theta(J_jg_j)\prod_\mu\Theta(\xi^\mu\sum_j J_j\xi_j^\mu/\sqrt{N}) \tag{10}$$

where:

$$Z = \int_{-\infty}^{\infty}\Pi_j dJ_j\delta(\sum_j J_j^2 - N)\Theta(J_jg_j) \tag{11}$$

This fractional volume is then averaged over a random choice of pattern vectors. For random, uncorrelated and unbiased stored patterns the averaged fractional volume is invariant under a change in the sign of g_j at a particular neuronal site j. Consequently, the optimal storage capacity is independent of the weight-sign bias. Thus the optimal storage capacity is the same even if all the weights are constrained to be positive, negative or a mixture of signs. The vanishing of the above fractional volume determines the optimal storage capacity of networks α_c, the number of stored patterns per neuron:

$$\alpha_c = 1 \tag{12}$$

This result may be compared with the optimal capacity of $\alpha_c = 2$ for neural networks with unconstrained weights [9,10]. If, rather than merely requiring that the patterns are correctly embedded, we require that the local fields are correctly aligned and stronger than a stability parameter κ, then equation (8) is replaced by:

$$\prod_{\mu,i} \Theta(\xi_i^\mu \sum_j J_{ij} \xi_j^\mu - \kappa\sqrt{N}) = 1; \quad \prod_{i,j} \Theta(J_{ij} g_{ij}) = 1 \tag{13}$$

In this case the optimal capacity α_c is related to κ by :

$$\alpha_c = [2 \int_{-\infty}^{\kappa} e^{-t^2/2}(\kappa - t)^2 dt / \sqrt{2\pi}]^{-1} \tag{14}$$

for networks with sign-constrained weights. This capacity is again half the corresponding value for unconstrained networks storing patterns with the same value of the stability, κ. Equation (14) expresses the fact that as the value of κ is increased, the maximal number of patterns that can be stored decreases.

4. DYNAMICS OF ATTRACTOR NEURAL NETWORKS WITH SIGN CONSTRAINED WEIGHTS

Attractor neural networks are neural networks whose updated neuronal states are fed back to the input for subsequent updating. Its dynamics can be either synchronous or asynchronous. For synchronous dynamics, all the neuronal states are updated according to:

$$S_i^{t+1} = \text{sign}(h_i^t); \quad h_i^t = \sum_j J_{ij} S_j^t \tag{15}$$

while for asynchronous dynamics, a neuron is randomly chosen at each time step and its state updated according to (15).

This iterative dynamics will result in movement of state configurations of the network towards an attractor. The attractor may be a retrieval state i.e., the attractor configuration is identical to, or highly correlated with, a stored pattern. Alternatively it may be a spurious non-retrieval state with the attractor configuration having no correlation with the stored patterns. For attractor neural networks with sign constraints and a net weight sign bias it is also possible to have uniform attractors in which the neuron states are all 1's or all −1's simultaneously. Furthermore, if the weight-sign bias is large (for example, if all the weights are positive), then in a uniform attractor the local fields, h_i^t, are very strongly aligned corresponding to strongly attracting states. Here we shall focus on the solvable case of dilute networks, in which each neuron is fed by C other neurons with $1 \ll C \ll \ln N$. In this case the dynamics is solvable because the extreme dilution ensures that state correlations between different time steps are negligible [11].

The storage properties of dilute networks are similar to those of the fully connected networks presented in sections 2 and 3. The storage capacity, however, is now scaled by C instead of N, i.e. for dilute networks:

$$\alpha_c = p_c/C = 1 \tag{16}$$

For dilute networks it is possible to calculate the dynamical evolution of an input state S_i^t having a macroscopic overlap with one of the stored patterns (ξ_i^1, say). If m^t is the overlap between this input state and ξ_i^1 then the local field at neuron i is a Gaussian variable of mean $m^t \sum_j J_{ij}\xi_j^1$ and width $\sqrt{(1-(m^t)^2)C}$. Thus the dynamics of a dilute network with an arbitrary set of weights $\{J_{ij}\}$ is determined by the aligning fields $\Lambda_i^\mu \equiv \xi_i^\mu \sum_j J_{ij}\xi_j^\mu / \sqrt{C}$. Consequently a knowledge of the aligning field distribution:

$$\rho(\Lambda) \equiv \langle \overline{\delta(\Lambda - \Lambda_i^\mu)} \rangle \tag{17}$$

determines the dynamical equation (the overbar represents averaging over the fraction of weight space which stores the patterns and complies with the sign constraints, and $\langle \ldots \rangle$ represents averaging over the stored patterns). For synchronous dynamics the overlap m^{t+1} is therefore [12,13]:

$$m^{t+1} = f(m^t) = \int d\Lambda \rho(\Lambda)\mathrm{erf}\left(\frac{m^t\Lambda}{\sqrt{2(1-(m^t)^2)}}\right) \tag{18}$$

Here we are interested in a state configuration not only having a macroscopic overlap m^t with a stored pattern (ξ_i^1, say) but also having an overlap a^t with a uniform state ($S_i = +1$ for all i, say). The two relevent dynamical variables are therefore the overlap m^t with ξ_i^1 and the overlap a^t with the positive uniform state (i.e. the activity):

$$m^t = N^{-1}\sum_i \xi_i^1 S_i^t; \quad a^t = N^{-1}\sum_i S_i^t \tag{19}$$

Equivalently, the dynamics can be described by two alternative parameters:

$$m_\pm^t = N_+^{-1} \sum_{i \in \{\xi_i^1 = \pm 1\}} \xi_i^1 S_i^t \tag{20}$$

i.e. $m_\pm^t$ are the overlaps on those neurons having $\xi_i^1 = \pm 1$ respectively ($N_\pm$ are the number of neurons for which $\xi_i^1 = \pm 1$). Assuming that the stored patterns have independent components and a random distribution, m^t and a^t are related to $m_\pm^t$ by:

$$m^t = \frac{1}{2}(m_+^t + m_-^t); \quad a^t = \frac{1}{2}(m_+^t - m_-^t) \tag{21}$$

For input overlaps $m_\pm^t$, the local field at neuron i is now a sum of two Gaussian variables, of means $m_\pm^t \sum_{\xi_j = \pm 1} J_{ij} \xi_j^1$ and widths $\{[1 - (m_\pm^t)^2] C_{i\pm}^1\}^{1/2}$ ($C_{i\pm}^1$ is the number of neurons feeding neuron i for which $\xi_j^1 = \pm 1$). The dynamics is now determined by the two aligning fields $\Lambda_{i\pm}^\mu \equiv \xi_i^\mu \sum_{\xi_j^\mu = \pm 1} J_{ij} \xi_j^\mu \{C_{i\pm}^\mu\}^{-1/2}$, and a knowledge of the double field distributions:

$$\rho_\pm(\Lambda_+, \Lambda_-) \equiv \langle \delta(\Lambda_+ - \Lambda_{i+}^\mu) \delta(\Lambda_- - \Lambda_{i-}^\mu) \rangle_{\xi_i^\mu = \pm 1} \tag{22}$$

determines the dynamical equations which, for synchronous dynamics, become:

$$m_\pm^{t+1} = f_\pm(m_+^t, m_-^t) = \int_{-\infty}^\infty d\Lambda_+ d\Lambda_- \rho_\pm(\Lambda_+, \Lambda_-) \mathrm{erf} \left[\frac{m_+^t \Lambda_+ / \sqrt{2} + m_-^t \Lambda_- / \sqrt{2}}{\sqrt{2 - (m_+^t)^2 - (m_-^t)^2}} \right] \tag{23}$$

The corresponding dynamical equation for asynchronous dynamics is:

$$dm_\pm/dt = f_\pm(m_+^t, m_-^t) - m_\pm \tag{24}$$

For sign-constrained networks with no bias in the sign distribution (i.e. the number of positive and negative synapses feeding a neuron are the same), the two double field distributions $\rho_\pm(\Lambda_+, \Lambda_-)$ are identical. In other words, the double field distribution is independent of the output bit of the stored patterns. In this case, the dynamical equation (23) is identical to the previously stated equation (18) for $m_+^t = m_-^t$.

However, the most interesting case is for networks with a bias in the sign-distribution given by:

$$\sum_j g_{ij} = \frac{B}{\sqrt{C}} \qquad \forall i \tag{25}$$

i.e. the number of excitatory (inhibitory) synapses feeding neuron i is $(1 \pm B/\sqrt{C})/2$. With this asymmetry in the sign distribution, the two double field distributions $\rho_\pm(\Lambda_+, \Lambda_-)$ are no longer identical. Using the replica method outlined in ref. [3] it is possible to show that the double field distributions are given by:

$$\rho_{\pm}(\Lambda_+, \Lambda_-) = \int d\Lambda \rho(\Lambda)\delta\left[\frac{1}{\sqrt{2}}\left(\Lambda \pm \frac{B}{\sqrt{\pi}}\right) - \Lambda_+\right]\delta\left[\frac{1}{\sqrt{2}}\left(\Lambda \mp \frac{B}{\sqrt{\pi}}\right) - \Lambda_-\right] \tag{26}$$

where $\rho(\Lambda)$ is the aligning field distribution defined in (17). In the limit of maximally stable storage,

$$\rho(\Lambda) = [1 + \text{erf}(\kappa/\sqrt{2})]\delta(\Lambda - \kappa)/2 + \Theta(\Lambda - \kappa)\exp(-\Lambda^2/2)/\sqrt{2\pi} \tag{27}$$

where the maximal stability κ can be found for a given α via (14).

Substituting (26,27) in (23) the dynamical equations become:

$$f_{\pm}(m_+^t, m_-^t) = \int_{-\infty}^{\infty} d\Lambda \rho(\Lambda)\text{erf}\left[\frac{(m_+^t + m_-^t)\Lambda/2 \pm (m_+^t - m_-^t)B/2\sqrt{\pi}}{\sqrt{2[1 - (m_+^t)^2/2 - (m_-^t)^2/2]}}\right] \tag{28}$$

5. ATTRACTOR AND TRANSIENT BEHAVIOURS

5.1 Attractor types and stability For $\alpha \leq 1$ these dynamical equations possess three types of attractor:

(1) $(m_+, m_-)=(1,1)$ and $(-1,-1)$ are stable attractors corresponding to the retrieval attractor $(m = 1)$ and its anti-correlated counterpart $(m = -1)$.

(2) $(m_+, m_-)=(1,-1)$ and $(-1,1)$ are uniform attractors $(a = \pm 1)$ and they are stable with finite basins of attraction for any non-zero B.

(3) $(m_+, m_-)=(0,0)$ can also act as an attractor corresponding to a spurious non-retrieval state with no overlap with the uniform or retrieval attractors.

In order to understand the stability of these attractors we note that a fixed point of the dynamics must satisfy:

$$g_{\pm}(m_+, m_-) = f_{\pm}(m_+, m_-) - m_{\pm} = 0 \tag{29}$$

Furthermore $g_{\pm}(m_+, m_-)$ must be stable to perturbations in m_+^t and m_-^t. Thus for a stable attractor the two eigenvalues of the matrix $\partial g_{\pm}/\partial m_{\pm}$ must both be negative. From this requirement we find that the retrieval and uniform attractors are always stable below the storage capacity for positive B (though the uniform attractor is not stable if $B = 0$). On the other hand the non-retrieval attractor is a stable fixed point provided:

$$\int_{-\infty}^{\kappa} e^{-t^2/2}(\kappa - t)dt \leq \pi \quad \text{or} \quad \alpha \geq \alpha^* = 0.21 \tag{30}$$

and for B positive:

$$B \le B^* = \pi/\sqrt{2} = 2.22 \tag{31}$$

α^* is half the corresponding value for unconstrained networks [12], the difference arising from the weight-sign constraints.

In Figs. 1-5 the above dynamical equations are illustrated at various values of α and B. For convenience we have chosen a^t and m^t to define the axes, and restricted the figures to the first quadrant (since the other quadrants are symmetrically similar). The labels R,U and NR represent the retrieval, uniform and non-retrieval fixed points respectively. The arrows on the diagrams show the direction of movement but not the magnitude of each step.

In the figures the bold lines indicate basin boundaries. In order to study the transient behaviour of the dynamics, we have also plotted the curves $dm/dt = 0$ and $da/dt = 0$ as the thin lines on the figures. The curve $dm/dt = 0$ demarcates the regions in the space of (a^t, m^t) where m^t increases or decreases in one time step and $da/dt = 0$ similarly separates the regions where a^t increases or decreases after one time step. In general there are four types of transient states: (1) both m^t and a^t decrease, which will be called *non-retrieval transients*; (2) m^t increases and a^t decreases — *retrieval transients*; (3) m^t decreases and a^t increases — *uniform transients* and finally (4) both m^t and a^t increase — *active transients*.

5.2 $B = 0$. When $B = 0$ and $\alpha < \alpha^*$ any state with a small (positive) macroscopic overlap with the stored pattern will inevitably move towards the $(a^t, m^t)=(0,1)$ retrieval fixed point since this is the only attractor present.

For $\alpha > \alpha^*$ a non-retrieval attractor appears centred at $(0,0)$. (See Fig. 1.) In the neighbourhood of the retrieval state, there is a transient motion towards the retrieval state, while in the neighbourhood of the non-retrieval attractor the transient movement is towards the non-retrieval state. Finally there also exist two separated regions of retrieval transients where there is a small transient increase in m^t before moving towards the non-retrieval state.

5.3 $0 \le B \le 2.22$. When B is positive and non-zero, uniform attractors appear. In Fig. 2(a) there exist both retrieval and uniform attractors and correspondingly regions of retrieval and uniform transients. In addition there exists a region of non-retrieval transient states in both the retrieval and uniform basin near their common basin boundary, where there is an initial transient decrease in a^t and m^t before the states move towards either $(0,1)$ or $(0,1)$.

In Fig. 2(b) the non-retrieval state $(0,0)$ becomes an attractor for $\alpha > \alpha^*$. In this case there exist retrieval, non-retrieval and uniform transients. As in Fig. 2(a) there are regions of non-retrieval transients lying within the basins of both the retrieval and uniform attractors.

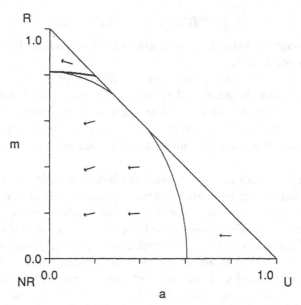

Fig.1. The basin boundary (bold line) and transient boundaries (thin line) for $\alpha = 0.3$ and B=0.

5.4 $B \geq 2.22$. For B greater than 2.22 the non-retrieval state (0,0) is no longer an attractor. (See Fig. 3.) As expected there exist regions of retrieval and uniform transients. There also exists a region of active transient states (Fig. 3(a)) in which both m^t and a^t increase initially, but eventually move towards the retrieval or uniform fixed points depending on which side of the basin boundary they lie. In addition there exists a region of non-retrieval transients on both sides of the basin boundary (Fig. 3(b)); these states will eventually fall into the retrieval or uniform transient regions.

5.5 $B \leq 0$. For negative B the behaviour of synchronous (23) and asynchronous (24) dynamics is quite different. For synchronous updating we see that if $B \to -B$ then m_+^{t+1} and m_-^{t+1} in (28) are interchanged. Consequently the uniform state $(a^t, m^t) = (\pm 1, 0)$ will be a cyclic attractor of period 2 and the approach to this attractor will involve oscillations about the $a^t = 0$ axis, giving states of all 1's and all -1's at each alternative time step. The retrieval and non-retrieval attractors still exist as in the case of positive bias B, but the approaches to these attractors will also involve oscillations about the $a^t = 0$ axis. Apart from this oscillatory behaviour the basin boundaries for synchronous dynamics are identical to their positive B counterparts. Thus there are no non-retrieval attractors for $B \leq -2.22$ and $\alpha \geq 0.21$.

For negative B and asynchronous dynamics the uniform attractors disappear. Thus only the retrieval and non-retrieval attractors exist and the latter state is stable even for $B \leq -2.22$. Figs. 4 and 5 illustrate asynchronous dynamics with negative B whose boundaries of transience $dm/dt = 0$ are identical to those of Figures 2 and 3 respectively. In Figs 4(a) and 5(a) where α is below α^*, all states move towards the retrieval state, although non-retrieval transience is present in the region far from the retrieval fixed point. In Figs. 4(b) and 5(b) interference due to other stored patterns stabilizes the non-retrieval attractor when α exceeds the critical value α^*.

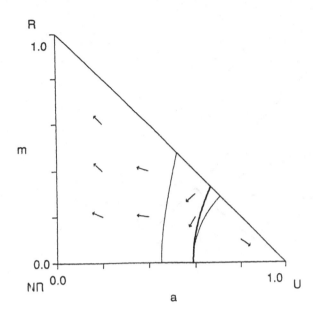

Fig.2(a). Basin and transient boundaries for $\alpha = 0.2$ and B=2.0.

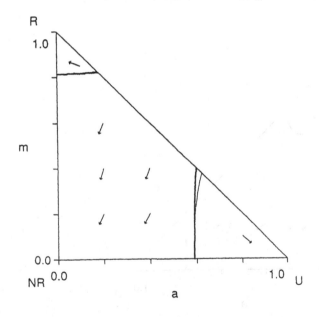

Fig.2(b). Basin and transient boundaries for $\alpha = 0.3$ and B=2.0.

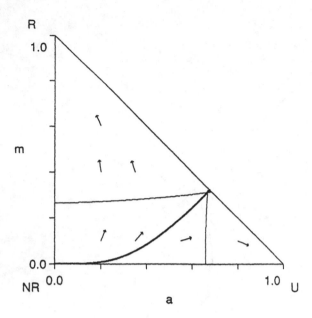

Fig.3(a). Basin and transient boundaries for $\alpha = 0.1$ and B=2.5.

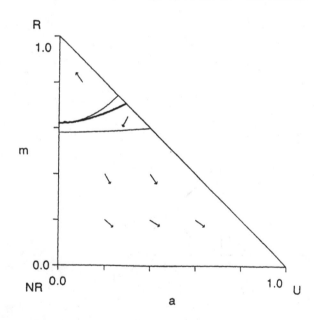

Fig.3(b). Basin and transient boundaries for $\alpha = 0.25$ and B=2.5.

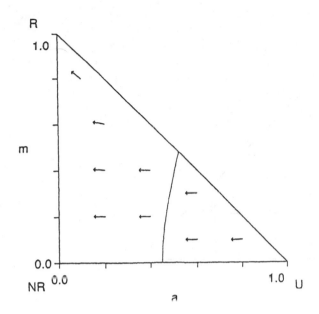

Fig.4(a). Basin and transient boundaries for $\alpha = 0.2$ and B=-2.0.

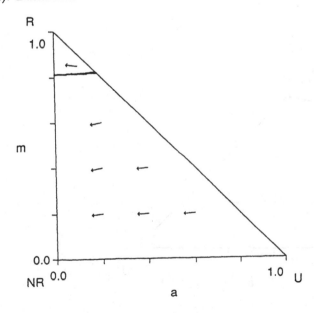

Fig.4(b). Basin and transient boundaries for $\alpha = 0.3$ and B=-2.0.

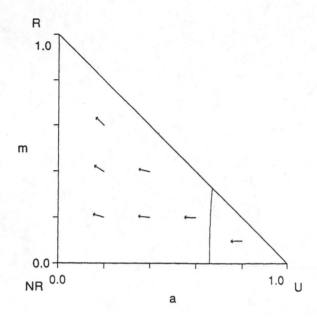

Fig.5(a). Basin and transient boundaries for $\alpha = 0.1$ and B=-2.5.

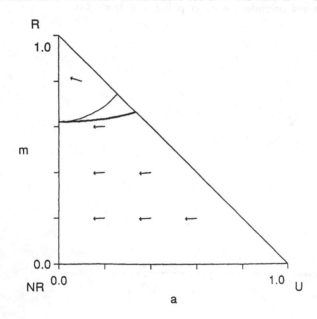

Fig.5(b). Basin and transient boundaries for $\alpha = 0.25$ and B=-2.5.

5.6 Summary. In summary, we have found three types of attractors in dilute networks with a bias in weight signs: retrieval, non-retrieval and uniform attractors. Their size and stability are determined by two factors: storage level α and weight bias B. Roughly speaking, low α favours the retrieval attractor, and high B favours the uniform attractor. When the storage level increases, interference from other stored patterns causes the retrieval basin to narrow, allowing for the existence of the non-retrieval attractor. However, when the weight bias B also increases, the uniform attractor enlarges and encroaches on the non-retrieval attractor. Thus the non-retrieval attractor exists for $\alpha \geq 0.21$ and $B \leq 2.22$ only.

We found interesting transient behaviours. The normal behaviour within the retrieval basin is that the m-component of the network state motion should increase and the a-component decrease. However, in all cases with $B \neq 0$ presented here there exists, alongside with this normal behaviour, transient motion towards the non-retrieval fixed point in the vicinity of the retrieval basin boundary, i.e. the m-component of the network state first moves away from the retrieval fixed point before eventually approaching it. This effect, which is absent in the dynamics of network states with zero activity, is most marked for states with increasingly high activities. For sufficiently large B, transient motion towards the uniform fixed point also exist in the vicinity of the basin boundary, as illustrated in Fig. 3(a-b).

Similarly, the normal behaviour within the uniform basin is that the m-component of the network state should decrease and the a-component increase. However, in all cases presented here, there exists a region in the vicinity of the uniform attractor basin boundary where the a-component of the network state first moves away from the uniform fixed point before eventually approaches it. For sufficiently large B, transient motion towards the retrieval fixed point also exist in the vicinity of the basin boundary, as illustrated in Fig. 3(a-b).

6. DISCUSSION

In this paper we have used ± 1 quantisation for the dynamical state vector S_i^t. If instead we use the quantisation $V_i^t = (0,1)$ we would obtain two populations of neurons for networks satisfying Dale's rule ($J_{ij}g_j \geq 0$), one in which the neurons are excitatory or quiescent and another in which the neurons are inhibitory or quiescent. This would mirror biological networks more closely since neurons are exclusively excitatory or inhibitory. However the use of (0,1) quantisation does not substantially alter any of the results of this paper and it can be straightforwardly handled using the replacement $h_i^t = \sum_j J_{ij}S_j^t \rightarrow 2\sum_j J_{ij}V_j^t - \sum_j J_{ij}$ and replacement of the signum updating function with a step function throughout.

Though biological neurons appear constrained to being solely excitatory or inhibitory, both the number and function of excitatory and inhibitory neurons appears to be quite different in networks such as the human cortex. For example, learning appears to take place at excitatory efferent synapses and not inhibitory ones [14] (though the learning rules in Section 2 can be readily adapted to effect learning at excitatory (positive) efferent synapses only). More biologically detailed neural network models, along the lines of our study, are expected to mirror the complex and delicate interplay between excitation and inhibition found in biological networks.

REFERENCES

[1] Dale H H, *Proc. R. Soc. Med.* **28** (1935) 319.

[2] Shinomoto S, *Biol. Cybern.* **57**(1987) 197.

[3] Wong K Y M and Campbell C, "The Dynamics of Neural Networks With Sign Constrained Weights", 1990, in preparation.

[4] Minsky M L and Papert S A, "Perceptrons" (MIT Press, 1969).

[5] Rumelhart D E, Hinton G E and Williams R J, *Nature* **323**(1986) 533.

[6] Amit D J, Wong K Y M and Campbell C, *J. Phys.* **A22** (1989) 2039.

[7] Campbell C and Karwatzki J M, "Error Correcting Learning Algorithms for Neural Networks with Sign Constrained Weights", University of Bristol Preprint (1990).

[8] Amit D J, Campbell C and Wong K Y M, *J. Phys.* **A22** (1989) 4687.

[9] Gardner E, *J.Phys.* **A21** (1988) 257; Gardner E *Europhys. Lett.* **4** (1987) 481; Gardner E and Derrida B *J.Phys.* **A21** (1988) 271.

[10] Cover T M, *IEEE Trans. Electron. Comput.* **EC - 14** (1965) 326.

[11] Derrida B and Pomeau Y, *Europhys. Lett.* **1** (1986) 45; Derrida B, Gardner E and Zippelius A, *Europhys. Lett.* **4** (1987) 167.

[12] Gardner E, *J. Phys.* **A22** (1989) 1969.

[13] Kepler T and Abbott L F, *J. Physique* **49**(1988) 1657, Krauth W, Nadal J-P and Mezard M, *J. Phys.* **A21** (1988) 2995.

[14] Treves A and Amit D J, *J. Phys.* **A22** (1989) 2205; Kohring G A 1989 "Coexistence of Global and Local Attractors in Neural Networks" (Bonn University preprint).

THE NEURAL BASIS OF THE LOCOMOTION OF NEMATODES

Paul Erdös and Ernst Niebur*

Institute of Theoretical Physics, University of Lausanne, BSP Université,
CH - 1015 Lausanne, Switzerland

Abstract: A model of electrotonic neurons is presented which allows computer simulation of a physiologically realistic neural network, such as that found in nematodes. The undulatory locomotion of *Caenorhabditis elegans* is investigated by solving the equations of motion of a segmented model of the discretized body taking into account all internal and external forces. The spatio-temporal muscle excitation patterns which produce locomotion are determined, and it is concluded that these are most probably generated by stretch receptor cells and signals which globally turn on or off the neural circuitry governing forward or backward motion. The results are illustrated in the form of a computer generated videotape and compared with the observed motion of a nematode.

I. Introduction

The purpose of this work is to contribute towards an understanding of neural networks occurring in nature. Our first consideration was to determine an appropriate starting point. Fig. 1 shows three nerve cells: (a) of the nematode *Caenorhabditis elegans* (*C.e.*), (b) of the shark, (c) of man. Note, that for obvious reasons, the synapses, where the cell transmits information to another cell, are only shown for *C.e.* Besides the increase of dendritization clearly visible in Fig. 1, at some stage of evolution towards species with more and more complex nervous systems, probably two important transitions have taken place: the transition from electrotonic neurons to a combination of electrotonic and spiking neurons with electrochemical compensation of signal attenuation; and the transition from a completely deterministic, genetically pre-wired nervous system to a net with an, at least, partially random connectivity. As a result of the increased complexity of the nervous systems of advanced organisms as noted above, we have chosen, as a starting point, to consider the nematode *C.e.* which has a very simple nervous system.

Lecture Notes in Physics, Vol. 368
L. Garrido (Ed.)
Statistical Mechanics of Neural Networks
© Springer-Verlag Berlin Heidelberg 1990

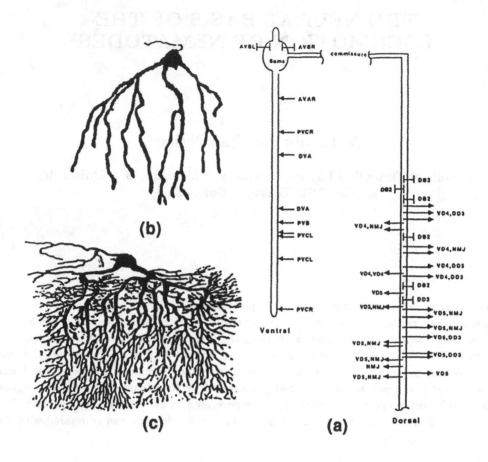

Fig. 1. Three nerve cells. (a) schematic diagram of a nerve cell of the nematode *Caenorhabditis elegans*. Arrows indicate in- or outgoing chemical synapses, *H*-symbols denote electrical synapses. The alphanumeric code identifies other neurons, to which the synapses connect. (b) nerve cell of shark. (c) nerve cell of man.

The nematode *C.e.* is an animal for which the wiring diagram of the whole nervous system is known [1]. *C.e.* is a worm of $\sim$ 1mm length and $\sim$ 0.08mm diameter. Each individual has 302 neurons. Each neuron has, on the average, 20 synapses. Through laser ablation experiments (i.e. selective killing of neurons), the function of many of the neurons is known. We have reasons to believe that the neurons are electrotonic, i.e. transmit graded electric signals which are attenuated along the axons, and not spikes.

Related to its simple nervous system is the simple lifestyle of *C.e.* Its only observable activities are locomotion, feeding, digesting (bacteria), and reproduction. This makes the study of its nervous system more tractable than that of higher species, for instance vertebrates, which may engage in activities other than those listed above.

Our second consideration, therefore, was to select the neural basis for one specific activity of *C.e.* to investigate. Several reasons led us to choose locomotion. Among these, the primary reason was the fact that the innervation by 69 motor neurons and

10 connecting- and interneurons of the 95 muscle cells is clearly established. Also, we were intrigued by the possibility of being able to explain, in terms of basic physical considerations, the nature of the complicated spatio-temporal pattern of excitation of the muscle cells that are needed for undulatory motion.

II. Undulatory Locomotion

Before we were able to understand the functioning of the motor-neural system of *C.e.*, two preliminary questions had to be answered. First, what type of muscular activity is required to produce the creeping wavelike motion of vermiform, legless animals such as a nematode; and, second, how is such a muscular excitation wave produced in the organism ? This first question has been answered essentially by the British zoologist J. Gray [2]. To understand how muscular contraction will produce a forward thrust, we use the principle of virtual work, as applied to a curved segment of the animal's body. The segment (2) is shown in full in Fig. 2.

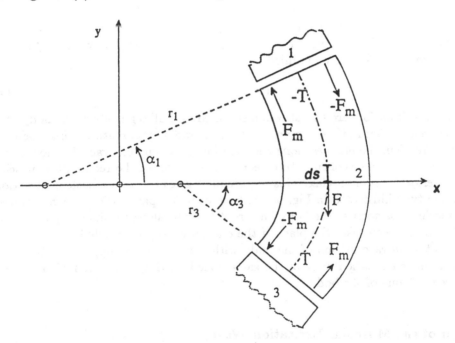

Fig. 2. Illustration of the origin of forward thrust generated by the contraction of muscles in a vermiform body. For explanation, see text.

When the muscles inside the exterior arc of the segment contract, they produce the forces F_m and $-F_m$ shown just inside that arc. Because of the incompressibility and viscosity of the body fluid, inside of the segment pressure forces will build up creating a second pair of forces $-F_m$ and F_m shown adjacent to the inner arc delimiting the body. The two forces which act at the cross-section adjacent to segment 1 may be combined to a form couple $-T$, whereas the two other forces yield a couple T adjacent to segment 3.

If segment 2 undergoes a virtual displacement ds in the direction of the body axis (dash - dotted line) and F is the frictional force exerted by the support on the body, the principle of virtual work states that, in equilibrium or uniform motion, the sum of the infinitesimal work elements by all the couples and forces is zero. From Fig. 2 it follows that

$$-Td\alpha_1 + Td\alpha_3 + F\,ds = 0. \tag{1}$$

Here, $d\alpha_1$ and $d\alpha_3$ are the infinitesimal angular displacements related to ds by

$$d\alpha_1 = \frac{ds}{r_1}, \qquad d\alpha_3 = \frac{ds}{r_3}, \tag{2}$$

where r_1 and r_3 are the radii of curvature of the body at the two ends of the segment. Therefore,

$$F = T\left(\frac{1}{r_3} - \frac{1}{r_1}\right). \tag{3}$$

It follows, that the muscular contraction along the outer arc is equivalent to a force $-F$, which is precisely the axial thrust necessary to move the body in the direction of its axis in order to produce a speed which, in turn, gives rise to the frictional force F.

It is clear, that for forward movement (in Fig. 2 from segment 3 towards segment 1) the following condition has to be fulfilled :

$$r_1 > r_3. \tag{4}$$

To produce a resultant forward thrust, when summed over all segments of the body, *it is necessary, therefore, that in those segments where the radius of curvature is increasing in the forward direction, the muscles along the exterior arc should contract.* In those parts of the body where the radius of curvature decreases in the forward direction, the muscles along the exterior arc should relax. This requirement gives rise to a muscle contraction pattern of the type illustrated in Fig. 6a. This contraction pattern is *momentary*, and has to be modified as soon as the body has crept forward along its sinuous path so far that another muscle cell takes the place of the one previously contracted.

Hence, to the *forward* motion of the body with speed v on its support, there corresponds a muscular excitation wave of the same speed moving *backwards* with respect to the reference frame of the body.

III. Origin of the Muscular Excitation Wave

In this section we answer the second preliminary question raised in Section II, i.e., how is a muscular excitation wave of the type described above produced in $C.e$? Experiments [3] fix the speed v of this wave as

$$v \cong 0.2cm\,s^{-1}.$$

Several hypotheses have been put forward to explain the production of this muscular excitation wave. These are as follows:

(a) the nerve ring ("brain") produces pulses at $\cong 0.25\,s$ intervals. These pulses travel down the length of the body in the interneurons with a speed identical to that of the muscular excitation wave and activate the contraction wherever they pass;

(b) Some local nerve circuits act as pacemakers and trigger the contraction of the muscles to which they are attached [4]; and

(c) there are stretch receptor cells in the cuticle which sense that the skin at that point is stretched because it is on an arc of strong curvature. The stretch receptors excite the muscle cells at about 1/4 body wavelength behind the stretch receptors, initiating contraction of those cells. The process repeats itself giving rise to the required muscle contraction pattern.

The next Section discusses why we did not accept hypotheses (a) and (b).

IV. Signal Propagation in the Nerves of C.e.

In the absence of synaptic interactions, and if the membranes are passive, the intracellular voltage $V(x,t)$ in the neural process (the branches of the neuron, such as dendrites, are called processes), is the solution of the equation [5]

$$\lambda^2 \frac{\partial^2 V}{\partial x^2} - \tau \frac{\partial V}{\partial t} - V + V_R = 0. \tag{5}$$

The parameters λ and τ characterize the process and are related to the diameter D, the transmembrane capacity per unit area C, the volume resistivity of the cytoplasm R, and the transmembrane resting conductivity per unit area g by

$$\lambda = \sqrt{4Rg/D}, \qquad \tau = C/g. \tag{6a, b}$$

V_R is the intracellular resting potential. Generally speaking, the nerve voltage decreases by a factor $e \cong 2.7$ after propagating a distance λ, or during a time τ.

For a neural network with synaptic connections, eq. (5) should be modified to include terms which express the coupling to the other neurons, and there are as many equations as there are neurons, each with its own set of parameters. These equations, which we do not write out explicitly here, are the ones used in this work.

The nematode C.e. is not large enough to permit the measurement of these parameters. However, on the larger nematode Ascaris lumbricoides (A.l.), these parameters have been measured.

To test our equations, we first calculated the speed v of nerve signals in A.l. and found $v = 22\,cm\,s^{-1}$, which is in good agreement with the experimental values of $v = 21 - 38$ cm s^{-1} [7]. We then appropriately scaled all parameters such that they would be applicable to C.e.. These parameters are shown in Table 1.

To study the neural activity, individual neurons with both chemical synapses and gap junctions (i.e. electrical synapses) were investigated. The distribution of these synapses along the neuron was taken from experiment (see Fig. 1). The partial differential equation (5), including the synaptic terms (not shown in eq. 5), was solved numerically both for A.l. and for C.e.. The shape of the *voltage vs time* curve obtained by the numerical simulation for A.l. agreed well with the experimental results. For C.e., the speed of propagation of neural excitations for the cases (a), (b) and (c) detailed in the Table varied between

$$v \cong 2 - 31 \, cm \, s^{-1}.$$

Since the speed of the muscular excitation wave never exceeds $0.2 \, cm \, s^{-1}$ [8], it is clear that the hypothesis (a) is not appropriate.

	Interneurons	Motorneurons
$D(m)$	10^{-6}	0.33×10^{-6}
$1(m)$	10^{-3}	0.5×10^{-3}
$R(\Omega m)$	1	
$C(Fm^{-2})$	0.75×10^{-2}	
$V_R(mV)$	-35	
$V_E(mV)$	15	
$g_{exc}(\Omega^{-1}m^{-2})$	10^4	
$g_{el}(\Omega^{-1}m^{-2})$	10^4	
$g_R(\Omega^{-1}m^{-2})$	(a) 0.05	(a) 0.25
	(b) 0.5	(b) 1.0
	(c) 5.0	(c) 10

Table 1. Parameters of *Caenorhabditis elegans* neurons. D: cell diameter, l: process length, R: cytoplasm resistivity, C: transmembrane capacity, V_R: intracellular resting potential, V_E: reversal potential for excitatory synapses, g_{exc}: transmembrane conductivity for excitatory synapses, g_{el}: transmembrane conductivity for electrical synapses, g_R: transmembrane resting conductivity. (a), (b), and (c) yield different velocities in the range $2 - 31$ cm s^{-1}.

The hypothesis (b) concerning auto-oscillatory circuits is also not acceptable since in the very primitive neural system of C.e., there is no evidence of such circuits. In order to produce the correct phase shift between oscillators which activate different muscle cells, further neurons would have to be present. However, there are none which could serve this purpose. Also, chains of coupled oscillators would make the observed separate circuits for forward and backward movement unnecessary. There remains for consideration the last hypothesis (c) involving stretch receptors.

Stretch receptors are known to exist in many animals. The existence of stretch receptors in nematodes was postulated in the 1970's by R. Russel [9] based on the observation of long, undifferentiated prolongations of some somatic excitatory motor neurons (classes DA, DB, VA and VB in C.e., using the notation of White et al., [1]). As we shall see, a stretch receptor has to be about 1/4 wavelength λ of the sinusoidal body shape posterior to the muscle it controls for motion in the anterior direction and vice versa. The wavelength of the body shape is approximately equal to the length of the body. It follows that the neural processes which connect the stretch receptors to the muscle must be of a length equal to 1/4 of the body length, which is indeed the case for the cells which we assume to perform this function.

V. Forward and Backward Motion

Observation through a microscope reveals, that when a *C.e.* in forward movement touches an obstacle with its head, it reverses its direction of motion. When it touches an obstacle with its tail creeping backwards, it starts to move forward. Of course, it can also stop and start at any time. The laser microbeam ablation experiments show that the nematode has two distinct neural circuits, one for forward motion and one for backward motion. [10].

In light of what was said above about the stretch receptors, it is logical to assume that each of these two circuits is activated and deactivated at the proper time by signals from the nerve ring ("brain") transmitted through the interneurons. The role of the nerve ring in governing the motion is, thus, only that of a switch, the sinusoidal muscle contraction wave itself being produced by the concerted action of the stretch receptors and trigger cells activated by the latter.

These ideas were subject to two tests. The first test was designed to see if the interneurons are indeed capable of carrying a signal consisting of a graded and attenuated potential pulse to all motor neurons of the forward- or backward-moving circuit. The second test was designed to see if a computer simulation of the locomotion of the nematode governed by its neural excitation could be developed that would replicate the actual motion of a nematode.

VI. Interneuron Signal Propagation

The neurons of *C.e.* have all been classified by an alfanumeric code [1]. We shall not discuss this code except to say that V stands for ventral, D for dorsal, L for left and R for right, and that $AVBL$ and $AVBR$ denote two interneurons. Both of these interneurons are located in the ventral cord, and receive their input from 50 chemical synapses in the nerve ring. Each interneuron is coupled by electrical synapses to 11 ventral motor neurons of the class VB and to 7 dorsal motor neurons of the class DB. In addition, the 7 dorsal motor neurons are connected to their own ventral processes by so-called commissures.

The system of partial differential equations which describe the spatio-temporal variation of the excitation potential in these neurons has already been described in Section IV. It was solved by discretizing the variable x which is the distance from the nerve ring along the neural cord, and the time t measured from the moment of excitation by the nerve ring.

Fig. 3 shows one snapshot of the situation after 5ms. Those parts of the nerve cells, where the potential exceeds -32 mV, corresponding to full excitation, are drawn with a full line. Parts with lower potential are indicated by dotted lines. One sees that the excitation almost completely invaded the motor neurons VB1-6 (middle diagram), and partially invaded the motor neurons DB1-7 (top diagram).

The numerical simulation proved that the excitation of all relevant motor neurons was possible within a time span needed to initiate coherent muscle action. The numerical solution of the system of *nonlinear* partial differential equations is discussed briefly in the Appendix.

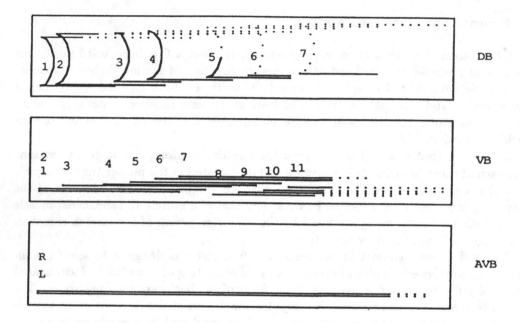

Fig. 3. Calculated electrical potential distribution 5ms after excitation by a signal from the nerve ring in the head of *Caenorhabditis elegans* in the motor neurons *VB*1-11, *DB*1-7 and in the interneurons AVBL and AVBR. The number of each neuron is printed above its head-side (left) end. Full lines indicate neural regions where the potential exceeds −32 mV (full excitation), dotted lines represent regions of lesser potential. The curved arcs are the commissures.

VII. Simulation of the Nematode Locomotion

In order to achieve a quantitative understanding of the neural and physical basis of the undulatory locomotion of *C.e.* the following program has to be carried out:

(1) Identify and mathematically formulate all forces acting on the body of the nematode;

(2) Formulate the equations of motion of the body taking into account (1);

(3) Define certain spatio-temporal muscle excitation patterns, and solve the equations of motion corresponding to these patterns; and

(4) Compare the computer-simulated movement with the motion of the animal as filmed through a microscope.

Some general remarks concerning this program are in order:

It is clear that (1) is entirely defined by the physiology of the animal and by the mechanical properties of the support on which it moves. This point will be discussed shortly. On the other hand, we have considerable liberty in the choice of the muscular excitation pattern (3), since this pattern is not directly observable. We made use of this liberty, and tested different patterns, among them one which gave the nematode the highest speed of locomotion.

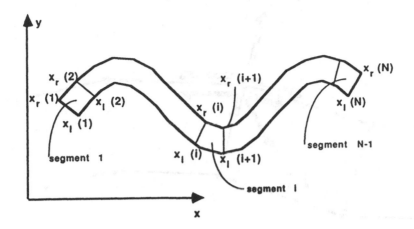

Fig. 4. Division of the body in N-1 segments for the purpose of simulation. Usually $N = 20$ or 50. The corners of the segments have the position vectors $\mathbf{x}_r(t)$ and $\mathbf{x}_l(t)$, where r and l refer to right and left, respectively, and $i - 1, \ldots, N$.

The position of the body is defined by the projection onto the plane of motion of the outline of the body. This closed curve is divided into 19 segments by lines perpendicular to the body axis. The curved parts of the outline of each segment are replaced by straight line sections. The resulting diagram is shown in Fig. 4. The momentary position of the body will be described by the totality of the positions $\mathbf{x}_s(i)$ ($i = 1, \ldots, 20$; s=r(ight), l(eft)) of the corners of the segments. The forces acting on the body will be reduced to forces acting on the points $\mathbf{x}_s(i)$. We now describe in detail the four steps of our program.

1. Identification of the Forces

The following forces act on $\mathbf{x}_s(i)$ (see Fig. 5):

(a) Force produced by the interior pressure of the liquid in the body cavity (pseudo-coëlome);

(b) Forces of the elastic cuticle;

(c) Muscle forces;

(d) Frictional forces; and

(e) Inertial forces.

(a) The pressure force $\mathbf{F}_s^p(i)$ acting outward from the point $\mathbf{x}_s(i)$ is

$$\mathbf{F}_s^p(i) = p(t)r[\mathbf{x}_s(i+1) - \mathbf{x}_s(i-1)]_\perp, \qquad i = 2, \ldots, N - 2. \tag{7}$$

Here, $\mathbf{x}_\perp$ denotes a vector of length $|\mathbf{x}|$, and perpendicular to $\mathbf{x}$, r is the body radius and $p(t)$ is the interior pressure. The latter is calculated from

$$p(t) = \frac{[p(0)V(0)]^k}{V(t)}, \qquad k = \text{positive integer.} \tag{8}$$

This expression takes into account the incompressibility of the liquid filling the body. Using $k = 4, 6$ or 8, we always found that during the simulation the volume V remained

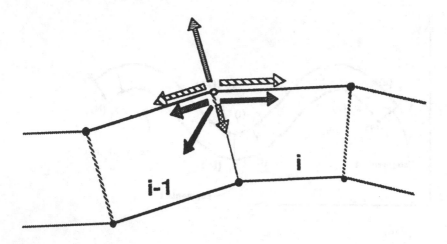

Fig. 5. Forces acting on the point $\mathbf{x}_r(i)$ (empty circle). Arrow marked by horizontal lines: interior pressure force $\mathbf{F}_r^p(i)$; cross-hatching: component $\mathbf{C}_r^e(i)$ of the elastic force of the cuticle acting between contralateral points; oblique lines: component $\mathbf{L}_s^e(i)$ of the elastic force of the cuticle acting between ipsilateral points; fine cross-hatching: muscle forces $\mathbf{F}_r^m(i-1,i)$ and $\mathbf{F}_r^m(i,i+1)$; full dark interior: frictional force $\mathbf{F}_r^f(i)$ exerted by the support.

constant within 1%. (This is a necessary condition to assure that the calculation does not lead to a disintegration of the simulated nematode.)

The pressure forces on the first and last segments ($i = 1$ and $N - 1$) are given by somewhat different formulas.

(b) The force $\mathbf{F}_s^e(i)$ exerted by the elastic cuticle has been decomposed into two parts: one force, $\mathbf{C}_s^e(i)$, which is member of a pair of equal and opposite forces acting between contralateral points, and another force, $\mathbf{L}_s^e(i,i+1)$, which is partner in a pair of equal and opposite forces between ipsilateral points.

The former force is put in the form

$$\mathbf{C}_r^e(i) = -\mathbf{C}_l^e(i) = k_C \left(\frac{d(i)}{2r} - 1\right)[\mathbf{x}_e(i) - \mathbf{x}_r(i)]. \tag{9}$$

Here, $d(i) = |\mathbf{x}_l(i) - \mathbf{x}_r(i)|$ is the momentary diameter of the segment. Observation shows that the diameter of the segments varies little compared to the variation of their length; therefore, the elastic constant k_C of diametral stretch must be large compared to the elastic constant k_L involved in the longitudinal stretch. We used the relationship

$$k_C = 10^3 k_L.$$

The ipsilateral force has a more complicated character due to the elastic properties of the cuticle:

$$\mathbf{L}_s^e(i,i+1) = -\mathbf{L}_s^e(i+1,i) = (e_s(i) + 2p(0)r^2)\mathbf{t}(i,i+1). \tag{10}$$

Here $\mathbf{t}(i,i+1)$ is a unit vector pointing from $\mathbf{x}_s(i+1)$ to $\mathbf{x}_s(i)$. The function $e_s(i)$ depends on the distance between the two points $\mathbf{x}_s(i+1)$ and $\mathbf{x}_s(i)$, i.e. on the length of the cuticle covering the segment i. This function, whose analytical form we do not explicitly write down here, is so designed that the length of the segment can be easily changed when this length is close to the equilibrium length, but the deformation is more difficult when the segment is either strongly stretched or strongly compressed. The function $e_s(i)$ is proportional to the longitudinal elastic constant k_L valid for small deformation. The equilibrium body radius is r.

The elastic forces defined by these equations are so chosen that they are in equilibrium with the pressure forces generated by an interior pressure $p(0)$ if all segments have the length $l = L/19$ (L=length of the nematode) and if the body diameter is everywhere $2r$.

(c) The muscle forces are modeled to take into account the observation that the somatic musculature of C.e. is strictly longitudinal. Therefore, these muscles exert forces between neighboring ipsilateral points. The muscular force $\mathbf{F}_s^m(i, i+1)$ acting on the point $\mathbf{x}_s(i)$ will be written in the form

$$\mathbf{F}_s^m(i, i+1) = f_s(i) k_L lt(i, i+1), \qquad i = 2, \dots, N-2. \tag{11}$$

The muscle forces in the first and last segments are zero. The set of numbers

$$f_s(2), \dots, f_s(N-2) \tag{12}$$

defines the **excitation pattern**. This excitation pattern is imposed on the muscles by the nervous system, and, as will be seen, leads to locomotion.

(d) The frictional forces exerted by the support on the nematode have not been measured directly, but in model experiments glass fibers of the same diameter as small nematodes have been pulled over an agar surface [11]. The force was found to be in the range

$$F^f = (17 - 72) \times 10^{-6} kgm\ s^{-2}.$$

The animal creates a shallow groove on the surface, in part due to the surface tension of the water film which usually covers its body. For this reason the frictional force which impedes lateral slipping is greater than the force counteracting longitudinal motion. Therefore, we will assume that the frictional force can be written as

$$\mathbf{F}_s^f(j) = c_L \mathbf{v}^L(j) - c_N \mathbf{v}^N(j), \tag{13}$$

where the longitudinal velocity is defined as

$$\mathbf{v}^L(j) = [\mathbf{v}(j) \cdot \mathbf{t}(j)]\mathbf{t}(j), \qquad j = 2, \dots, N-1, \tag{14}$$

and the velocity normal to the body axis is

$$\mathbf{v}^N(j) = \mathbf{v}(j) - \mathbf{v}^L(j). \tag{15}$$

The vector $\mathbf{t}$ has been defined previously. The velocity is

$$\mathbf{v}_s(j) = \frac{d}{dt}\mathbf{x}_s(j), \qquad j = 2, \dots, N-1,\ s = r, l, \tag{16}$$

and the dot in Eq. (14) denotes the scalar product. Since $\mathbf{v}^L(j)$ and $\mathbf{v}^N(j)$ are practically independent of s, i.e. on which side of the body they are calculated, we omitted this subscript in eqs (14,15). For $j = 1$ and $j = N$ somewhat modified formulas hold.

For the reason indicated above, we have used

$$c_N \gg c_L.$$

(e) The inertial forces due to the acceleration of the body are negligibly small.

2. Equations of Motion

In uniform motion the sum of the forces acting on all points $\mathbf{x}_s(i)$ which define the position of the body have to vanish. Hence

$$\mathbf{F}_s^p(i) + \mathbf{C}_s^e(i) + \mathbf{L}_s^e(i, i-1) + \mathbf{L}_s^e(i, i+1) + \mathbf{F}_s^m(i, i-1) + \mathbf{F}_s^m(i, i+1) + \mathbf{F}_s^f(i) = 0,$$

$$i = 2, \ldots, N-1, \qquad s = r, l. \tag{17}$$

The last term contains the time derivatives of the position vectors. The two terms preceding the last one contain the muscular excitation pattern imposed. The equations are coupled to each other through the terms which contain the indices of two segments, e.g., i and $i + 1$. The motion will be completely described when these equations have been integrated with respect to the time variable to obtain $\mathbf{x}_s(i, t), i = 1, \ldots, N; s = r, l$.

At $t = 0$ we assume that the body is in a position which corresponds to one wavelength of a sine-function whose amplitude is about equal to the body diameter - as observed in vivo. The initial interior pressure is 10^4 Pa [12]. The longitudinal friction coefficient $c_L = 18 \times 10^{-5} kgs^{-1}$ is so chosen as to give the observed speed of $10^{-3} m\, s^{-1}$. A computer program permitted the visualization of the body outline on the screen at any stage of the simulation.

In a preliminary simulation the muscles were not excited, i.e. the set of coefficients (12) were equal to zero. In this case the simulated body relaxed to a straight shape, but did not move either forward or backward.

In a second step, muscle excitation patterns were imposed – these will be discussed in the next Section. In this case the simulated nematode progressed, but after a while the body straightened out and the progression stopped. It was clear at this point that something was missing in the simulation.

VIII. Eppur' si muove

Even when its body is at a stillstand, one observes that the nematode moves its head left and right. When it progresses, this head movement is always present, and we concluded that it is this head movement which guides the animal along a sinuous (instead of straight) path. As was shown in Section II, such a path is indispensable if the muscle contraction is to result in a thrust.

Therefore, we have to assume that some neural circuit exists in the head which produces this oscillatory motion. In addition, to direct the backward motion, such a circuit has to be present in the tail as well. There are many nerve cells in the nerve ring situated in the head which may be responsible for the guidance of the nematode. In fact, the head muscles receive a more detailed innervation than the musculature in the bulk of the body. There are also some nerve cells in the tail, whose function is unknown, and we assume that they serve the purpose mentioned. It would be interesting to have this assumption tested by microbeam laser ablation experiments.

For the mathematical simulation, this guidance has been built into the equations by turning the first (or last) segment of the body along a direction which corresponds to the next position along a sinusoidal trajectory. For details we refer to [6].

We continue now the discussion of the four-point program exposed in Section VII.

3. Muscle Excitation Patterns

(a) Muscle excitation pattern controlled by stretch receptors.

The relative situation of the stretch receptors with respect to the muscles to be excited has been discussed in Section IV (i.e. 1/4 λ). Applying this excitation pattern yielded a steady forward (or backward) movement. Fig. 6 (a) shows a snapshot of the body with the contracted muscles shown in heavy lines. Fig. 7 represents graphically the motion. The maximal excitations were $f_s^{max}(i) = 0.05$.

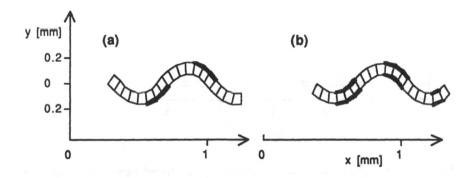

Fig. 6. Picture of the segmented model nematode body in motion, as photographed on the computer screen. Regions of momentary muscular excitation are indicated by heavy lines. (a) The excitation pattern is generated by stretch receptors. (b) Muscle excitation pattern producing fastest motion for given maximal muscle forces. The pattern is obtained by an optimization process.

(b) Muscle excitation pattern yielding maximal velocity.

With a suitable optimization algorithm the contraction pattern of the 38 simulated muscle cells which gives the fastest motion can be found. Fig. 6 (b) shows this pattern, and in Fig. 7 the distance *vs* time diagram obtained with this pattern is also shown. Again, $f_s^{max}(i) = 0.05$. Possibly some animals use this type of pattern.

4. Comparison of Experiment and Theory

A videotape (filmed from the screen of a Cray 2 computer) was made which compares the locomotion of a live *C.e.* with that of the simulated nematode. Among other phenomena, the videotape shows the nematode creeping forward and touching with its head an obstacle. In consequence, the neural circuitry for forward motion is switched off, and the circuity for backward motion is activated: the nematode then creeps backward.

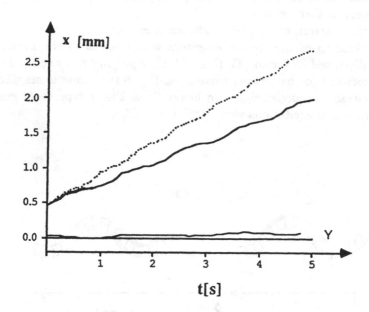

Fig. 7. Coordinate x [mm] of the center of gravity of the simulated nematode motion as a function of time $t[s]$. The full curve is the result of the algorithm based on stretch-receptor induced excitation. The dotted curve is obtained by the algorithm which creates fastest movement. For both curves $f_s^{max}(i) = 0.05$. The y-coordinate as a function of t is also shown. It remains practically zero.

It should be emphasized, that *the only external force* which has been introduced to act on the animal in this simulation is the frictional force exerted by the support, and this force is *opposite to the direction of motion*. Yet the simulation results in a progression, which makes it likely that not only the nematode, but we are also on the right track.

Appendix

The partial differential equations which yield the intracellular voltage $V(x,t)$ as a function of the position coordinate x within the neuron and at the time t cannot be solved analytically. Integral transformations have often been used [13] assuming that the equations are linear in V. We cannot make this assumption, because in a chemical synapse the transmembrane impedance depends on V, which renders the equations non-linear. Having in mind future work with active, e.g. Hodgkin-Huxley-type membranes it was, in any case, preferable to develop a method which is generally applicable to neural nets.

Discretizing the equations with respect to both x and t leads to so called "stiff" difference equations, whose solution by means of "explicit" (e.g. Runge-Kutta-type) algorithms leads to divergences. Therefore, implicit algorithms were tried, such as the "backward differentiation" and the Crank-Nicholson method. Finally we settled on Gear's algorithm [14]. A comparison of the results using different algorithms may be found in [6].

Acknowledgments

The support of the Swiss National Science Foundation through Grant 2.000-5.295 is gratefully acknowledged. Our thanks are due to Dr. R.C. Herndon for his critical reading of the manuscript, and to Prof. L. Garrido for the excellent organization of this conference.

References

* Present address: Division of Biology 216-76, California Institute of Technology, Pasadena, CA 91125, U.S.A.

[1] J.G. White, E. Southgate, J.N. Thomson and S. Brenner, Phil Trans. R. Soc. Lond. Ser. B **314**, 1 (1986).
[2] J. Gray, Quart. J. Micr. Sci. **94**, 551 (1953).
[3] E. Niebur and P. Erdös, in *Computer Simulation in Brain Science*, R. Cotterill, ed., Cambridge Univ. Press, Cambridge (1988).
[4] A.O.W. Stretton, R.E. Davis, J.D. Angstadt, J.E. Donmoyer and C.D. Johnson, Trends Neurosci. **8**, 294 (1985).
[5] W. Rall, in *Handbook of Physiology, The Nervous System*, Vol. 1, Section 1, E.R. Kandel (ed.), American Physiological Society, Bethesda (1977).
[6] E. Niebur, *Thesis*, Faculté des Sciences, Université de Lausanne, (1988).
[7] C.D. Johnson and A.O.W. Stretton. In *Nematodes as Biological Models* vol. 1, B.M. Zuckermann (ed.), Academic Press, New York (1980).
[8] J.P. Walrond and A.O.W. Stretton, J. Neurosci. **5**, 16 (1985).
[9] R. Russel, personal communication.
[10] M. Chalfie, J.E. Sulston, J.G. White, E. Southgate, J.N. Thomson and S. Brenner, J. Neurosci. **5**, 956 (1984).
[11] H.R. Wallace, Nematologica **15**, 67 (1969).
[12] D.L. Lee and H.J. Atkinson, *Physiology of Nematodes*. The Macmillan Press Ltd, London (1976).
[13] C. Koch and T. Poggio, J. Neurosci. Meth. **12**, 303 (1985).
[14] *The NAG Fortran Library Manual-Mark 12*, The Numerical Algorithms Group Ltd, Oxford, UK (1987).

REVERSIBILITY IN NEURAL PROCESSING SYSTEMS

K. Gustafson

Department of Mathematics and Optoelectronic Computing Center

University of Colorado, Boulder, CO, USA 80309-0426

Abstract

Reversibility (or the lack of it) is one of the most important properties any physical system, model, or theory may have. Here I explore a number of reversibility (irreversibility) aspects of neural processing systems, with reversibility taken in the widest possible sense. The format is nonstandard: far more questions than answers.

Outline

1. Issues of Reversibility (General)
   - Question: is there any reversibility at all?
   - Locality and Globality
   - Linearity and Nonlinearity
   - The Counter Problem
2. Reversibility Features in Neural Processing Systems
   - Question: is there any reversibility in neural processing systems?
   - Temporal and Spatial Features
   - Some Things We Have Done
   - The Neuron as Counter
3. One from Partial Differential Equations Who Flew into the Cuckoo's Net
   - Question: is the problem well posed?
   - Initial Conditions and Boundary Conditions
   - The Hodgkin-Huxley Approach and Reversibility
   - A Sigmoid Calculus
4. Possibilities
   - Question: are (really) new algorithms to be expected?
   - Little Machines, Big Machines, and Memory
   - Long Term Potentiation, Handshaking, and Glia
   - Computing, Cognition, and Content

Acknowledgements

References

1. Issues of Reversibility (General)

I am going to put forth here some ideas that are quite speculative. I will stress an inquiring view, rather than a developmental view, toward a better understanding of neural processing. For example, that is why I chose the title: *reversibility in neural processing systems*. Almost everyone who is knowledgeable in the subject will jump up and say "there is no reversibility in neural processing!" I have good evidence for this reaction, having produced it in 4 out of 4 trials when casually presenting the proposition to knowledgeable persons during stops on my voyage to this conference.

• Question: is there any reversibility at all?

In other words, to the knowledgeable person who jumps up with the correct immediate reaction, I reply "right - but is there any reversibility at all?" My goal here is not to become enmeshed in the morass of such epistemological considerations for their own sake. Rather, I want to point out that when dealing with the very real, the very minute, and the very large, imperfect reversibility (and imperfect irreversibility) considerations should be allowed and dealt with to better understand what is really going on. Less important than an absolute group invariance will be some imperfect group reversal such as a feed back mechanism invoked by a processing system when it reaches a certain threshold. Less important than the fact that a model acts as a dissipative dynamical system may be the manner in which the decay products are returned (or not returned) to the system, or in what larger environment this system resides (and therefore interacts).

I must interject a personal note. In my university physics courses long ago, as a student I was always uneasy with the physicists "clean" arguments using Hamiltonians and the like. I was always seeing some ignored side effects which it seemed to me should still be included. Friction for example was usually neglected. It particularly bothered me that some nice oscillation was glibly modelled without regard for the reaction it might be having with its environment. Another example is the Taylor problem of flow between concentric cylinders. I always saw the ends of the cylinders as affecting the flow characteristics. Only in the last ten years have researchers admitted this problem. Only recently have I begun to realize that I was right, I had been aware of natures' everpresent symmetry breaking all along. Awareness of such complexity, is, however, a handicap to quick clean results.

• Locality and Globality

Time and Space are the natural contexts in which we usually consider reversibility aspects of a physical system or model. I believe that it is useful to also consider reversibilities in the separate contexts of locality and globality.

For example, when does a baby conceive of time? I am sure that this has been much studied by medical and psychological researchers. But here we see the distinction between a local (the baby's) sense of time, and global time as we envision it within our universe. Personally, I don't consciously remember anything prior to my ability to walk, eg. nothing from my first year. Possibly my unconscious remembers some experiences while I was crawling. Maybe even earlier there is some effect still within my brain electrochemistry of a grasp at a fallen rattle I wanted back into my hand. Probably my sense of time developed with activity, which involved simultaneously the formation of a (local) spatial sense.

Is this local time of ours hardwired? Is it softwired? When I try to exercise memory recall going back to my earliest impressions as a baby, what mutes, and eventually blocks, that reversibility? How are such memory recall mechanisms conducted by our neural processing system? If we succeed in building biologically motivated or even biological computers, how will such time reversibility mechanisms be built in? These are important questions, for, as I will come back to at the end of this paper, many judgment and other higher cognitive decisions based upon experience require selective recall, i.e., selective local time reversal.

One may ask the same questions of space: when does a baby conceive of space? Is it hardwired? Is it softwired? How is it connected to the input/output devices of seeing, feeling, touching? How are grabbing, reaching, crawling, walking learned? How much is genetic preprogramming? How are such reversibility (eg., feedback) spatial mechanisms conducted by our neural processing systems?

How are our local neural processing systems connected to the more global neural processing systems of family, society, nature? How are our localities in space and time synchronized to the global systems of the earth and the universe? How much of that is hardwired biologically through evolution and survival, psychologically to remain sane, sociologically in order to function beyond an animal state?

In his pioneering book Wiener (1948) cast both communication theory and control theory in the contexts of a passage from Newtonian to Gibbsian time.

> "The relation of these mechanisms to time demands careful study. It is clear of course that the relation input-output is a consecutive one in time, and involves a definite past-future order. What is perhaps not so clear is that the theory of the sensative automata is a statistical one."

From Wiener (1948) we also have

"Thus the brain, under normal circumstances, is not the complete analogue of the computing machine, but rather the analogue of a single run on such a machine."

This insightful statement contains a tacit sense of a local time, the sequence of our brain's operations over our lifetime. Our local processing unit may have little to do with the next person's local (in space and time) neurocomputer, and little to do with the global space and time in which we envision ourselves. Yet we synchronize into a collective society which could be viewed as a very large parallel computer.

• Linearity and Nonlinearity

One is accustomed to reversibility - nonreversibility considerations in linear equations. For example, either the equation has a unitary inverse under group action, or it doesn't. However, I believe that much confusion has been generated by mixing up questions of reversibility vs irreversibility, locality vs nonlocality, linearity vs nonlinearity, and combinations of these six concepts.

In particular, just the presence of nonlinearity should trigger a changed thinking on our part about reversibility. To talk about broken symmetries in nonlinear systems is quite a different notion than in linear systems. For example, if $u(x)$ is a solution to a homogeneous cubic nonlinear problem, then $-u(x)$ may also be, which may not be the case at all in a quadratic problem. Such considerations do not arise at all in linear problems.

In their most recent book, Nicolis and Prigogine (1989) do a pretty good job of exposing at a reasonable level some of the issues of irreversibility and nonlinearity, so let me just refer to that book for further references, etc. However, I believe the linear vs nonlinear dichotomy is still under-understood.

For example, nonlinear dynamical systems are radically different from linear dynamical systems. In the latter, only $u \equiv 0$ is an attractor, and it is totally unstable. As another example, the stability of a self organized structure far from equilibrium is a completely nonlinear concept, tracing back to the principle of exchange of stabilities in fluid dynamics (see Chandrasekhar, 1961).

As a final instance of keeping well in mind the linear-nonlinear dichotomy, in (Goodrich, Gustafson, Misra, 1986) we answered affirmatively the conjecture of Misra and Prigogine (1980) that an irreversible Markov semigroup

$$W_t = PU_tP$$

formed by coarse graining a unitary evolution U_t will be positivity (eg, probability) preserving only if the U_t carries an underlying K-flow dynamics. That is, U_t must be of the form $(U_tf)(x) = f(T_tx)$ where T_t

is an underlying deterministic measure preserving dynamics in the physical phase space X with K-flow dynamical structure, eg, T_t could be a horseshoe, Baker, or Bernoulli transformation. See the book Nicolis and Prigogine (1989) for more details. The point I want to make here is that our arguments in the state space $L_2(X)$ are all completely linear operator theory with not much chance, in my opinion, for unlocking truly nonlinear details in the phase space dynamics.

I will use this Kolmogorov flow dynamics example one time later to illustrate an important property vis a vis initial conditions in reversibility considerations.

Quantum Mechanics remains the last bastion of linearity. More than any other linear theory, it seems to approximate the physics, i.e., in its csae, atomic spectra, exactly. Recently there are new attempts to question its linearity, see Weinberg (1989), and the discussion by Thompson (1989). Quantum mechanics is firmly based on the linear matrix mechanics of Heisenberg, the linear partial differential equation of Schrödinger, the linear Hilbert space setting of Von Neumann. But its paradoxes hint of a deeper level of understanding, where one can speculate that it will be nonlinear. This will perhaps not be seen in terms of small corrections.

It remains, then, that locality, reversibility, and linearity, are all idealizations. We use them in lieu of something better.

• The Counter Problem

Some years ago (Gustafson, 1975) I was working with B. Misra at Colorado on what we called the (quantum mechanical) Counter Problem. This problem concerns the collapse of the wave packet during the act of measuring. From my point of view it includes a number of very interesting operator theoretic questions about operator limits, viz,

$$s - \lim_{n \to \infty} (PU(t/n)P)^n$$

for unitary evolutions $U(t)$. Misra left Colorado and shortly later combined forces with Chiu and Sudarshan to exposite what they call the Zeno's paradox in quantum theory, see Misra and Sudarshan (1977). This has recently induced renewed experimental interest, eg., see the discussion by Pool (1990).

I mention this here for two reasons. First, even though on the quantum level, it illustrates the general elusiveness of real reversibility. The constantly watched particle does not decay. This violates the prediction of its (localized, isolated) reversible Hamiltonian, would it not be watched. It illustrates how the placing of an idealized reversible structure into another environment leads to contradictions, and to limits in our understanding.

Secondly, and in a completely different context, I want to return to interpretations of the neuron as a counter, exhibiting analogous difficulties to our understanding of its microfunction.

2. Reversibility Features in Neural Processing Systems

Almost all modelling of neural processing systems has built-in orderings. For example, neural nets glean and store information from their inputs, whether supervised or not, according to an input order. Some algorithms such as Boltzmann schemes actively employ the second law of thermodynamics, and hence a definite ordering in time. So Neural Processing Systems, just like Digital Computers, are to a very large extent operated in a fashion without regard to any potential reversibility features. Nonetheless I want to bring out a few of the latter.

• Question: is there any reversibility in neural processing systems?

The first reversibility process to come to mind is memory. How do we do recall? I have already discussed several facets of this in the first section. Most learning algorithms, etc., do not concern themselves with temporal recall. Most emphasis is placed upon the forward store operation, or on the building up of an associative store.

Pao (1989) gives a nice exposition (Chapter 6 therein) of the state of the art of neural net or similar methods for associative recall, see also Kohonen (1977). There are the linear matrix associative memories, holographic memories, Hopfield associative memories. Although there has been of course some success in obtaining response for partial cues (Hopfield memories come to mind), it seems to me that much less is understood about recall (or read) operations of neural nets, as compared to the record (or write) operation. This lack of reversibility speaks of a need for much more research into reversible read-write (recall). Here I should admit to some confusion of terms: recall, read, reverse, which should be carefully distinguished when considering these memory questions.

This limited situation concerning temporal reversibility of artificial neural processing systems should be contrasted to the rather good temporal recall that biological neural nets (eg, our brain) exhibit. Is the latter high quality performance due to a massive capacitance of the 10^{12} neurons with their 10^4 connections? Is distributed (recallable) memory really of necessity very very distributed? Are the low capacities of the artificial nets and the frustrating nonoverlayabilities in holograms real limits?

One should ask many more 'reversibility' questions about neural processing systems. It would be fruitful. It reminds of 'inverse problems', usually more difficult than 'forward problems'.

● Temporal and Spatial Features

I have brought out some temporal aspects of reversibility already. A spatial aspect was just mentioned: the needed *extent* in a neural processing system to store information in such a way that it can react to recall cues.

Another spatial feature is found in parallel processing. Biologically, why is seeing apparently so parallel (in space), whereas hearing is apparently far more serial? Does this mean in our artificial neural nets we should have a lot of neural "pixels" in a single layer for visual pattern recognition? If so, how is interconnect complexity kept down or handled?

The write-read issue mentioned above has led to recent attempts to understand nature's coding of neural signals. See for example Richmond, Optician, Podell, Spitzer (1987), and the discussion by Vaughn (1988). These investigations, still controversial, present geometric patterns (specifically, Walsh functions) to primates (monkeys in this case) and attempt to follow those patterns to a temporally coded signal sent from eye to brain. There is some evidence that geometrically related patterns are temporally coded in an additive (linearity, again) manner. The samples are small but this is an important and difficult problem.

Even asynchronous Hopfield nets give rise to concepts of short term memory and long term memory as distinct temporal features. A spatial feature is the notion of an attractor as found in chaotic dynamical system theory. Such features quickly become spatial-temporal as they are viewed as evolving dynamical systems.

● Some Things We have Done

In Goggin, Johnson, Gustafson (1990) we analyzed primacy and recency effects for backpropagation algorithms and holographic recall. Using a conventional momentum parameter, a primacy effect occurs: the current values of the weights are biased toward the first presentations in the sequence of training patterns. To produce a recency effect, we introduced a different momentum parameter. A method was devised for selecting momentum parameters based upon the degree of the effects desired: primacy or recency. The key to this method is a momentum function, for example, for $\alpha < 1$,

$$\frac{(1 - \alpha^{t+1-n})}{1 - \alpha}$$

whose derivative w.r.t. n decreases as n approaches the current presentation t.

In Gustafson, Goggin, Johnson (1990) we developed a connectionist overrelaxed nonlinear Gauss-Seidel algorithm (CNLOR) that converges much faster than Backpropagation. From a reversibility sense, it needs far less feedback than Backpropagation. Moreover it provides a comparison methodology possessing known convergence criteria, and also it enables an analysis of appropriate values for hidden units. A key insight into this scheme is provided by the weight update formula at a $j\underline{th}$ output unit. Whereas the weight update for a generalized delta rule scheme is (assuming the sigmoid nonlinearity)

$$\Delta w_{ji} = \eta(t_j - o_j)(o_j)(1 - o_j)o_i$$

it turns out that the weight update for CNLOR is

$$\Delta w_{ji} = \frac{\eta(t_j - o_j)o_i}{o_j(1 - o_j)}.$$

It is useful to think of o_i as 1 and the target t_j as 1. Then, whereas the GDR weight change slows down as o_j gets nearer and nearer to 1, in CNLOR the quotient $(t_j - o_j)/(1 - o_j)$ cancels and the weight change approaches the learning constant. This much more rapidly drives the sigmoid to saturation.

In Goggin, Johnson, Gustafson (1991) we have developed a wedge/ring detector second order architecture for spatial translation, rotation, and scale invariant processing of 2-D images with n input units. These may be interpretted as matters of spatial reversibility. This scheme has complexity $O(n)$ weights as opposed to $O(n^3)$ weights in a third order rotation invariant architecture. The reduction in complexity is due to the use of discrete frequency information.

• The Neuron as Counter

Some time ago I wondered (Gustafson, 1989) about regarding the Neuron as a Counter. I was led to this from the quantum mechanical counter question mentioned above. Recently I have noticed that early on, over twenty years ago, early researchers (eg., Jack Cowan) had mentioned this interpretation, although it appears to have not been developed. It is surely a natural idea that has occured to others.

I think it is worth a little further exposure, especially under the further view: the synapse latency dynamics (potential building, firing, dead period) as an analogue to the collapse of the wave package dynamics (not well understood) in the quantum mechanical case, which I described above.

Let me remark here without going into detail that in the analogy of collapse of wave packets, if one considers the more general problem of the arrival of an impulse, particle, signal, what have you, there is always a question of the extent of the incoming entity as it impacts the counter. Often overlooked, it raises intriguing questions, for example, as to the degree in which the interaction is elastic or inelastic, whether both entities are "scattered" in the micros, and so on. One is immediately forced to look at such phenomena over a range from the classical to the inclusion of possible quantum effects.

Most actual counters used in practice "lock up" immediately upon the "registration" of an incoming impulse. During the ensueing latent or dead period, no counting will take place and any arriving impulses will be missed. Biological neurons possess a similar recovery period before they can again react. A similar reciprocity failure may be found in certain photorefractive materials. I am working on such questions with my coworkers (K. Johnson, K. Gustafson, S. Goggin, 1991).

3. One from Partial Differential Equations Who Flew into the Cuckoo's Net

Those of us inured to a partial differential equations approach to problems of science are trained to ask: is the problem well posed? I will make this term precise below, but an analogy I like is that mathematicians find themselves in the (somewhat thankless) role of accountants of science: can the theories be made rigorous, are they consistent, are all the items defined and accounted for, do the equations really have solutions (debits = credits)?

• Question: is the problem well posed?

The usual requirements for well posed differential equations are (1) existence of a solution (2) uniqueness (or exact counting) of solutions (3) stability (with respect to initial conditions) of solutions. In my book (Gustafson, 1987) I add three other requirements (1') construction of (exact) solution (2') regularity (eg., physically realistic) of solutions (3') approximation (eg., numerical) of solutions.

There was a lot of mathematical work on the equations of Hodgkin-Huxley and their simplifications due to Fitzhugh and Nagumo during the late 1960's and the 1970's. By way of gossip I mention that Gail Carpenter was one of our mathematics students at the University of Colorado (I didn't know her) and went on elsewhere to do a Ph.D. thesis in mathematics on such equations. Although some nice results were obtained in that period for simplified uncoupled equations or explicitly coupled equations, such as the existence of traveling wave solutions, I think that it became apparent that there was little hope of a rigorous theory of the general equations being modelled by Hodgkin, Huxley, and other biologists. Thus, although for very simplified model equations one could (1) prove the existence of traveling wave solutions (2) count their exact number in some cases (3) look at their dependence on initially assumed constants of motion, also (1') in a few special cases find an explicit solution (2') demonstrate sufficient differentiability, such results usually necessitated a reduction down to the level of ordinary differential equations, and very small and low order systems at that.

This is in fact the general state of the mathematics of systems of nonlinear partial differential equations. The estimates are too hard and too messy and too conservative to tell you much even when you can obtain them. Equations from physics (eg., fluid dynamics) are better posed and easier to work on: They come from physical conservation principles which when studied can lead you to the right estimates. Those from biology usually do not carry such physical-mathematical consistency, hence are harder to prove things about.

So it has been (3'): approximation, especially numerical. I don't attempt to list all of the numerical simulations. Even in the absence of analytic results, these numerical studies will continue to be very interesting. For example recently Babloyantz and Destexhe (1988, to appear) have numerically modelled rather large systems of neural partial differential equations and have been able to produce both stationary and periodic attractors. It should be noted that (private communication, A. Destexhe), thus far at least, in order to produce chaotic solutions, for which there appears to be ample physical evidence, a time delay is needed in the simulation equations.

Thus we can say that, by and large, these biological partial differential equation systems are not well posed. We really don't know very much about them, mathematically.

The like must be admitted concerning the stochastic analyses of models of neural activity. See Tuckwell (1989) for an excellent account of all of this work, and my review (Gustafson, 1990) of this book. Basically, one has had to settle for a systems input-output approach without much understanding of the detailed internal dynamics of the neural systems.

I should also address the Algorithms. Let's call them for simplicity the ART, Backprop, and the Boltzmann classes of algorithms. These are also not in good shape mathematically. The ART schemes (see Carpenter and Grossberg, 1987) use an assembly of rather simple nonlinear ordinary differential equations but my impression is that there is not a cohesive rigorous mathematics for their collective behavior. The Backpropagation Algorithm has renewed faith in neural net algorithms in general, but beyond the old perceptron convergence proof, little has been established rigorously. The Boltzmann schemes, being akin to linear statistical mechanics, can borrow rigor from that field. I don't want to sound too disparaging. Quite the contrary: there is much important mathematical work yet to be done, on the biological partial differential equations, the mathematical stochastic process theory, and the computational algorithms.

• Initial Conditions and Boundary Conditions

A proper understanding of wellposedness requires a careful consideration in each problem of acceptable initial conditions and the constraints imposed by boundary conditions. Sometimes I joke with my class in partial differential equations that physicists never consider boundary conditions, mathematicians never consider physics, and engineers must deal with both!

I don't have time or space here to do more than mention a couple of observations. We might just lump these, as concerns initial conditions, under the phrase "preparation of the state". Most scientific models just assume that you can do this, so that one starts with a nice clean initial state $u(0)$. Here is a question. Why does the neuron react in a sigmoid-like way? This experimentally observed fact (Hodgkin, Huxley 1952) probably depends on an unknown chemical "preparation of incoming state". A comparable artificial neural processing system with extensive interacting nodes would need to do the same. Possibly the ingredients to initial condition preparation would be reusables, possibly from constant exterior supply. In a second analogy, the K-flow dynamical systems mentioned earlier partition the states at time $t = 0$ into those which will eventually expand exponentially, and those which will not. The initial state selection is a fundamental part of irreversibility.

As concerns boundary conditions, they have been included with some success in "clamping" arguments in neural net mean field theories, for example. But if we think of our own brain, we should ask: what are the boundary conditions? This is a very complicated question, but for example, it would be nice to know when hyperbolic boundary conditions, which can sustain waves, or parabolic boundary conditions, which will damp waves, are around. To use the fluid dynamics analogy, the Navier Stokes equations tell really nothing unless they are accompanied by boundary conditions, which select from the otherwise infinity of potential solutions.

• The Hodgkin-Huxley Approach and Reversibility

There are two points to be made here. First, as concerns the preparation of the initial state, Hodgkin and Huxley (1952) propose a phenomenological model consisting of charged particles with a hypothesized affinity for Na. Then by also assuming Boltzmann statistics they are led to the sigmoid as describing the response ability of a nerve to undergo a change in Na permeability. This model supports their earlier finding of a sigmoid curve-fit to experimental data.

Secondly, their eventual circuit model equations become so complicated that simplifications with reversibility implications must be invoked. For example there is the equation

$$\frac{a}{2R_2}\frac{\partial^2 V}{\partial x^2} = I = C_M\frac{dV}{dt} + \bar{g}_k n^4(V - V_k) + \bar{g}_{Na}m^3 h(V - V_{Na}) + \bar{g}_l(V - V_l)$$

I don't attempt to detail the quantities. Because this PDE cannot be dealt with, a wave (a reversible) relationship $\partial^2 V/\partial x^2 = \Theta^{-2}\partial^2 V/\partial t^2$ is assumed to reduce the above equation to an ODE, amenable to numerical solution. Thus a reversibility has been interspersed into the analysis. I wonder how many researchers today have the joint capability to accurately deal with such joint mathematical-biological considerations?

• A Sigmoid Calculus

I am accumulating (Gustafson, to appear) what I call a sigmoid calculus. But I want to expose here an interesting little wrinkle which carries reversibility connotations.

Suppose that we retreat from PDE (short for partial differential equations) to ODE (short for ordinary differential equations) but that we retain the sigmoid as a basic synapse response mechanism. In looking at it as a solution to a differential equation, one remembers how to solve the latter, by the two most elementary techniques, variable separation and partial fractions. Thus from the general differential equation initial value problem

$$y'(x) = ay(x) - by^2(x), \quad y(x_0) = y_0 \neq a/b,$$

we arrive at the solution

$$y(x) = a(b + (ay_0^{-1} - b)exp(-a(x - x_0)))^{-1}.$$

Next, regarding the solution as a cumulative probability distribution function, we must of necessity restrict attention to the case $a = b$. Imposing the additional symmetry condition that $y_0 = 1/2$, we thus obtain the two parameter sigmoid distribution function

$$y(x) = (1 + e^{-a(x-x_0)})^{-1}$$

with threshold x_0 and gain a. Its derivative $y'(x)$ is a probability density function satisfying the differential equation

$$y'(x) = ay(x)(1 - y(x))$$

with mean x_0 and spread a. Let $x_0 = 0$ without loss of generality (or replace $x - x_0$ by x) and for simplicity let $a = 1$. I then took the derivative of $y(x)$ directly, from which

$$y'(x) = (1 + e^{-x})^{-2} e^{-x}$$
$$= (1 + e^{-x})(1 + e^{x})$$
$$= y(x) \cdot y(-x).$$

The original differential equation is now seen to be a difference-differential equation!

It is easily checked that the sigmoid $y(x) = (1 + e^{-x})^{-1}$ is the only symmetric probability distribution function satisfying this difference-differential equation. I wondered, what about that other neural net favorite for soft threshold modelling,

$$y(x) = \frac{1}{2}(1 + \tanh x).$$

It is easily checked that it satisfies the difference-differential equation

$$y'(x) = 2y(x) \cdot y(-x).$$

It is the only symmetric probability distribution function doing so (the proof is the same as above) and is thus seen to be the sigmoid $y(x) = (1 + e^{-2x})^{-1}$ with gain 2.

More generally (same proof) one finds the sigmoid of mean 0 and gain a characterizied by the difference-differential equation

$$y'(x) = ay(x) \cdot y(-x), \qquad y(0) = 1/2.$$

This provides, among other things, a different way to look at the Δw_{ji} update in the generalized delta rule algorithms, and reveals a simple spatial reversibility in the action of the net cummulative input sums to a given summing node. Also, for example, the error $(t_j - o_j)$ in the Δw formula can be thought of as $o_j(-x)$ when $t_j = 1$, and of course as $-o_j(x)$ when $t_j = 0$.

For a nonzero threshold x_0, the sigmoid satisfies the difference-differential equation

$$y'(x) = ay(x) \cdot y(2x_0 - x)$$

and the same considerations go through.

As a next related point in this section, I want to argue how important it is to leave the thresholds unspecified, i.e., so that they may change. From the delta rule formula above,

$$\Delta w_{ji_0} = \eta \cdot (t_j - o_j) \cdot a \cdot o_j(x) \cdot o_j(2x_0 - x) \cdot o_{i_0}$$

applied to a threshold modelled as an additional bias node i_0 of constant "on" output 1, one sees that the threshold weight x_0 is "always changing". This reflects its importances as the "most random" or "most sensitive" among the weights, depending on the application.

4. Possibilities

Within the speculative context of this paper, it would be hardly fair to not put forth some intriguing possibilities for neural net processing systems. Leaving Neuroscience as a whole to others, here I want to propose some conceptual ideas toward future implementations within Neurocomputing systems. In other words, I will emphasize a specific point of view: the search for (really) new algorithms worthy of experimental implementations.

• Question: are (really) new algorithms to be expected?

By now the attentive reader would be led to predict an initial "no" answer to this last leadinquestion. But to the contrary, I expect neural processing system research to open new doors to (really) new algorithms. Even though in the meantime I expect most algorithmic development, no matter how clever, to follow already existing lines and known concepts, we should keep beating on the walls looking for hidden ways to new visions. That, in fact, is more or less what I am doing in this paper.

No doubt we will continue to make improvements in algorithms by use of existing techniques. The effective solution of the traveling salesman problem by simulated annealing and the reemergence of Ising model and spin glass methods in the Hopfield net are examples of dramatic improvements. For a nice review of the resistance Hopfield encountered in that interdisciplinary endeavor, see the recent account by Philip Anderson (Anderson, 1990). But physicists are good mathematicians and deeply involved in computing, so they are able to connect these ingredients. In another example, our introduction of nonlinear overrelaxation (Gustafson, Goggin, Johnson, 1990) shows how the connecting of existing methods from numerical analysis with those of neural algorithms such as Backpropagation can greatly improve and clarify the latter.

But real neural processing will not be numerical analysis, real learning is not optimization, real memory is not statistical mechanics, the higher mental functions are not feedback control mechanisms. So the really new algorithms should be expected not from engineering, physics, or mathematics, but more from biology, psychology, the soft sciences.

Time and Space have run out so I will quickly finish the paper's content, with just a few hints of how one might want to think about finding such new algorithms from the soft sciences. Elaboration will be given elsewhere.

- ### Little Machines, Big Machines, and Memory

The competition in current parallel computation architectures, whether they be classical digital systems or connectionist networks, is between "a lot of Chinese coolies" or a few linked supercomputers. In any of these configurations a key question, alluded to earlier, for next generation computing, is: how much memory (and saved earlier memory experiences) is needed? This may be viewed as a hardwiring question. I mention an intensely personal example from my very distant past. As a highschool student in Boulder, one of my schoolmates was named Atanasoff. Little did I know that his father would some day be recognized as one of the principal pioneers of digital computing. With all due respect to my friend John, my friends and I were more impressed with John's older sister, a cheerleader at the school. She was a knockout, I remember images of her today - why, and how?

- ### Long Term Potentiation, Handshaking, and Glia

The recent findings (Davies et al., 1989), see commentary by Stevens (1989), approach at a foundational level the fundamental biological - electrical - chemical couplings that may underly learning and memory. In particular, the temporal reversibility aspects ("handshaking" to check levels) indicate possible new schemes for neural processing systems. The report by Weiss (1990) on the new findings on the possible roles of the glia cells is another hint as to how algorithms could operate to imitate brain functioning.

- ### Computing, Cognition, and Content

In current work I am investigating algorithms which might enable the forming of *Content,* eg., as needed to make reasonably accurate subjective judgment calls.

Acknowledgements

Although the viewpoint and for the most part the ideas and questions exposed here are my own, formulated as concerns neurocomputing during the last two years, I would like to express my thanks to a number of individuals for very helpful related collective scientific working experiences and dialogue: my colleagues Professor Kristina Johnson and Shelly Goggin of the Optoelectronic Computing Systems Center in Boulder, Dr.

Jakob Bernasconi at ABB Rersearch in Baden, and Professor Ilya Prigogine and Baidyaneth Misra and Ioannis Antoniou at the Free University of Brussels.

Elements of this research were partially supported by NSF Grant CDR 8622236.

References

1. Norbert Wiener, *Cybernetics*, Wiley & Sons, New York (1948)

2. G. Nicolis and I. Prigogine, *Exploring Complexity*, Freeman & Co., New York (1989)

3. S. Chandrasekhar, *Hydrodynamic and Hydromagnetic Stability*, Oxford Press, Oxford (1961)

4. R. Goodrich, K. Gustafson, B. Misra, On K-flows and Irreversibility, *J. Stat. Physics* 43 (1986), 317-320

5. B. Misra and I. Prigogine, On the Foundations of Kinetic Theory, *Supplement of the Progress of Theoretical Physics* 69 (1980), 101-110

6. S. Weinberg, *Phys. Rev. Letters* 62 (1989), 485-488

7. R. Thompson, *Nature* 341 (1989), 571-572

8. K. Gustafson, The Counter Problem (*unpublished notes*, 1975)

9. B. Misra and E.C.G. Sudarshan, The Zenos Paradox in Quantum Theory, *J. Math. Physics* 18 (1977), 756-763

10. R. Pool, Quantum Pot Watching, *Science* 246 (1990), p888

11. Y.H. Pao, *Adaptive Pattern Recognition and Neural Networks*, Addison-Wesley Pub. Co (1989)

12. T. Kohonen, *Associative Memory: A System-Theoretic Approach*, Springer-Verlag (1977)

13. B.Richmond, L. Optician, M. Podell, H. Spitzer, Temporal Encoding of Two-Dimensional Patterns by Single Units in Primate Interior Temporal Cortex I. Response Characteristics, *J. Neurophysiology* 57 (1987), 132-146. II. Quantification of Response Waveform, *J. Neurophysiology* 57 (1987), 147-162. III. Information Theoretic Analysis, *J. Neurophysiology* 57 (1987), 163-178

14. C. Vaughn, *Science News* 134 (1988), 58-60

15. S. Goggin, K. Johnson, K. Gustafson, Primacy and Recency Effects Due to Momentum in Back-Propagation Learning, *Progress in Neural Nets* (1990, to appear)

16. K. Gustafson, S. Goggin, K. Johnson, Iterative Methods for Neural Net Architectures, *SIAM J. Sci. Stat. Computing* (1990, to appear)

17. S. Goggin, K. Johnson, K. Gustafson, A Second-Order Rotation and Scale Invariant Neural Network (1991, to appear)

18. K. Gustafson, The Neuron as Counter (*unpublished notes*, 1989)

19. K. Johnson, K. Gustafson, S. Goggin (1991, to appear)

20. K. Gustafson, *Partial Differential Equations*, 2nd Edition, John Wiley & Sons, New York (1987)

21. A. Babloyantz and A. Destexhe, *Biological Cybernetics* 58 (1988) and a paper in preparation

22. H. Tuckwell, *Stochastic Processes in the Neurosciences*, SIAM Publications, Philadelphia (1989)

23. K. Gustafson, Review of: Stochastic Processes in the Neurosciences, *SIAM Review* (June, 1990, to appear)

24. G. Carpenter and S. Grossberg, Art2: Self-organization of stable category recognition codes for analogy input patterns, *Applied Optics* 26 (1987), 1-23

25. Hodgkin and Huxley, *J. Physiology* 116, 117 (1952)

26. K. Gustafson, *to appear*

27. Philip W. Anderson, Spin Glass VII: Spin Glass as Paradigm, *Physics Today*, March, 1990, p9

28. S. Davies, R. Lester, K. Reymann, G. Collingridge, Temporally distinct pre- and post-synaptic mechanisms maintain long-term potentiation, *Nature* 338 (1989), 500-503

29. S. Stevens, Strengthening the Synapses, *Nature* 338 (1989), 460-461

30. R. Weiss, Hints of Another Signaling System in Brain, *Science News* 137 (1990), 54

Lyapunov functional for neural networks
with delayed interactions
and statistical mechanics of temporal associations

Zhaoping Li[1] and Andreas V.M. Herz[2]

[1] Fermi National Laboratory, P.O. Box 500, Batavia, IL 60510, USA

[2] Sonderforschungsbereich 123 an der Universität Heidelberg, 6900 Heidelberg, FRG

1. Introduction

Computational functions of associative neural structures may be analytically studied within the framework of attractor neural networks such as proposed by Little [1] and Hopfield [2], where *static* patterns are stored as attractors of the system dynamics. Environmental input to a neurobiological network, however, always provides *information in both space and time*. It is therefore desirable to extend the original approach and to search for a joint representation of static patterns and temporal sequences.

Signal delays are omnipresent in the brain and play an important role in biological information processing [3]. Their incorporation into theoretical models seems to be mandatory. In particular, the distribution of delay times involved should be taken into account. Kleinfeld [4] and Sompolinsky and Kanter [5] proposed models for temporal associations, but they used only a *single* delay. Tank and Hopfield [6] presented a feedforward architecture for sequence recognition based on multiple delays, but they just considered information relative to the *very end* of each given sequence. Besides these deficiencies, both approaches lack the important quality to acquire knowledge through a true learning mechanism: They both require that synaptic efficacies be calculated by hand, which is certainly not satisfactory, neither from a neurobiological point of view nor for applications in artificial intelligence.

This drawback has been overcome by a careful analysis of the formal aspects of Hebbian learning [7]. Implemented in a neural network with a *broad* delay distribution, Hebb's rule gives rise to a surprisingly fault-tolerant mechanism for the autoassociative recall of stationary patterns *and* temporal sequences — learnt by the *same* principle [8].

In the present contribution we concentrate on parallel dynamics and generalize the idea of "duplicated" spins [9]. We introduce a Lyapunov functional H for the temporal

[1] Address after September 1, 1990: School of Natural Sciences, Institute for Advanced Study, Princeton, NJ 08540, USA   [2] Address after October 1, 1990: Division of Chemistry and Biology, California Institute of Technology, Pasadena, CA 91125, USA

evolution in the deterministic case and prove that under a stochastic Glauber dynamics the system relaxes to a Gibbs distribution generated by H. Next, we focus on the thermodynamics of networks with finitely many stored memories. Finally, we discuss the storage capacity, based on numerical simulations and replica symmetric calculations — extending the studies of Fontanari and Köberle [10] to networks with signal delays.

2. Neural dynamics

Throughout what follows, we describe a neural net as a collection of N formal neurons with activities $S_i(t) = \pm 1$, connected by modifiable synapses with efficacies $J_{ij}(\tau_{ij})$, where τ_{ij} denotes the time delay for the information transport from j to i . In the present paper we consider a model where there is for each pair of neurons a *large* number of axons whose delays τ are independent of i and j. (Other architectures having only a single link between each pair of cells have been considered elsewhere [8]; see also [11] and [12].) External stimuli are fed into the system via receptors with activities $\sigma_i(t) = \pm 1$ [13]. The local fields $h_i(t)$ are then given by

$$h_i(t) = (1 - \gamma) \sum_{j,\tau} J_{ij}(\tau) S_j(t - \tau) + \gamma \sigma_i(t) , \tag{1}$$

where the parameter γ measures the system's input sensitivity. The stochastic nature of biological processes involved in the time evolution of a single neuron may be described by a Glauber dynamics at a finite inverse temperature $\beta = T^{-1}$. In that case, the probability for S_i to take the values ± 1 at time $(t + \Delta t)$ is

$$p[S_i(t + \Delta t) = \pm 1] = \frac{1}{2}[1 \pm \tanh(\beta h_i(t))] , \tag{2}$$

which reduces in the deterministic limit, $\beta \to \infty$, to

$$S_i(t + \Delta t) = \text{sgn}[h_i(t)] . \tag{3}$$

Updating is either performed for *all* spins in parallel, or sequentially in that only a *single* spin may be flipped at each time step Δt, which is adapted accordingly ($\Delta t \propto N^{-1}$). In the present contribution, we study parallel dynamics with $\Delta t = 1$. Consequently, the signal delays take nonnegative integer values.

3. Synaptic dynamics

During a learning session the synaptic strengths may change according to the Hebb rule [7]. In order to understand the implications for a network with parallel dynamics and delayed interaction, we focus our attention on a connection with delay τ between a specific pair of neurons, say j and i. Following Hebb's principle, the efficacy $J_{ij}(\tau)$ will be increased if .cell j takes part in *firing* cell i. In its neurophysiological context, this rule was originally formulated for excitatory synapses only, but as usual [2], we apply it to all synapses in a network with formal ± 1-neurons.

Taking the transmission delay τ into account, the synaptic strength will therefore change at each timestep t by an amount proportional to the product of $S_j(t - \tau)$ and $S_i(t+1)$ since due to the delay τ in (1) and the parallel dynamics (2) it takes $\tau + 1$ time steps until neuron j actually influences the *state* of neuron i. Starting with a *tabula rasa* ($J_{ij}(\tau) = 0$) we obtain after P single learning sessions, labeled by μ and each of duration D_μ,

$$J_{ij}(\tau) = \varepsilon(\tau) \cdot N^{-1} \sum_{\mu=1}^{P} \sum_{t_\mu=1}^{D_\mu} S_i(t_\mu + 1) S_j(t_\mu - \tau) \equiv \varepsilon(\tau) \tilde{J}_{ij}(\tau) . \tag{4}$$

Here N^{-1} is a scaling factor useful for the theoretical analysis of a fully connected network. The parameters $\varepsilon(\tau)$ are nonnegative and normalized such that $\sum_{\tau=0}^{\tau_{\max}} \varepsilon(\tau) = 1$. They are additional a priori weights for the synaptic strengths $J_{ij}(\tau)$ which take morphological characteristics of the different transmission lines into account and may be used to define various network architectures.

According to (4), synapses measure and store correlations of the external stimuli in space (ij) and time (τ) — as "seen" by the network via (1) and (2). We would like to emphasize that the synaptic efficacies are in general *asymmetric*, $J_{ij}(\tau) \neq J_{ji}(\tau)$.

In spite of the asymmetry, an analysis of the network capabilities in terms of statistical mechanics becomes possible if during the learning sessions we take $\gamma = 1$ in (1) and $T = 0$ in (2). This amounts to a clamped learning scenario where the system evolves strictly according to the environmental inputs, $S_i(t_\mu) = \sigma_i(t_\mu - 1)$. Furthermore, we focus on temporal associations with a *cyclic* structure. For example, each of the P learning sessions may corresponds to teaching a cycle of n_μ static patterns ψ_{ip}^μ, $1 \leq p \leq n_\mu$, each of duration Δ_μ with $\Delta_\mu n_\mu = D_\mu$. During such a learning session we thus impose upon the network a *time-dependent* external stimulus of the form $\sigma_i(t_\mu) = \psi_{ip(t_\mu)}^\mu$ with $p(t_\mu) = p(\bmod n_\mu)$ for $(p-1)\Delta_\mu \leq t_\mu \leq p\Delta_\mu$. For a discussion of the general case see [8]. In passing we would like to note that we should offer the sequences $\tau_{\max}$ time steps before allowing for synaptic plasticity according to (4) so that both S_i and S_j are in well defined states during the actual learning sessions.

For the time being we work with unbiased random patterns where $\psi_{ip}^\mu = \pm 1$ with equal probability. A generalization to the low activity regime requires an appropriate change of the learning rule (4) as e.g. in [14].

4. A new kind of synaptic symmetry

Using formulae (1) - (4) one may derive equations of motion for macroscopic order parameters, see e.g.[8], and investigate the stability of special solutions and their dependence upon the weight distribution $\varepsilon(\tau)$ [15]. However, this kind of study only applies to the case where $\sum_{\mu=1}^{P} n_\mu \ll \log N$, and does not allow any predictions about the extensive case with nonzero storage level $\alpha \equiv \lim_{N \to \infty}(P/N)$.

As a first step towards an analysis in terms of statistical mechanics we now show that there is a valuable relation between the efficacies of specific pairs of synapses for a certain class of network architecture — if we demand that all cycles to be stored have equal length D. In that case we get from the definition (4)

$$\tilde{J}_{ij}(\tau) = \tilde{J}_{ji}(D - (2 + \tau)) , \tag{5}$$

true for all τ. As in the following, temporal arguments of the synaptic couplings should always be understood *modulo D*. For all networks whose weight distributions obey

$$\varepsilon(\tau) = \varepsilon(D - (2 + \tau)) \tag{6}$$

we have thus found an "extended" synaptic symmetry,

$$J_{ij}(\tau) = J_{ji}(D - (2 + \tau)) . \tag{7}$$

Notice that for a sequential dynamics (3) and $N \to \infty$, one obtains $\tilde{J}_{ij}(\tau) = \tilde{J}_{ji}(D - \tau)$. Both formulae generalize Hopfield's symmetry assumption [2] in a natural way and reduce to it for learning sessions with static patterns only. Equation (7) will be the essential ingredient to establish a Lyapunov functional for parallel dynamics.

5. The Lyapunov functional

We now show that there is a Lyapunov functional for the noiseless retrieval dynamics, i.e. a bounded scalar functional which decreases under the deterministic evolution of the system. We do so for the case where the network evolves freely, taking $\gamma = 0$ in (1). The general case [13] is discussed elsewhere [15]. Extending a proposal of Li [16] we take

$$H(t) \equiv -\frac{1}{2} \sum_{i,j=1}^{N} \sum_{a,\tau=0}^{D-1} J_{ij}(\tau) S_i(t + 1 + a - D) S_j(t + 1 + (a - \tau - 1)\%D - D) , \tag{8}$$

where $a\%b \equiv a \bmod b$. The functional H depends on *all* network states between $(t + 1 - D)$ and t so that solutions with constant H *need not be static fixpoints*.

Using the extended symmetry (7) and the definition of the postsynaptic fields (1) for $\gamma = 0$, the difference $\Delta H(t) \equiv [H(t) - H(t - 1)]$ may after some algebra be written

$$\Delta H(t) = -\sum_{i=1}^{N}[S_i(t) - S_i(t-D)] h_i(t-1) - \frac{\varepsilon(D-1)}{2N} \sum_{\mu=1}^{P} \sum_{t_\mu=1}^{D_\mu} \{\sum_{i=1}^{N} \psi_{ip(t_\mu)}[S_i(t) - S_i(t - D)]\}^2.$$

Due to the dynamics (3), the first term on the right hand side will only take nonpositive values. Since $\varepsilon(\tau) \geq 0$, the same is true for the second term. For reasons to be explained when we discuss the statistical mechanics of the network, we nevertheless set that term to zero. This is easily achieved, if we demand that $\varepsilon(D - 1)$ vanishes. Due to (6), this choice does not affect other efficacies. Furthermore, it will also allow a Lyapunov functional for systems with couplings more general than (4), see [15]. In other words,

$$J_{ij}(D - 1) \equiv 0 , \tag{9}$$

and we finally get (the complete derivation (8) $\to$ (10) will be given in section 6)

$$\Delta H(t) = -\sum_{i=1}^{N}[1 - S_i(t)S_i(t - D)] |h_i(t - 1)| \leq 0 . \tag{10}$$

For finite N, H is bounded from below and ΔH has to vanish after finitely many steps. This can be realized only if the system settles in a state with $S_i(t) = S_i(t - D)$ for all i.

We have thus discovered two rather important facts: a) the retrieval dynamics is governed by a Lyapunov functional, and b) the system relaxes to static states or limit cycles with $S_i(t) = S_i(t - D)$ — oscillatory solutions with the *same* period as that of the taught cycles (or equal to an integer fraction of D) . This periodicity allows for a new formalism and finally paves the way to a description of temporal associations in terms of equilibrium statistical mechanics.

6. An equivalent static problem

In order to proceed with the analytical treatment, we follow a little detour and turn to a problem of *static* pattern storage, of which we shall show later on that it is equivalent to our original problem of storing *temporal* associations with a cyclic structure.

Consider a two-dimensional network with D columns and N rows of ± 1-neurons whose activities will be denoted by $S_{ia}(t)$, $1 \leq i \leq N$ and $0 \leq a \leq D-1$. Suppose we want to store P patterns $\xi_{ia}^{\mu 0}$, $1 \leq \mu \leq P$, and for each of them also $D-1$ "shifted copies" $\xi_{ia}^{\mu \alpha}$, $1 \leq \alpha \leq D-1$, duplicates of $\xi_{ia}^{\mu 0}$, cyclically translated along the a-direction, i.e.,

$$\xi_{ia}^{\mu \alpha} \equiv \xi_{i,(a-\alpha)\%D}^{\mu 0} \, . \tag{11}$$

We may try to do so by constructing synaptic efficacies similar to those of the Hopfield model,

$$J_{ij}^{ab} \equiv \mathcal{E}_{ab} N^{-1} \sum_{\mu=1}^{P} \sum_{\alpha=0}^{D-1} \xi_{ia}^{\mu \alpha} \xi_{jb}^{\mu \alpha} \, . \tag{12}$$

The matrix elements $\mathcal{E}_{ab}$ characterize the architecture of the network, i.e. the connectivity in the a-direction. They are normalized to assure the correct scaling behavior,

$$\sum_{a=0}^{D-1} \mathcal{E}_{ab} = \sum_{b=0}^{D-1} \mathcal{E}_{ab} = 1 \, , \tag{13}$$

and are supposed to be symmetric,

$$\mathcal{E}_{ab} = \mathcal{E}_{ba} \, , \tag{14}$$

and invariant under cyclic transformations,

$$\mathcal{E}_{ab} = \mathcal{E}_{(a+c)\%D,(b+c)\%D} \, . \tag{15}$$

Equations (11) - (15) lead couplings that are manifestly symmetric,

$$J_{ij}^{ab} = J_{ji}^{ba} \, , \tag{16}$$

and translation invariant with respect to the upper indices,

$$J_{ij}^{(a+c)\%D,(b+c)\%D} = J_{ij}^{ab} \, . \tag{17}$$

Due to (16), there is a well defined Hamiltonian for this new system,

$$H = -\frac{1}{2} \sum_{i,j=1}^{N} \sum_{a,b=0}^{D-1} J_{ij}^{ab} S_{ia} S_{jb} \, , \tag{18}$$

formally equivalent to that of a Hopfield network of size ND. It might be used to describe the statistical mechanics of the net once we have specified the dynamical rules and proven that they do lead to an invariant density governed by this Hamiltonian.

At first, however, we want to establish the equivalence with the delay network storing temporal associations. The reader will be convinced that this is indeed possible if he or she applies the following "translation rules",

$$\xi_{ia}^{\mu a} = \psi_{i,p(a,\alpha)}^{\mu} \quad , \quad p(a,\alpha) = \text{int}\left(\left(\frac{a-\alpha}{\Delta_{\mu}}\right)\%n_{\mu}\right) + 1 \ , \tag{19}$$

and

$$\mathcal{E}_{ab} = \varepsilon\left((b-a-1)\%D\right) \ , \tag{20a}$$

and conversely

$$\varepsilon(\tau) = \mathcal{E}_{a,((a+\tau+1)\%D)} \ . \tag{20b}$$

For the synaptic couplings, this prescription leads to

$$J_{ij}^{ab} = J_{ij}\left((b-a-1)\%D\right) \ , \tag{21a}$$

and conversely

$$J_{ij}(\tau) = J_{ij}^{a,((a+\tau+1)\%D)}. \tag{21b}$$

Next, we set

$$S_{ia}(t) = S_i(t-a) \ . \tag{22}$$

The energy functional (18), evaluated at time t, will be identical to (8) if the diagonal terms of $\mathcal{E}$ vanish,

$$\mathcal{E}_{aa} = 0 \ . \tag{23}$$

By (20), this implies (9). The analogy between the two kinds of networks is complete if we now formulate the original dynamics (1) - (3) in terms of the new variables. We define postsynaptic potentials by

$$h_{ia}(t) = h_i(t-a) \ , \tag{24}$$

and arrive at

$$h_{ia}(t) = \sum_{j=1}^{N} \sum_{b=0}^{D-1} J_{ij}^{a,((b+a+1)\%D)} S_{j,b+a}(t) \ . \tag{25}$$

There are two different but equivalent ways to visualize the updating procedure in this network of size ND. One either concentrates on *single* time steps from t to $(t+1)$ where the dynamics corresponds to a shift for all neurons except those with $a = 0$,

$$S_{ia}(t+1) = S_{i,a-1}(t) \ , \tag{26}$$

satisfying (22), and a true update for this last column according to Equation (3),

$$S_{i0}(t+1) = \text{sgn}(h_{i0}(t)) \ . \tag{27}$$

The second alternative explicitly refers to the new network of size ND: D consecutive sweeps of the original dynamics are grouped together to form one single step for the dynamics of the enlarged system. The resulting updating rule has a pseudo sequential characteristic: synchronous within single columns and sequentially ordered — with decreasing index a corresponding to increasing time t — with respect to these columns. In what follows we will mainly work with the first interpretation (26)-(27).

We cannot exploit the formal similarity to the Hopfield model in a naive way when proving (10). However, the correct derivation only takes a few steps. We take (23) into account and evaluate H (18) at t and $(t-1)$ so as to obtain

$$H(t) = -\frac{1}{2}\sum_{i,j=1}^{N}\sum_{a,b=1}^{D-1} J_{ij}^{ab} S_{ia}(t)S_{jb}(t) - \sum_{i,j=1}^{N}\sum_{b=1}^{D-1} J_{ij}^{0b} S_{i0}(t)S_{jb}(t)$$

$$= -\frac{1}{2}\sum_{i,j=1}^{N}\sum_{a,b=0}^{D-2} J_{ij}^{ab} S_{ia}(t-1)S_{jb}(t-1) - \sum_{i=1}^{N} S_{i0}(t)h_{i0}(t-1) ,$$

$$\equiv H_R(t-1) - \sum_{i=1}^{N} S_{i0}(t)h_{i0}(t-1) , \qquad (28)$$

$$H(t-1) = -\frac{1}{2}\sum_{i,j=1}^{N}\sum_{a,b=0}^{D-2} J_{ij}^{ab} S_{ia}(t-1)S_{jb}(t-1) - \sum_{i,j=1}^{N}\sum_{b=0}^{D-2} J_{ij}^{D-1,b} S_{i,D}(t)S_{jb}(t-1) ,$$

$$= H_R(t-1) - \sum_{i=1}^{N} S_{iD}(t)h_{i0}(t-1) .$$

Using (27) we then get an expression identical with (10),

$$\Delta H(t) = -\sum_{i=1}^{N}\left[1 - S_{i0}(t)S_{iD}(t)\right]\left|h_{i0}(t-1)\right| \leq 0 , \qquad (29)$$

and have thus finished our proof that $H(t)$ is indeed a Lyapunov functional for the noiseless dynamics (3). Furthermore, due to the equivalence with the static network, we may apply equilibrium statistical mechanics so as to analyze the behavior of the system in the present of noise as will be shown in the next section.

Let us finally note that for $D=2$ the above considerations straightforwardly lead to well known results about the Little model [1]. Due to (23) only couplings with zero delay enter (18) in this case. Equation (17) leads to

$$J_{ij}^{ab} = J_{ij}^{ba} \equiv J_{ij} . \qquad (30)$$

For static patterns, i.e. $\Delta = 2$, these couplings differ from those of the original Little model by a factor of 2 due to the sum $\sum_{t=1}^{2}$ in (4). Notice, however, that there is also the possibility of explicitly storing 2-cycles with $\psi_1^\mu \rightleftharpoons \psi_2^\mu$. Equation (18) may be written in terms of the "duplicated" spins $S_i \equiv S_{i1}$ and $\sigma_i \equiv S_{i0}$ [9],

$$H = -\sum_{i,j=1}^{N} J_{ij}\sigma_i S_j , \qquad (31)$$

or, inserting the dynamics (25) and (27), in terms of S_i only,

$$H = -\sum_{i=1}^{N}\left|\sum_{j=1}^{N} J_{ij}S_j\right| , \qquad (32)$$

as first discovered by Peretto [17]. The general analysis (10) immediately tells us that there are two classes of solutions, static states and limit cycles of length 2.

7. Equilibrium statistical mechanics

In the preceding sections we have demonstrated that the zero temperature dynamics (3) is governed by a Lyapunov functional H. If we switch to a stochastic dynamics à la Glauber (2), the important question arises whether H also determines the equilibrium states of the network. It will now be shown that this is indeed the case.

Statistical properties of a stochastic dynamical system may be computed by averaging observables over all possible states A of the system [18]. The corresponding probability measure will be denoted by the normalized distribution $\rho(A(t))$,

$$\text{Tr}_A \, \rho(A(t)) = 1 \,. \tag{33}$$

For a homogeneous Markov chain, the time evolution of $\rho(A(t))$ may be described in terms of time independent single step transition probabilities $W(B|A)$,

$$\rho(B(t+1)) = \text{Tr}_A W(B|A)\rho(A(t)) \,. \tag{34}$$

Equation (33) results in

$$\text{Tr}_B W(B|A) = 1 \,. \tag{35}$$

The probability distribution will tend to a *unique* equilibrium distribution $\rho_{eq}(A)$ with

$$\rho_{eq}(B) = \text{Tr}_A W(B|A)\rho_{eq}(A) \tag{36}$$

if the system is ergodic, i.e. if there exists some finite number M such that from an arbitrary state A all others are accessible in at most M steps of the dynamics (34),

$$W^M(B|A) > 0 \qquad \text{for all } A, B \,, \tag{37}$$

with

$$W^M(B|A) \equiv \text{Tr}_{\{C_i\}} W(B|C_{M-1})\dots W(C_2|C_1)W(C_1|A) \,. \tag{38}$$

Turning to the problem at hand, we have to define the network states for a neural assembly with a broad distribution of transmission delays. These are given by all neuron states which influence the future dynamics, so that we may set

$$A(t) \equiv \{S_{ia}(t); \ 1 \le i \le N, \ 0 \le a \le (D-1)\} \,. \tag{39}$$

The $S_{ia}(t)$ corresponding to a global state $A(t)$ will henceforth also be written $S_{ia}(A(t))$. Since (39) corresponds to a system with size ND, there are 2ND different network states. We will frequently use the notation

$$A(t) = (A_0(t), A_1(t), \dots, A_{D-1}(t)) \quad , \quad A_a(t) \equiv \{S_{ia}(t); 1 \le i \le N\} \,. \tag{40}$$

During retrieval only the first component A_0 results directly from the dynamics (2), the other components, "the past", are shifted copies of their identical predecessors at earlier times, see Equations (26) and (27). The single-step transition probability thus takes the form

$$W(B(t+1)|A(t)) = W(B_0(t+1)|A(t)) \prod_{a=1}^{D-1} \delta\{B_a(t+1), A_{a-1}(t)\} \,, \tag{41}$$

with

$$\delta\{B_a(t+1), A_{a-1}(t)\} \equiv \prod_{i=1}^{N} \delta_{S_{ia}(B(t+1)), S_{i,a-1}(A(t))} \tag{42}$$

corresponding to the shift operation. Parallel dynamics implies independent transitions for all neurons so that the probability of $B_0(t+1)$ given $A(t)$ can be read off from (2),

$$
\begin{aligned}
W\big(B_0(t+1)|A(t)\big) &= \prod_{i=1}^{N} \frac{1}{2}\{1 + \tanh[\beta S_i(B_0)h_{i0}(A_0\ldots A_{D-1})]\} \\
&= \frac{\exp[\beta\sum_i S_i(B_0)h_{i0}(A_0\ldots A_{D-1})]}{\prod_i 2\cosh[\beta h_{i0}(A_0\ldots A_{D-1})]} .
\end{aligned}
\tag{43}
$$

In section 5, couplings with delay $\tau = D - 1$ were explicitly set to zero — see (9). Consequently, h_{i0} does *not* depend on A_{D-1}. We may therefore rewrite (28) and obtain for the case of an arbitrary state $X \equiv (X_0 X_1 \ldots X_{D-1})$

$$H(X_0 X_1 \ldots X_{D-1}) = H_R(X_1 \ldots X_{D-1}) - \sum_{i=1}^{N} S_i(X_0)h_{i0}(X_1 \ldots X_{D-1}) , \tag{44}$$

where neither H_R nor h_{i0} depend on $S_i(X_0)$. Summing over all substates X_0 we get

$$\mathrm{Tr}_{X_0} \exp[-\beta H(X)] = \exp[-\beta H_R(X_1 \ldots X_{D-1})] \prod_{i=1}^{N} 2\cosh[\beta h_{i0}(X_1 \ldots X_{D-1})] . \tag{45}$$

Inserting the last two identities with $X = (B_0 A_0 \ldots A_{D-2})$ into (43) we obtain

$$W\big(B_0(t+1)|A(t)\big) = \frac{\exp[-\beta H(B_0 A_0 \ldots A_{D-2})]}{\mathrm{Tr}_{B_0} \exp[-\beta H(B_0 A_0 \ldots A_{D-2})]} . \tag{46}$$

By (46), the transition probability (41) explicitly satisfies the normalization (35). In passing we note that due to (17), H is constant under cyclic permutations,

$$H(X_0 \ldots X_{D-2} X_{D-1}) = H(X_{D-1} X_0 \ldots X_{D-2}) . \tag{47}$$

We are now prepared to prove that the equilibrium distribution is a Gibbs distribution generated by H,

$$\rho_{eq}(A) = Z^{-1} \exp[-\beta H(A)] , \tag{48}$$

with

$$Z \equiv \mathrm{Tr}_A \exp[-\beta H(A)] . \tag{49}$$

First, we notice that the Glauber dynamics (2) assures strictly positive probabilities at the single neuron level for all nonzero temperatures β^{-1}. According to (41) and (43), this allows for transitions between arbitrary states A and B after at most D steps, and ergodicity is thus established with $M = D$ in (37).

Second, using the expression (48) for $\rho_{eq}(A)$, and inserting (41), (42) and (48) into the right-hand side of the equilibrium condition (36), we get

$$Z \cdot \text{Tr}_A W(B|A)\rho_{eq}(A) =$$

$$= \text{Tr}_{A_0} \ldots \text{Tr}_{A_{D-1}} \frac{\exp[-\beta H(B_0 A_0 \ldots A_{D-2})]}{\text{Tr}_{B_0} \exp[-\beta H(B_0 A_0 \ldots A_{D-2})]} \prod_{a=1}^{D-1} \delta\{B_a, A_{a-1}\} \exp[-\beta H(A_0 \ldots A_{D-1})]$$

$$= \prod_{a=0}^{D-2} \text{Tr}_{A_a} \exp[-\beta H(B_0 A_0 \ldots A_{D-2})]\delta\{B_{a+1}, A_a\} \frac{\text{Tr}_{A_{D-1}} \exp[-\beta H(A_0 \ldots A_{D-2} A_{D-1})]}{\text{Tr}_{B_0} \exp[-\beta H(B_0 A_0 \ldots A_{D-2})]}$$

The last factor is unity due to the cyclic invariance (47). We are left with a trivial summation and obtain

$$\text{Tr}_A W(B|A)\rho_{eq}(A) = Z^{-1} exp[-\beta H(B_0 B_1 \ldots B_{D-1})] = \rho_{eq}(B) .$$

The stochastic dynamics therefore evolves towards a Gibbs distribution generated by the Lyapunov functional H, which justifies the application of techniques originally developed for the study of neural networks with instantaneous interactions [19-22].

Stepping back for a moment, let us point out that in proving (48) we did *not* use the principle of detailed balance,

$$W(A|B)F(B) = W(B|A)F(A) , \tag{50}$$

where F is some function of the state variables. As known from the theory of dynamical systems [18], the stochastic process (34) has to converge towards an equilibrium distribution $\rho_{eq}(A) = F(A)$ if the relation (50) is satisfied for all pairs of states. Showing that this equation holds, with $\rho_{eq}(A)$ given by (48), is therefore a common method to establish the applicability of equilibrium statistical mechanics. But detailed balance is a sufficient condition, stronger than what is actually needed, and is violated in our case. In general, $W(A|B)$ vanishes for nonzero $W(B|A)$, due to the memory effect visible in (41). For a detailed discussion see [15].

Finally, let us once again discuss the case without delays, i.e. $D = 2$. The Hamiltonian (31) leads to an equilibrium distribution (48) with

$$Z = \text{Tr}_A \exp[\beta \sum_{i,j=1}^{N} J_{ij} S_{i0}(A) S_{j1}(A)] , \tag{51}$$

and we obtained a description of the Little model in terms of "duplicated" spins [9]. The connection to an expression derived by Peretto [17] — compare also with (32) — is easily achieved if we use the reduced probability distribution,

$$\tilde{\rho}_{eq}(A_1) \equiv \text{Tr}_{A_0} \rho_{eq}(A_0 A_1) = Z^{-1} \exp[-\beta \tilde{H}(A_1)] , \tag{52}$$

with

$$\tilde{H}(A_1) \equiv -\beta^{-1} \ln\{\text{Tr}_{A_0} \exp[-\beta H(A)]\} = -\beta^{-1} \sum_{i=1}^{N} \ln[2 \cosh(\beta \sum_{j=1}^{N} J_{ij} S_j(A_1))] . \tag{53}$$

Having presented a rather extensive introduction to the fundamentals of our approach, we will in what follows just give an outline of the quantitative analysis. Details can be found elsewhere [15].

8. Finitely many cycles

We study networks with a finite number of stored cycles as $N \to \infty$, and use a general formalism introduced to the field of neural networks by van Hemmen and Kühn [19], based on the theory of large deviations. We define "partial" overlaps $m_a^{\mu\alpha}$,

$$m_a^{\mu\alpha} \equiv \frac{1}{N} \sum_i \xi_{ia}^{\mu\alpha} S_{ia} , \tag{54}$$

which measure the similarity between a network state $\mathbf{S}$ and the pattern $\xi^{\mu\alpha}$ in a specific column a. In terms of these macroscopic order parameters, the Hamiltonian H may be written

$$-\beta H_{ND}(\mathbf{m}) \equiv ND \cdot F(\mathbf{m}) , \tag{55}$$

with

$$F(\mathbf{m}) \equiv \frac{\beta}{2D} \sum_{a,b} \varepsilon(a,b) \sum_{\mu,\alpha} m_a^{\mu\alpha} m_b^{\mu\alpha} . \tag{56}$$

The free energy density f,

$$f = -\beta^{-1} \lim_{N \to \infty} (ND)^{-1} \ln[\mathrm{Tr}\,\exp(-\beta H_{ND})] , \tag{57}$$

is at its saddle-points given by

$$f = -\beta^{-1} \{ F(\mathbf{m}) - \mathbf{m} \cdot \partial F(\mathbf{m}) + c[\partial F(\mathbf{m})] \} - \beta^{-1} \ln 2 , \tag{58}$$

where $c(\mathbf{t})$ denotes the cumulant generating function of the random variables $NDm_a^{\mu\alpha}$ given by

$$c(\mathbf{t}) = \frac{1}{D} \sum_a \left\langle \ln \cosh\left[D \sum_{\mu,\alpha} t_a^{\mu\alpha} \xi_a^{\mu\alpha}\right] \right\rangle . \tag{59}$$

Here $\langle \ldots \rangle$ represents an average over the random patterns. We thus obtain

$$f = \frac{1}{2D} \sum_{a,b} \varepsilon(a,b) \sum_{\mu,\alpha} m_a^{\mu\alpha} m_b^{\mu\alpha} - \frac{1}{\beta D} \sum_a \left\langle \ln 2\cosh\left[\beta \sum_b \varepsilon(a,b) \sum_{\mu,\alpha} m_b^{\mu\alpha} \xi_a^{\mu\alpha}\right] \right\rangle. \tag{60}$$

From the general fixed-point equations for the partial overlaps,

$$\mathbf{m} = (\partial c) [\partial F(\mathbf{m})] , \tag{61}$$

we get

$$m_a^{\mu\alpha} = \left\langle \xi_a^{\mu\alpha} \tanh\left[\beta \sum_b \varepsilon(a,b) \sum_{\nu,\gamma} m_b^{\nu\gamma} \xi_a^{\nu\gamma}\right] \right\rangle . \tag{62}$$

The local stability of a specific solution is guaranteed if

$$\lambda_{\min}(\mathbb{1} - A(\mathbf{m})B(\mathbf{m})) > 0 , \tag{63}$$

with

$$A_{a,b}^{\mu\alpha,\nu\beta}(\mathbf{m}) = D\delta_{ab} \left\langle \xi_a^{\mu\alpha} \xi_a^{\nu\beta} \cosh^{-2}\left[\beta \sum_d \varepsilon(a,d) \sum_{\tau,\delta} m_d^{\tau\delta} \xi_a^{\tau\delta}\right] \right\rangle , \tag{64}$$

and

$$B_{a,b}^{\mu\alpha,\nu\beta} = \beta D^{-1} \mathcal{E}_{ab} \delta_{\mu\nu} \delta_{\alpha\beta} \,. \tag{65}$$

The preceding equations are formally rather similar to those of the Hopfield model [19]. In our case, however, they contain averages over the nontrivially *correlated* patterns $\xi^{\mu\alpha}$, describing complete cycles of temporal associations. This problem may be solved if we switch back to the individual *uncorrelated* patterns ψ_p^μ [15].

For the rest of this section, we assume a "uniform" distribution,

$$\mathcal{E}_{ab} = (D-1)^{-1}(1 - \delta_{ab}) \,, \tag{66}$$

focus on cyclic associations with $n_\mu \Delta_\mu = D$, and study the case $D \gg 1$. Neglecting corrections of order D^{-1}, see however [15], we arrive at the following results. Above a certain critical temperature T_c,

$$T_c = \max_\mu(\Delta_\mu) \equiv \Delta_{\max} \,, \tag{67}$$

there exists only the paramagnetic state, $\mathbf{m} = 0$. Below the critical temperature T_c there are two different cases.

If $\Delta_{\max} = 1$, all cycles and copies are δ-correlated, $< \xi_a^{\mu\alpha} \xi_a^{\nu\beta} >= \delta_{\mu\nu} \delta_{\alpha\beta}$, so that all PD stored patterns can be treated on equal footing. For *retrieval* states, one may introduce the "total" overlaps $\hat{m}^{\mu\alpha} \equiv D^{-1} \sum_a m_a^{\mu\alpha}$. Both the fixed-point equations for $\hat{m}$ and the stability conditions become identical with their counterparts for the Hopfield model and there is no formal difference between "intercyclic mixtures" — coherent oscillations of different sequences, "intracyclic mixtures" — combinations of shifted copies of the same cycle, and mixtures of both kinds. The analysis of Amit et al. [20] can be taken over and it shows in particular that only true retrieval states of the form $\hat{m}^{\mu\alpha} = m\delta_{\mu 1}\delta_{\alpha 0}$ are stable at all $T < T_c$.

The scenario is changed however, if we consider $\Delta_{\max} > 1$. The first nontrivial stable solutions branch off from the paramagnetic state in the direction of a static n_ν-symmetric intracyclic mixture state of a cycle ν with $\Delta_\nu = \Delta_{\max}$, i.e. $m_a^{\mu\alpha} = \delta_{\mu\nu} m$, indicating the recognition of this specific *cycle* without discerning its individual *patterns*. Symmetric mixtures of other cycles μ appear as soon as the temperature T falls below the corresponding duration Δ_μ, and we obtain an ordering of these static solutions. In the case usually studied, the cycles are degenerate with respect to Δ and symmetric intercyclic mixtures appear too [15].

The fixed-point equation for a n_μ-symmetric intracyclic mixture is identical to that for a n_μ-symmetric state of the Hopfield model, but the stability conditions are changed. The analysis reveals that stability depends on three eigenvalues only. First, we have to consider a single eigenvalue expressing the stability against a uniform increase in the amplitude m,

$$\lambda_1 = 1 - \beta \Delta_\mu [1 - q_{EA} + (1 - n_\mu)Q] \tag{68}$$

where q_{EA} denotes the corresponding Edwards-Anderson order-parameter,

$$q_{EA} = \langle \tanh^2[\beta \Delta_\mu m \sum_{p=1}^{n_\mu} \psi_p^\mu] \rangle \,, \tag{69}$$

and Q is defined by

$$Q \equiv -\left\langle \psi_1^\mu \psi_2^\mu \cosh^{-2}\left[\beta\Delta_\mu m \sum_{p=1}^{n_\mu} \psi_p^\mu\right]\right\rangle . \tag{70}$$

Next there is an eigenvalue whose degeneracy is equal to the number of (other) cycles whose single patterns last for $\Delta_{\max}$ time steps. It describes the stability against the appearance of solutions correlated with these cycles,

$$\lambda_2 = 1 - \beta\Delta_\mu[1 - q_{EA}]\frac{\Delta_{\max}}{\Delta_\mu} . \tag{71}$$

Finally, we get a two-fold degenerate eigenvalue which measures the stability against oscillations with period D — i.e. stability against the onset of the true retrieval of the stored cycle,

$$\lambda_3 = 1 - \beta\Delta_\mu[1 - q_{EA} + Q] \cdot Z_{D,n_\mu} , \tag{72}$$

with

$$Z_{D,n_\mu} \equiv \Delta_1^{-2}\frac{1 - \cos(\frac{2\pi}{n_\mu})}{1 - \cos(\frac{2\pi}{D})} . \tag{73}$$

All three eigenvalues correspond closely to their counterparts determining the stability of a n_μ-symmetric mixture state in the Hopfield model — up to the additional factors $\Delta_{\max}/\Delta_\mu$ and Z_{D,n_μ} for λ_2 and λ_3 respectively. But those change the picture completely: Directly below the critical temperature, the first eigenvalue is positive as in the Hopfield model. However, λ_2 is negative unless $\Delta_\mu = \Delta_{\max}$ — so only the first solution to appear below T_c is locally stable in terms of thermodynamics. Finally, the last eigenvalue is positive for all $n_\mu < D$. In other words, for $\Delta_\mu = \Delta_{\max} \neq 1$ the n_μ-symmetric intracyclic mixture is stable near T_c for arbitrary $n_\mu < D$ — contrary to the case with $\Delta_{\max} = 1$. At very low temperatures, $q_{EA} = 1$ and $Q = 0$ for odd n_μ (up to exponentially small corrections), and all three eigenvalues equal unity. For even n_μ, however, both λ_2 and λ_3 are proportional to $-\beta$. These results indicate that static mixture states of cycles with $\Delta_\mu = \Delta_{\max}$ and an odd number of patterns are locally stable at all temperatures; the even-n_μ ones loose their stability below a certain temperature, given by $\lambda_3 = 0$, as could be verified by numerical simulations for the dynamics of finite networks with $\varepsilon(\tau) = D^{-1}$, corresponding to the limit $D \gg 1$ considered above [15].

In closing we would like to mention that for a network storing static patterns only, i.e. $\Delta = D$, retrieval states are stable by the above analysis — in accordance with results on the dynamics of analog networks with a single delay [23]. See also [12].

9. Extensively many patterns and storage capacity

In the present section, we extend the analysis of the static Little model by Fontanari and Köberle [10], and study our model at finite storage level $\alpha = \lim_{N\to\infty}(P/N) > 0$. As in the replica-symmetric theory of Amit et al. [21], we single out finitely many, say s, cycles of the original system and assume that the network is in a state highly correlated with these spatio-temporal objects. The remaining, extensively many cycles are described as a noise term in the context of the replica method. For the equivalent network of size ND, we thus concentrate on sD static patterns — the original s cycles and their shifted copies — and treat the other as "uncondensed" [21] patterns. For an alternative approach along the lines developed by Gardner [24] see the contribution of Bauer [25].

The general analysis is technically rather involved and can be found elsewhere [15]. In what follows, we focus on some results for the case $\Delta = 1$. The relevant order parameters for a n-fold replicated network, $\mathbf{m}$ and $\mathbf{q}$, are defined as

$$m_{\sigma a}^{\mu \alpha} \equiv N^{-1} \sum_i \xi_{ia}^{\mu \alpha} S_{ia}^{\sigma} , \qquad 1 \leq \sigma \leq n , \tag{74}$$

and

$$q_{ab}^{\rho \sigma} \equiv N^{-1} \sum_i S_{ia}^{\rho} S_{ib}^{\sigma} , \qquad 1 \leq \sigma, \rho \leq n . \tag{75}$$

Let us consider retrieval solutions for cycles with $\Delta = 1$. The replica-symmetric ansatz

$$m_{\sigma a}^{\mu \alpha'} = m^{\mu} \delta_{\alpha',0} \tag{76}$$

and

$$q_{ab}^{\rho \sigma} = \delta_{ab} [\delta_{\rho \sigma} (1 - q) + q] , \tag{77}$$

leads to

$$m_\mu = \left\langle\!\!\left\langle \xi^{\mu 0} \tanh[\beta \{ \sum_{\mu'} m_{\mu'} \xi^{\mu' 0} + \sqrt{\alpha r} z \}] \right\rangle\!\!\right\rangle \tag{78}$$

and

$$q = \left\langle\!\!\left\langle \tanh^2[\beta \{ \sum_{\mu'} m_{\mu'} \xi^{\mu' 0} + \sqrt{\alpha r} z \}] \right\rangle\!\!\right\rangle , \tag{79}$$

with

$$r \equiv q \sum_{k=1}^{D} \frac{[\lambda_k(\mathcal{E})]^2}{[1 - \beta(1 - q)\lambda_k(\mathcal{E})]^2} . \tag{80}$$

The double angular brackets represent an average with respect to both the condensed cycles and the normalized Gaussian random variable z. The $\lambda_k(\mathcal{E})$ are eigenvalues of the weight matrix $\mathcal{E}$. At the fixed-points given by (78) - (80), the free energy is

$$-\beta f(\beta) = -\frac{\beta}{2} \sum_\mu (m_\mu)^2 - \frac{1}{2} \alpha \sum_k \{ \ln[1 - \beta(1 - q)\lambda_k(\mathcal{E})] - \frac{q\beta\lambda_k(\mathcal{E})}{1 - \beta(1 - q)\lambda_k(\mathcal{E})} \}$$

$$+ \frac{1}{2} \alpha \beta^2 (q - 1) r + \left\langle\!\!\left\langle \ln 2 \cosh[\beta \{ \sum_\mu m_\mu \xi^{\mu 0} + \sqrt{\alpha r} z \}] \right\rangle\!\!\right\rangle . \tag{81}$$

The preceding equations closely resemble their counterparts for the Hopfield model [22] and become identical with them for a uniform distribution (66) in the limit $D \to \infty$. For the same distribution but finite D, there is at zero temperature a first-order phase transition between the retrieval state and a spin-glass phase as in the Hopfield case. The critical storage level α_c, however, and the corresponding overlap m_c are changed.

Notice that each cycle consists of D independent patterns so that the storage capacity for *single patterns* is $\tilde{\alpha}_c = D\alpha_c$. During the recognition process, however, each of them will trigger the cycle it belongs to, and cannot be retrieved as a static pattern.

As a second measure for the network performance, we include the information content $I_R(\mathcal{E})$, measured *per synapse* and relative to that of the Hopfield (or Little) model,

$$I_R(\mathcal{E}) \equiv \frac{I(\mathcal{E})}{I(Hopfield)} = \frac{D \cdot \alpha_c(\mathcal{E})}{d \cdot \alpha_c(Hopfield)}, \tag{82}$$

where d denotes the number of different delay lines between each pair of neurons, $1 \leq d \leq D - 1$. A numerical solution of the saddle-point equations at $T = 0$ leads to the following table for networks with a uniform weight distribution:

D	α_c	m_c	I_R	
2	0.100	0.93	1.45	
3	0.110	0.95	1.20	(I)
4	0.116	0.96	1.12	
5	0.120	0.96	1.09	

These results agree well with simulations on large finite systems as shown by two examples. For $D = 3$, we found $\alpha_r = 0.120 \pm 0.015$, and for $D = 4$, $\alpha_r = 0.125 \pm 0.015$.

The influence of the weight distribution may be demonstrated by some choices of $\varepsilon(\tau)$ for $D = 4$:

τ	$=$	0	1	2	3	α_c	m_c	I_R	
$\varepsilon(\tau)$	$=$	1/3	1/3	1/3	0	0.116	0.96	1.12	
$\varepsilon(\tau)$	$=$	1/2	0	1/2	0	0.100	0.93	1.45	(II)
$\varepsilon(\tau)$	$=$	0	1	0	0	0.050	0.93	1.45	

The uniform distribution leads to the largest α_c but smallest I_R. The other two networks have the same value of I_R as the (unique) $D=2$ – system due to the particular structure of their eigenvalue spectrum. Furthermore, it turns out that we get $I_R = 1.45$ independently of D for all networks with a "minimal connectivity", where only one synapse links two neurons. However, these single-delay networks have lower associative capabilities, as discussed in [8]. For systems with a uniform distribution, I_R approaches 1 for $D \to \infty$. Simulation data show once again slightly higher values — possibly indicating effects of replica symmetry breaking as known from the Hopfield model [21].

10. Conclusions and Outlook

The present study has been motivated by the desire to find a joint representation for static patterns and temporal sequences in neural networks — and keeping analytic tractability at the same time. It has been shown that this is indeed possible for certain network architectures and learning paradigms. We introduced a Lyapunov functional for deterministic dynamics in neural networks with delayed interactions and could successfully develop the statistical mechanics for cyclic associations.

Within this article we could not touch upon many related topics — nonsymmetric mixture states, the role of external fields, and extensions to low activity, to name only a few — and there remain open questions as well. The present approach is based on parallel dynamics, and one would like to apply it to the sequential case too. However, the corresponding Lyapunov functional is not known yet, but we do hope that our study will also initiate some research towards the understanding of this kind of neural network.

Acknowledgements

Not aware of each others complementary work, both authors met at Caltech in December 1989. This event triggered the present study. We are therefore grateful to John Hopfield for his hospitality and to the Studienstiftung des Deutschen Volkes for a generous travel grant. The present paper is a first report about research that has been done meanwhile — in collaboration with Wulfram Gerstner and Leo van Hemmen, and based on the previous studies with Reimer Kühn and Bernhard Sulzer. It is a great pleasure to thank all of them for many stimulating and helpful discussions, and, in particular Leo van Hemmen and Reimer Kühn for critically reading the manuscript.

References

1. W.A. Little, Math. Biosci. **19**, 101-120 (1974)
2. J.J. Hopfield, Proc. Natl. Acad. Sci. USA **79**, 2554-2559 (1982)
3. V. Braitenberg, in: *Lecture Notes in Biomathematics*, M. Conrad, W. Güttinger, and M. Dal Cin (eds.), (Springer, Berlin, Heidelberg, New York, 1974) pp 290-298
4. D. Kleinfeld, Proc. Natl. Acad. Sci. USA **83**, 9469-9473 (1986)
5. H. Sompolinsky and I. Kanter, Phys. Rev. Lett. **57**, 2861-2864 (1986)
6. D.W. Tank and J.J. Hopfield, Proc. Natl. Acad. Sci. USA **84**, 1896-1900 (1987)
7. D.O. Hebb, *The Organization of Behavior* (Wiley, New York, 1949) p 62
8. A.V.M. Herz, B. Sulzer, R. Kühn, and J.L.van Hemmen, Europhys. Lett. **7**, 663-669 (1988); Biol. Cybern. **60**, 457-467 (1989)
9. J.L. van Hemmen, Phys. Rev. **A 34**, 3435-3445 (1986)
10. J.F. Fontanari and R. Köberle, Phys. Rev. **A 36**, 2475-2477 (1987); J. Phys. France **49**, 13-23 (1988); J.Phys.A: Math.Gen. **21**, L667-L671 (1988)
11. A.C.C. Coolen and C.C.A.M. Gielen, Europhys. Lett. **7**, 281-285 (1988)
12. M. Kerszberg and A. Zippelius, to appear in Physica Scripta (1990)
13. A.V.M. Herz, in: *Connectionism in Perspective*, R. Pfeiffer, Z. Schreter, F. Fogelman-Soulié, and L. Steels (eds.), (North-Holland, Amsterdam, 1989)
14. W. Gerstner, J.L. van Hemmen, and A.V.M. Herz, preprint TU München (1990)
15. A.V.M. Herz, PhD thesis (Heidelberg, 1990); A.V.M. Herz, Z. Li, W. Gerstner and J.L. van Hemmen, to be published
16. Z. Li, unpublished (1987)
17. P. Peretto, Biol. Cybern. **50**, 51-62 (1984)
18. K. Binder, in *Monte Carlo Methods in Statistical Physics*, K. Binder (ed.), (Springer, Berlin, 1979) pp 1-45
19. J.L. van Hemmen and R. Kühn, Phys. Rev. Lett. **57**, 913-916 (1986); see also: J.L. van Hemmen, Phys. Rev. Lett. **49**, 409 (1982)
20. D.J. Amit, H. Gutfreund, and H. Sompolinsky, Phys. Rev. **A 32**, 1007-1018 (1985)
21. D.J. Amit, H. Gutfreund, and H.Sompolinsky, Phys.Rev.Lett. **55**, 1530-1533 (1985)
22. J.L. van Hemmen and V.A. Zagrebnov, J.Phys.A: Math.Gen. **20**, 3989-3999 (1987)
23. C.M. Marcus and R.M. Westervelt, Phys. Rev. **A 39**, 347 (1989)
24. E. Gardner, J. Phys. A: Math. Gen. **21**, 257-270 (1988)
25. K.Bauer, private communication, these proceedings, and to be published

Semi-local Signal Processing in the Visual System

André J. Noest

Dept. Medical and Physiological Physics, Utrecht University
Princetonplein 5, NL-3584-CC Utrecht, The Netherlands

1 Introduction

Studying neural networks from the viewpoint of statistical physics has certainly led to progress in our understanding of some instructive toy-problems. For example, there is now a reasonable comprehension of the dynamical behaviour of long-range associative pattern reconstruction networks with fixed Hebb-style synapses [1,2]. Simple models of learning are also being analysed succesfully [3]. One may even entertain the hope that our present theoretical machinery can be generalised so as to apply to much more natural and complex networks than have been analysed sofar. Nevertheless, the sobering fact remains that we are still quite far from linking our models to psychophysical or physiological experiments, as well as to the independently developed theories pertaining to specific areas of brain function such as vision, hearing, or locomotion.

In this paper, I will suggest one of the directions in which one could work towards establishing such a link. The direction I choose is determined by the (far-away) goal of understanding the way in which visual information is processed in the brains of, say, primates. Let me stress from the outset that the size of my step in this direction is tiny. I will first discuss some basic aspects of the geometric structure inherent in visual data, as well as the related gross architecture of networks in the visual system. Then I sketch how the local structure of the data seems to be extracted in the initial stages of processing. Finally, I present a network model (still very much a toy) which illustrates how global patterns made up of textured patches can be reconstructed even if the network connectivity is merely semi-local. In fact, restrictions on the range of the connections are *necessary* if one wants to obtain such functional behaviour at all. At the same time, one avoids the 'wiring explosion' that occurs when long-range networks are scaled up. Thus, the model presented below is at least compatible with the basic restrictions of a biological or electronic implementation.

2 Geometric structure in data and networks

At the spatiotemporal scales relevant to an animal's survival, the world contains a considerable amount of geometric structure. Causality and cohesion forces tend to produce correlations between physical properties found at neighbouring locations. The primary function of the visual system is to convey sufficient information about this structure, so as to allow perception of the world as a collection of (possibly nested) coherent objects. However, some of the structure is lost, hidden, or disturbed by noise during the image formation, the detection or the encoding of visual signals. Thus, the neural machinery of the visual system must extract every piece of structure remaining in the visual signals, filter out the incoherent components, fill in any obliterated parts, and segment the data so as to obtain a reconstructed set of coherent objects. These requirements suggest that the geometric structure of the data must be related to the kinds of operations that occur in visual system networks. If, in addition, the brain must implement these operations with minimal wiring, then one expects that the abstract structure in the data and the operations should map onto the anatomy of the neural networks in a smooth way. This is often found experimentally [4,5].

In order to set the stage for the more quantitative discussion that follows, it seems appropriate to sketch a (brutally simplified) picture of the functional architecture of the initial stages of the visual system [4].

At the coarsest level of discrimination, one can distinguish a set of processing 'streams', dealing more or less separately with attributes defining form, colour, motion, etc. Their distinctions can be seen from very early stages (retina and LGN) onwards, and although there are some mutual interactions along the way, their main output signals are fused again only after several stages of attribute-specific extraction of local structure, as well as its integration into more globally coherent signals.

Each stage of processing tends to occur in a layer, with half a dozen or so layers superimposed on each other in large contiguous areas of the visual cortex. On a coarse scale, these areas tend to be smooth maps of the visual field. Much wiring is saved by such a relation between the anatomy and the fundamental requirement of having interactions between neurons receiving signals that contain correlations induced by the geometric structure of the visual world. On a finer scale one finds functional substructure along the lateral dimensions of areas (and their layers). Each substructure is a module ('column') in which one finds neurons that receive data from roughly the same local patch of visual field.

In discussing the functional architecture of the networks at the level of areas and smaller, it is then useful to distinguish:

- Feedforward connections, mapping local data in one layer onto neurons in the next layer (possibly in another area). These are responsible for the 'receptive fields' of neurons, which I will discuss in the next section.

- Within-column connections. These specify the local networks that can compute quantities like local isophote curvature, etc. Again, see the next section for more details.

- Lateral connections between neurons in neighbouring columns. These provide the possibility of linking up locally measured data so as to fill in gaps, or segment the scene into a collection of coherent parts. In Section 4, I define and analyse a rudimentary model of such processes.

- Feedback connections to previous layers. Their structure and rôle are still very enigmatic, and I simply ignore them for now.

3 Receptive fields and data representations

Limiting the discussion to the processing of (quasi-)static images defined only by luminance, one can roughly say that the initial layers (retina and LGN) provide the visual cortex with a smooth map of the visual image, sampled densely by neurons whose activity encodes local luminance with a compressed dynamic range, and with reduced sensitivity to components having low spatial frequencies. The next few layers (in area V1) analyse the local luminance structure in this 'image' $I(z)$. For simplicity, I shall often take the spatial coordinates $z = (x, y)$ to be continuous.

   The connections that feed the image data forward to V1 define the kernels $R(z)$ of convolutions $O(z) = R * I(z)$. These linear functionals then form the input to V1 neurons. The kernels are local, in the sense of having fast-decaying tails, and one finds specific substructure in the central part of their supports. This structure is responsible for the experimentally well-studied 'receptive fields' of these cells. It is striking that most of the experimentally found receptive field structures [4] agree very well with what they should be according to an analysis which starts from simple and natural first principles [6,7].

   The basic assumptions are that the image contains useful structure over a wide range of scales, and that this structure is statistically isotropic. One also requires that the image gradually looses (and never gains) in fine-structure when a larger-scale version of the kernels is used. Then one finds that the kernels must belong to the family of Gaussians of all widths, and their derivatives. Nature seems content with a range of scales of about 30, with orders of differentiation of 1 to 4, and with a lack of mixed partial derivatives. Instead of mixed partials, it uses a quasi-continuous set of directional derivatives with all orientations. Note the virtual absence of the zeroth order. As a consequence, very smooth luminance structure cannot actually be seen, but must be reconstructed from the information at edges, allowing all manner of brightness illusions to occur. Similar effects occur in the temporal domain.

   If, say, x is the direction into which derivatives are taken, the receptive field kernels become

$$R_n(z, \sigma) = (\partial_x)^n \frac{1}{2\pi\sigma^2} \exp(-\frac{z^2}{2\sigma^2}).$$

Rotation of these kernels over all angles α generates the full family $R_n^\alpha(z, \sigma)$. One can also view the convolution with such kernels as a decomposition into Hermite functions of semi-local parts of the image, viewed through a Gaussian aperture. This becomes apparent if one writes the x-dependent factor of, say, $R_n(z, \sqrt{\frac{1}{2}})$ as

$$(\partial_x)^n \exp(-x^2) = (-1)^n \phi_n(x) A(x),$$

with 'aperture' $A(x) = \exp(-\frac{1}{2}x^2)$, and Hermite functions

$$\phi_n(x) = H_n(x) \exp(-\frac{1}{2}x^2) \ ; \ <\phi_n(x)\phi_m(x)> = 2^n n! \sqrt{\pi} \, \delta_{n,m}.$$

The orthogonality of the $\phi(x)$ functions is an advantage in many further analyses, whereas the interpretation of the convolution as measuring terms in a Taylor-expansion of the image around the center of the receptive field suggests at once that differential-geometric characteristics of local shape can be found from the local measurements. A simple example is the local curvature of an 'edge' (say, tangent to the x-axis), which is formally $\kappa = -\partial_x^2 I(z)/\partial_y I(z)$. The network version of this is obtained when one interprets the formal differentials as convolutions with similarly differentiated Gaussian kernels. Note that the network need not actually do a division (with an obvious problem when $\partial_y I = 0$). It is entirely in the spirit of neural computation to represent a quantity by an ordered tuple of lower-level quantities that suffice to define it.

In general, further processing in the next layers of the system uses the representation of the image as it appears on the output of the neurons whose receptive fields we have just described. One can consider the action of the whole set of receptive fields as mapping the scalar (luminance) image to a scalar (activity) distribution defined on a multidimensional support, being the product of the original $D = 2$ space and the local 'feature'-space spanned by the outputs of the set of fuzzy derivative operators. In turn, one may expect that this representation acts as a new 'image', to be sampled by the receptive fields of neurons in the next layer. Thus, the network may end up representing purely spatial structure in the input as geometric structure in a high-dimensional space of visual pattern attributes. Although theoretical as well as experimental results on these matters are as yet very scarce, one can at least identify one class of operations on these data-representations for which lateral interactions between columns would be needed. Such operations are modelled in the remainder of this paper.

4   Reconstruction of patchy patterns

In this section, I present a model that takes as its input the signals from an early preprocessing stage as sketched above. Thus, the semi-local image structure is represented by the activity of many neurons with different receptive fields, such as occur in each cortical module. Connectivity of the network is restricted to a range much smaller than the network diameter, but is still much larger than the distance between neighbouring neurons [8,9,10].

The task of the network is to reconstruct, fill-in, and segment the data into coherent patches. The task definition thus incorporates the assumption that realistic data can be viewed as a patchwork of textured areas, each of which is rather homogeneous. The set of texture types is very much smaller than the set of patchwork patterns that may be constructed out of them. The essential differences between the model presented here and long-range associative models are that my model locally 'knows' only the texture codes, and that it reconstructs a global patchwork pattern by finding a compromise between coarsening the patch structure and adhering to the data which comes in continuously. This allows an enormously large set of patchwork patterns to be reconstructed: capacity grows faster than any power of N, the network size. Several other benificial properties of the network will also emerge, such as the ability to get rid of mixture states by means of nucleation of pure phases.

In ref.9, I sketched the construction of the perhaps simplest model that can do this using interactions with restricted range, and I gave a brief overview of the rich phenomenology of domain behaviour in this model. Here, I set up a version with slightly less artificial assumptions, and I analyse in a bit more detail the domain-wall phenomena that are responsible for the functional properties of the model.

4.1 Construction of the model

4.1.1 Network connectivity

The architecture in these models can best be defined by identifying the neuron indices i with the position vectors of a D-dimensional unit-spaced lattice of sidelength L, with $D = 2$ as the most practically relevant case. Randomness in the positions of the neurons could be incorporated without serious trouble, but I avoid the distracting complexity that such a choice would entail.

In defining the order parameters, the encoding of the local texture, and the input of external data onto the network, we employ a modular element of structure that corresponds to the neurophysiological 'columns'. These modules can be defined most simply by covering the i-lattice with a coarse-grained (block) lattice of spacing $\rho \gg 1$.

Each neuron has Z incoming connections with other neurons, sampled at random from a spherically symmetric connectivity kernel G of effective radius $r \gg \rho$, and with tails that decay at least exponentially fast. I shall give explicit results for G kernels that are either a Gaussian, or a 'pillbox', i.e. uniform in a sphere. However, the qualitative behaviour of the model does not depend on this as long as $r \ll L$. Overall, we now have the hierarchy of spatial scales $1 \ll \rho \ll r \ll L$.

The ZN (ordered) i,j-pairs so selected from G are given the usual Hebb-style synaptic strength

$$T_{ij} = \sum_{k=1}^{P} \xi_i^k \xi_j^k,$$

in which the $\xi_i^k \in \{-1, +1\}$ are the neural states that represent the k-th of P textures. Note that the same texture may or may not have the same representation within different modules. This does not affect the derivation that follows, as long as there is (quasi-)orthogonality between the representations of different textures on the scale of the modules. One could generalize the model to allow sparse-coding of the textures, but I avoid this here for the sake of simplicity. The major unrealistic feature that results from this is the spin-inversion symmetry, which is broken in the sparse-coded version.

4.1.2 Neural dynamics and external data

There are $N \to \infty$ neurons $s_i \in \{+1, -1\}$, updated asynchronously according to

$$P[s_i(t + \delta t) = \pm 1] = \frac{1}{2}\{1 \pm F_b[h_i(t)]\}.$$

To get the most compact results, on should take $F_b(h) = \mathrm{erf}(h/\sqrt{2b})$, which can also be motivated by observing that it describes precisely the outcome of a threshold operation $\text{sgn}(h_i + g)$, where g is a zero mean Gaussian noise of variance b. Most other sigmoids will however produce qualitatively the same model behaviour.

The local fields have contributions from external input as well as lateral interactions:

$$h_i(t) = h_i^0(t) + Z^{-1} \sum_j T_{ij} s_j(t) \, .$$

The external inputs $h_i^0(t)$ are

$$h_i^0(t) = \eta_i(t) + \sum_{k=1}^{P} H_k(i) \xi_i^k.$$

This represents the patch-structured data to be processed by the system. Note that it contributes to the neural local fields persistently, not merely through the initial conditions as assumed in more classical models. The term $\eta_i(t)$ is a Gaussian noise with zero mean and variance c. The $H_k(i)$ are the weights with which the code for texture k occurs locally. I shall drop the k-index when the P-component vector of weights is meant. Other models [11,12] for the way in which the data resembles members of the texture set could be used with little effect on the results, but at a cost in added complexity.

The initial conditions are most naturally defined as $s_i(0) = \text{sgn}(h_i^0(0))$, and one is interested in how a set of order parameters (measuring similarity between network states and exemplars) will evolve under the influence of rather weak external fields, in which the pattern to be recovered is seriously affected by noise, i.e., $H_k^2 < c < 1$.

4.2 Order parameter evolution equations

The appropriate order parameters for describing the functional behaviour of the model should also be defined on the semi-local scale of the modules. Labelling the modules by position vectors l, and denoting the sets of ρ^D neurons within module l by $I(l)$, one defines a P-component overlap $m(l) = \{m_k(l)\}, k = 1, \ldots, P$, where

$$m_k(l) = \rho^{-D} \sum_{I(l)} \xi_i^k s_i \, .$$

In order to derive exact evolution equations, one has to take the $N \to \infty$ limit such that the various spatial scales in the network grow in proper relation. Both ρ and r must diverge when $L = N^{1/D} \to \infty$, but such that $\rho = o(r)$ and $r = o(L)$. For example, in ref.9 I took $r = L^a$ and $\rho = L^{a'}$, with $0 < a' < a < 1$.

In addition, the spatial structure in the patchy data must be coarse-grained to at least the scale of the modules: $H_k(i)$ must be essentially constant on distances of $O(\rho)$, but may vary arbitrarily over distances of $O(r)$. It is then useful to define rescaled spatial coordinates $x = i/r$, which will become continuous as $N \to \infty$.

Finally, the degree of connectivity Z and the size of the texture set P have to be fixed. Clearly, Z is limited by the number of neurons within range, which is of $O(r^D)$. In ref.9, I assumed very sparse connections $Z = o(\log N)$, in which case one may extend the formalism of ref.1 to the present model. For heavily loaded

$(P = O(Z))$ networks, this is the only connectivity regime in which one can compute the dynamics exactly. Larger Z lead to an unmanagable explosion of correlations due to feedback loops. Although the semi-local connectivity in the visual system is far from complete, it tends to be not so sparse as to prevent finite-size loops. On the other hand, the dimensionality of the set of textural features measured locally by the receptive fields seems to be only between 10 and 20. Thus, P/Z is very small in practice, and –as will appear later– it is of minor importance in determining the size of the set of global patterns that the network can reconstruct. This happy circumstance enables one to derive exact evolution equations for range-restricted networks with any Z. The most general way of doing this has been presented at this conference by Ton Coolen [10]. In this paper, I only need to apply a simplified version of his formalism, since the sampling kernel G that determines my connectivity structure is translation and rotation invariant. In addition, one can show that restricting the dynamics to states that have semi-local overlap components $m_k(x) = O(1)$ only for a finite set of k, while all others are of $O(\rho^{-D/2})$ allows one to do the derivation for moderate loadings $P - o[Z/\ln(Z)]$

In any of these cases, one obtains

$$\partial_t m_k(x) = -m_k(x) + \langle\langle \xi_i^k F_d \left[\sum_{k=1}^n \xi_i^k \{H_k(x) + G * m_k(x)\} \right] \rangle\rangle, \qquad (1)$$

where $*$ denotes a convolution, and $\langle\!\langle \rangle\!\rangle$ is an average over the n 'active' patterns $\{\xi_i^k\}, k = 1, \ldots, n; n \leq P$, i.e. those that may have $m_k = O(1)$. The width parameter $d = b + c$ of the sigmoid can be seen as incorporating the effects of noise from the stochastic neural dynamics and from the external data. The rather simple form of eq.1 is due to the choice of an error function for F_d ; other sigmoids produce a more complicated expression, but do not generally affect the qualitative behaviour of the model. Note that eq.1, like its simpler counterpart for the classical long-range models, is a deterministic evolution equation. If, as in reality, one can have very large but finite N, L, r, ρ, and Z, then fluctuations have to be taken into account.

It is worth noting that eq.1 differs from the evolution equation of the corresponding infinite-range network only by the occurrence of the convolution $G * m_k(x)$ instead of m_k. Thus, the 'flat' (x-independent) solutions of eq.1 are simply the same as in the infinite-range model. These have been rather widely studied [13].

In summary, $2P$ pure (say k=1) stable solutions $m_1 = M \equiv F_d(M)$ appear as $d < d^*$ with $H = 0$. In the regime of practical importance ($d < 0.1$), these states are exponentially close to the exemplar patterns: $1 - M \approx \sqrt{2d/\pi} \exp[-1/(2d)]$. In addition, mixture states ($n \geq 3$) of increasingly complex composition become stable as $d \to 0$. Both the pure and the mixed states are also stable under finite H perturbations, as long as these are uniformly small. The phenomena that are most characteristic of restricted range models all involve solutions which are spatially inhomogeneous. It is this set of solutions that we study in the remainder of this paper.

4.2.1 Relation with cubic spin models

In trying to understand the behaviour of eq.1, it is helpful first to study the critical regime $d \approx d^* = 2/\pi$, although the important regime for applications is $d < 0.1$, say. As always, the critical point d^* is where the trivial solution $m(x) = 0$ first looses its stability. It is easy to show that, for small H, the solutions then quickly relax to small amplitudes $|m(x)| \ll 1$, and smooth spatial structures. This enables one to approximate the convolution as $G * m(x) \approx m(x) + \gamma \nabla^2 m(x)$, and the sigmoid as $F_d(h) \approx uh - vh^3$, with $u = 1 + (d^* - d)/4$ and $v = \pi/4$ for the favoured choice of $F_d(h) = \mathrm{erf}(h/\sqrt{2d})$. Thus, for $d \to d^*$ and $|H(x)| \ll |m(x)|$, one gets: $\partial_t m_k(x) =$

$$= uH_k(x) + u\gamma\nabla^2 m_k(x) + (u - 1)m_k(x) - 3vm_k(x)\sum_k m_k^2(x) + 2vm_k^3(x). \quad (2)$$

This is a dynamic Landau-Ginzburg equation for an n-vector spin model with axial cubic symmetry breaking and an inhomogeneous bias field. In many cases, we want to study solutions that are either time-independent, or that evolve very slowly. For small H, such solutions of eq.2 can be constructed with relative ease in the case $D = 1$. For $D > 1$, symmetry reduction methods [14] can be used. I only discuss the simplest $D = 1$ solutions of eq.2, as a way of 'warming-up' for the analysis of the practically important regime.

First put $H = 0$. Then $m(x) = 0$ is the only stable solution for $d > d^*$, and it becomes unstable for $d < d^*$. In the latter regime, the cubic term in eq.2 which breaks the continuous rotation invariance in the k-space of patterns, forces the solutions (locally) towards one of the axes of this space. All of these axial directions are equivalent because of the k-permutation invariance of eq.1. The corresponding modes become unstable on all spatial scales larger than a critical wavelength which has been computed by Ton Coolen in his contribution to this conference [10]. After the initial fast growth phase of the modes (in $O(1)$ time), the system will enter a phase of quasi-stationary behaviour. For the simplest case of global overlap on only a single pattern (say with $k = 1$), the quasi-stationary solutions have local structures that are easily obtained in closed form. These come in two types:

- 'Flat' solutions: $m_k(x) = m_k = \pm\delta_{k,1}\sqrt{(u - 1)/v}$.

- 'Kinks': $m_k(x) = \pm\delta_{k,1}\sqrt{(u - 1)/v}\,\tanh[x\sqrt{(u - 1)/(2u\gamma)}]$.

Note that in $D > 1$, the x now signifies only the coordinates normal to the (planar) kink. The position of the kink along the x-coordinate is arbitrary because of translation invariance of eqs.1 and 2. Accordingly, the kink has a neutrally stable (Goldstone) mode, $\nabla m(x)$, which corresponds to infinitesimal translations of the kink. Since an external field component H_1 couples to this mode, the kinks will move under the influence of such a bias. The same goes for the mutual effect of two adjacent kinks. Since such kinks can only be of opposite sign, they attract and eventually annihilate each other. The exponential tails of the kinks cause this collaps to take an exponentially long time when the initial separation is large compared to the effective kink-width $W = \sqrt{2u\gamma/(u - 1)}$. More complicated domain effects will be discussed under the more realistic assumption that the model is well below its critical point.

4.3 Domains and Walls at low d

In view of the intended application of segmenting textured patches, the most interesting behaviour of the model is to be found in the regime where $d < 0.1$, say. In this regime, the evolution of the model leads in $O(1)$ time to a collection of domains that conforms to the spatial structure in the data $H_k(x)$, up to details of at least an $O(1)$ size. With very noisy, imcomplete or smeared data, much of the domain structure so generated will differ substantially from the intended reconstruction, especially in the finer structural details. As already mentioned, either pure or mixed ordered states may be found in the domains. Inside pure domains, only exponentially few reconstruction errors exist away from the walls, and we will see that the domain walls become sharp, with effective width $W = O(\sqrt{d})$. Later stages in the evolution of the model are best analysed in terms of the motion of the domain walls, as driven by mutual interaction, differing domain state stability, data input, or curvature.

4.0.1 Walls between pure domains

The simplest domain walls are those that separate two pure domains. These come in just two types, since eq.1 is invariant under permutation of k-indices and under sign inversion. The types are:

- $(-|+)$ walls, between two domains with opposite overlaps on any single pattern ($k = 1$, say), and

- $(1|2)$ walls, between domains with overlap on any two distinct patterns ($k = 1$ and $k = 2$, say).

One can check that both types of wall only couple exponentially weakly to external field components H_k with $k > 2$. Thus, one can restrict $H(x)$ to lie in the $k = 1, 2$ subspace, and study the structure and behaviour of the two types of wall by analysing the solutions of eq.1 with $n = 1$, respectively $n = 2$. In fact, one can reduce the $n = 2$ case to that of $n = 1$, and thus obtain the solutions describing both wall types in one go. This is due to the possibility of *decoupling* the $n = 2$ case into the evolution of two *independent* modes, despite the non-linearity of eq.1. The modes are $Q_{\pm}(x) = m_1(x) \pm m_2(x)$, and their evolution is governed by

$$\partial_t Q_{\pm}(x) = -Q_{\pm}(x) + F_d[H_{\pm}(x) + G * Q_{\pm}(x)], \tag{3}$$

with $H_{\pm}(x) = H_1(x) \pm H_2(x)$. Thus, each of the Q-mode evolutions (eq.3) is identical in form to the $n = 1$ case of eq.1, and we only need to solve this simple case to be able to construct any $n = 2$ solution by superposition.

Let us first put $H = 0$, and look for a stationary, isolated, planar kink in one mode $Q(x) = -Q(-x) \neq 0$. This kink must then satisfy $Q(x) = F_d[G * Q(x)]$. For d just below d^*, we already obtained a kink solution in closed form in the previous subsection. In fact, such kinks occur at any $d < d^*$, and they are unique and stable except for the neutral stability mode $\nabla Q(x)$ associated with infinitesimal translation. In the important low-d regime the kink becomes a sharp feature:

$$Q(x) \approx \text{erf}[Mf(x)/\sqrt{2d}], \text{ for } |x| < 1,$$

where $f(x) = \text{erf}(x/\sqrt{2})$ for a Gaussian G in any D. For a 'pillbox' G, one finds D-dependent terms $f(x) = x$ in $D = 1$, $f(x) = 2[x\sqrt{1-x^2} + \arcsin(x)]/\pi$ in $D = 2$, $f(x) = (3x - x^3)/2$ in $D = 3$, etc.. Since the sigmoid has a sharp transition from -1 to +1 when d is small, the influence of alternative choices for G on the shape of the kink profile is actually very small. The effective width of a kink is $O(\sqrt{d})$, and the tails outside this region decay extremely fast. For example,

$$Q(x)/M \approx 1 - \sqrt{2d/\pi}x^{-1} \exp[-x^2/(2d)], \text{ for } \sqrt{d} \ll x < 1, \text{ in D=1.}$$

As a consequence, a pair of parallel (necessarily opposite-signed) kinks do not interact on practical timescales if their separation is at least $1 + W$ for a pillbox G, or a few times that distance for Gaussian G. However, once the kinks come within distance 1, they annihilate each other in unit time.

To complete this subsection, we specify how the Q-kinks make up the domain walls, and note how this leads to simple rules governing what may be called the wall "chemistry", i.e. a symbolic notation for the phenomena that occur when two walls come to within unit distance of each other.

Each of the independent Q-modes which make up a general $n = 2$ solution (with $n = 1$ as a special case) can be either flat, $Q(x) = \pm M$, or have a kink of either polarity. A $(1|2)$ type wall corresponds to those cases where only one of the Q-modes has a kink at the wall position, while the other mode is flat there. Obviously, there are 8 different forms of this wall, corresponding to which mode carries the kink, and how the polarities are chosen. In addition, any other pair of unequal pattern indices may be taken. A $(-|+)$ type wall is obtained when both Q-modes have a kink at the same position. Because the modes are independent and neutrally stable to translation, one sees at once that, say, a $(-1|+1)$ wall is just an accidental degeneracy of a pair of walls like $(-1|2|+1)$. Clearly, the domain intervening between these walls could belong to any $k \neq 1$, and could have arbitrary polarity. Thus, any such wall pairs do not interact at all: They can travel through each other when external data forces the individual walls to move, and any $(-1|+1)$ wall will be split into such a $(-1|\pm k|+1)$ pair by arbitrarily weak k-state perturbations caused by H_k or other walls. Wall pairs of the types $(1|2|1)$ or $(-|+|-)$ can interact because they involve kinks of opposite polarity in the same Q-mode(s). Since these kinks annihilate each other in $O(1)$ time when they approach to within distance 1, one gets the symbolic wall 'reactions':

$$(1|2|1) \rightarrow (1), \text{ and } (-|+|-) \rightarrow (-).$$

A slightly more complicated reaction (via an intermediate state) occurs between walls of unequal type:

$$(1|-2|+2) \rightarrow (1|-2|1|+2) \rightarrow (1|+2).$$

This classification exhausts the possibilities of the $n = 2$ case, but in practical situations one will also be confronted by interacting wall pairs like (1|2|3). Unfortunately, the $n \geq 3$ case of eq.1 can no longer be decoupled; instead one has to solve a system of n coupled equations, describing actitivity of 2^{n-1} subsets of the neurons that have the same pattern-dependent contribution to their local field. For the $n = 3$ case of the (1|2|3) pair of walls at separation < 1, one finds again a decay to a (1|3) wall in $O(1)$ time, but via an intermediate mixture state. The fate of larger, longer lived mixture domains will be analysed in section 4.3.3. of this paper.

4.3.2 Walls driven and pinned by data and curvature

Sofar, only planar walls have been analysed in systems without data $(H(x) = 0)$. In order to analyse the phenomena of greatest interest to the intended applications, one has to study how curvature and data influence the evolving shapes of walls.

For walls between pure domains, and away from the $D - 2$ dimensional intersections of different walls, the problem reduces to that of a curved Q-kink in a field $H \neq 0$. Let us first study only the influence of the data on planar kinks. In the limit $d \to 0$, one can then solve eq.3 with, e.g., $H_+ < 0$, to obtain a simple closed form solution for a moving Q_+ kink. Its profile is now skew-shaped, with a longer tail trailing in the wake of a sharp front, moving with a velocity V. In co-moving coordinates $y = x - Vt$, one finds:

$$Q_+(y < 0)/M = -1 + 2\exp(y/V), \ Q_+(y \geq 0) = +M.$$

This kink profile is obtained in any D, and for any allowed form of G, but the exact relation between the velocity V and H does depends on D and G, albeit only in quantitative details. For a pillbox G, one gets the simple relation $V[1-\exp(-1/V)] = -H_+$ in $D = 1$, with more complicated forms for larger D. With a Gaussian G in arbitrary D, the relation is $\exp[1/(2V^2)] \operatorname{erfc}[1/(\sqrt{2}V)] = -H_+$. The common result for any G and D is that V is proportional to H when $H \ll 1$, and that $V \to \infty$ as $H \to 1$.

In $D > 1$, walls are generally curved because the data represented in the initial conditions can have an almost arbitrary patchstructure, and because of fluctuations in the local overlaps due to input noise. In qualitative terms, curvature drives domain walls towards the concave side, unless there is a sufficiently strong bias field. Thus, large domains bordering on several domains with different pattern overlap tend to develop smooth walls, while domains that are embedded in a larger one will become spherical and shrink. More quantitatively, a wall with local mean curvature κ (radius $R = \kappa^{-1}$ in $D = 2$) will travel inward with velocity $V = O(\kappa)$ when $H \ll \kappa$. Thus, an embedded domain of size $S(0) \gg 1$ contracts to size $S(T) \approx 1$ in time $T = O(S^2)$, and then collapses in unit time. The result is that, after some time $T \gg 1$, one has lost all embedded domains of initial size up to $O(\sqrt{T})$, while the larger ones have not yet shrunk by an appreciable fraction of their initial size. The coarsening of the wall shape of the larger domains proceeds similarly; it takes time of $O(S^2)$ to essentially remove undulations of size S.

In general, the coarsening process sketched above would continue indefinitely under conditions where $H = 0$, i.e. without persistently presented data. The order in the system would eventually lose all similarity to the data component in the initial conditions. Note that this differs fundamentally from the behaviour of infinite range models. However, with the more realistic assumption of persistent non-zero H terms, the domain walls will quickly become pinned in configurations where the curvature-driven coarsening is balanced by the earlier analysed forces exerted on the walls by the data. The condition of balance $H_k(x') = -G * m_k(x')$ implicitly defines the positions x' along the centerline of the pinned walls. For reasonably coarse patch structures, it is easy to relate H to the local shape of the walls in the final, pinned configuration. For example, to prevent the collaps of a $k = 1$ spherical domain of radius $R \gg 1$, embedded in a $k = 2$ domain, the data must contain a term $H_1 \geq 1/(3R)$ in $D = 2$ (and $\pi/(4R)$ in $D = 3$) within the corresponding patch. Likewise, local input data of this strength across the boundary of two adjacent textured patches will pin the walls between the corresponding pure domains so as to follow the boundaries up to details with local mean curvature $\kappa = O(H)$.

4.3.3 Destruction of mixture domains

Finally, we have to deal with phenomena involving domains within which a state with mixed overlaps occurs. In infinite range models, these mixture states are a serious nuisance, since their aggregated basins of attraction dominate those of the pure states at low d. This problem will appear to be largely abolished in restricted-range models. As a result, the latter networks can disambiguate inputs in which substantial superposition of texture information has occurred. I shall deal explicitly with the most robust symmetric mixtures, the $n = 3$ states that have $m_k \to 1/3$ as $d \to 0$.

The basic observation is that pure domains rapidly invade mixed domains. Suppose, for example, that a small pure ($k = 1$) 'droplet' arises within an arbitrarily large mixed domain. This may be due to a small patch of unambiguous data in an otherwise ambiguous context. Analysis of the $n = 3$ case of eq.1 for the $(1|1, 2, 3)$ wall of the droplet at low fields H and $d \to 0$ leads to the conclusion that the droplet wall will move outwards (locally) if more than $1/3$ of the volume of a pillbox G centered at the wall lies inside the droplet. For the simple case of a spherical droplet, this implies the existence of a small critical nucleation radius for growth of the droplet. As a supercritical droplet grows, the expansion of its wall speeds up to a large limiting velocity V when the wall curvature becomes $\ll 1$. For example, $V[1 - \exp(-1/V)] = 1/3$ in $D = 1$, with a pillbox G. Independently of G or D, the profile of the invading $(1|1, 2, 3)$ wall in comoving coordinates $y = x - Vt$ then becomes

$$m_1(y) = 1 - m_2(y) \; ; m_2(y) = m_3(y) = \exp(y/V), y < 0 \; ; m_k(y \geq 0) = 1/2.$$

It would take unrealistically strong mixed data fields $H_k \geq 1/4$ to stop this invasion of mixture domains by expanding pure domains. The nucleation size of droplets with opposite polarity (giving $(-1| + 1, 2, 3)$ walls) or with overlap on another (say $k = 4$)

pattern is the same, although the structure and the speed of their invasion walls are more complicated. It takes even smaller pure droplets to invade a mixture state of higher complexity ($n \geq 5$). In any event, the speed of the invasion process is $O(1)$, so mixture domains of size S are destroyed in a timescale of $O(S)$.

In $D = 1$, mixture states are completely destroyed, while in higher dimensions, there will remain very small mixed domains at the intersections of walls. In fact, such small domains arise in $O(1)$ time wherever more than two different pure domains meet in a roughly symmetric arrangement. If the parent domains have a size of at least a few times the unit scale, then these 'decorated' intersections have large lifetimes, and they can evolve by sliding motions, related to those of the adjoining domain walls. This motion can be driven or pinned by data much like the planar wall motion analysed before, except that the 'free' motion of a decorated corner is such as to produce symmetric configurations of three adjoining parent domains. The sliding process need not settle quickly into global equilibrium since the overall structure of the webb of walls will in general show a sequence of topological transitions. In the practical situations of our intended application however, such complicated long-time coarsening effects are prevented from occuring by the already mentioned pinning of walls by the data, which happens on $O(S)$ timescales for size S domains.

5   Conclusions

The analysis of the domain-wall behaviour indicates that the restricted-range model could be useful for the intended application of reconstructing and segmenting patterns consisting of medium-scale textured patches. The total amount of wiring scales as $rNZ = O(N^\beta)$, with $1 + a/D \leq \beta \leq 1 + a + a/D$. The limits of the range of β correspond to very sparse, respectively complete, semi-local connectivity. The dynamical behaviour of the network produces a rich phenomenology, which can be summarised as follows.

In $O(1)$ time, the pattern is reconstructed using information on a unit spatial scale. Pure domains have very few remaining errors with respect to the pattern that was locally closest to the data, as long as one is not inside a wall, which anyway is very thin in the proper regime of low d. However, if signal/noise is very low or the data is ambiguous, one may initially get mixture domains, or many small pure domains, or fine-grained undulations in the domain walls that are unrelated to the signal content in the data. These problems are remedied at later times, by a variety of effects. In $O(S)$ time, size S mixture domains are destroyed by invasion of neighbouring pure domains, even when the latter are initially very small. Smooth domain walls that are not near a corresponding patch boundary in the signal are moved in the proper direction with a speed proportional to the signal. Thus, the system can also deal with moving scenes. In $O(S^2)$ time, but limited to a scale $S = O(1/H)$, size S embedded domains collaps, and undulations in walls of large domains flatten out. Finally, even weak and noisy data will pin the walls near to the patch boundaries, up to details with minimum radius of curvature $R = O(1/H)$. Thus, the network does an automatic trade-off, sacrificing more details as signal/noise becomes worse.

Normally, this requires some sort of outside intervention. One should only match the scale of the smallest details of interest to the unit-scale of the network (= interaction range), below which domains cannot be stabilised by realistic signals.

The capacity of the network in terms of different global patterns that can be faithfully reconstructed is enormous despite the very low number of different texture types encoded in the connections. For finite H, one can recover any combination of patches of typical diameters $> 1/H$. Thus, the global pattern capacity A can be estimated as $\ln(A) = O(N^{1-a})$. It seems that this conforms much better to the observation that real networks can deal with a vastly larger repertoire of global patterns than the ones that they have been ever exposed to, or that can be coded globally in long-range connected architectures.

Putting these results back into the wider perspective of the problems that were discussed in the first part of this paper shows how small is the step that has been made. However, one can still hope that at least the direction was right.

Acknowledgement. This work was funded by SPIN-SNN, the Dutch Foundation for Neural Network Research.

References

1] B.Derrida, E.Gardner, and A.Zippelius, Europhys.Lett. 4, 167 (1987).
2] A.C.C. Coolen and Th.W.Ruijgrok, Phys.Rev.A 38, 4253 (1988).
3] W.Kinzel, these Proceedings.
4] D.C.van Essen, in 'Visual Cortex', A.Peters & E.Jones, Eds., Plenum 1985.
5] E.I.Knudsen, S.du Lac and S.D.Esterly, Ann.Rev.Neurosci. 10, 41 (1987).
6] J.J.Koenderink and A.J.van Doorn, Biol.Cybern. 55, 367 (1987).
7] J.J.Koenderink, Biol.Cybern. 58, 163 (1988).
8] A.Canning and E.Gardner, J.Phys.A 21, 3275 (1988).
9] A.J.Noest, Phys.Rev.Lett. 63, 1739 (1989).
10] A.C.C.Coolen, these Proceedings, and Utrecht Univ. PhD thesis 1990.
11] A.Engel, H.English and A.Schutte, Europhys.Lett. 8, 393 (1989).
12] A.Rau and D.Sherrington, Europhys.Lett. 11, 499 (1990).
13] D.J.Amit, H.Gutfreund, and H.Sompolinsky, Phys.Rev.A 32, 1007 (1985),
        Phys.Rev.Lett. 55, 1530 (1985), Ann.Phys.(N.Y.) 173, 30 (1987).
14] M.Skierski, A.M.Grundland, and J.A.Tuszynski, J.Phys.A 22, 3789 (1989)

STATISTICAL MECHANICS AND
ERROR-CORRECTING CODES

Nicolas Sourlas

Laboratoire de Physique Théorique de l'Ecole Normale Supérieure
24, rue Lhomond - 75231 PARIS Cédex 05 - FRANCE

Recent developments in the theory of spin-glasses are a major breakthrough in statistical mechanics and made it applicable to a new class of problems, well outside its original field. Good examples of this are the topics covered in this workshop. I could add combinatorial optimisation, theoretical models of the immune system, models of prebiotic evolution etc. Here I am going to present some recent applications to the theory of communication[3].

A typical case considered in communication theory is the case where a source "produces" some information at some point A and that information is to be used at some other point B. Points A and B are connected by the "communication channel". There are several examples of this: The example of a telephone line, communication with radio waves in space or through the atmosphere, storage of data on a computer disk, storage of music on a compact disk, etc. One may eventually want to use the data in the same place (spatial transportation is not necessary), but later in time, as in the case of the computer disk. I will assume for simplicity that the information data are represented in digital form, i.e. they consist of a sequence A of "bits", a_i, $i = 1, ..., N$, equal to zero or to one. Let us consider the ensemble $\mathcal{A}$ of the messages the source can generate and call p_i the probability for a_i to be one.

A central role in the mathematical theory of communication is played by the notion of information introduced by Shannon in his seminal work[1,2]. The information content of the message is, according to Shannon, given by

$$H(A) = -\sum_i p_i \, log_2 p_i \tag{1}$$

where A represents, as before, the whole bit sequence.

During the transmission of any information data, errors may occur due to the presence of noise. The exact nature of noise depends on the particular circunstanses. It can be thermal noise in an electronic apparatus or in the wires of a transmission line, absorption by the atmosphere or interference with other sources of radiation in the case of the transmission of radio waves through the atmosphere, interference between neighbouring points on a computer disk, etc.

To be more specific, we will assume that the transmission proceeds as follows. An electric signal v_i is sent during the time interval τ. $v_i = v$ if $a_i = 1$, and $v_i = -v$ if

$a_i = 0$. Upon reception, instead of the v_i signals, one rather gets $u_i = v_i + y_i$ because of the noise. In an ideal case the y_i's, which are the noise contribution, are independent random variables with zero mean gaussian distribution and variance w^2 . (This is called the memoryless binary gaussian channel). Let us call $p(u_i|a_i)$ the conditional probability to receive a signal u_i when a_i is the input information bit and $p(a_i|u_i)$ the probability for the input to be a_i when one observes an output u_i. Because of the noise there is a loss of information during transmission. This loss is

$$L(U|A) = - \sum_{\{u\}} \sum_i p(u_i)\, p(a_i|u_i)\, log_2\, p(a_i|u_i) \qquad (2)$$

where U represents the output sequence. So the maximum of information which can be transmitted by the channel per time interval τ is

$$C = \lim_{N \to \infty} \frac{1}{N} \ max\ (\ H(A) - L(U|A)\ ) \qquad (3)$$

where the maximum is taken over all possible pairs (A, U) of inputs-outputs of length N. C is called the capacity of the channel and plays a very important role in communication theory. It denotes the maximum possible information transfer rate over the channel. The capacity of the gaussian channel has been computed by Shannon and is given by

$$C = \frac{1}{2}\, log_2\, (1 + \frac{v^2}{w^2}) \qquad (4)$$

v^2/w^2 measures the signal to noise ratio.

One wants to recover the information as error-free as possible. One possibility is to increase the signal to noise ratio to decrease the loss of information during transmission. The other possibility is to try to spread the information content over a longer sequence of bits (make the message redundant) to compensate for the loss in the transmission. This procedure is called coding. At the end one has to reconstruct the original message. This is the decoding procedure. The ratio of the number of information bits over the number of transmitted bits is called the rate R of the code and measures its redundancy.

It is customary[4,5] to measure the performance of a code by measuring the error probability per information bit P_e as a function of the ratio of the signal energy ($v^2\tau$ in our example) over the noise energy ($2w^2$) per information bit, i.e., $v^2\tau/(2w^2R)$. Naturally a code is better if it achieves a smaller error for a given signal to noise ratio. An alternative way of measuring the performance of a code is the following. The same P_e which is obtained by using the code with an energy per information bit E_c ($v^2\tau/R$ in our example) can be obtained without using any code provided that the energy per bit fed to the channel is adjusted to an appropriate value. Lets call E_{un} this value. The gain of the code is defined as the ratio E_{un}/E_c and is usually measured in decibels. Of course better codes have higher gain.

How good a code can possibly be? If we want to recover the information error-free, we must use a code with $R \le C$, otherwise, we would transmit more information than the channel capacity allows. So the channel capacity is an upper bound on the rate of error-free coding. There is a very remarkable theorem, first proven by Shannon, that there exist codes with a rate which can approach C as close as we want, such that the error probability is zero. (They saturate the channel capacity).

The idea of the proof is the following. We called $\mathcal{A}$ the ensemble of the 2^N possible messages A of length N; it corresponds to the vertices of the hypercube in N dimensions. Each original message is transformed into a message of length $K = N/R$ by the coding procedure. So the code can be considered as a map of the hypercube in N dimensions to the hypercube in K dimensions and any such map defines a code. Let's call $\mathcal{B}$ the ensemble of the vertices of the Kth dimensional hypercube whose inverse map is $\mathcal{A}$. $\mathcal{B}$ is only a fraction of the Kth dimensional hypercube whose size depends on the rate of the code. Because of the noise, the output of the transmission channel will in general not be a point in $\mathcal{B}$ but will be some point x in the Euclidian space $\mathbf{R}^K$. The probability of the distance of x from a point in $\mathcal{B}$ obviously depends on the noise, i.e. on the probabilities $p(u_i|a_i)$. Now we consider the code corresponding to a random map and the following decoding algorithm. We consider a sphere of radius r centered at x. If one and only one point of $\mathcal{B}$ falls inside this sphere we decide that this was the transmitted message. Otherwise we decide that there is a decoding error. This is a theoretical decoding algorithm, very complicated to implement. We want to compute the error probability using this code and the previous decoding algorithm. This is impossible. But what is possible to compute is the average error probability $\mathbf{P}_e$, the average being taken over all random maps. $\mathbf{P}_e$ is obviously a function of the radius r and of the transition probabilities $p(u_i|a_i)$. If the rate $R < C$ it can be shown that there exists a choice of r such that $\mathbf{P}_e$ goes to zero when the length of the message N goes to infinity. Observe now that $\mathbf{P}_e$ is an upper bound on the error probability which can be achieved, i.e. in this ensemble of random codes there are good codes and bad codes and the good codes must have an error probability smaller than that of the average. In this way we obtain a lower bound on the possible rates for error-free codes and the difference between the upper and the lower bounds goes to zero as N goes to infinity.

This argument shows that there exists a large number of codes which asymptotically saturate the channel capacity (since the "average" code does) but does not construct them explicitly, nor provides an easy decoding algorithm. On the other hand it is a fact that the explicitly known codes with easily implementable decoding do not saturate the channel capacity.

For the case of the gaussian channel this implies that there exist codes such that it is possible to communicate without any error if and only if $R \leq \frac{1}{2}\log_2(1 + \frac{v^2}{w^2})$ If $v/w \ll 1$ this bound is $R \leq (1/\ln 2)v^2/2w^2$.

Let me now, to be concrete, give two examples of explicit codes[4,5]. The first example is a trivial one. We send every input bit twice, so there are two outputs u_i and w_i for every information bit a_i. Decoding proceeds as follows. If $u_i + w_i > 0$, we decide that the input was one, otherwise we decide it was zero. The rate of this code is $1/2$. What is its gain? The error probability when we dont use any code is

$$P_e^{uncoded} = \int_{v/w}^{\infty} \frac{dx}{(2\pi)^{1/2}} \exp(-x^2/2) \tag{5}$$

The signal to noise ratio per information bit for this code is $v^2/(2w^2 R) = v^2/w^2$, while the error probability is given by eq. (5) with v/w replaced by $2v/(w\sqrt{2}) = (v\sqrt{2})/w$.

So we obtain the same error without using the code provided we take $v^{uncoded} = v\sqrt{2}$, i.e. the gain of this simple code is one (or zero if expressed in decibels).

We take as a second example one of the Viterbi codes. Formally we can represent the bit sequence by the polynomial $G(x) = \sum a_k x^k$ whose coefficients are 0 or 1. From $G(x)$ we construct two new polynomials by multiplication with $g_1(x) = 1 + x + x^2$ and $g_2(x) = 1 + x^2$. (g_1 and g_2 are called the generating polynomials of the code.)

$$G_1(x) = g_1(x)\, G(x) = \sum_k (a_k + a_{k-1} + a_{k-2})_{mod\ 2}\ x^k \tag{6}$$

$$G_2(x) = g_2(x)\, G(x) = \sum_k (a_k + a_{k-2})_{mod\ 2}\ x^k \tag{7}$$

where the addition of the a's is modulo 2 (i.e. $0 + 0 = 1 + 1 = 0$, $0 + 1 = 1 + 0 = 1$). The coefficients of $G_1(x)$ and $G_2(x)$ are again 0 or 1.

Instead of transmitting the original information bits, i.e. the coefficients of $G(x)$ one transmits the coefficients of $G_1(x)$ and $G_2(x)$. This is the coding operation. In this example one transmits twice as many bits as there are information bits. Here again $R = 1/2$. Decoding proceeds as follows. If u_i is received as an output, the probability p_i for the input a_i to be 1 is given by the Bayes formula

$$p_i = \frac{p(u_i|1)}{p(u_i|1) + p(u_i|0)} = \frac{\exp(-(u_i - v)^2/2w^2)}{\exp(-(u_i - v)^2/2w^2) + \exp(-(u_i + v)^2/2w^2)} \tag{8}$$

(Remember that we assume a zero means gaussian noise distribution with variance w^2). So it is possible to compute the probability of every possible input sequence of the a_i's, when a given sequence of u_i's is observed as the channel output. Given the observation of the u_i's there exists an algorithm, the Viterbi algorithm, to find the most probable input sequence. (We will come back later to the Viterbi decoding algorithm.) There is a whole family of Viterbi codes, every one in the family being specified by its rate and by the generating polynomials. Their performance increases with the degree ν of the generating polynomials and they are very widely used.

I will now show that there is an equivalence between Viterbi codes and theoretical models of spin glasses. Spin-glasses[6] are usually modelled with the Edwards-Anderson Hamiltonian

$$H_{EA} = - \sum_{i,j=1,N} C_{ij} J_{ij} \sigma_i \sigma_j \tag{9}$$

We first consider the case where the σ_i's are Ising spins. We call C_{ij} the connectivity matrix. Its elements are 1 if the two spins interact and 0 otherwise. The J_{ij}'s which give the strength of the two spin interaction are independent random variables with known probability distribution. Let J_0 denote the mean and ΔJ^2 the variance of the J_{ij}'s.

Two results of spin-glass theory will be of interest for us.

a) At zero temperature (and for not a too weak connectivity) there is a phase transition. For $J_0/\Delta J$ greater than certain critical value there is a spontaneous magnetisation $m = \frac{1}{N}\sum_i \sigma_i \neq 0$, while in the opposite case $m = 0$ (N is the number of spins).

b) The configuration space of the spins is invariant under the transformation $\sigma_i \rightarrow \sigma_i \epsilon_i$ where the ϵ_i's are an arbitrary configuration of Ising spins. As $\epsilon_i^2 = 1$ the Hamiltonian is also invariant under this transformation and the simultaneous tranformation of the J_{ij}'s $J_{ij} \rightarrow J_{ij}\epsilon_i\epsilon_j$. (This is called gauge invariance).

The above two properties are also true for the following Hamiltonian, a straightforward generalisation of the previous one

$$H = -\sum_p \sum_{i_1,...,i_p=1,..,N} C_{i_1,...,i_p}^{(p)} J_{i_1,...,i_p} \sigma_{i_1}...\sigma_{i_p} \tag{10}$$

Let again $J_0^{(p)}$ and $\Delta J_{(p)}^2$ be the mean and the variance of the J's.

In order to show the equivalence between Viterbi codes and spin glass models, first make the trivial correspondence $\sigma_i = 1 - 2a_i$ between Ising spins and information bits. Modulo 2 addition is equivalent to spin multiplication. (More generally there is a correspondence between group multiplication for the groups $\mathbf{Z}_n$ and modulo n addition.) With this correspondence, the coefficients of the polynomials $G_i(x)$ in eqs. (6) and (7) can be written as $(a_k + a_{k-1} + a_{k-2})_{mod\,2} \rightarrow J_{k,k-1,k-2}^{input} = \sigma_k^0\sigma_{k-1}^0\sigma_{k-2}^0$, the σ^0's corresponding to the input information bits, and similarly for $J_{k,k-2}^{input}$. To decode one has to find the most probable configuration of the σ's given the occurrence of J^{output}'s. According to eq. (8) the probability of a σ configuration is

$$\mathbf{P} = \prod_k \mathcal{P}_k^1 \mathcal{P}_k^2$$

$$\mathcal{P}_k^1 = \frac{\exp(-(J_{k,k-1,k-2} - \sigma_k\sigma_{k-1}\sigma_{k-2})^2/2w^2)}{\exp(-(J_{k,k-1,k-2} - 1)^2/2w^2) + \exp(-(J_{k,k-1,k-2} + 1)^2/2w^2)} \tag{11}$$

$$\mathcal{P}_k^2 = \frac{\exp(-(J_{k,k-2} - \sigma_k\sigma_{k-2})^2/2w^2)}{\exp(-(J_{k,k-2} - 1)^2/2w^2) + \exp(-(J_{k,k-2} + 1)^2/2w^2)}$$

(J obviously stands for J^{output} in this equation). One can equally well maximize $\log \mathbf{P}$

$$\log \mathbf{P} \sim \sum_k J_{k,k-1,k-2}\,\sigma_k\sigma_{k-1}\sigma_{k-2} + J_{k,k-1,k-2}\,\sigma_k\sigma_{k-2} + const. \tag{12}$$

where $const.$ means σ independent. The right hand side of eq. (12) is a one dimensional spin glass Hamiltonian with three spin nearest neighbour and two spin next nearest neighbour interactions (one should add a minus sign in front, for what follows). The configuration of the σ's which is its ground state, is the same one which maximizes the probability $\mathbf{P}$. The Viterbi decoding algorithm is nothing else than the zero temperature transfer matrix algorithm used in statistical mechanics to solve one dimensional problems!

Before realizing this precise correspondence between Viterbi codes and one dimensional spin glasses, I suggested the following family of codes[3]. Let a_i be the N bits of information and $\epsilon_i = 1 - 2a_i$ the spins associated with them. The coded message, i.e. the input to the transmission channel, are the matrix elements $J_{i_1,...,i_p}^0 = \epsilon_{i_1}...\epsilon_{i_p}$. Due to the presence of noise, the output of the channel will be $J_{i_1,...,i_p} = J_{i_1,...,i_p}^0 + \Delta J_{i_1,...,i_p}$ where ΔJ is the noise contribution. Decoding is finding the ground state of H given by eq. (10). In the absence of noise our Hamiltonian is a "pure gauge", i.e. a generalisation of the Mattis model. All the couplings can be made ferromagnetic by a gauge transformation. Its ground state is obtained from the ferromagnetic ground state by the same gauge transformation. The above is not true in the presence of a non infinitesimal noise and the problem is to find how many errors are induced by the presence of the noise.

Any particular code in the family of codes defined above, is specified by the connectivity matrices C. To estimate the performance (i.e. the error probability for a given noise level) it is enough to consider the case where all the input bits equal one. (The general case is the same because of the gauge symmetry.) In this case the number of errors is the number of spins equal to -1, i.e. the bit error probability is $P_e = (1-m)/2$, where m is the ground state magnetisation.

We assume now for simplicity that there is only one value of p in eq. (10), and that the coordination number $z_i = \sum_{j_2,\dots,j_p} C_{i,j_2,\dots,j_p} = z$ is independent of i.

The rate R of the code equals the number of spins devided by the number of independent terms appearing in the Hamiltonian. Any particular spin appears in z different terms, but every term is connected to p spins: so $R = p/z$. So we established the following dictionary between error correcting codes and statistical mechanics.

$$
\begin{array}{rcl}
Coding & \Longleftrightarrow & Construct\ a\ pure\ gauge\ Hamiltonian \\
Generating\ polynomials & \Longleftrightarrow & Connectivity \\
Decoding & \Longleftrightarrow & Find\ a\ ground\ state \\
Signal\ to\ noise & \Longleftrightarrow & J_0^2/\Delta J^2 \\
Error\ probability\ per\ bit & \Longleftrightarrow & Ground\ state\ magnetisation
\end{array}
$$

The error probability per bit P_e can be computed in two cases.

A) When the noise is very weak. In this case almost all the σ's are 1. P_e equals the probability for the sum of all the couplings to a particular spin to be negative:

$$
P_e = \int_{v\sqrt{z}/w}^{\infty} \frac{dy}{\sqrt{2\pi}} \exp\left(-\frac{y^2}{2}\right) \tag{13}
$$

According to this calculation, in the weak noise limit, the gain of our code is $\mathcal{G} = 10\log_{10} p$ db, independent of R. This suggests it is advantageous to use large p.

B) When the connectivity is extensive, i.e. $z \sim \binom{N-1}{p-1}$. It is shown[6] that only the first two moments of the noise distribution are relevant, and one has to rescale the couplings, $J_0 = (p!/N^{p-1})j_0$, and $\Delta J^2 = (p!/N^{p-1})\Delta j^2$, to get a sensible thermodynamics. We set $\Delta j = 1$ and $\tau = 1$ to fix the normalisation. The signal to noise ratio is then

$$
J_0^2/\Delta J^2 = j_0^2\, p! \,/\, N^{p-1} \tag{14}
$$

If $p = 2$ or $p \to \infty$, and $p/N \to 0$ the spin-glass model is soluble. For $p \to \infty$ it is Derrida's random energy model (or REM)[7]. B. Derrida remarked that, in this case, if two configurations of the spins are macroscopically distinguishable, their energies are uncorrelated. This remark allowed him to solve the model for $j_0 = 0$. He showed that there is a phase transition in this model. For $T > T_c = 1/2\sqrt{\ln 2}$ the system is paramagnetic and the free energy is $F_p = -T\ln 2 - 1/4T^2$ while for $T \leq T_c$ the system is in the spin-glass phase and $F_{sg} = -\sqrt{\ln 2}$ and the entropy vanishes. For $j_0 \neq 0$

an additional term appears in the Hamiltonian, $J_0 \sum \sigma_{i_1} \ldots \sigma_{i_p}$, which can be written as $N j_0 m^p$ plus non leading terms in N, where m is the magnetisation. When $p \to \infty$ this term is zero except when $m = 1$. In this latter case, the entropy vanishes and $F_f = -j_0$. There is a transition from the paramagnetic to the ferromagnetic phase when $F_p = F_f$, or $j_0 = T \ln 2 + 1/4T$. This transition line stops when it meets the separation line between the paramagnetic and the spin-glass phase, $T = 1/2\sqrt{\ln 2}$. This happens for $j_0 = j_{0,c} = \sqrt{\ln 2}$. So we have shown that for $j_0 > j_{0,c} = \sqrt{\ln 2}$ the spontaneous magnetisation $m = 1$, while for $j_0 < j_{0,c}$ $m = 0$. Translated into the coding language, this means that for $j_0 > \sqrt{\ln 2}$ one can recover the input without error. Remarking that $v^2/w^2 = j_0^2 p!/N^{p-1}$ and that the rate is $R = p!/N^{p-1}$, the signal to noise ratio per information bit is j_0^2, i.e. the REM, used as an error correcting code, saturates Shannon's famous bound for the Gaussian channel. Of course this happens in the assymptotic limit $p \to \infty$. For the more interesting case of finite but large p, one can use the replica method[8,9] and show that $j_{0,c}$ decreases with p and that for $j_0 \sim \sqrt{\ln 2}$

$$ m - 1 - \frac{\exp(-pj^2)}{\sqrt{p}} \, c(j) \qquad c(\sqrt{\ln 2}) = .087 \qquad (15) $$

i.e. the approach to asymptotia is very fast. (The details of the computations will be published elsewhere.)

The $p = 2$ case is the Sherrington–Kirkpatrick (or SK) model[10]. The SK model develops a spontaneous magnetisation for $j_0 > j_{0,c} = 1/2$ (our normalisation for j_0 differs by a factor of 2 from the usual one) which has been computed analytically[11] and numerically[10]. For $j_0 = \sqrt{\ln 2}$ $m_{SK} = .85$.

So the REM is, at least in theory, an ideal error correction code; for $j_0 > \sqrt{\ln 2}$ one can recover the input without error, the approach to asymptotia is that of eq. (15). By solving the REM we rediscover, in a completely new way, Shannon's famous $\ln 2$.

No exact algorithm is known to minimise H, except for the one dimensional case and one has to use heuristics, like simulated annealing[12,13] which provides only an approximation to the ground state. If we have converged to the correct ground state, we know what the value of the energy should be. Because the energy is gauge invariant, we can make a gauge transformation so that all the σ's are one. Then

$$ H = -\sum_p \sum_{i_1,\ldots,i_p=1,\ldots,N} C^{(p)}_{i_1,\ldots,i_p} J_{i_1,\ldots,i_p} = -\mathcal{N} + O(\sqrt{\mathcal{N}}) \qquad (16) $$

where $\mathcal{N} = \sum_{p,i} C^{(p)}_{i_1,\ldots,i_p}$. So it is possible to monitor the convergence to the ground state by measuring the energy.

We will now introduce some more general codes. The information message, instead of being a sequence of bits, will now be a sequence of symbols a_i, $i = 1,\ldots,N$ each of which can take l values (the a_i's are "letters from an input alphabet of length l"). A symbol contains now $\log_2 l$ bits. Let G be a $m \times m$ matrix representation of some group and H a finite subgroup of G with l elements taken in the same representation. I consider the following Hamiltonian

$$ H = -\sum_p \sum_{i_1,\ldots,i_p=1,\ldots,N} C^{(p)}_{i_1,\ldots,i_p} Re \, Tr\{J_{i_1,\ldots,i_p} h_{i_1} \ldots h_{i_p}\} \qquad (17) $$

C is the connectivity matrix as before, the h_i's are elements of $\mathbf{H}$, Tr is the trace of the product matrix and the J's are now $m \times m$ matrices constructed in such a way that the Hamiltonian is again a pure gauge. More explicitly

$$J_{i_1,...,i_p} = (h^0_{i_p})^{-1}...(h^0_{i_1})^{-1} \qquad (18)$$

where h^0_i is the input sequence. Again the J's are the input to the transmission channel. There are several ways to transmit the J's, one is sending its matrix elements. In this case, for example, one sends a signal $+v$ for $(\sigma_3)_{11}$, $-v$ for $(\sigma_3)_{22}$ and nothing for $(\sigma_3)_{12}$ and $(\sigma_3)_{21}$. We consider now an alternative possibility. (The error probability is the same in both cases). An input J is, by construction, an element of $\mathbf{H}$, say the element k. We send then a signal proportional to $\phi_k(t)$, where the $\phi_i(t)$, $i = 1,...,l$ form an orthonormal base on the interval $[0, \tau]$

$$\int_0^\tau dt\, \phi_i(t)\phi_j(t) = \delta_{ij} \qquad (19)$$

The channel output $f(t)$, because of the noise, is not an element of $\mathbf{H}$ any more, but we decompose it on the same ϕ_i base

$$f^k = \int_0^\tau dt\, \phi_k(t)f(t) \qquad (20)$$

To decode we have to find the ground state of the Hamiltonian

$$H = -\sum_p \sum_{k=0}^l \sum_{i_1,...,i_p=1,..,N} C^{(p)}_{i_1,...,i_p}\, f^{p,k}_{i_1,...,i_p}\, \mathrm{Re}\, \mathrm{Tr}\{h_k h_{i_1}...h_{i_p}\} \qquad (21)$$

If the connectivity is such that H corresponds to the Hamiltonian of a one dimensional system, we can decode using Viterbi decoding in this case also. The performance of these codes can be computed in the weak noise limit. (Because of the gauge symmetry we can again consider only the case where all h^0_i's are one.) The error probability at site i is then equal to the probability that a configuration with $h_i \neq 1$, (while at all other sites $h_j = 1$) has lower energy than the configuration with $h_i = 1$. I computed it for the following cases.

$\mathbf{G} = U(1)$ and $\mathbf{H} = Z_n$. In this example $m = 1$, i.e. the h_i's are not matrices and the case $n = 2$ is the Ising case studied before. It turns out that $n = 3$ is marginally better than $n = 2$ and all other cases are less interesting.

$\mathbf{G} = SU(2)$ and $\mathbf{H}$ is a finite subgroup of $SU(2)$. In this example $m = 2$. It turns out that the most interesting case is when $\mathbf{H} = \mathbf{Q}$, the quaternion group which has eight elements (it is the smallest nonabelian subgroup of $\mathbf{G}$). It is composed by the unit matrix ± 1, plus the Pauli matrices $\pm i\sigma_j$, $j = 1, 2, 3$. Any particular "spin" h_i has an energy $E = -\sum_k F_k e_k$ where

$$F_k = \sum_p \sum_{i_2,...,i_p} C^{(p)}_{i,i_2,...,i_p}\, f^{p,k}_{i,i_2,...,i_p} \qquad (22)$$

$$e_k = \mathrm{Re}\, \mathrm{Tr}\{h_k h_i h_{i_2}...h_{i_p}\} \qquad (23)$$

The F_k's are independent random variables with known mean and variance, which can be computed through eq. (22) and the noise characteristic of the channel. The probability P_s to get a symbol error is, in the weak noise limit, given by

$$1 - P_s = \int_{-F^0/\Delta F}^{\infty} \frac{du}{\sqrt{2\pi}} \exp(-u^2/2) \left\{ \int_{-u}^{u} \frac{dv}{\sqrt{2\pi}} \exp(-v^2/2) \right\}^3 \tag{24}$$

$$P_s \sim \exp(-\frac{1}{4}(F^0/\Delta F)^2) \tag{25}$$

where F^0 is the average of F_0 and ΔF^2 the variance of the F_k's.

It is straightforward to generalise to the case $G = SU(2) \otimes SU(2)$ where $\otimes$ is the tensor product and $H = Z_2 \otimes \overline{Q} \otimes \overline{Q}$ where $\overline{Q} = Q/Z_2$. More generally I consider the case $G = [SU(2)]^k$ and $H = Z_2 \otimes [\overline{Q}]^k$. The length of the input alphabet is now $l = 2 \times 4^k$. It can be shown that, in this case, the error probability per symbol in the weak noise limit is given by

$$P_s \sim (l-2) \exp(-\frac{1}{4}(F^0/\Delta F)^2) \tag{26}$$

The number of information bits per input letter is $\log_2 l = 1 + 2k$ The error probability per bit is the error probability per symbol times $l/2(l-1)$. Combining all this, we get, in the weak noise limit, that the gain of this code is

$$\mathcal{G} = 10 \, log_{10}[p(2k+1)/2] \quad \text{db} \tag{27}$$

This result is to be compared with the corresponding result for the Ising code

$$\mathcal{G} = 10 \, log_{10}p \quad \text{db} \tag{28}$$

I assumed again, for simplicity, that a single term (a single p) appears in eq. (17). We see that there is a substantial gain by using non abelian codes with large groups.

Let us now discuss the important question of decoding. We saw that when the connectivity matrix is such that H is a one dimensional Hamiltonian, one can use Viterbi decoding. This is an exact algorithm to produce the most probable input sequence, given the output of the channel. (It is the transfer matrix algorithm of statistical mechanics at zero temperature). One gets better codes as the range of the interaction ν (or equivalently the degree of the generating polynomials) increases, but, on the other hand the algorithmic complexity of decoding, i.e. the number of elementary operations an ideal computer has to perform, increases as l^ν (2^ν in the case of Ising spins). So, in practice, one is limited to relatively small values of ν.

In the more general case where H is not one dimensional, no exact decoding algorithm is known. We proposed to use simulated annealing for decoding. This algorithm will hopefully find, after some reasonable number of steps, an approximate ground state. It will be a fixed point of the zero temperature Monte-Carlo dynamics. Two issues are important here. a) What is the size of the basin of attraction of the ground state fixed

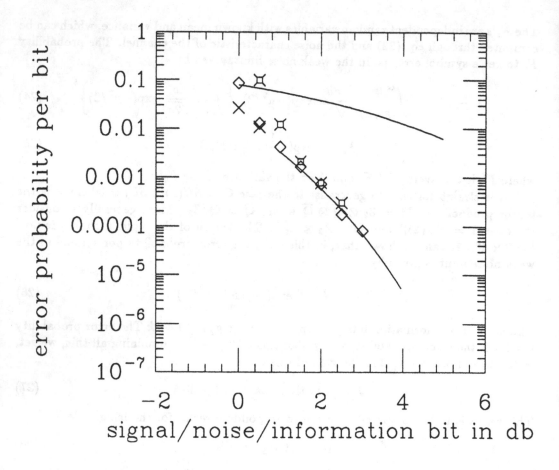

Figure 1

point, or, in other words, how easy is it to find it by the simulated annealing algorithm, and b) the approximate ground state one finds, how close is it to the real ground state.

The answer to these questions is not known a priori, nor, a fortiori, is it possible to give an analytic estimation of the algorithmic complexity for the simulated annealing algorithm to converge to the ground state. It is only possible to answer eventually these questions by numerical experiments. I performed preliminary numerical simulations for several of the codes proposed in this paper. The first simulations concerned the Ising codes and $p = 2$, $R = 1/3$ and $R = 1/4$. The results have been published already[3] and are pretty encouraging. One finds a very good approximation of the ground state, down to a signal to noise ratio per information bit of 2 db and the basin of attraction is large. But we saw that these $p = 2$ models are not very performant codes. I also tried $p = 4$ Ising models, which are better codes, but I found that the basin of attraction of their fixed points is small. Coming now to the case of the non abelian codes, it is clear from eq. (27) that it is possible to achieve a large performance in two different ways. One by choosing a large p and the other by choosing a large group. So I next tried non abelian models with $p = 2$ hoping to get at the same time good performance codes and large basin of attraction of the simulated annealing algorithm. The results are shown in the figures and seem very promising.

The models I simulated have the connectivity of a thre dimensional cubic lattice, and there are only two "spin" interactions, i.e. they all have $p = 2$ and rate $R = 1/3$. In all the figures the upper line shows the error probability per bit as a function of the signal to noise ratio for the uncoded message, for the purpose of comparison, while the lower line shows the error probability per bit in the weak noise limit, as computed analytically.

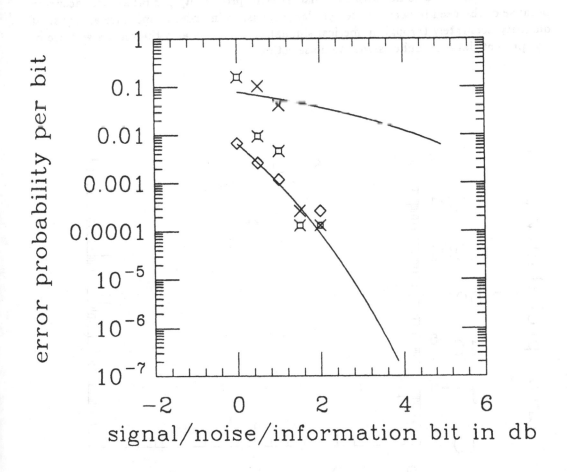

Figure 2

Figure 1 shows the results for the discrete group $\mathbf{H} = Z_2 \otimes [\overline{\mathbf{Q}}]^2$ whose input alphabet has length $l = 32$ while figure 2 shows the results for the case $\mathbf{H} = Z_2 \otimes [\overline{\mathbf{Q}}]^4$ with $l = 128$. Figure 3 shows, for comparison the results for the Ising code with $R = 1/2$ and $\nu = 7$ and Viterbi decoding. Different points in figures 1 and 2 correspond to different annealing schedules. In figure one the points marked with a square were obtained using 440 iterations per information bit, the points with a diamond using 900 iterations and the points with a cross using 1200 iterations per information bit. In figure 2 the points with a cross correspond to 1500 iterations, those with a square to 2100 iterations and the points with a diamond correspond to 4300 iterations per information bit. The points with the smallest error probability per bit are less accurate, because of the need to accumulate very large statistics in order to measure a very small quantity accurately through numerical simulations. This is also the reason why we do not present data for higher signal to noise ratios.

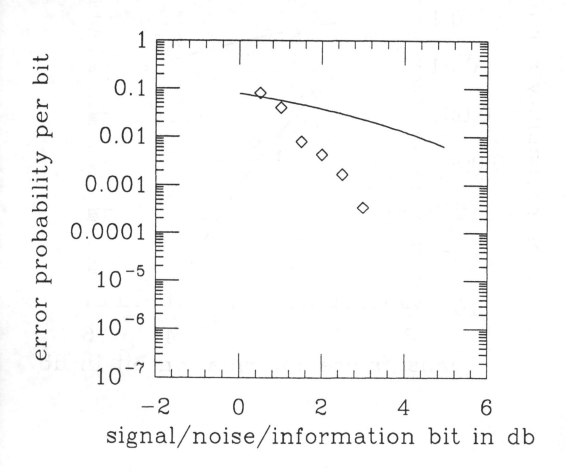

Figure 3

We first remark that our points fall on the curve obtained analytically in the weak noise limit, at least when the signal to noise ratio is not very small. This means that it is possible to reach by simulated annealing the ground state of our Hamiltonian, or a very good approximation of it. Secondly the algorithmic complexity, i.e. the number of iterations needed to find the ground state, depends on the noise level. The convergence is much faster for weaker noise. This is the opposite of what happens with deterministic decoding algorithms of the Viterbi type, where the algorithmic complexity of decoding is the same, irrespectively of the noise level. Finally we remark that better codes obtain better performance but at the cost of a more complex decoding.

The $\nu = 7$ Viterbi code illustrated in figure 3 is one of the most performant codes used in practice. Our codes illustrated in figure 1 and particularly the one of figure 2 obtain much smaller error for the same signal to noise ratio, but at the cost of a longer decoding. The difference is particularly significant for weak signals (zero to one decibels) where the traditional algorithms fail to achieve strong gains.

Because the Monte-Carlo algorithm can be made parallel, custom circuits, of the neural network type, can be used for decoding. Neural network models for error correction codes have already been considered by Platt and Hopfield[14].

The simulations show that simulated annealing works very well as a decoding algorithm and one may wonder why, because it is known that finding the ground state of spin-glasses is a NP-complete optimisation problem. The reason is simple. We are interested here in the ferromagnetic phase and not in the spin-glass phase. Furthermore at the noise levels we are interested in, it can be shown that the error does not depend on the exact determination of the ground state; replica symmetry breaking, which signals the appearance of quasidegenerate metastable states, occurs at very low temperature and it is known that with a very good approximation, the magnetisation does not vary below that temperature.

Clearly the central issue for practical applications is the performance of approximate decoding and a more systematic numerical study is needed.

References

1) Shannon, C. E., *Bell Syst. Tech. J.* 27, 379 and 623 (1948)
2) Shannon, C. E. and Weaver W., A Mathematical Theory of Communication, (Univ. of Illinois Press, 1963)
3) Sourlas, N., *Nature* 339, 693 (1989)
4) McEliece, R. J., The Theory of Information and Coding, (Encyclopedia of Mathematics and its Applications, Addison-Wesley, 1977)
5) Clark, G. C. and Cain, J. B. , Error-Correction Coding for Digital Communications, (Plenum Press, 1981)
6) Mézard, M., Parisi, G., and Virasoro, M. A., Spin Glass Theory and beyond, (World Scientific, 1987)
7) Derrida, B. *Phys. Rev.* B24, 2613-2626 (1981)
8) Gross, D. J. and Mézard, M. *Nucl. Phys.* B240, 431-452 (1984)
9) Gardner, E. *Nucl. Phys.* B257, 747-765 (1985)

330

10) Sherrington, D. and Kirkpatrick, S. *Phys. Rev. Lett.* 35, (1975) 1792-1796; *Phys. Rev.* B17, 4384-4403 (1978)
11) Vannimenus, J., Toulouse, G., and Parisi, G., *J. Physique(Paris)* 42, 565-571 (1981)
12) Kirkpatrick, S., in Lecture Notes in Physics 149, (Springer Verlag, 1981)
13) Kirkpatrick, S., Gelatt, C. D., and Vecchi, M. P., *Science* 220 671-680 (1983)
14) Platt, J. C., and Hopfield, J. J., in Neural Networks for Computing, (ed. Denker, J.) (AIP 1986)

SYNERGETIC COMPUTERS - AN ALTERNATIVE TO NEUROCOMPUTERS

H. Haken

Institut fur Theoretische Physik und Synergetik

Universitat Stuttgart

Pfaffenwaldring 57/IV D-7000 Stuttgart

1. Why various approaches?

Many readers probably know the story of a drunken man who went home but then he
realized that he has lost his key. So he started searching for his key under a
lantern. Just by chance a friend of his came along the street and asked the
drunken man: "Why do you look for your key under the lantern? Are you sure that
you have lost it here?" So the drunken man replied: "No, but it is the only
place where there is light." I think that this story bears of truth to the field
of neurocomputers. Let me first talk about the by now classical "lantern", which
was established by McCulloch and Pitts [1] nearly fifty years ago. They
conceived the brain as a neural net in which each neurone was modeled as a
system having two states "on" and "off" or "active" or "non-active". The
activation was achieved by the sum of inputs from the other two-level neurones.
When the sum exceeded a certain threshold, the neurone was activated and could
send out its signal. As was shown by McCulloch and Pitts such a network was
capable of doing all the logical operations of a Boolean algebra once the
connections between the neurones were weighed adequately. A first simple
realization of such a network was made by Rosenblatt [2] by means of his
perceptron. Later on this approach has been refined by a number of research
workers, e.g. by replacing the threshold by means of a sigmoid curve and by
using three layers instead of one. The approach made under the McCulloch and
Pitts lantern may be called bottom up, because here first the properties of
individual elements are fixed and then their connections are determined in such
a way that the network performs a specific task. A number of theoretical
physicists have contributed to the theoretical understanding, where I wish to
mention names such as Amari [3], Caianiello [4], Hopfield [5], and Amit,
Gutfreund, and Sompolinsky [6]. At the same time it is a remarkable effect that
over the years a shift of emphasis has occurred. While first the network was
conceived to perform logical operations, in its later interpretation it is
rather considered from the point of view of dynamical systems theory and here
the attractor states of the system are studied. While there are beautiful
theoretical approaches to the study of such networks, it must not be

overlooked that there are a number of serious drawbacks. The first of all is that the network possesses attractor states which are not wanted, i.e. they represent for instance patterns that are not stored and are not to be retrieved by the network. Time consuming methods are necessary to get the system out of these unwanted states for instance by simulated annealing. The same is true for learning processes which again may be time consuming and may lead to states which are not the optimal ones or even wrong. From the more theoretical sight one should mention that there is no general theory for the behavior of such a network. This has led some people to speak of some mysticism which has occurred with respect to the use of these networks. So in spite of some spectacular successes of this approach for instance by the NET talk system of Sejnowski [7], there is certainly a need either to improve the theory of these networks and of its applications or to look for different construction principles. This leads me to the discussion of the second lantern which is based on my own field of synergetics.

In synergetics [8] so far the spontaneous, i.e. self-organized, formation of spatial, temporal or functional structures in systems far from thermal equilibrium has been studied. This approach has been described in a number of text books so that I shall not present this field here. I just wish to mention a few features which are essential for our approach in the field of computers and pattern recognition by them. Consider a specific system, namely a fluid heated from below. Beyond a critical value of the temperature difference between its lower and upper surface, the fluid may spontaneously form rolls of up- and down-welling fluid. When the vessel is circular, the orientation of the rolls may be arbitrary depending on initial fluctuation or on a partly ordered initial state. In other words, the fluid shows multistability, and which stable state is reached depends on its preparation or in still other words, the fluid may form a totally ordered pattern out of a partially ordered pattern. Thus it acts as an associative memory. This may be used to formulate an algorithm which acts as follows: First a set of specific patterns are stored in the computer, then an algorithm is established which pulls an initially given partly ordered state into one of the prescribed patterns, provided the initially given pattern is similar to one of the stored patterns. It is then pulled to the one to which it had been most similar in the beginning.

Before I describe this formalism and derive its mathematical properties I wish to make a few general statements, how this formalism may be used.

1. It can be implemented as an algorithm on a serial computer in order to perform the recognition of patterns, such as faces, city-maps, a.s.o..

2. It can be realized by a parallel network where the individual nodes or neurones have properties which are rather different from those of the hitherto known neurones of McCulloch and Pitts.

3. The model may be used as a model of the functioning of the brain when patterns are recognized.

4. This model can be used to explain psychophysical results for instance on the recognition of ambivalent patterns where it is known from psychophysics that oscillations of the perception occur.

2. The algorithm of the synergetic computer [9]

a) The concept of associative memory. When an incomplete set of data is given, the associative memory must be able to complement it.

b) We wish to construct a dynamical process by which pattern recognition is done. To this end we wish to construct a potential landscape in which a fictitious particle, which describes the patterns, moves. The particle is pulled into one of its attracting states provided an initial condition is set so that the symmetry may be broken. In other words, the pattern is recognized once it is within its basin of attraction.

c) We wish to treat the system as a synergetic system according to the following idea: A fluid in which part of the rolls have been formed, may generate its order parameter which competes with the other order parameters of the system. Because of the special preparation of the initial state with partially ordered subsystems, the order parameter belonging to that specific order wins the competition and, eventually, enslaves the total system so that now the total system is in the ordered state. In pattern recognition we wish to employ the same mechanism. Once a set of features is given, they may form their order parameter which competes with still other order parameters. Eventually one order parameter, which was initially supported the most, wins and forces the system to exhibit the features which have been still lacking with respect to the special pattern. Thus we see that there is a complete correspondence between the complementation process during pattern formation and associative memory during pattern recognition.

In order to show how these ideas can be cast into a mathematical form, we first briefly discuss the selection of features. For our present purposes it will be sufficient to illustrate our procedure by means of a specific example. When we have a photograph of a face, we may put a grid over it in order to digitize the whole image. We label the individual cells or pixels by numbers 1, 2, ... N. To each pixel we attribute a number indicating its tone of grey. We then form the vector

$$\underset{\sim}{v} = \begin{matrix} v_1 \\ v_2 \\ . \\ . \\ v_N \end{matrix} \qquad (2.1)$$

Because we wish to store a whole set of different faces in the computer, we distinguish them by a label k so that $\underset{\sim}{v}$ is now replaced by the prototype vectors

$$\underset{\sim}{v}_k = \begin{matrix} v_{1k} \\ v_{2k} \\ \\ v_{Nk} \end{matrix} \qquad (2.2)$$

We shall assume that v_{kj} is real. The label k may adopt the values

$$k = 1, \ldots , M \qquad (2.3)$$

where M is the number of stored patterns. We shall assume that the number of patterns is smaller than or equal to the number of features

$$M \leq N . \qquad (2.4)$$

If not stated otherwise we subject the vectors $\underset{\sim}{v}_k$ to the condition

$$\sum_\ell v_{k\ell} = 0 \qquad (2.5)$$

which can always be achieved by forming

$$\underset{\sim}{v}_k = \tilde{\underset{\sim}{v}}_k - \frac{1}{N} \sum_\ell \tilde{v}_{k\ell} \qquad (2.6)$$

where $\tilde{\underset{\sim}{v}}_k$ are the "raw vectors" and N is the number of components.

For what follows we need the transposed vector which is defined by

$$\bar{\underset{\sim}{v}}_k = (v_{k1}, v_{k2}, \cdots v_{kN}). \qquad (2.7)$$

We shall assume throughout this paper that the normalization

$$(\bar{\underset{\sim}{v}}_k \underset{\sim}{v}_k) = \sum_{j=1}^N v_{kj}^2 = 1 \qquad (2.8)$$

holds which can always be achieved by dividing a "raw vector $\tilde{v}_k$" by $(\tilde{v}_k \, \tilde{v}_k)^{1/2}$. Because the vectors v_k are not necessarily orthogonal to each other, we need the adjoint vectors

$$v_k^+ = (v_{k1}^+, \ v_{k2}^+, \ \ldots \ v_{kN}^+) \ . \tag{2.9}$$

which obey the orthonormality relations

$$(v_k^+ \, v_{k'}) = \delta_{kk'} \ . \tag{2.10}$$

We shall represent the adjoint vectors v_k^+ as a superposition of the prototype vectors v_k

$$v_k^+ = \sum_{k'} a_{kk'} \, \bar{v}_{k'} \ , \tag{2.11}$$

where $a_{kk'}$ are constants and $\bar{v}_{k'}$ denotes the transposed of $v_{k'}$. As we shall see later on, (2.11) will help us considerably in doing the numerical calculations.

2.1 Construction of the dynamics

As indicated at the beginning of this paper, our goal is this: Let a test pattern, which we denote by the vector q, be given. Then we wish to construct a dynamics which pulls the test pattern via intermediate states $q(t)$ into one of the prototype patterns v_{k_0}, namely the one to which $q(0)$ has been closest, i.e. in whose basin of attraction it has been lying

$$q(0) \rightarrow q(t) \rightarrow v_{k_0} \ . \tag{2.12}$$

The thus required equation reads

$$\dot{q} = \sum_k \lambda_k (v_k^+ \, q) \, v_k - B \sum_{k' \neq k} (v_{k'}^+ \, q)^2 \, (v_k^+ \, q) \, v_k$$

$$- C(q^+ \, q) \, q + F(t) \tag{2.13}$$

Note that the constant B can be made dependent on k and k'

$$B \rightarrow B_{kk'} \tag{2.14}$$

and then has to be taken under the sum. But for the moment being such a generalization is not needed.

Let us briefly discuss the meaning of the individual terms on the right-hand side of (2.11). λ_k are called attention parameters. A pattern can be recognized only if the corresponding attention parameters are positive otherwise it will not be recognized. The expression $\underset{\sim}{v}_k \cdot \underset{\sim}{v}_k^+$ acts as a matrix. As one can see, first $\underset{\sim}{v}_k^+$ is multiplied by a column vector $\underset{\sim}{q}$ so that a scalar is generated. Finally, the vector $\underset{\sim}{v}_k$ becomes effective which again acts as a column vector. Thus by the whole process, a column vector is transformed into a new column vector or, in other words, $\underset{\sim}{v}_k \cdot \underset{\sim}{v}_k^+$ acts as a matrix. This matrix has occurred in a number of early publications by a number of authors and is called learning matrix. The next term serves for the discrimination between patterns. Because the first term, at least when λ is positive, will lead to an exponential growth of $\underset{\sim}{q}$, we need a factor which limits that growth. This is achieved by the third term on the right-hand side. Finally $\underset{\sim}{F}$ are fluctuating forces which we shall use occasionally, but which we shall drop when not otherwise noted.

We shall decompose the vector $\underset{\sim}{q}$ into the prototype vectors (2.2) and into a residual vector $\underset{\sim}{w}$

$$\underset{\sim}{q} = \sum_{k=1}^{M} \xi_k \underset{\sim}{v}_k + \underset{\sim}{w} \tag{2.15}$$

and require that

$$(\underset{\sim}{v}_k^+ \underset{\sim}{w}) = 0 \quad \text{for all } k=1, \ldots, M \tag{2.16}$$

holds. We now define $\underset{\sim}{q}^+$ which appears in $C(\underset{\sim}{q}^+ \underset{\sim}{q}) \underset{\sim}{q}$ in (2.13), by means of the relations

$$\underset{\sim}{q}^+ = \sum_{k=1}^{M} \xi_k \underset{\sim}{v}_k^+ + \underset{\sim}{w}^+ \tag{2.17}$$

where $\underset{\sim}{w}^+$ obeys the orthogonality relations

$$(\underset{\sim}{w}^+ \underset{\sim}{v}_k) = 0 \quad \text{for all } k=1, \ldots, M . \tag{2.18}$$

One readily convinces oneself that the relation

$$(\underset{\sim}{v}_k^+ \underset{\sim}{q}) = (\underset{\sim}{q}^+ \underset{\sim}{v}_k) \tag{2.19}$$

holds. Namely by inserting (2.15) on the left-hand side of (2.19), we obtain

$$(\underset{\sim}{v}_k^+ \underset{\sim}{q}) = (\underset{\sim}{v}_k^+ (\sum_{k'=1}^{M} \xi_{k'} \underset{\sim}{v}_{k'} + \underset{\sim}{w})) \tag{2.20}$$

which by means of the orthogonality relations (2.10), (2.16) can be written as

$$(\underset{\sim}{v}_k^+ \underset{\sim}{q}) = \xi_k . \tag{2.21}$$

The same result is obtained when we insert (2.17) on the right-hand side of (2.19) and utilize the orthogonality relations (2.10), (2.18). (2.19) allows us to express the right-hand side of (2.13) either by means of the left-hand side of (2.19) everywhere or of the right-hand side of (2.19) everywhere so that only the variables $\underset{\sim}{q}$ or $\underset{\sim}{q}^+$ are involved. One then readily may convince oneself that the formal equations

$$\dot{\underset{\sim}{q}} = - \frac{\partial V}{\partial \underset{\sim}{q}^+} \, , \tag{2.22}$$

$$\dot{\underset{\sim}{q}}^+ = - \frac{\partial V}{\partial \underset{\sim}{q}} \tag{2.23}$$

hold, in which V plays the role of a potential function which is explicitly given by

$$V = - \frac{1}{2} \sum_{k=1}^{M} \lambda_k \, (\underset{\sim}{v}_k^+ \, \underset{\sim}{q})^2 + \frac{1}{4} B \sum_{k \neq k'} (\underset{\sim}{v}_k^+ \, \underset{\sim}{q})^2 \, (\underset{\sim}{v}_{k'}^+ \, \underset{\sim}{q})^2$$

$$+ \frac{1}{4} C \, (\underset{\sim}{q}^+ \, \underset{\sim}{q})^2 \, . \tag{2.24}$$

We now wish to derive the order parameter equations belonging to equation (2.13). To this end we multiply (2.13) from the left by $\underset{\sim}{v}_k^+$. Using the orthogonality relations between $\underset{\sim}{v}_k^+$, $\underset{\sim}{v}_k$, $\underset{\sim}{w}$, and $\underset{\sim}{w}^+$, and the definition (2.21), we obtain

$$\dot{\xi}_k = \lambda_k \, \xi_k - B \sum_{k' \neq k} \xi_{k'}^2 \, \xi_k - C \, (\sum_{k'=1}^{M} \xi_{k'}^2 + (\underset{\sim}{w}^+ \, \underset{\sim}{w})) \, \xi_k \, . \tag{2.25}$$

Note that the sum over $k' \neq k$ runs over all values $k' = 1 \ldots M$ except for the value k which appears at the coefficient of ξ_k on the left-hand side of (2.25). When we multiply (2.13) by vectors $\underset{\sim}{u}_\ell^+$ which belong to the space orthogonal to the prototype pattern vectors $\underset{\sim}{v}_k$ and sum up over the individual components, we obtain the equations

$$\dot{\underset{\sim}{w}} = - C \, (\sum_{k'=1}^{M} \xi_{k'}^2 + (\underset{\sim}{w}^+ \, \underset{\sim}{w})) \, \underset{\sim}{w} \, . \tag{2.26}$$

In the derivation of (2.26), the following equations have been used:

$$\underset{\sim}{w} = \sum_{\ell=M+1}^{N} f_\ell \, (t) \, \underset{\sim}{u}_\ell \tag{2.27}$$

$$\underset{\sim}{w}^+ = \sum_{\ell=M+1}^{N} f_\ell \, (t) \, \underset{\sim}{u}_\ell^+ \tag{2.28}$$

$$(\underline{u}_\ell^+ \, \underline{u}_\ell) = \delta_{\ell\ell'} \tag{2.29}$$

$$(\underline{u}_\ell^+ \, \underline{v}_k) = (\underline{v}_k^+ \, \underline{u}_\ell) = 0 \; . \tag{2.30}$$

Because the factor in front of $\underline{w}$ on the right-hand side of (2.26) is negative everywhere, we find

$$|\underline{w}| \to 0 \text{ for } t \to \infty \tag{2.31}$$

where the norm $|\underline{w}|$ is defined by

$$|\underline{w}| = (\underline{w}^+ \, \underline{w})^{1/2} \; . \tag{2.32}$$

Thus the dynamics reduces the problem to the prototype vector space

$$\dot{\xi}_k = \lambda_k \, \xi_k - B \sum_{k'\neq k} \xi_{k'}^2 \, \xi_k - C \, (\sum_{k'=1}^M \xi_{k'}^2) \, \xi_k \; . \tag{2.33}$$

The order parameters obey the initial condition

$$\xi_k(0) = (\underline{v}_k^+ \, \underline{q}(0)) \tag{2.34}$$

which follows directly from (2.20). Quite clearly, the right-hand side can be derived from a potential function

$$\dot{\xi}_k = - \frac{\partial \tilde{V}}{\partial \xi_k} \tag{2.35}$$

where $\tilde{V}$ is given by

$$\tilde{V} = - \frac{1}{2} \sum_{k=1}^M \lambda_k \, \xi_k^2 + \frac{1}{4} B \sum_{k'\neq k} \xi_{k'}^2 \, \xi_k^2 + \frac{1}{4} C (\sum_{k'=1}^M \xi_{k'}^2)^2 \; . \tag{2.36}$$

Note that the sum over $k' \neq k$ is now a double sum where k' and k each run from 1 ... M, with the only exception $k' = k$. The role played by the individual terms in (2.24) or in (2.36) can be easily visualized when we look at a plot of the potential V in a two-dimensional feature space, where the two axes are spanned by the two prototype vectors $\underline{v}_1$ and $\underline{v}_2$. Close to the origin, the terms quadratic in $\underline{q}$ dominate and we see that the first sum in (2.24) has a negative sign provided the attention parameters λ_k are positive. This decrease of the potential is stopped when $\underline{q}$ is further increasing because then the last term in (2.24) takes over and, eventually, increases much more quickly than the first term decreases. This interplay between the first and last term generates the valleys. The middle term on the right-hand side of (2.24) finally generates the ridges which define the basins of attraction and thus allows for the discrimination between different patterns.

2.2 Important properties of $V(\xi_k)$

In this section we shall assume that all attention parameters are equal and positive

$$\lambda_k = \lambda > 0 . \tag{2.37}$$

and we put $\lambda = C$.

We summarize the results. The stable fixed points are at $\underline{q} = \underline{v}_k$, i.e. at the prototype patterns, and there are no other stable fixed points. The stable fixed points are equivalently characterized by $\xi_k = 1$, all other ξ's $= 0$. At $\underline{q} = 0$ there is the only unstable fixed point. There are saddle points which are situated at $\xi_{k_1} = \xi_{k_2} = \ldots = \xi_{k_m} = 1$, all other ξ's $= 0$. Here k_1, k_2, $\ldots$ k_m may be any selection out of 1, $\ldots$, M. If $|\xi_{k_0}|$ is initially bigger than any other $|\xi|$, the dynamics pulls it to the stable fixed point $\xi_{k_0} = 1$, all other ξ's $= 0$. If $|\xi_{k_1}| = |\xi_{k_2}| = \ldots = |\xi_m|$ are initially bigger than any other ξ_k, the dynamics terminates at the corresponding saddle point, from which only a fluctuation can drive the system into any of the fixed points belonging to k_1, k_2, $\ldots$ k_m.

References

[1] W.S. McCulloch, and W.H. Pitts, A logical calculus of the ideas immanent
     in nervous activity, Bulletin of Mathematical Biophysics, 5, 115-133
     (1943)

[2] R. Rosenblatt, Principles of neurodynamics, Spartan Books, New York
     (1962)

[3] S. Amari, Characteristics of randomly connected threshold element
     networks and network systems, Proc. IEEE, 59, 35-47 (1971)

[4] E.R. Caianiello, J. Theor. Biol. 1, 209 (1961)

[5] J.J. Hopfield, Neural networks and physical systems with emergent
     collective computational abilities, P.Nat.Acad.Sci. U.S.A., vol. 79,
     2445-2458 (1982)

[6] D.J. Amit, H. Gutfreund, and H. Sompolinsky, Spin-glass models of neural
     networks, Phys. Rev. A2, 1007-1018 (1986)

[7] T.J. Sejnowski, and C.R. Rosenberg, Complex systems, 3, 145 (1987)

[8] H. Haken, Synergetics, An Introduction, 3rd ed., Springer, Berlin (1983)

[9] H. Haken, Synergetic computers for pattern recognition and associative
     memory. In: H. Haken (ed.), Computational systems, natural and
     artificial. Springer, Berlin, Heidelberg, New York (1987)

DYNAMICS OF THE KOHONEN MAP

Tamás Geszti, István Csabai, Ferenc Czakó and Tamás Szakács
Department of Atomic Physics, Eötvös University, H-1088 Budapest, Hungary,

Roger Serneels
Limburgs Universitair Centrum, Universitaire Campus, 3610 Diepenbeek, Belgium,

Gábor Vattay
Department of Solid State Physics, Eötvös University, H-1088 Budapest, Hungary

1. INTRODUCTION

Kohonen's feature map [1] is a kind of neural computer, combining explicit feed-forward and simulated feedback elements in an efficient way. It is a screen composed of neuron-like elements (Figure 1.) of which only one fires at a time. Which one: it depends on the actual value of a continuous (possibly vectorial) input. Then one says that the input (a "feature") is mapped onto the firing neuron. This is meant to mimic the mapping of visual input from a given direction, received by our retina, onto a neuron or a small group of them in the visual cortex.

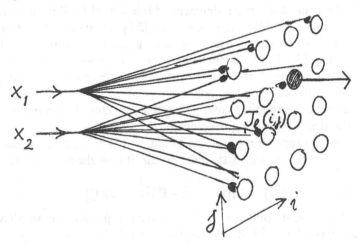

Figure 1. Kohonen's feature map: input vector $\vec{x}$ is fed forward to all neurons to select a 'winner' that will fire alone; feedforward connections learn to choose neighbouring winners for nearby inputs.

Kohonen's feature map is also a learning machine: it learns to associate the firing of neighbouring neurons to neighbouring inputs, like an infant's brain learns how a true image of the visual input can be generated in the visual cortex by neural computation, without using linses or mirrors.

Let us regard the two parts of the job, viz. creating the localized mapping and learning topological faithfulness separately. Concerning the first, in our brain output localization is achieved by lateral inhibitory connections creating a dynamical competition between neurons receiving the same multiplexed inputs: after a short while only the winner remains firing. In Kohonen's implementation this is simulated by an algorithm to select a formalized "winner" among neurons in a screen. The common inputs x_l ($l = 1, 2, \ldots$) reach different neurons labelled by an index vector $\mathbf{i} = i, j, \ldots$ through different feed-forward connections of strength $J_l(\mathbf{i})$. Inputs and corresponding connection strengths on a neuron can be written as vectors in another space: $\vec{x}$ and $\vec{J}(\mathbf{i})$ respectively. Then Kohonen's algorithm is this: winner is neuron $\mathbf{i}^*$ for which $\vec{J}(\mathbf{i})$ is closest to the actual input $\vec{x}$, i.e.

$$\mathbf{i}^*: \quad |\vec{J}_{\mathbf{i}^*} - \vec{x}| \text{ is minimum of } |\vec{J}_\mathbf{i} - \vec{x}|. \tag{1}$$

This is abstraction from neural algorithms assuring large output from a neuron $\mathbf{i}$ for which $\sum_l J_l(\mathbf{i})x_l \equiv \vec{J} \cdot \vec{x}$ is large: if both $\vec{J}$ and $\vec{x}$ are normalized to the same dimensionless number, the biggest scalar product is equivalent to the smallest Euclidean distance. As a further generalization, the thing seems to work with any reasonable definition of the distance.

The second part of the job, to make the input-to-winner mapping topologically faithful, is a task for an unsupervised, self-organizing network: no one fixes which neuron should respond to which input range, the only requirement being that neighbourhood relations should be respected.

Our brain solves the task by a refinement of the lateral inhibition architecture into what is called "Mexican hat" lateral connections [2,3]: excitation for close neighbours, inhibition for more remote ones. That forces neighbouring neurons to fire together, which is gradually fixed by Hebbian learning in the feed-forward connections $J_l(\mathbf{i})$. This creates a shift of those connection strengths in a way to establish a faithful mapping of neighbourhood relations among input features.

The Kohonen version is again a compact algorithm. One should repeat the following two steps until the desired ordering is reached:

(i) generate a random input $\{x_l\}$ and use Equation (1) to find the winner $\mathbf{i}^*$;

(ii) modify connection strengths of the winner and its neighbours according to

$$J_l(\mathbf{i}) \rightarrow J_l(\mathbf{i}) + \lambda(\mathbf{i} - \mathbf{i}^*)\left[x_l - J_l(\mathbf{i})\right], \tag{2}$$

where $\lambda(\vec{r})$ usually depends only on $|\vec{r}|$, being sharply peaked for small values of the distance r, giving largest amplitude of adaptation to $\mathbf{i}^*$, then next largest to its closest neighbours, etc. This modification brings connection strengths in a neighbourhood closer to the common $\vec{x}$, which is an efficient way to bring them closer to one another. This establishes topological faithfulness of the mapping from $\vec{x}$ to $\mathbf{i}$.

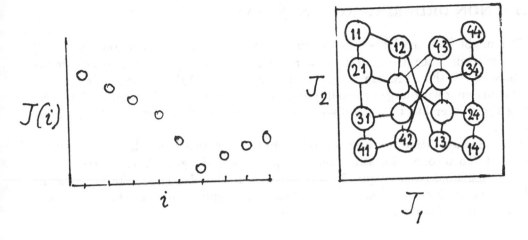

Figure 2. Partially ordered Kohonen maps with defects: *a)* a breaking point in 1 dimension; *b)* a twist in two dimensions. In *b)* a circle represents the connection vector $\vec{J}$ of a neuron; those belonging to neighbouring neurons are connected by 'chords'.

To make the mapping one-to-one (apart from discretization), topology is not sufficient: the components of $\vec{J}(i)$ must become monotonous functions of i.

This takes time: initially the mapping – although neighbourhood relations are established – contains breaking points (in 1 dimension) or twists (in more dimensions), see Figure 2. However, it is the miracle of the Kohonen learning process that if sufficient time is given and the input vector $\vec{x}$ has no more dimensions than the output screen index i (for the opposite see [1,4]), a mapping not only topologically faithful but also free of twists is obtained. The creation of this twistless order is the subject of the following sections.

Once that happened, the mapping can be smoothed out by continuing to run the Kohonen algorithm with a learning function $\lambda(r)$ gradually reducing both in amplitude and in width [1]. Care must be taken not to do this reduction too fast, and too fast is any decay function of a finite integral [4].

2. THE NON-ORDERING KOHONEN MAP

One might wonder why is it necessary to include into the modification process (2) the winner itself, since it is closest to the random input anyway. Closest is not always close however: if there are large voids in the distribution of the connection strengths, an input arriving with large probability into a void will help to fill it out. Therefore, shifting the winner has the important role to assure a uniform representation of the input set.

Let us regard a degenerate version of the Kohonen model in which *only* the winner is modified. Then connection strengths $\vec{J}$ are just atoms of a gas, filling out the 'vessel' defined by the input range (Figure 3), and diffusing indefinitely. Their diffusion constant can be measured, if periodic boundary conditions are imposed and an 'atom' is allowed to move on freely from one image vessel to the other.

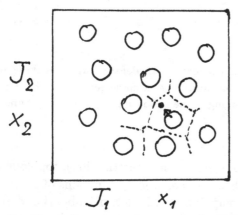

Figure 3. Connection strength vectors of a Kohonen map are positions of quasi-atoms filling out the input range

If successive diffusion steps were not correlated, it would be easy to calculate the mean square displacement in one step and subsequently the diffusion constant. Approximating the Voronoi cell by a sphere (which sounds reasonable if the dimensionality is not too high), one obtains

$$D \approx D_0 = \frac{\left(\Gamma(\tfrac{d}{2})\tfrac{d}{2}\right)^{2/d}}{1 + \tfrac{2}{d}} \frac{(\lambda(0))^2}{n^{2/d}}. \tag{3}$$

Dynamical correlation effects can change this simple picture. However, in our numerical study no strong deviations have been observed from the simple behaviour (Figure 4).

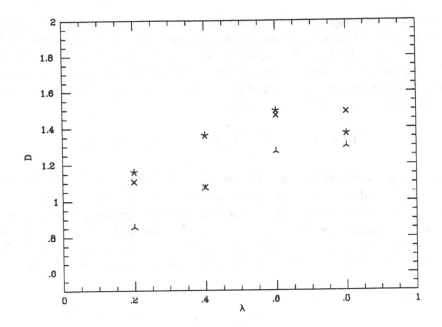

Figure 4. Reduced diffusion constant D/D_0 for a non-ordering Kohonen map (see text) as a function of the single coupling strength $\lambda = \lambda(0)$, averaged over different densities; dimensions: ⅄ 3, ✕ 4, ✶ 5.

3. FRAMEWORK OF A DYNAMICAL DESCRIPTION

It is easy to see how the Kohonen process brings neighbours in the spatial index **i** together in connection strengths $J_l(\mathbf{i})$. It is more difficult to see how twistless order is created.

The global reason is this [1,5]: once the monotonous order is established, the Kohonen process is unable to change it. Therefore such ordered states are attractors of the dynamics, and the diffusion-like random process sooner or later drives the system to one of those (symmetry-breaking) attractors.

The stability of the ordered phase depends on the shape of the learning function $\lambda(r)$. In one dimension the property trivially holds if $\lambda(r)$ is a monotonously decreasing function. Indeed, let us assume that $J(i)$ is already monotonously ordered. Then the closer neuron i is to winner i^*, the closer is $J(i)$ to input x; consequently, the bigger is its relative displacement towards x: the order of two connection strengths cannot be reversed.

In more dimensions a formal proof of the stability of the twistless ordered state seems quite difficult; the monotonous decreasing property of $\lambda(r)$ is by all means a necessary condition.

On the borderline of stability the system becomes critical which means an effective decoupling of remote points. This brings a major simplification into the dynamics of fluctuations, as described below in Section 5.

To obtain more insight to the nature of the ordering process, let us observe that $\vec{J}(\mathbf{i}, t)$ is a vector field on a lattice of nodes $\mathbf{i}$, changing in time steps of the Kohonen process (1) – (2), and try to find an approximate equation of motion for that field.

Following the usual procedure, we split the evolution into and average and a random noise term. For the average evolution of the connection strengths $\vec{J}(\mathbf{i}, t)$ of N neurons we obtain a coupled N-dimensional mapping:

$$\vec{J}(\mathbf{i}, t+1) = \vec{J}(\mathbf{i}, t) + \sum_{\mathbf{j}} p_{\mathbf{j}} \lambda(\mathbf{i} - \mathbf{j}) [\vec{x}_{\mathbf{j}} - \vec{J}(\mathbf{i}, t)] \tag{4}$$

where $p_{\mathbf{j}}$ is the probability that neuron $\mathbf{j}$ is the winner (this happens if $\vec{x}$ is in the Voronoi cell around $\vec{J}(\mathbf{j})$), and $\vec{x}_{\mathbf{j}}$ is the conditional mean value of $\vec{x}$ in the same case.

Equation (4), although easy for a numerical study, is very complicated for a theoretical treatment, mainly because of the need of evaluating the Voronoi cell weights and averages (see the complications arising in a similar but much simpler case [6]). Therefore, at least for later stages of ordering when the field is already smooth although not monotonous yet, it is promising to turn to a continuum approximation.

Then $\vec{J}(\mathbf{i}, t)$ becomes a continuous field $\vec{J}(\mathbf{r}, t)$. Expanding this field around $\mathbf{r} = \mathbf{i}$, it turns out that the winner selection probability is proportional to the Jacobian $|\partial \vec{J} / \partial \mathbf{r}|$. The average $\vec{x}_{\mathbf{j}}$ is a weighted average of the corners of the Voronoi cell that can be approximated by $\vec{J}(\mathbf{j}, t)$ (neglecting second derivative corrections). Then first derivatives cancel from the sum in (4) if $\lambda(\mathbf{r})$ is inversion symmetric. If it is even isotropic then – turning to continuous time – the mean evolution equation becomes

$$\partial_t \vec{J}(\mathbf{r}, t) = W(\mathbf{r}) \, |\partial \vec{J} / \partial \mathbf{r}| \, S \, \Delta \vec{J}(\mathbf{r}, t) \tag{5}$$

with the stiffness constant

$$S = \frac{1}{2} \sum_{\mathbf{j}} \lambda(\mathbf{i} - \mathbf{j}) \cdot (\mathbf{i} - \mathbf{j})^2. \tag{6}$$

This looks like a nonlinear diffusion equation, with a strange complication lumped into the innocent-looking factor $W(\mathbf{r})$. It hides the fact that while ordering is not monotonous yet, different groups of neurons having approximately equal connection strengths $\vec{J}$ are competing for the input. $W(\mathbf{r})$ is meant to express the share of neurons around $\mathbf{r}$ in this competition.

4. NUMERICAL RESULTS AND THEIR QUALITATIVE DISCUSSION

Without solving the above equations, a qualitative picture of the ordering process emerges.

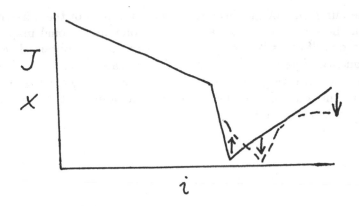

Figure 5. The final stage of topological ordering in one dimension

At least in one dimension, there is a systematic (although very weak) driving force to sweep out defects of monotonous ordering, even without any built-in asymmetry.

The qualitative reason is in the asymmetry between the two sides of the almost-ordered configuration. A fluctuational displacement of the boundary is followed by a slow rearrangement of the two ordered regions. The displacement can happen in both directions, however if it goes towards the nearby end of the sample then it is accompanied by a larger curvature, causing somewhat faster response and reducing the probability to step back before the long-range rearrangement might be completed.

In two dimensions the situation is less clear, since curvatures in one direction can be compensated by an opposite curvature in the other direction, therefore configurations like $J_1 \propto \mathbf{a} \cdot \mathbf{r}$, $J_2 \propto \sin(kx)\sinh(ky)$ can survive a long time, being steady for the mean evolution and evolving only by fluctuations.

For a connection strength varying as $\exp(-r^2/2\sigma^2)$, Equation (6) suggests a time scale $\propto \sigma^{-2}$ for the evolution. Instead, our simulations reveal an ordering rate of $\mathcal{O}(\sigma^{5.4})$, over three orders of magnitude in the rate. This clearly suggests that the ordering process is fluctuation dominated, and its nature remains to be understood.

5. CRITICAL DYNAMICS IN ONE DIMENSION

To see what might happen if the stability condition is violated, for a one-dimensional chain of N neurons, with $\lambda(0)$ and $\lambda(1) = \lambda(0) + \epsilon$ being the only non-vanishing connections (nearest-neighbour learning), we measured the fluctuations of the disorder parameter [1] $\Delta = \sum_{i=2}^{N} |J(i) - J(i-1)| - |J(N) - J(1)|$ starting from an ordered state with $\Delta = 0$. A fluctuating steady state was established with $\overline{\Delta} \propto \epsilon^{\xi}$ with $\xi \approx 2$ for $\epsilon > 0$.

In the critical point $\epsilon = 0$ ($\lambda(1) = \lambda(0) = \alpha$) there is no inversion yet, however for $\alpha > 1/2$ a strong tendency is observed for two neighbouring $J(i)$s to come close to each other. We carried out measurements of the distribution function $\rho(d)$ of the difference $d(i) = J(i) - J(i - 1)$ in the middle of an already ordered one-dimensional map. A Weierstrass-Mandelbrot-type self-affine [7] steady-state distribution was obtained (Figure 6), consisting of a sequence of peaks centered at d values of ratios about $1 - \alpha$. For $\alpha > 1/2$ the distribution function diverges at $d \to 0$, whereas for $\alpha < 1/2$ it goes to 0. The self-affine distribution function multiplies by a constant if its argument d is rescaled to $d' = (1 - \alpha)d$ (see Figure 6. for $\alpha = 0.8$).

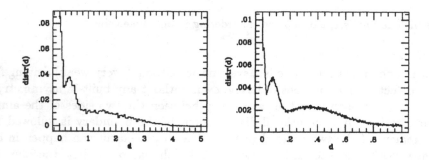

Figure 6. The distribution function of the distance d between connection strengths of two neighbouring neurons in an ordered one-dimensional Kohonen map with marginally stable connection strengths $\lambda(0) = \lambda(1) = \alpha = 0.8$, plotted on two different scales of d to illustrate self-affinity.

To understand the origin of this behaviour let us observe that $d = J(i) - J(i - 1)$ changes to

$$d' = (1 - \alpha)d \quad \text{if } i \text{ or } i - 1 \text{ is winner.} \tag{7}$$

On the other hand,

$$d' = d + \frac{\alpha}{N}y \quad \text{if } i + 1 \text{ or } i - 2 \text{ is winner,} \tag{8}$$

where y can be regarded as a random variable of a fixed distribution $p(y)$ in a mean-field-like approximation, neglecting correlations between neighbours except for the couple of i and j. In this approximation it becomes apparent that the competition of muliplicative shrinking and additive random stretching, each of which happens with probability 1/2, determines the steady distribution. For large α the shrinking dominates, causing the diverging accumulation at the origin. The simple one-dimensional random mapping (7) – (8) reproduces faithfully the distribution of d obtained from the full Kohonen map.

The steady distribution of (7) – (8) can be evaluated analytically from the integral equation

$$\rho(d) = \frac{1}{2}\frac{1}{(1-\alpha)}\rho\left(\frac{d}{1-\alpha}\right) + \frac{1}{2}\int p(y)\rho\left(d - \frac{\alpha}{N}y\right)dy. \tag{9}$$

From the point of view of the behaviour around the origin one can neglect the second term, since $\rho(x) = 0$ for $x < 0$ if the Kohonen map is not unstable. Then for the purely multiplicative process the integral equation has a solution of form $\rho(d) \propto d^\delta$, where

$$\delta = -1 - \frac{\ln 2}{\ln(1-\alpha)}. \tag{10}$$

In accordance with our measurements on the Kohonen map, $\delta < 0$ causing divergence for $\alpha > 1/2$ and vice versa.

6. OUTLOOK

The above results are only the beginning of a true understanding of the Kohonen map. The $\sigma^{-5.4}$ law points to a dominating role of fluctuations in the dynamics of ordering, and advances in theory are of clear practical importance. The self-affine distribution observed in the marginally stable one-dimensional map are rather for the theory-minded reader, unless it turns out in the future that it is good for your brain or your computer to operate in the marginally stable range.

REFERENCES

1. T. Kohonen, *Self-Organization and Associative Memory*, 2nd Edition (Springer, Berlin 1988).
2. S. Grossberg, *Biol. Cybern.* **21**, 145; **23**, 121 (1976).
3. D. J. Willshaw and C. von der Malsburg, *Proc. Roy. Soc.* B **194**, 431 (1976).
4. H. Ritter and K. Schulten, *Biol. Cybern.* **60**, 59 (1988).
5. M. Cottrell and J. C. Fort, *Biol. Cybern.* **53**, 405 (1986).
6. L. A. Bunimovich and Ya. G. Sinai, *Nonlinearity* **1**, 491 (1988).
7. T. Tél, *Z. Naturforsch.* **43a**, 1154 (1988).

EQUIVALENCE BETWEEN CONNECTIONIST CLASSIFIERS AND LOGICAL CLASSIFIERS

Laurent Bochereau, Paul Bourgine and Guillaume Deffuant

CEMAGREF, BP 121, 92185 Antony, France

Tel. (1) 40 96 61 21 / Fax (1) 40 96 60 36

Abstract : The aim of this paper is to show that connectionist and logical classifiers are functionally equivalent in the sense that they can always be derived from one another. Connectionist classifiers are defined as associative memories followed by a decision function. The associative memory models discussed in this paper are multilayer perceptrons and Hopfield models. By using a decision function, the associative memory can always be transformed into a connectionist classifier. On the other hand, logical classifiers are sets of logical rules which map an input space into an output or class space. These logical classifiers can be seen as the knowledge base of an expert system. Methods for deriving an equivalent logical classifier from a multilayer perceptron followed by a decision function are also presented. They include (i) defining a validity domain for the connectionist classifier, (ii) detecting invariance among the input patterns and (iii) extracting the equivalent logical rules.

Introduction

In a classification problem, the input units are pieces of information or signs that can be relevant or irrelevant for the classification task. There are different ways of coding the information on both the input and the output layers. In our work, we chose to use a localist coding scheme for these layers. Thus, each unit on the input and ouput layers corresponds to the presence/absence of a specific piece of information or the membership/non-membership to a given class. If an input parameter is to vary over an interval [a,b], then this parameter can always be encoded in p exclusive booleans representing the membership to an interval $[a_i, a_{i+1}]$, $a = a_0 < .. a_i ... < a_p = b$.

Connectionist classifiers and logical classifiers will be defined in the first section. It can be easily shown that building a connectionist classifier implementing a set of logical rules is always possible [3]. Therefore, the main contribution of this paper is to show that there always exists a logical classifier that is functionally equivalent to a connectionist classifier (§2). This being the case, the result has a very interesting consequence: the connectionist classifier is no longer seen as a black box but may provide an explicit explanation.

The third section discusses the concept of validity domain for a connectionist classifier and describes methods for defining it. The fourth section concerns specifically connectionist classifiers constructed as the composition of a multilayer perceptron and a decision function. Methods for extracting a set of logical rules equivalent to such a connectionist classifier are discussed. This includes techniques for delimiting a validity domain and detecting invariance in the input space.

1. <u>Connectionist classifiers and logical classifiers</u>

1.1. <u>Associative memories and decision functions</u>

The <u>associative memories</u> considered in this paper are neural networks that map an input space into an output space. The output space may be seen as classes, prototypes or hypotheses space. Two associative memories models will be developped: the <u>Hopfield model</u> and the <u>multilayer perceptron</u>. The <u>decision functions</u> are functions that take arguments on a set of hypotheses and map them into a boolean vector of classes.

1.1.1. <u>Hopfield models</u>

The Hopfield model is a neural network where all the neurons are boolean and interconnected [1,11,15]. Hopfield proved that such a network was able to memorize patterns or prototypes as attractors and was also able to recall them, when given an initial state, after stabilization of the network.

The neurons have boolean activations $X \in B^m$; the n prototypes are given as $X^j = (x_1^j, .. x_m^j)$ and the transition function is defined by :

$$T(x_i) = 1 \quad \text{if } \Sigma_j w_{ij}x_j \geq 0 \quad \text{where } w_{ij} \text{ is the weight between neurons i and j}$$
$$T(x_i) = 0 \quad \text{if } \Sigma_j w_{ij}x_j < 0$$

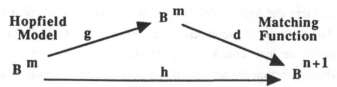

<u>Figure 1</u>: Construction of an Hopfield model followed by a matching function

For each prototype X^j, one can define the attraction basin B^j as

$$B^j = \{X \in B^m / \exists\, k \in N,\ T^k(X) = X^j\}$$

We can now define the output space B^{n+1}, with $n+1$ exclusive booleans defined as:

- $(X \in B^j)$ for $j <= n$,

- "unknown" that is set to TRUE when all $(X \in B^j)$ are FALSE.

The decision function attached to Hopfield models is simple : it selects the prototype that has emerged after stabilization and tries to match the result with one of the prototypes. If this procedure succeeds, the decision function returns the prototototype's name; in the alternative case, the answer is "unknown".

1.1.2. Multilayer Perceptrons

Multilayer perceptrons are multilayer neural networks with complete connections between adjoining layers [9,10]. The number of hidden layers between the input and output layer is arbitrary. The input-output relationship of each unit (except for those belonging to the input layer) is given by :

$$y = f[\,net\,] \qquad \text{where}\quad net = \sum_{i=1}^{n} w_i\, x_i - \theta$$

where y represents the output of the unit, n the number of inputs x_i, w_i the connection weights and θ the threshold. f is chosen to be the sigmoid function for each unit's output function. We are given a set of examples $\{X^k, Y^k\}$ where X^k takes its values in the hypercube B^m and Y^k in the hypercube B^n. The network calculates a mapping g (figure 1) from B^m to I^n by using the back propagation algorithm [13,16].

Multilayer perceptrons provide a continuous answer for each hypothese, that can be seen as a measure of membership to each class. This is clearly not a mecanism for eliminating hypotheses or rejecting classes. In order to derive a classifier or a decisional machine, one has to fit the output vector of the associative memory into a decision function that will select some hypotheses and eliminate the nonselected hypotheses (see figure 2).

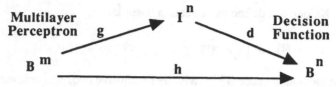

: Construction of a multilayer perceptron followed by a decision function

There is a large number of possible decision functions; therefore, below are some selected examples:

(i) the maximum rule keeps one hypothese among a set of exclusive hypotheses.

(ii) a threshold rule allows a selection of a variable number of hypotheses.

(iii) the rule that keeps the p bests, ordered or not.

The class of decisional machines can be further extended by introducing the costs c_{ij} that are defined as the cost of choosing hypothese i when j is the right answer (second order risk). This can be illustrated by considering a medical doctor that has to make a diagnosis and hesitates between a malignant and a benign disease. In such a case, the costs c_{ij} and c_{ji} may be very different. One way for the doctor to limit the riskiness of this decision is to adopt the following rule:

(iv) choose the hypothese i minimizing $\Sigma_j y_j c_{ij}$.

This expression becomes an average cost if the y_j are probabilities. This can be achieved if during the training phase a penalty term is introduced stating that $\Sigma_i y_i = 1$.

(v) other types of decisional functions may also be used.

1.2. Connectionnist classifiers as abductive machines

A connectionist classifier is a decisional machine defined by the composition between an associative memory g and a decision function d (figure 3).

Since g maps the input hypercube B^m into E and d maps E into the ouput hypercube B^n, it follows that a connectionist classifier is a mapping between the input hypercube B^m and the ouput hypercube B^n.

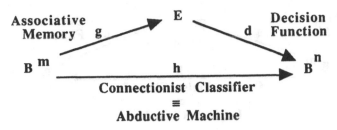

Figure 3: Construction of a connectionist classifier

The associative memories considered in figure 3 are neural networks mapping an input space into an output space. We have developed two models of associative memories : the multilayer perceptron and the Hopfield model. However, our claim is that the previous framework can be extended to any associative memory [8]. Decision functions take arguments on a set of hypotheses and map them into a boolean vector of the classes.

The connectionist classifier behaves as an <u>abductive machine</u> because it is able to select a few hypotheses and eliminate the rest of them [14].

1.3. Logical classifiers

Let us first define some notations that will be used in the following. $X = [x_i]i \in [1,m]$ represents the vector of boolean inputs and $Y = [y_j]j \in [1,n]$ the vector of classes or hypotheses.

A <u>logical classifier</u> is defined as a set of clauses with a unique positive output boolean.

$$(x_{i1} \wedge x_{i2} \wedge x_{i3} \wedge ... \Rightarrow y_i) \wedge (x_{j1} \wedge ... \Rightarrow y_j) \wedge (x_{k1} \wedge ... \Rightarrow y_k) \wedge ...$$

where $x_{ij} \in X$ are positive or negative boolean inputs,
                   $y_i \in Y$ are positive boolean outputs.

Logical rules as defined above can easily be included in the knowledge base of an expert system [7].

2. Equivalence between a connectionist classifier and a logical classifier

In this section, we show the equivalence between a connectionist classifier and a logical classifier according to the following diagram :

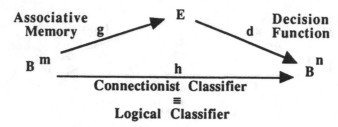

Figure 4: Equivalence diagram between a connectionist classifier and a logical classifier

Theorem : A connectionist classifier is functionally equivalent to a logical classifier.

Proof : Each boolean of B^n is a boolean function of B^m. It can then be expressed as a disjunction of conjunctions:

$$[(x_{i1} \wedge x_{i2} \wedge x_{i3} \wedge ...) \vee (x_{j1} \wedge x_{j2} \wedge x_{j3} \wedge...) ...] \Rightarrow y_j$$

Therefore,

$$[(x_{i1} \wedge x_{i2} \wedge x_{i3} \wedge ... \Rightarrow y_j) \wedge (x_{j1} \wedge x_{j2} \wedge x_{j3} \wedge... \Rightarrow y_j) \wedge ..]$$

Since this is true for each boolean of B^n, it follows that the connectionist classifier is equivalent to a logical classifier.

These results allow us to look at the classifier in two different ways that are functionally equivalent, either as the composition of an associative memory and a decision function, or as a logical machine. Furthermore, the connectionist classifier can also explain, if necessary, its own computation in a logical way.

The choice of a decision function determines a set of logical rules. Changing the decision function (i.e. modifying **d** into **d'**) will then lead to a change in the derived set of rules.

When the connectionist classifier is a multilayer perceptron followed by a decision function, it will tend to behave poorly when presented cases far from the training base. The rules extracted corresponding to those cases will then not be relevant. In order to overcome this problem, we propose to consider a set of connectionist classifiers trained on the same examples and to compare their results.

3. Validity domain construction

The validity domain concept is related to the idea that there exists a training base neighborhood on which the generalization works well [5]. The construction of such a validity domain can be performed by various methods depending on both the considered problem and the probability of occuring for each example in the training base as mentioned earlier [2].

3.1. Upper bound of the validity domain

An upper bound D of the validity domain can be determined from a priori knowledge, either related to the data coding or formulated by an expert, or statistical regularities observed from the training base. We will then derive constraints on the input hypercube B^m of the neural network. This allows one to reduce the initial cardinality of the domain (2^m).

If we assume that the input vector is a vector of booleans, the constraints about the domain can then be written as mathematical and logical formulas. We can also interpret a set of booleans as a characteristic function of a subset and derive constraints as properties of these subsets, i.e. their cardinality. We describe below a few common cases of domain restriction with the corresponding cardinality calculation rules :

a) A simple method is to use the Hamming distance with respect to a $\{X_1^k\}_{k \in K}$ where X_1 is a subset of X. We could then define a measure of membership to the validity domain by the following definition : an example x is said to belong to the domain validity D of the connectionist classifier with a similitude coefficient $c = 1 - s/m \in [0,1]$ if :

$$\exists \, y \in \{X_1^k\}_{k \in K} \, / \, x \in \text{hypersphere of center y and of radius s}$$

b) D consists of booleans $(a_1, ..., a_n)$ where m among them are TRUE and $m1 \leq m \leq m2$,

$$D = \{ A' \subset A \, / \, m1 \leq \text{card} (A') \leq m2 \} \quad \text{with card(D)} = \sum_{m \in [m1,m2]} C(n,m)$$

b') When $m1 = m2 = m$, D consists of booleans $(a_1, ..., a_n)$ where m are TRUE.

$$D = \{ A' \subset A \, / \, \text{card} (A') = m \} \quad \text{with card(D)} = C(n,m)$$

b") For m = 1, D consists of exclusive booleans $\{a_1, ..., a_n\}$ and card(D) = n

c) D consists of a cartesian product of domains $D_1, ..., D_n$,

$$D = D_1 \times ... \times D_n \quad \text{with card(D)} = \text{card}(D_1) \times ... \times \text{card}(D_n)$$

Rules a, b, b' and b" are descriptions of leaf domains. Rule c recursively calls rules a, b, b', b" and c. Thus, a recursive language can be used to describe constraints on the neural network validity domain. This language may be extended by considering other constructions [6].

3.2. Construction by μ-stability

Let us then assume that we are building in an relatively different manner a large number of PAC (Probably Almost Correct) connectionist classifiers that are trained on the same learning data base with a given probability distribution. According to Valiant's framework [17], those connectionist classifiers h_i (each defined by a couple (g_i, d_i)) can classify correctly, with a probability of $1-\delta$, at least $(1-\varepsilon)$ of the examples of a test data base K (with the same distribution) [2]. If we consider that the decision function d is the maximum function, we can then write:

$$\text{Probability} \, [\, E < \varepsilon \,] \, > \, 1 - \delta$$

$$\text{where} \quad E = (1/\text{card}(K)) \, \Sigma_{k \in K} \, \, d\text{Hamming} \, [Y^k, \, h(X^k)]$$

The definition of PAC connectionist classifiers can be extended to the context of statistical pattern classification. In this case, it has been shown that the minimum possible probability of misclassification is achieved by the Bayes rule. Kanaya [12] defines a network that gives valid generalization from a m-length training data as a network such that there exists $0 < \eta, \varepsilon < 1$ satisfying:

$$\text{Probability} \, [\, E < E_0 + \varepsilon \,] \, > \, 1 - \delta$$

$$\text{where} \quad E = (1/\text{card}(K)) \, \Sigma_{k \in K} \, \, d\text{Hamming} \, [Y^k, \, h(X^k)]$$
$$\text{and} \quad E_0 \text{ is the minimum attainable error computable with the Bayes rule.}$$

In the Valiant's framework, the classification function is well known; therefore, $E_0 = 0$.

Let us now assume that we are building in a relatively different manner a large number (greater than 20) of connectionist classifiers $h_i = dog_i$, $i = 1, ..., p$, that are trained on the same learning data base.

For each point x of the input space B^m, we can consider

$$m_p(x) = (1/p) \, \Sigma_{i \in [1,p]} \, \, h_i(x)$$

We can then define the p-averaged connectionist classifier h as:

$$h : B^m \rightarrow [0, 1]^n$$
$$x \rightarrow m_p(x)$$

For each $x \in B^m$, the ith component of $h(x)$ is noted $[h(x)]_i$

The p-averaged connectionist classifier h will said to be μ-stable in $x \in B^m$ if there exists $i \in [1,p]$ such that:

$$[h(x)]_i \geq \mu \qquad \mu \in [0, 1]$$

When p becomes large, each component $[h(x)]_i$ can be seen as a measure of accepting hypothcoo i.

We can now define a domain of μ-validity as the set of points in D where the p-averaged connectionist classifier is μ-stable.

$$D\mu = \{ x \in D / h \text{ is } \mu\text{-stable in } x \}$$

Therefore, the set of μ-relevant logical rules will be defined as the rules derived from the domain of μ-validity of a p-averaged connectionist classifier. This idea will be further developped below.

As illustrated on figure 5, the domain of μ-validity decreases as μ increases. Similarly, the performance of the p-averaged connectionist classifier measured on the domain of μ-validity will increase as μ increases.

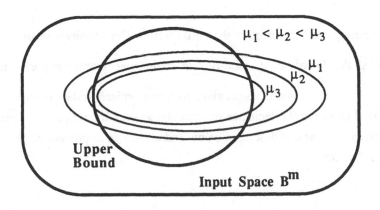

Figure 5: Construction of the validity domain

When $\mu = 1$, the error rate ε of the p-averaged connectionist classifier is not guaranted to be equal to 0 since all the connectionist classifiers may be wrong on one point of the domain of 1-validity. On the other hand, there is also a non-zero lower bound for μ below which the domain of μ-validity is equal to $\mathbb{D}$ (i.e. when d is the maximum rule and the lower bound is 1/n).

In order to construct the validity domain, we propose to link the choice of μ to the performance level desired; the performance level can be estimated on the test data base K by:

$$E(\mu) = (1/N) \Sigma_{k \in K \cap D\mu} \, d_{Hamming} [Y^k, h(X^k)]$$
$$\text{where} \quad N = \text{card}(K \cap D\mu)$$

The performance level desired depends on the problem; however, $E_0 < E(\mu) < E_0 + \varepsilon$, and the choice of E allows to determine μ.

The problem of extracting the equivalent rules is NP-complete [4]. By choosing μ close to 1, we improve the rules relevance and we reduce the complexity of the computation. Other means for reducing the computation complexity will now be discussed.

4. Derivation of a logical classifier from a connectionist classifier

The connectionist classifiers considered from now on are multilayer neural networks followed by a decision function. However, extracting logical rules from an Hopfield model is also possible [18]. Techniques for deriving a logical classifier from a multilayer neural network followed by a decision function are now described.

4.1. Construction of the μ-stability domain

This idea has already been discussed in the third section. The validity domain is reduced by calculating in $\mathbb{D}$ the domain of μ-stability. For multilayer neural networks, the relative difference in the networks construction is given by a randomized initial choice of the weights and a randomized choice in the sequence of examples at each training cycle. This method is very complementary to the one described below and allows an improvement in computation time and rules relevance.

4.2. Invariance detection

We are now considering transformations of a neuron $f[w_0 + \Sigma_j \, w_j \, x_j]$ that keep its transfer function invariant, but allow better semantic properties.

4.2.1. Smoothing of the weights

When the number of x_j's is large, it is possible to perform a histogram of the weights w_j and cluster the weights in several classes. When the classes are difficult to distinguish, one can use statistical methods such as discriminant analysis. By choosing an average coefficient w_k for each class, we can derive an equivalent neuron $f[w_0 + \Sigma_k \, w_k (\Sigma_{jk} \, x_{jk})]$ where x_{jk} belongs to class k.

This can also be used during the training period. When weights are observed to move in the same direction, they can be averaged and then constrained. By reducing the number of degrees of freedom, this procedure forces the network to extract semantic features and also allows speeding up the training process.

4.2.2. Appearance of intermediate numerical variables

If the neurons x_{jk}s of the same class k are or behave as boolean units (this is always the case on the input layer), then the summation $\Sigma_{jk} \, x_{jk}$ can be interpreted as the number of x_{jk} equal to 1. We can then introduce an intermediate numerical variable defined as the number of elements of class k. This method allows a better interpretation and simplifies the rule extraction task. Constraints such that "the number of class k elements is contained between n_1 and n_2" can then expressed with the Hamming distance:

$$n_1 \leq d_{Hamming}[X, X_0] \leq n_2$$

where   X is vector defined by the elements x_{jk} of class k $[x_1, ..., x_{jk}]$.
and X_0 is the null vector $[0, ..., 0]$.

Such constraints can then be easily taken into account in the implicit enumeration techniques that can be used to extract logical rules.

4.3. Extraction of logical rules

We can always, by enumeration, extract a disjonctive normal form, by superposing all the cases with the OR connector. We can afford to use an explicit enumeration algorithm since the number of variables has been reduced as explained in the previous paragraph. In any case, it is always possible to improve the implicit enumeration by using heuristic methods such as propagation/extraction techniques [6].

The introduction of a validity domain for the connectionist classifier (§3) allows speeding up the rule extraction process. When using explicit enumeration, the number of cases to be considered is greatly reduced. On the other hand, the mathematical and logical functions describing the validity domain can also be introduced in the implicit enumeration methods (i.e. constraints propagation/extraction) and increase their efficiency.

The methods presented above have been tested with networks whose hidden layers consist of less than 10 units. The limitation on the number of input neurons is obviously related to the existence of regularities in the coefficient.

Conclusion

This paper showed that connectionist and logical classifiers are functionally equivalent in the sense that they can always be derived from one another. Methods for deriving an equivalent logical classifier from a connectionist classifier involve several steps : the detection of invariance among the input patterns, the definition of a validity domain in the input space and the extraction of the equivalent logical rules.

The composition of a multilayer perceptron with a decision function allows a selection/elimination process. For this reason, the connectionist classifier can be said to behave as an abductive machine in the sense that such a machine is able to focus on a limited number of likely hypotheses that can then be passed down to a more precise problem solver.

The set of equivalent rules that can be extracted from the connectionist classifier can be used as a first knowledge base when building an expert system. The expert neural network is then able to provide an explicit explanation of its computation.

References:

[1] Amit D. J., Gutfreud H. and Sompolinsky H., "Storing infinite numbers of patterns in a spin-glass model of neural network", Phys. Rev. Lett., vol. 50, pp 1110-1112, 1985.

[2] Baum E.B., " Are k-nearest neighbor and back propagation accurate for feasible sized sets of examples?", Lecture Notes in Computer Science, 412, pp 2-26, Springer Verlag, 1990.

[3] Bochereau L. and Bourgine P., "Implémentation et extraction de traits sémantiques sur un réseau neuromimétique", Proceedings of Neuro-Nîmes 89, pp 125-143.

[4] Bochereau L. and Bourgine P., "Extraction of semantic features and logical rules from a multilayer neural network", Proceedings of IJCNN 90, Washington DC, 15-19/01/90.

[5] Bochereau L. and Bourgine P., "Validity domain and extraction of rules on a multilayer neural network", Proceedings of IJCNN 90, San Diego, 17-21/06/90.

[6] Bourgine P., "PROMAT language and his virtual machine", Proceedings of RFIA 89, AFCET/INRIA, Paris.

[7] Clancey W., "Heuristic Classification", Artificial Intelligence, 1985, 27, 289-350.

[8] Deffuant G., "Neuron units recruitment algorithm for generation of decision trees", Proceedings of IJCNN 90, San Diego, 17-21/06/90.

[9] Feldman S.E. and Ballard D.H., "Connectionist models and their properties", Cognitive Science, 1982, 6, pp 205-254.

[10] Fogelman-Soulie F., Gallinari P., Lecun Y., Thiria S., "Automata networks and artificial intelligence" in Automata Networks in Computer Science", F. Fogelman-Soulié, Y. Robert, M. Tchiente eds, Manchester University Press, 1987, pp133-186.

[11] Hopfield J.J., "Neural networks and physical systems with emergent collective computational abilities", Proceedings of the National Academy of Sciences, 1982, 79, pp 2554-2558.

[12] Kanaya F., "On the Bayes Statistical Behavior and Valid Generalization of Pattern Classifying Neural Networks", Cognitiva 90, Madrid, 20-23/11/90.

[13] Le Cun Y., "A learning procedure for asymmetric networks", Proceedings of Cognitiva 85, Paris, 1985, pp599-604.

[14] Peirce, 6-530.

[15] Personnaz L., Guyon I. and Dreyfus G., "Collective computational properties of neural networks : new learning mechanisms", Phys. Rev. A, Vol. 34, pp 4217-4228, 1986.

[16] Rumelhart D.E., Hinton G.E., Williams R.J., "Learning internal representations by error propagation" in Parallel Distributed Processing, D.E. Rumelhart & J.L. MacClelland eds, MIT Press, Volume 1: pp 318-369, 1986.

[17] Valiant L.G., "A theory of the learnable", Communications of the ACM V27, Nº 11 pp 1134-1142, 1984.

[18] Victorri B., "Modéliser la polysémie", TA Informations, Vol. 29, N° 1.2, pp 21-42, 1988.

On Potts-glass neural networks with biased patterns

D. Bollé [*]and P. Dupont[†]
Instituut voor Theoretische Fysica
Universiteit Leuven
B-3030 Leuven, Belgium

Abstract

Neural networks of the q-state Potts type are considered for a finite
number of biased patterns. The existence and stability properties of
the Mattis states and symmetric states are discussed at zero temperature for all q and arbitrary bias and near the "critical" temperature
for $q = 3$ and two classes of representative bias parameters. Some of
these properties are illustrated through the solution of the dynamical
equations governing the time evolution of the macroscopic overlap between the learned patterns and the instantaneous microscopic state of
the network.

1   Introduction

Potts-glass models of neural networks have been introduced in [1]. There,
the capacity of storage and retrieval of information in this type of networks
has been discussed, mostly concentrating on the limit of zero temperature.
In [2],[3] the general temporal development for these models, extended to
asymmetric synaptic couplings has been considered by deriving evolution

---

[*]Onderzoeksdirecteur NFWO, Belgium; e-mail: FGBDA18@ BLEKUL11.BITNET
[†]Onderzoeker IIKW, Belgium; e-mail: FGBDA22@ BLEKUL11.BITNET

equations describing the overlap between the learned patterns and the instantaneous microscopic state of the network. These equations have been solved numerically for a small number of Potts states q and a small number of patterns.

In this contribution we report on some further progress, mostly based upon analytic methods, in the study of Potts networks. In particular we present the first results we have obtained on the stability of the Mattis and symmetric states and on the time evolution of the overlap when a finite number of biased patterns are considered.

In section 2 we describe the model and its dynamics. Using the mean-field theory approach [4] we find in section 3 that the positive Mattis states are stable at zero temperature for all q and arbitrary bias. Near the "critical" temperature T_c, defined as the maximum temperature for which a positive Mattis state exist, they are also stable for $q = 3$ and two representative classes of the bias. This is different from the Hopfield model where the Mattis states become unstable near T_c when the bias parameter a satisfies $a^2 > \frac{1}{3}$ [5]. Concerning the symmetric states, we have studied the crossing of the energy levels in function of the bias at $T = 0$ for $q = 3$. For some choice of the bias parameter we find that the free energy of the Mattis states is lower than the free energy of the symmetric states for all allowed values of that bias. For other choices the free energy of the Mattis state becomes higher than the free energies of the symmetric states. The stability of the symmetric states in function of the bias parameters is discussed in some detail. In section 4 we present the dynamical equations describing the time evolution of the overlap [3]. We solve them numerically obtaining the overlap flow diagrams, to illustrate some of the stability properties found in section 3. A more systematic and complete discussion will appear in future work.

2   The model

Consider a system of N neurons. We assume that every neuron can occupy q discrete states by viewing it as a q-state Potts spin. The instantaneous configuration of all the spin variables at a given time describes then the state of such a network. The neurons are interconnected with all the others by a synaptic matrix of strength $J_{ij}^{\alpha\rho}$ which determines the contributions of a signal fired by the j-th presynaptic neuron in state ρ to the post-synaptic potential which acts on the i-th neuron in state α. This contribution can either be positive (excitatory synapse) or negative (inhibitory synapse).

The potential h_{σ_i} of neuron i which is in a state σ_i is the sum of all postsynaptic potentials delivered to it in a time unit, i.e.

$$h_{i,\sigma_i} = -\sum_{\substack{j=1 \\ j \neq i}}^{N} \sum_{\alpha,\rho=1}^{q} J_{ij}^{\alpha\rho} m_{\sigma_i,\alpha} m_{\sigma_j,\rho} \qquad (1)$$

with

$$m_{\sigma_i,\rho} = q\delta_{\sigma_i,\rho} - 1. \tag{2}$$

We assume that the synaptic couplings satisfy

$$J_{ij}^{\alpha\rho} = J_{ji}^{\rho\alpha}. \tag{3}$$

The dynamics of this q-state Potts model is the following. At zero temperature the state of the neuron in the next time step is fixed to be the state which minimizes the induced local field (1). The stable states of the system are those configurations where every neuron is in a state which gives a minimum value to $\{h_{i,\sigma_i}\}$. If the relation (3) holds, this stability is equivalent to the requirement that the configurations $\{\sigma_i\}$ are the local minima of the anisotropic Hamiltonian

$$H = -\frac{1}{\eta} \sum_{\substack{i \neq j}}^{N} \sum_{\alpha,\rho=1}^{q} J_{ij}^{\alpha\rho} m_{\sigma_i,\alpha} m_{\sigma_j,\rho}. \tag{4}$$

In the presence of noise there is a finite probability of having configurations other than the local minima. This can be taken into account by introducing an effective temperature $T = \frac{1}{\beta}$. We assume that the temporal development of the system occurs via sequential updating. The transition probability of neuron i to go from a state σ_i into another state α_i is given by

$$w(\sigma_i \rightarrow \alpha_i) = \frac{1}{Z_i} \exp\left[-\frac{\beta}{2}(h_{i,\alpha_i} - h_{i,\sigma_i})\right] \tag{5}$$

where

$$Z_i = \sum_{\rho=1}^{q} \sum_{\substack{\gamma=1 \\ \gamma \neq \rho}}^{q} \exp\left[-\frac{\beta}{2}(h_{i,\gamma} - h_{i,\rho})\right]. \tag{6}$$

At this point we remark that (3) is a sufficient condition for w to satisfy detailed balance [6].

To build in the capacity of learning and memory in this network, its stable configurations must be correlated with the p patterns $\{k_i^a\}$, $a = 1, ..., p$ fixed by the learning process. The latter are allowed to be biased i.e. the k_i^a are chosen as independent random variables which can take the values $1, ..., q$ with probability

$$P(\alpha) = \frac{1 + B_\alpha}{q}, \quad \alpha = 1, ..., q, \tag{7}$$

where the $\{B_\alpha\}$ are the bias parameters. Analogous to the Hopfield model [5], we therefore propose the learning rule

$$J_{ij}^{\alpha\rho} = \frac{1}{q^2 N} \sum_{a=1}^{p} \left(m_{k_i^a,\alpha} - B_\alpha\right)\left(m_{k_j^a,\rho} - B_\rho\right). \tag{8}$$

We note that the bias parameters are independent of the patterns and hence the latter are learned in the same way. From the fact that $0 \leq P(\alpha) \leq 1$ and

$\sum_{\alpha=1}^{q} P(\alpha) = 1$ we deduce the further properties

$$-1 \leq B_\alpha \leq q - 1, \qquad \sum_{\alpha=1}^{q} B_\alpha = 0. \tag{9}$$

Using the learning rule (8), both the postsynaptic potential and the Hamiltonian can be rewritten as

$$h_{i,\sigma_i} = -\frac{1}{N} \sum_{\substack{j=1 \\ j \neq i}}^{N} \sum_{a=1}^{p} \left(m_{k_i^a,\sigma_i} - B_{\sigma_i} \right) \left(m_{k_j^a,\sigma_j} - B_{\sigma_j} \right), \tag{10}$$

$$H = -\frac{1}{2N} \sum_{\substack{i,j=1 \\ i \neq j}}^{N} \sum_{a=1}^{p} \left(m_{k_i^a,\sigma_i} - B_{\sigma_i} \right) \left(m_{k_j^a,\sigma_j} - B_{\sigma_j} \right). \tag{11}$$

where we have employed $(\rho, \gamma = 1, ..., q)$

$$\frac{1}{q} \sum_{\alpha=1}^{q} m_{\rho,\alpha} m_{\alpha,\gamma} = m_{\rho,\gamma}. \tag{12}$$

In the sequel we study the biased Potts neural network for finite p and q and in the limit $N \rightarrow \infty$. An important role is played by the order parameter R_a defined by

$$R_a = \frac{1}{N} \sum_{j=1}^{N} \left(m_{\sigma_j,k_j^a} - B_{\sigma_j} \right). \tag{13}$$

It can be interpreted as a measure for the macroscopic overlap with pattern a. When we have total overlap, R_a becomes in the limit $N \rightarrow \infty$

$$R_a = q - 1 - \frac{1}{q} \sum_{\alpha=1}^{q} B_\alpha^2. \tag{14}$$

For random configurations with probability distribution (7), we have

$$R_a = -\frac{1}{q} \sum_{\alpha=1}^{q} B_\alpha^2. \tag{15}$$

3 Stability properties

Starting from the Hamiltonian (11) and applying standard techniques (linearization and the saddle-point method [4]), the ensemble averaged free-energy is given by

$$f = \frac{1}{2} \sum_{a=1}^{p} R_a^2 - \frac{1}{\beta} \left\langle\!\left\langle \ln \left(\sum_{\rho=1}^{q} \exp \left[\beta \sum_{a=1}^{p} (m_{k^a,\rho} - B_\rho) R_a \right] \right) \right\rangle\!\right\rangle. \tag{16}$$

The double brackets $\langle\!\langle . \rangle\!\rangle$ stand for averaging over the distribution of all learned patterns $\{k_i^a\}$. The saddle-point equations for the order parameters

R_a are given by

$$R_a = \left\langle\!\left\langle \frac{\sum_{\rho=1}^{q} (m_{k^a,\rho} - B_\rho) \exp\left[\beta \sum_{b=1}^{p}(m_{k^b,\rho} - B_\rho)R_b\right]}{\sum_{\rho=1}^{q} \exp\left[\beta \sum_{b=1}^{p}(m_{k^b,\rho} - B_\rho)R_b\right]} \right\rangle\!\right\rangle. \qquad (17)$$

The following type of solutions of (17) will be distinguished. First of all, there are the Mattis states, having only one non-zero overlap which we denote by M. If $M > 0$, they are correlated with one of the p learned patterns. Since all patterns are treated in the same way the index of the pattern can be chosen arbitrarily so that we have p solutions of this type.

Secondly, we have the n-symmetric states, being solutions with n ($1 \leq n \leq p$) non-zero overlaps with equal magnitude S_n. When $S_n > 0$, they can be seen as states which mix the n corresponding patterns. We remark that S_1 is the Mattis state M.

In the sequel we discuss stability properties of both the Mattis states and the n-symmetric states. The possible existence of asymmetric solutions is not considered here.

3.1   Mattis states

The saddle-point equation (17) for a Mattis state can be written more explicitly

$$M = \sum_{\alpha=1}^{q} \left(\frac{1 + B_\alpha}{q}\right) \frac{\sum_{\rho=1}^{q}(m_{\alpha,\rho} - B_\rho) \exp\left[\beta M(m_{\alpha,\rho} - B_\rho)\right]}{\sum_{\rho=1}^{q} \exp\left[\beta M(m_{\alpha,\rho} - B_\rho)\right]}. \qquad (18)$$

In the case of $T = 0$ the positive solution of (18) is given by

$$M_0^+ = q - 1 - \frac{1}{q}\sum_{\alpha=1}^{q} B_\alpha^2, \qquad (19)$$

so this means complete retrieval since, as we will see later, M_0^+ is stable. There exists also a negative solution when $T = 0$ given by

$$M_0^- = -\sum_{\alpha=1}^{q} \frac{1 + B_\alpha}{q}(1 + B_\alpha^{max}) \qquad (20)$$

with

$$B_\alpha^{max} = \max\{B_\rho | \rho \neq \alpha\}. \qquad (21)$$

In order to study the local stability of the Mattis states, we look at the stability matrix:

$$A_{ab} \equiv \frac{\partial^2 f}{\partial R_a \partial R_b}. \qquad (22)$$

The explicit form of the matrixelements is given by

$$A_{ab} = \delta_{ab} - \beta \left\langle\!\!\!\left\langle \frac{\sum_{\rho=1}^{q}(m_{k^a,\rho} - B_\rho)(m_{k^b,\rho} - B_\rho)U_\rho}{\sum_{\rho=1}^{q} U_\alpha} \right. \right.$$
$$\left. \left. - \frac{\sum_{\rho=1}^{q}\sum_{\alpha=1}^{q}(m_{k^a,\rho} - B_\rho)(m_{k^b,\alpha} - B_\alpha)U_\rho U_\alpha}{\left(\sum_{\rho=1}^{q} U_\rho\right)^2} \right\rangle\!\!\!\right\rangle$$

(23)

where

$$U_\rho = \exp\left[\beta \sum_{d=1}^{p}(m_{k^c,\rho} - B_\rho)R_d\right].$$

(24)

It is clear that A is diagonal. Furthermore, there are only two different eigenvalues. A non-degenerate one, viz.

$$\lambda_1 = 1 - \beta \sum_{\alpha=1}^{q}\left(\frac{1+B_\alpha}{q}\right)\left\{\frac{\sum_{\sigma=1}^{q}(m_{\sigma,\alpha} - B_\sigma)^2\exp[\beta M(m_{\sigma,\alpha} - B_\sigma)]}{\sum_{\sigma=1}^{q}\exp[\beta M(m_{\sigma,\alpha} - B_\sigma)]}\right.$$
$$\left. - \left[\frac{\sum_{\sigma=1}^{q}(m_{\sigma,\alpha} - B_\sigma)\exp[\beta M(m_{\sigma,\alpha} - B_\sigma)]}{\sum_{\sigma=1}^{q}\exp[\beta M(m_{\sigma,\alpha} - B_\sigma)]}\right]^2\right\}$$

(25)

and a $(p-1)$-times degenerate one, i.e.

$$\lambda_2 = 1 - \beta \sum_{\alpha=1}^{q}\sum_{\rho=1}^{q}\left(\frac{1+B_\alpha}{q}\right)\left(\frac{1+B_\rho}{q}\right)$$
$$\left\{\frac{\sum_{\sigma=1}^{q}(m_{\sigma,\rho} - B_\sigma)^2\exp[\beta M(m_{\sigma,\alpha} - B_\sigma)]}{\sum_{\sigma=1}^{q}\exp[\beta M(m_{\sigma,\alpha} - B_\sigma)]}\right.$$
$$\left. - \left[\frac{\sum_{\sigma=1}^{q}(m_{\sigma,\rho} - B_\sigma)\exp[\beta M(m_{\sigma,\alpha} - B_\sigma)]}{\sum_{\sigma=1}^{q}\exp[\beta M(m_{\sigma,\alpha} - B_\sigma)]}\right]^2\right\}.$$

(26)

The sign of λ_1 and λ_2 determine the stability of the solutions of (18).

In order to see if the Mattis states (19) and (20) are stable, we have to calculate the eigenvalues (25) and (26) in the limit $T \to 0$. For M_0^+ they are both equal to one and hence M_0^+ is always stable near $T = 0$. The situation for M_0^- is more complicated: it represents a stable solution if and only if all bias parameters B_α, $\alpha = 1, ..., q$ are different or $p = 1$ or $q = 2$ (see [7] for more details).

The study of the Mattis states at finite temperatures is more involved. In the following we restrict ourselves to positive ones since they correspond with "recognizing" the learned patterns. First, we define two temperatures which are of interest, i.e.

$$T_0 = inf\{T|(T, M = 0) \text{ is a stable solution of } (18)\} \tag{27}$$
$$T_c = sup\{T|\exists M > 0 : (T, M) \text{ is a solution of } (18)\}. \tag{28}$$

Using the expressions (25) and (26), the temperature T_0 can be found from

$$T_0 = inf\{T|\lambda_1(T, M = 0) > 0 \text{ and } \lambda_2(T, M = 0) > 0\}. \tag{29}$$

This leads to

$$T_0 = q - 1 - \frac{1}{q}\sum_{\rho=1}^{q} B_\rho^2. \tag{30}$$

Above this temperature the state with all overlaps equal to zero becomes stable.

The calculation of the temperature T_c and the question of stability near this temperature is much more difficult. Only for $q = 2$ analytic results are known [5]. The result in that case is $T_c = T_0 = 1 - a^2$ where $B_1 = a$ and $B_2 = -a$. Furthermore, the Mattis states become unstable near T_c if $a^2 > \frac{1}{3}$.

For $q > 2$, it is convenient to introduce a new variable z

$$z = \beta M. \tag{31}$$

The saddle-point equation (17) then becomes ($\mathbf{B} = (B_1, ..., B_q)$)

$$\beta = \frac{z}{F(z, \mathbf{B})}, \tag{32}$$

where

$$F(z, \mathbf{B}) = \sum_{\alpha=1}^{q} \left(\frac{1 + B_\alpha}{q}\right) \frac{F_1(z, \mathbf{B}, \alpha)}{F_0(z, \mathbf{B}, \alpha)} \tag{33}$$

and

$$F_j(z, \mathbf{B}, \alpha) = \sum_{\rho=1}^{q} (m_{\alpha,\rho} - B_\rho)^j \exp[z(m_{\alpha,\rho} - B_\rho)]. \tag{34}$$

The temperature T_c is then given by

$$T_c = sup\left\{\left.\frac{F(z, \mathbf{B})}{z}\right| z \in (0, \infty)\right\}. \tag{35}$$

Equation (35) has been solved numerically for $q = 3$. Hereby two classes of bias parameters are of interest, viz.

$$\mathbf{B}_1 = (2a, -a, -a) \tag{36}$$
$$\mathbf{B}_2 = (a, 0, -a) \tag{37}$$

with $a \in [0, 1]$. The form (36) indicates that one state is privileged and the other two states have equal probability to appear. In the other case (expression (37)) all three states have different probability.

For both classes of bias the temperature T_c as a function of the parameter a is given in fig.1. For $a = 0$, we find back the value of the unbiased case [2]: $T_c(a = 0) \approx 2.18$. When a increases, T_c decreases. For $a = 1$ we find a T_c that is zero for the class $\mathbf{B}_1$ and non-zero for the class $\mathbf{B}_2$. The reason is the following: the bias $\mathbf{B}_1 = (2, -1, -1)$ corresponds with a probability distribution for the patterns where the lowest state has probability one. This

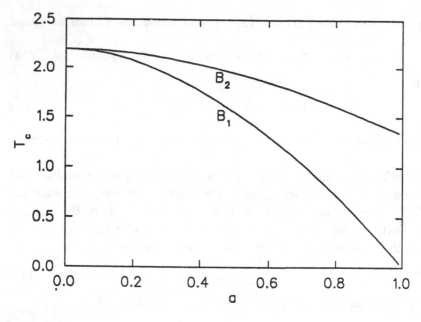

Figure 1

means that there is no freedom left for the neurons. The bias $\mathbf{B}_2 = (1, 0, -1)$ corresponds with a distribution where two states have a non-zero probability. Hence the neurons can still occupy different states.

The next step is to study the behavior of the eigenvalues (25) and (26) of the stability matrix. Using the fact that $T_c(z_c)$ is maximal, where z_c is the value of z for which $T = T_c$, and therefore $\frac{dT}{dz}(z_c) = 0$, we can easily verify that $\lambda_1(z_c) = 0$ for all $q \geq 2$. Since for temperatures below T_c, M decreases when T increases, we have for these temperatures that $z = z_c + \epsilon$ with ϵ a small positive parameter. Making an expansion of λ_1 around z_c, we find

$$
\begin{aligned}
\lambda_1 &= -\beta_c \sum_{\alpha=1}^{q} \left(\frac{1 + B_\alpha}{q}\right) \left[\frac{F_3(z_c, \mathbf{B}, \alpha)}{F_0(z_c, \mathbf{B}, \alpha)} - 3\frac{F_2(z_c, \mathbf{B}, \alpha)F_1(z_c, \mathbf{B}, \alpha)}{F_0{}^2(z_c, \mathbf{B}, \alpha)} \right. \\
&\qquad\qquad \left. +2\left(\frac{F_1(z_c, \mathbf{B}, \alpha)}{F_0(z_c, \mathbf{B}, \alpha)}\right)^3 \right] \epsilon + O(\epsilon^2) \\
&= C_1(z_c, \mathbf{B})\epsilon + O(\epsilon^2).
\end{aligned}
\tag{38}
$$

Numerical evaluation of $C_1(z_c, \mathbf{B})$ shows that it is always positive for both classes of bias (36),(37) so that λ_1 is positive just below T_c. Using (26) and (31) we find that $\lambda_2(z_c, \mathbf{B})$ is positive again for both classes of bias. Since λ_2 is an analytic function of z, we conclude that λ_2 is always positive just below T_c.

In conclusion, for the $q = 3$ state Potts model we find on the basis of these first results that the positive Mattis states are stable near T_c when the bias is of the form (36) and (37), independent of the value of a. This is different from the Hopfield model ($q = 2$) where the Mattis states become unstable near T_c when $a^2 > \frac{1}{3}$.

3.2 Symmetric states

In the case of symmetric states the positive solution of the saddle-point equation (17) becomes in the limit $T = 0$:

$$S_n = \frac{1}{n} \left\langle\!\left\langle \max_\rho \left(\sum_{a=1}^n m_{k^a,\rho} - nB_\rho\right) \right\rangle\!\right\rangle \tag{39}$$

and the corresponding free energy is given by

$$f_n = -\frac{n}{2} S_n^2. \tag{40}$$

Without bias it has been shown [4] for $q = 2$ that the following ordering of energy levels of the symmetric states occurs at zero temperature:

$$f_1 < f_3 < ... < f_4 < f_2. \tag{41}$$

This relation is no longer true in the case of general q. For large q we even have [7]:

$$f_1 \leq f_2 \leq f_3 \leq ... \tag{42}$$

In the case with bias we have for $q = 2$ that the Mattis states are no longer the ground states at zero temperature when the bias parameter exceeds a certain value [5]. For $q = 3$ with bias parameters B_1 (36) a similar statement can be made (see Fig.2 where the lowest energy levels are plotted against a). However, for $q = 3$ and bias parameters B_2 (37) the Mattis states remain the lowest in energy among the first three symmetric states and we expect this to be true when considering all the symmetric states (see Fig.3).

To study the local stability of the symmetric states we look at the stability matrix (22) which is now of the form

$$A = \begin{pmatrix} \mathcal{A}_n & 0 \\ 0 & \mathcal{D}_{p-n} \end{pmatrix} \tag{43}$$

where $\mathcal{A}_n$ is an $n \times n$-matrix with diagonal elements γ_1 and off-diagonal elements δ and $\mathcal{D}_{p-n}$ is a $(p-n) \times (p-n)$ diagonal matrix with elements γ_2. The quantities γ_1, γ_2 and δ are given by

$$\gamma_1 = 1 - \beta \left\langle\!\left\langle \frac{\sum_{\rho=1}^q (m_{k^1,\rho} - B_\rho)^2 U_\rho(n)}{\sum_{\rho=1}^q U_\rho(n)} \right.\right.$$
$$\left.\left. - \left(\frac{\sum_{\rho=1}^q (m_{k^1,\rho} - B_\rho) U_\rho(n)}{\sum_{\rho=1}^q U_\rho(n)}\right)^2 \right\rangle\!\right\rangle \tag{44}$$

$$\gamma_2 = 1 - \beta \left\langle\!\!\left\langle \frac{\sum_{\rho=1}^{q} (m_{k^{n+1},\rho} - B_\rho)^2 U_\rho(n)}{\sum_{\rho=1}^{q} U_\rho(n)} \right.\right.$$

$$\left.\left. - \left(\frac{\sum_{\rho=1}^{q} (m_{k^{n+1},\rho} - B_\rho) U_\rho(n)}{\sum_{\rho=1}^{q} U_\rho(n)} \right)^2 \right\rangle\!\!\right\rangle \tag{45}$$

$$\delta = -\beta \left\langle\!\!\left\langle \frac{\sum_{\rho=1}^{q} (m_{k^1,\rho} - B_\rho)(m_{k^2,\rho} - B_\rho) U_\rho(n)}{\sum_{\rho=1}^{q} U_\rho(n)} \right.\right.$$

$$\left.\left. - \frac{\left[\sum_{\rho=1}^{q} (m_{k^1,\rho} - B_\rho) U_\rho(n)\right]\left[\sum_{\rho=1}^{q} (m_{k^2,\rho} - B_\rho) U_\rho(n)\right]}{\left(\sum_{\rho=1}^{q} U_\rho(n)\right)^2} \right\rangle\!\!\right\rangle \tag{46}$$

where

$$U_\rho(n) = \exp\left[\beta S_n \left(\sum_{s=1}^{n} m_{k^s,\rho} - n B_\rho\right)\right]. \tag{47}$$

The matrix A has three different eigenvalues: $\lambda_1 = \gamma_1 - \delta$ which is non-degenerate, $\lambda_2 = \gamma_1 + (n-1)\delta$ with degeneracy $n-1$ and $\lambda_3 = \gamma_2$ with degeneracy $p-n$. It is convenient to rewrite the bias $\mathbf{B} = (B_1, ..., B_q)$ in the following way:

$$\mathbf{B} = a(c_1, ..., c_q), \quad a \in [0, 1) \tag{48}$$

and we call a the bias amplitude and $(c_1, ..., c_q)$ the bias structure.

The stability of the symmetric states at $T = 0$ depends on both the structure and the amplitude of the bias. A detailed study of (44)-(46) reveals the following preliminary results. If the structure is such that all componenets of $\mathbf{B}$ are different, we have stable states except for a finite number of values of the bias amplitude a. E.g. the state S_3 for $q = 3$ and bias $\mathbf{B}_2 = a(1, 0, -1)$ is stable unless $a \in \{0, \frac{1}{2}, 1\}$.

In the other case, two or more components of $\mathbf{B}$ are equal. Then for $S_n > 0$ two possibilities have to be considered. Firstly, if there are at least two components which are the smallest, the symmetric states $(n > 1)$ are always unstable independent of the value of a. Since for bias (36) the energy levels of the unstable symmetric states and the stable Mattis states cross for a certain value of $a = a_0$ (see fig.2) there must be another solution of (17) which is the global minimum of the free energy for $a > a_0$. Secondly, if there is only one component which is the smallest, there are only some regions for the bias amplitude a where we have stability. Results for $S_n < 0$ can be found in [7]. E.g. in the case without bias one has that for T small the state with $S_n < 0$ and $n = p \geq q$ is unstable and for T near T_0 this state is stable.

4 Dynamical behavior

In order to find out how an initial state of the network changes in time and to what extent one of the built-in patterns is approached, we look at the equations for the submagnetizations $\mu_k^\alpha(\sigma)$ derived in [2],[3]. These submag-

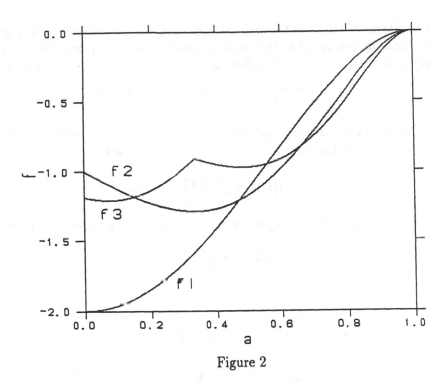

Figure 2

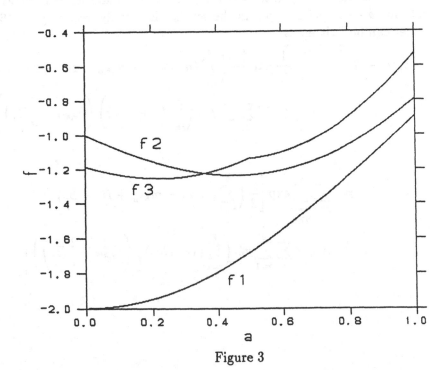

Figure 3

netizations are defined through the following procedure. Using the partition method [8], the set of all indices is divided into subsets labeled by the multi-index $\mathbf{k} = (k_1, \ldots, k_p)$ which depend on the built-in patterns as follows:

$$\{i \leq N\} = \bigcup_{\mathbf{k}} I_{\mathbf{k}}, \quad I_{\mathbf{k}} = \{i \leq N | \mathbf{k} = \mathbf{k}_i\}, \tag{49}$$

where $\mathbf{k}_i$ is the p-dimensional vector $\{k_i^1, k_i^2, \ldots, k_i^p\}$ for neuron i. We remark that the average number of indices in the set $I_{\mathbf{k}}$ is given by

$$\langle |I_{\mathbf{k}}| \rangle = q^{-p} N \prod_{a=1}^{p} (1 + B_{k^a}). \tag{50}$$

We then introduce submagnetizations for neurons restricted to the set $I_{\mathbf{k}}$

$$\mu_{\mathbf{k}}^\alpha(\boldsymbol{\sigma}) = \frac{1}{|I_{\mathbf{k}}|} \sum_{i \in I_{\mathbf{k}}} m_{\alpha, \sigma_i} \tag{51}$$

with $\boldsymbol{\sigma} = \{\sigma_1, \ldots, \sigma_N\}$.

Starting from the master equation for the probability to observe the network in a state $\boldsymbol{\sigma}$ at time t, one derives [2],[3]

$$\frac{d\mu_{\mathbf{k}}^\alpha}{dt} = -\frac{1}{q} \sum_{\rho=1}^{q} (1 + \mu_{\mathbf{k}}^\rho) \sum_{\nu=1}^{q} w_{\mathbf{k}}(\rho \to F^\nu \rho)(m_{\alpha,\rho} - m_{\alpha, F^\nu \rho}) \tag{52}$$

where F is the spin operator $F\alpha = [(\alpha \bmod q) + 1]$ and $w_{\mathbf{k}}$ is the transition rate for $\rho \to F^\nu \rho$ for a spin in the set $I_{\mathbf{k}}$. Explicitly, recalling (5),(6), the latter can be written as

$$w_{\mathbf{k}}(\alpha \to F^\nu \alpha) = \frac{1}{Z_{\mathbf{k}}} \exp\left[\frac{\beta}{2} \left(\sum_{a=1}^{p} (m_{k^a, F^\nu \alpha} - m_{k^a, \alpha} + B_\alpha - B_{F^\nu \alpha}) \right. \right.$$
$$\left. \left. \times q^{-p} \sum_{\mathbf{k}'} \sum_{\gamma=1}^{q} \mu_{\mathbf{k}'}^\gamma \left(\prod_{c=1}^{p} (1 + B_{k'^c}) \right) \left(\delta_{\gamma, k'^a} - \frac{1}{q} B_\gamma \right) \right) \right] \tag{53}$$

$$Z_{\mathbf{k}} = \sum_{\substack{\rho=1 \\ }}^{q} \sum_{\substack{\eta=1 \\ \eta \neq \rho}}^{q} \exp\left[\frac{\beta}{2} \left(\sum_{a=1}^{p} (m_{k^a, \eta} - m_{k^a, \rho} + B_\rho - B_\eta) \right. \right.$$
$$\left. \left. \times q^{-p} \sum_{\mathbf{k}'} \sum_{\gamma=1}^{q} \mu_{\mathbf{k}'}^\gamma \left(\prod_{c=1}^{p} (1 + B_{k'^c}) \right) \left(\delta_{\gamma, k'^a} - \frac{1}{q} B_\gamma \right) \right) \right]. \tag{54}$$

The equations (52)-(54) describe the evolution of the submagnetizations and hence also the evolution of the overlap parameters R_a through the relation $(N \to \infty)$

$$R_a = q^{-(p+1)} \sum_k \left(\prod_{a=1}^p (1 + B_{ka}) \right) \sum_{\alpha=1}^q \mu_k^\alpha(\boldsymbol{\sigma})(m_{\alpha,ka} - B_\alpha). \qquad (55)$$

We have solved these equations numerically to illustrate some of the stability properties found in section 3. For $q = 2$ and $a \neq 0$ we present in figs.4 and 5 the flow lines describing the evolution of the macroscopic overlap parameters R_1 and R_2 with two learned patterns. For $a = 0.2$ the Mattis states must be stable near T_c. This is clearly seen in fig.4 ($T = 0.95$, $T_c = 0.96$) where these states show up as attractors. For $a = 0.7$ however, the Mattis states must become unstable near T_c. Also this situation is found back using the dynamics of the system as is shown in fig.5 ($T = 0.5$, $T_c = 0.51$). The flow lines first go towards an (unstable) Mattis state but then bend away and the system evolves to an attractor on the diagonal.

In figs.6 and 7 we illustrate the result on the negative symmetric states mentioned at the end of section 3 for $q = 3, p = 3$. Starting from a state with 3 overlaps that are almost equal $\mathbf{R} = (-0.385, -0.388, -0.392)$ we see in fig.6 that at T=0.5 this state evolves to an asymmetric state. In fig.7 the asymmetric state $\mathbf{R} = (-0.237, -0.166, -0.392)$ evolves at $T = 1.5$ to a symmetric attractor.

References

[1] I. Kanter, Phys. Rev. **A37** (1988), 2739

[2] D. Bollé and F. Mallezie, J. Phys. **A22** (1989),4409

[3] D. Bollé, P. Dupont and F. Mallezie, in *Neural networks and spin glasses* eds. W.K. Theumann and R. Koeberle (World Scientific Singapore, 1990),151

[4] D. Amit, H. Gutfreund and H. Sompolinsky, Phys. Rev. **A32** (1985), 1007

[5] D. Amit, H. Gutfreund and H. Sompolinsky, Phys. Rev. **A35** (1987), 2293

[6] P. Peretto, Biol. Cybern. **50** (1984),51

[7] D. Bollé, P. Dupont and J. van Mourik, *Stability properties of Potts neural networks with biased patterns*, preprint KUL-TF90

[8] J.L. van Hemmen, D. Grensing, A. Huber and R. Kühn, Z. Phys. **B65** (1986), 53

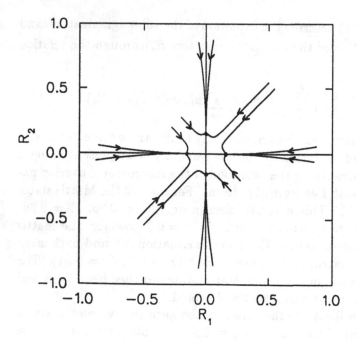

Figure 4

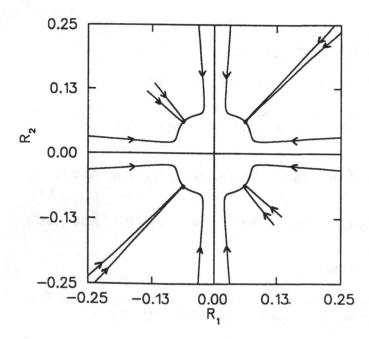

Figure 5

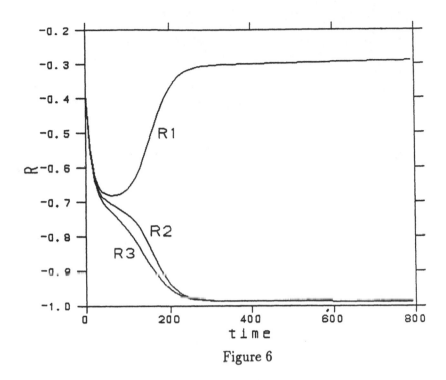

Figure 6

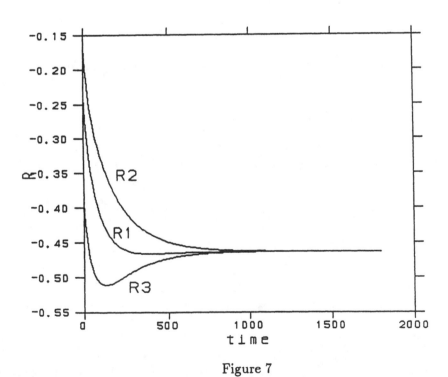

Figure 7

ISING-SPIN NEURAL NETWORKS WITH SPATIAL STRUCTURE

A.C.C. Coolen [†]

Dept. of Medical and Physiological Physics, University of Utrecht

Princetonplein 5 NL-3584CC Utrecht, The Netherlands

We study Ising spin models for neural networks with spatial structure, in which patterns are stored according to Hebb's rule. From the microscopic dynamics given by the master equation we derive a partial differential equation for the position- and time-dependent correlations between the system state and the stored patterns. This equation can be used to study networks with finite range connections (not necessarily symmetric), the behaviour of domain boundaries and information transport. For systems with finite range connections we can compute the range of macroscopic order.

I. INTRODUCTION

After Little [1] and Hopfield [2] published their papers in which they used the Ising model to describe the action of neural networks with patterns built in according to Hebb's rule [3], the methods of equilibrium statistical mechanics began to be widely applied to elucidate many properties of these networks [4,5,6,7]. The problem of treating non-symmetric couplings was also solved and it became possible to describe

† Present address:

    Dept. of Medical Physics and Biophysics, University of Nijmegen

    Geert Grooteplein Noord 21 NL-6525EZ Nijmegen, The Netherlands

the dynamical behaviour [8,9]. However, the question of how to describe networks in which the range of connections is restricted and in which the connection density is not uniform in space is still unanswered.

The purpose of this paper is to remedy this situation for the case where the range of the connections is large compared to the average distance between neighbouring neurons. The number of connections to and from each neuron will be assumed to be large as well. We introduce a macroscopic description in terms of correlation parameters which are both time- and position-dependent. From the microscopic dynamics a partial differential equation can be derived for large systems, which describes the evolution of these parameters in time. This equation also applies to non-symmetric systems where equilibrium statistical mechanics is not applicable. The macroscopic equation is used to study the onset of order in systems with finite range connections. In particular we will derive an analytical expression for the range of order that is given by the smallest wavelength in the Fourier spectrum of the correlations, as a function of the connection range and the noise level. This provides an answer to the question of which connection range one must choose in hardware realisations of attractor neural networks.

Equations similar to the ones we will derive have recently been given by Noest [10] for extremely diluted networks with symmetric couplings.

II. DEFINITIONS

We start with a network of N neurons, each of which can be in two states, $s_i = \pm 1$, $i=1,..,N$. The N-neuron network state will be written as $\vec{s}$. A finite number of special network states $\vec{\xi}^{(\mu)}$ ($\mu=1,..,p$) represent the patterns, which were built in using Hebb's rule [3] and which determine the connection matrix J_{ji} for the effect of neuron i on neuron j by

$$J_{ji} = \begin{cases} \dfrac{1}{S} \sum_{\mu=1}^{p} \xi_j^{(\mu)} \xi_i^{(\mu)} & \text{if } i \in S_j \\ 0 & \text{if } i \notin S_j \end{cases} \tag{1}$$

Here we have denoted by S_j the set of neurons which can send a signal to neuron j. S is the average number of neurons contributing to an input, i.e., $S = \frac{1}{N} \sum_j |S_j|$, where $|S_j|$ is the number of neurons in the set S_j.

As in statistical mechanics we will consider an ensemble of states, so we can speak about the probability $p_t(\vec{s})$ of finding the system at time t in the state $\vec{s}$. The evolution of this probability is now assumed to be governed by a stochastic process, in which the probability per unit time for neuron j to flip from s_j to $-s_j$ is some given function of the total state $\vec{s}$, which will be denoted by $w_j(\vec{s})$. The master equation for this process is

$$\frac{dp_t(\vec{s})}{dt} = \sum_j w_j(F_j\vec{s})p_t(F_j\vec{s}) - p_t(\vec{s}) \sum_j w_j(\vec{s}) \tag{2}$$

where F_j is an operator, defined by

$$F_j\Phi(s_1,\ldots,s_j,\ldots,s_N) = \Phi(s_1,\ldots,-s_j,\ldots,s_N)$$

For $w_j(\vec{s})$ we make the usual choice

$$w_j(\vec{s}) = \frac{1}{2}[1-\tanh(\beta s_j h_j(\vec{s}))] \tag{3}$$

where $\beta=1/T$ (the 'temperature' T being a measure of the amount of noise) and the input $h_j(\vec{s})$ acting upon neuron j is

$$h_j(\vec{s}) = \sum_{i \in S_j} J_{ji}s_i \tag{4}$$

In general it will be impossible to find the solution of Eq. (2). However, as in statistical mechanics, we are not interested in the microscopic details of a network state, but rather in the question whether the development of certain *macroscopic* features can be calculated. In the Appendix we analyse the evolution in time of linear order parameters Ω_μ:

$$\Omega_\mu(\vec{s}) = \sum_i \omega_i^{(\mu)} s_i \tag{5}$$

where $\mu \in \{1,..,n\}$. If the local fields $h_i(\vec{s})$ depend on $\vec{s}$ only through the values of the order parameters, $h_i(\vec{s}) = h_i[\vec{\Omega}(\vec{s})]$, and if the number n of order parameters is not too large, we find deterministic evolution equations for the order parameters in the limit $N \to \infty$:

$$\frac{d}{dt} \Omega_\mu(t) = \lim_{N \to \infty} \sum_j \omega_j^{(\mu)} \tanh(\beta h_j[\vec{\Omega}(t)]) - \Omega_\mu(t) \tag{6}$$

Here we assumed $\omega_i^{(\mu)} = O(|V_\mu|^{-1})$ for all i, μ. The support V_μ of Ω_μ is defined as $V_\mu = \{i | \omega_i^{(\mu)} \neq 0\}$, $|V_\mu|$ is the number of elements in V_μ (for simplicity we take all $|V_\mu|$ to be of the same order in N). The restriction on the number of order parameters is:

$$\lim_{N \to \infty} n / \sqrt{|V_\mu|} = 0 \quad \text{for all } \mu \tag{7}$$

III. LOCAL OVERLAP PARAMETERS

In order to describe the macroscopic dynamical behaviour of a network with a given spatial structure, as given by (1), we divide the system into a large number of non-overlapping clusters $\lambda = 1, 2, .., \Lambda$ of adjacent neurons. The indices of the neurons in cluster λ form the set I_λ, which contains $|I_\lambda| = N/\Lambda \gg 1$ indices. We now define macroscopic variables by

$$Q_{\lambda\mu}(\vec{s}) = \frac{1}{|I_\lambda|} \sum_{i \in I_\lambda} s_i \xi_i^{(\mu)} \tag{8}$$

This variable measures the correlation between the actual network state and pattern μ in cluster λ. A macroscopic state is now defined by the vector $\vec{Q} = (Q_{11}, ..., Q_{\Lambda p})$, where each $Q_{\lambda\mu}$ ranges between -1 and 1. In order to show that, in good approximation, the inputs $h_j(\vec{s})$ can also be considered as a function of $\vec{Q}(\vec{s})$, we write $\sum_{i \in S_j}$ in Eq. (4) as $\sum_{\lambda'} \sum_{i \in D_{j\lambda'}}$, where $D_{j\lambda'}$ is the intersection of S_j and I_λ, (see Figure 1). Then we obtain

$$h_j(\vec{s}) = \frac{1}{S} \sum_{\mu'} \xi_j^{(\mu')} \sum_{\lambda'} \sum_{i \in D_{j\lambda'}} s_i \xi_i^{(\mu')}$$

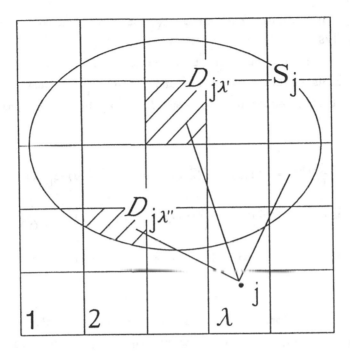

Figure 1:   The partitioning of the sum over all contributions to the input of neuron j; $D_{j\lambda'}$ denotes the intersection of cluster λ' and the set S_j of all neurons contributing to this input.

If S is taken sufficiently large, i.e. $S \gg N/\Lambda$, it is seen from the definition (8) that

$$h_j(\vec{s}) \simeq \sum_{\lambda'\mu'} \frac{|D_{j\lambda'}|}{S} Q_{\lambda'\mu'}(\vec{s})\, \xi_j^{(\mu')}$$

In the limit N→∞ this becomes an equality (if S_j is chosen such that the boundary ∂S_j coincides with boundaries of the sets I_λ then the equality holds without any restriction on S). Equation (6) can now be used to describe the evolution in time of $\vec{Q}$. However, in order to simplify the final result we will also assume that the number of connections from the set $D_{j\lambda'}$ to neuron j in cluster λ is the same for each $j \in I_\lambda$. We then find the density of connections from $I_{\lambda'}$, to any neuron in I_λ to be $n(\lambda,\lambda')=|D_{j\lambda'}|/S$.

We now take a continuum limit and replace the cluster labels λ and λ' by the continuous position vectors $\vec{x}$ and $\vec{y}$; for $Q_{\lambda\mu}(t)$ we write $q_\mu(\vec{x},t)$. By taking this limit, equation (6), applied to the order parameters $Q_{\lambda\mu}$, becomes

$$\frac{\partial \vec{q}(\vec{x},t)}{\partial t} = \vec{A}(\vec{x},\vec{q}) - \vec{q}(\vec{x},t) \quad \text{with} \quad \vec{q}(\vec{x},0)=\vec{q}_0(\vec{x}) \tag{9}$$

where

$$\vec{A}(\vec{x},\vec{q}) = < \vec{\xi}(\vec{x}) \tanh \beta\left[\vec{\xi}(\vec{x})\cdot \int n(\vec{x},\vec{y})\vec{q}(\vec{y},t)d\vec{y}\right] >_{\vec{\xi}(\vec{x})} \tag{10}$$

$\vec{q}$, $\vec{A}$ and $\vec{\xi}(\vec{x})$ are p-dimensional vectors. The average $<\ >_{\vec{\xi}(\vec{x})}$ is computed over the distribution of the pattern components at position $\vec{x}$; it therefore depends on their specific realisation. The function $n(\vec{x},\vec{y})$ represents the density of connections from position $\vec{y}$ to position $\vec{x}$. Since we assumed p to be fixed, the conditions for Eqs. (9,10) to hold are

$$\lim_{N\to\infty} \frac{N}{\Lambda S} = \lim_{N\to\infty} \frac{\Lambda^2}{S} = 0$$

One might choose $\Lambda = N^\epsilon$ and $S = N^\gamma$, such that $\epsilon < 1/2$ and $\gamma > 1-\epsilon$. If the pattern components $\xi_j^{(\mu)}$ do not vary within the clusters I_λ, Eq. (10) reduces to:

$$\vec{A}(\vec{x},\vec{q}) = \vec{\xi}(\vec{x}) \tanh \beta\left[\vec{\xi}(\vec{x})\cdot \int n(\vec{x},\vec{y})\vec{q}(\vec{y},t)d\vec{y}\right] \tag{11}$$

In this case we can make a decomposition: $\vec{q}(\vec{x},t) = q(\vec{x},t)\vec{\xi}(\vec{x}) + \vec{q}^\perp(\vec{x},t)$, where $\vec{q}^\perp(\vec{x},t)\cdot\vec{\xi}(\vec{x}) = 0$. It then follows from (9) that

$$\vec{q}^\perp(\vec{x},t) = \vec{q}^\perp(\vec{x},0)\ e^{-t}$$

$$\frac{\partial}{\partial t}q(\vec{x},t) = \tanh\left[\beta\int d\vec{y}\ \hat{n}(\vec{x},\vec{y})q(\vec{y},t)\right] - q(\vec{x},t) \tag{12}$$

where $\hat{n}(\vec{x};\vec{y}) = n(\vec{x};\vec{y})\ \vec{\xi}(\vec{x})\cdot\vec{\xi}(\vec{y})$ and $q(\vec{x},t)$ turns out to be simply the average activity (or magnetisation) at position $\vec{x}$ at time t. If, on the other hand, the components $\xi_j^{(\mu)}$ are drawn at random from $\{-1,1\}$ with equal probabilities, then:

$$\vec{A}(\vec{x},\vec{q}) = < \vec{\xi} \ \tanh \ \beta \left[\vec{\xi} \cdot \int n(\vec{x},\vec{y}) \vec{q}(\vec{y},t) d\vec{y} \right] >_{\vec{\xi}}$$ (13)

We will now discuss some general properties of (9) with randomly drawn patterns (13). We consider $n(\vec{x},\vec{y})$ to be a compact integral operator. Since every bounded $n(\vec{x},\vec{y})$ is compact, this is not a strong restriction. We define the inner product $<f|g> = \frac{1}{V} \int d\vec{x} \ f(\vec{x})g(\vec{x})$ $(V = \int d\vec{x})$ and norm $|f| = <f|f>^{1/2}$. Using $|\tanh(z)| \le |z|$ it follows that:

$$\frac{\partial}{\partial t} \ |\vec{q}^2| \ \le \ - \ 2 \ |\vec{q}^2| \ + \ 2\beta \ \frac{1}{V} \int\int d\vec{x}d\vec{y} \ n(\vec{x},\vec{y}) \ <|\vec{\eta}\cdot\vec{q}(\vec{x},t)| |\vec{\eta}\cdot\vec{q}(\vec{y},t)|>$$

$$\le \ - \ 2 \ |\vec{q}^2| \ + \ 2\beta \ \frac{1}{V} \int\int d\vec{x}d\vec{y} \ n(\vec{x},\vec{y}) \ < \vec{\eta}^2 \left[\vec{q}^2(\vec{x},t) \ \vec{q}^2(\vec{y},t)\right]^{1/2} >$$

$$\le \ - \ 2 \ |\vec{q}^2| \ + \ 2p\beta \ \frac{1}{V} \int\int d\vec{x}d\vec{y} \ \left[\vec{q}^2(\vec{x},t)\right]^{1/2} n(\vec{x},\vec{y}) \left[\vec{q}^2(\vec{y},t)\right]^{1/2}$$

By definition we have: $\frac{1}{V} \int\int d\vec{x}d\vec{y} \ n(\vec{x},\vec{y}) = 1$. Since $n(\vec{x},\vec{y})$ is compact, for all f: $|\int d\vec{y} \ n(.,\vec{y})f(\vec{y})| \le M|f|$ $(M \ge 1)$. Therefore:

$$|\int d\vec{y} \ n(.,\vec{y}) \left[\vec{q}^2(\vec{y},t)\right]^{1/2}| \ \le \ M| \left[\vec{q}^2\right]^{1/2}|$$

$$\frac{1}{V} \int\int d\vec{x}d\vec{y} \ \left[\vec{q}^2(\vec{x},t)\right]^{1/2} n(\vec{x},\vec{y}) \left[\vec{q}^2(\vec{y},t)\right]^{1/2} \ \le \ M| \left[\vec{q}^2\right]^{1/2}|^2 \ \le \ M|\vec{q}^2|$$

and finally: $\frac{\partial}{\partial t} \ |\vec{q}^2| \ \le \ - \ 2(1-\beta Mp) \ |\vec{q}^2|$. We can conclude that there is always a critical temperature $T_c \le Mp$, above which all order will eventually disappear. The upper bound Mp is probably not the best one can get. For translation invariant systems, i.e., $n(\vec{x},\vec{y}) = n(\vec{x}-\vec{y})$, we will show that $T_c = 1$.

For a symmetric connecting density, i.e, $n(\vec{x},\vec{y}) = n(\vec{y},\vec{x})$, the evolution in time corresponds to gradient descent on a free energy surface. The free energy functional $F[\vec{q}(.,t)]$ is:

$$F[\vec{q}(.,t)] = - \frac{1}{2} \int d\vec{x}d\vec{y} \ n(\vec{x},\vec{y}) \ \vec{q}(\vec{x},t)\cdot\vec{q}(\vec{y},t) - \frac{1}{\beta} \int d\vec{x} \ f(\vec{x},\vec{q}(\vec{x},t))$$ (14)

where $f(\vec{x},\vec{q})$ satisfies: $\vec{q} = - < \vec{\xi}(\vec{x})\tanh[\vec{\xi}\cdot\vec{\nabla}_{\vec{q}} f(\vec{x},\vec{q})] >_{\vec{\xi}(\vec{x})}$ for all $\vec{x},\vec{q}$. By using (9), (13) and (14) one can show that $\frac{d}{dt}F[q(.,t)] \le 0$, the stationary states being the equilibrium solutions of Eq. (9). In the case of non-symmetric connections, whether or not an equilibrium state will be approached will depend on the specific connections and the initial state.

If at t=0 the correlations are not position-dependent, i.e $\vec{q}(\vec{x},0)=\vec{q}_0$, and $\int d\vec{y}\, n(\vec{x},\vec{y})$ does not depend on $\vec{x}$, we recover the solution which is known for the fully connected network [9]:

$$\vec{q}(\vec{x},t) = \vec{q}(t)$$

$$\frac{d}{dt}\,\vec{q}(t) = <\vec{\xi}.\tanh[\beta\vec{\xi}\cdot\vec{q}(t)]>_{\vec{\xi}} - \vec{q}(t)$$

(15)

IV. SOME SIMPLE SOLUTIONS OF THE OVERLAP EQUATIONS

We will consider in more detail some simple choices for the spatial structure as determined by the connection density $n(\vec{x},\vec{y})$. The choice $n(\vec{x},\vec{y})=n(\vec{y})$, for instance, implies that any given neuron contributes equally to all inputs; it is to be expected that initial spatial variations in the correlations will vanish. This is indeed what happens, since from (9,13) it follows that:

$$\frac{\partial}{\partial t}\,[q_\mu(\vec{x},t)-q_\mu(\vec{x}',t)] = - [q_\mu(\vec{x},t)-q_\mu(\vec{x}',t)]$$

The system will evolve towards the homogeneous solution (15).

If we take $n(\vec{x},\vec{y})=\delta(\vec{x}-\vec{y})$ there is no communication between regions. This is reflected by the solution of (9,13):

$$\frac{\partial}{\partial t}\,\vec{q}(\vec{x},t) = <\vec{\xi}.\tanh[\beta\vec{\xi}\cdot\vec{q}(\vec{x},t)]>_{\vec{\xi}} - \vec{q}(\vec{x},t)$$

The position vector $\vec{x}$ has simply become a label. At each position $\vec{x}$ the correlations evolve in time independently, governed by the standard equation of simple fully connected networks [9] and the initial value $\vec{q}(\vec{x},0)$.

Since the connection density $n(\vec{x},\vec{y})$ need not be symmetric, we can also study feed-forward structures. A simple example is $n(\vec{x},\vec{y})=\delta(\vec{x}-\vec{y}-\vec{a})$, $\vec{a}\neq 0$. First we generalise the evolution equation by introducing transmission delays $\tau(\vec{x},\vec{y})$ which represent the time it takes for state changes at $\vec{y}$ to affect the input at position $\vec{x}$:

$$\frac{\partial}{\partial t}\,\vec{q}(\vec{x},t) = <\vec{\xi}.\tanh[\beta\!\int\!d\vec{y}.n(\vec{x},\vec{y})\vec{\xi}\cdot\vec{q}(\vec{y},t-\tau(\vec{x},\vec{y}))]>_{\vec{\xi}} - \vec{q}(\vec{x},t) \qquad (16)$$

In this example we choose $\tau(\vec{x},\vec{y})=|\vec{x}-\vec{y}|/v$, which corresponds to spike trains with equal effective velocity v. We find, writing $\vec{x}=(x,y,z)$ and choosing the z-axis in the direction $\vec{a}$:

$$\vec{q}(x,y,z,t) = \vec{\Psi}_{xy}(z,t)$$

$$\frac{\partial}{\partial t}\,\vec{\Psi}_{xy}(z,t) = <\vec{\xi}.\tanh[\beta\vec{\xi}\cdot\vec{\Psi}_{xy}(z,t-\tfrac{a}{v})]>_{\vec{\xi}} - \vec{\Psi}_{xy}(z,t)$$

The network decouples into independent parallel one-dimensional systems. If at $t=0$ we have $\vec{q}(x,y,z,t)=\vec{0}$ for $z>0$, with an input excitation in the direction of pattern ν at $z=0$ given by: $q_{\mu}(x,y,0,0)=\delta_{\mu\nu}I_{xy}$, then the above equations yield for $T=0$:

$$\Psi_{xy,\mu}(z,t) = \theta(vt-z)(1-e^{z/v-t}).\mathrm{sgn}(I_{xy})\delta_{\mu\nu}$$

The excitation propagates through the network with velocity v. If we choose $T>0$, the information transport will be slower.

If the connections are of finite range, domain structures may appear. We will give a simple example of a two-dimensional network to show how (9,13) gives rise to such behaviour. As connection density we choose:

$$n(\vec{x},\vec{y}) = \frac{1}{4R^2} \qquad \text{if } |x_1-y_1|<2R \text{ and } |x_2-y_2|<2R$$

$$n(\vec{x},\vec{y}) = 0 \qquad \text{elsewhere}$$

A neuron at position $\vec{x}$ receives input only from within a rectangle of size 2R centred at $\vec{x}$. If at $t=0$ we have a state like $q_{\mu}(\vec{x},0) = \delta_{\mu\nu}q(\vec{x})$,where $q(\vec{x}) > 0$ for $x_1>0$ and $q(\vec{x}) < 0$ for $x_1<0$, then Eq. (9,13) can easily be solved for $T=0$; the system evolves towards the equilibrium configuration $q_{\mu}(\vec{x},\infty)=\delta_{\mu\nu}\mathrm{sgn}[x_1]$. We find two domains with a boundary at $x_1=0$. We don't know whether or not this boundary is stable under a small deformation.

V. THE ONSET OF ORDER

We will now study the onset of order, i.e., the development of a non-trivial solution of Eqs. (9,13) from the trivial state $\vec{q}(\vec{x},t) = \vec{0}$. Expansion of (9) in powers of $\vec{q}(\vec{x},t)$ yields:

$$\frac{\partial}{\partial t} \vec{q}(\vec{x},t) = \beta \int d\vec{y}\ n(\vec{x},\vec{y})\vec{q}(\vec{y},t) - \vec{q}(\vec{x},t) + O(\vec{q}^3) \tag{17}$$

All bifurcations from the trivial solution must be eigenfunctions of the integral operator $n(\vec{x},\vec{y})$ with sufficiently large eigenvalue.

We will restrict ourselves to translation invariant systems: $n(\vec{x},\vec{y}) = n(\vec{x}-\vec{y})$. In this case we can apply a Fourier transform to Eq. (17) to find the solution of the linearised equation:

$$\vec{q}(\vec{x},t) = \int d\vec{k}\ \vec{A}_{\vec{k}}(t)\ e^{i\vec{k}\cdot\vec{x}-i\omega_{\vec{k}}t} \tag{18}$$

where:

$$\vec{A}_{\vec{k}}(t) = \left(\frac{1}{2\pi}\right)^d e^{[\beta\ \text{Re}\ \hat{n}(\vec{k})-1]t}\ \hat{\vec{q}}(\vec{k},0) \qquad \omega_{\vec{k}} = -\beta\ \text{Im}\ \hat{n}(\vec{k})$$

The Fourier transform $\hat{g}(\vec{p})$ of a function $g(\vec{x})$ is: $\hat{g}(\vec{p}) = \int d\vec{x}\ g(\vec{x})e^{-i\vec{p}\cdot\vec{x}}$. From (18) both the bifurcation temperature $T_{\vec{k}}$ for Fourier component $\vec{k}$ and the critical temperature T_c can be found:

$$T_{\vec{k}} = \int d\vec{x}\ n(\vec{x})\ \cos(\vec{k}\cdot\vec{x})$$

$$T_c = \max_{\vec{k}} T_{\vec{k}} = T_{\vec{0}} = 1$$

A very useful feature is that from (18) it follows that there exists a cut-off wavelength $\Lambda(T)$ in the Fourier spectrum, given by:

$$\Lambda(T) = 2\pi|\vec{k}^*(T)|^{-1} \tag{19}$$

where $\vec{k}^*(T)$ is defined as the largest solution of: $T_{\vec{k}^*} = T$. This $\Lambda(T)$ determines the minimum correlation length of the non-trivial solution that develops from the trivial solution. All Fourier components with $|\vec{k}| > |\vec{k}^*|$ will decay exponentially. If a network is to store and retrieve

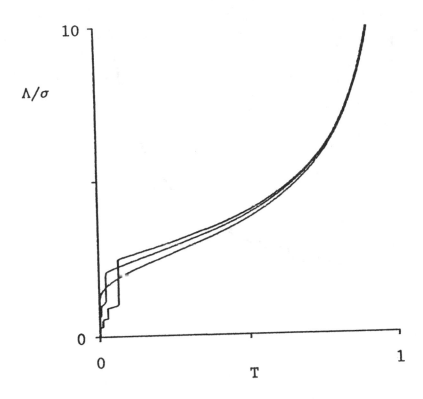

Figure 3:   The relative cut-off wavelength Λ/σ as a function of T (D=2)
Upper curve: a block-shaped interaction kernel
Middle curve: a triangular-shaped interaction kernel
Lower curve: a Gaussian-shaped interaction kernel

information, as do Hopfield networks, then one would like $\Lambda(T)$ to be as large as possible in order to guarantee that from local clues the system will reconstruct complete patterns only. It is possible to compute this $\Lambda(T)$ for all interactions of the form:

$$n(\vec{z}) = \sigma^{-d} f(|\frac{\vec{z}-\vec{a}}{\sigma}|) \tag{20}$$

Due to normalisation and because σ is the actual width the following conditions hold: $\int d\vec{z}\ f(|\vec{z}|) = \int d\vec{z}\ \vec{z}^2\ f(|\vec{z}|) = 1$. Using the Fourier transform of Eq. (20) we find that $\Lambda(T)$ is the smallest solution of:

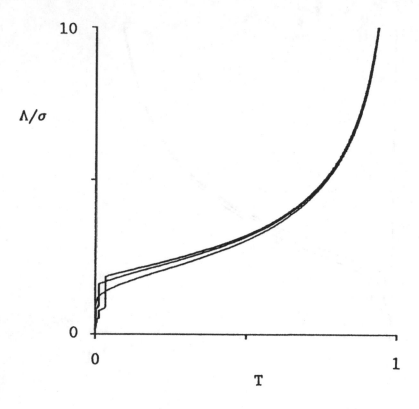

Figure 4:   The relative cut-off wavelength Λ/σ as a function of T (D=3)
             Upper curve: a block-shaped interaction kernel
             Middle curve: a triangular-shaped interaction kernel
             Lower curve: a Gaussian-shaped interaction kernel

D=1:
$$T = 2\cos\left(\frac{2\pi a y}{\Lambda(T)}\right) \int_0^\infty dy\ f(y)\ \cos\left(\frac{2\pi\sigma y}{\Lambda(T)}\right) \tag{21}$$

D=2:
$$T = 2\pi \int_0^\infty dy\ y\ f(y)\ J_0\left(\frac{2\pi\sigma y}{\Lambda(T)}\right) \tag{22}$$

D=3:
$$T = 4\pi \int_0^\infty dy\ y^2 f(y)\ \left(\frac{2\pi\sigma y}{\Lambda(T)}\right)^{-1}\sin\left(\frac{2\pi\sigma y}{\Lambda(T)}\right) \tag{23}$$

D is the physical dimension of the network; J_0 is the Bessel function of the first kind. General features of (21,22,23) are: $\Lambda(0) = 0$ and $\Lambda(1) = \infty$. If f(y) is taken to be Gaussian, the relations (21,22,23) can be inverted analytically. One obtains:

$$\Lambda(T)/\sigma = \pi\sqrt{2/D} \left[\log(1/T)\right]^{-1/2} \qquad (24)$$

In Figure 2 (D=2) and Figure 3 (D=3) the solutions of Eqs. (22,23) are shown for Gaussian interactions, as well as for block-shaped and triangular-shaped interaction functions. It seems that $\Lambda(T)/\sigma$ does not depend much on the actual choice of the interaction function.

In building a hardware realisation of an attractor neural network the major problem is to meet the requirement of huge connectivity. Full connectivity in artificial networks is, except perhaps in the case of optical realisations, out of the question. However, one will have to choose the connection range σ sufficiently large, i.e., such that $\Lambda(T) \geq L$ (where L is the physical size of the network). For a give noise level one can use equations (21-24) to calculate the connection range needed.

VI. CONCLUSIONS

The proposed description of the temporal development of the correlations $q_\mu(\vec{x},t)$, Eq. (9), enables us to study the phenomena that emerge in network models where there is some spatial structure in the connectivity or where connections have a finite range. In a way our analysis proceeds naturally along the lines of recent papers on systems with inhomogeneities and finite connectivity by Kanter [11] and Canning and Gardner [12] and the recent study of domain effects in diluted networks with symmetric connection density by Noest [10] (his analysis also leads to an equation like (9)). Our equation applies to non-diluted systems, and, since the connection density need not be symmetric, also to models where thermodynamic analysis is not applicable.

Using the evolution equation, we studied the onset of order, i.e., the emergence of non-trivial macroscopic states. There is a critical temperature below which all macroscopic order will eventually disappear. If the network is translation invariant it is possible to compute the critical temperature, independent of the actual shape of the interaction kernel. Furthermore, one can show that the correlation length of non-trivial solutions can be much larger than the range of the interactions. This enables one to retrieve complete patterns with limited-range interactions from local clues. For a large class of interactions the ratio between the minimum correlation length and the range of the connections can be computed as a function of the noise level, and, as a result, the connection range that is needed in hardware realisations of attractor neural networks.

ACKNOWLEDGEMENT

It is a pleasure to thank Prof. Dr. Th.W. Ruijgrok for many significant contributions to this study.

APPENDIX: THE EVOLUTION OF LINEAR ORDER PARAMETERS

In this section we derive flow equations for order parameters Ω_μ of the form

$$\Omega_\mu(\vec{s}) = \sum_i \omega_i^{(\mu)} s_i$$

The vector $\vec{\Omega}(\vec{s}) = (\Omega_1(\vec{s}),..,\Omega_n(\vec{s}))$ constitutes our macroscopic level of description. The support V_μ of Ω_μ is defined as: $V_\mu = \{i| \ \omega_i^{(\mu)} \neq 0 \}$. In order to find deterministic equations we assume $\omega_i^{(\mu)} = O(|V_\mu|^{-1})$ for all i,μ ($|V_\mu|$ is the number of elements in V_μ). For simplicity we assume all $|V_\mu|$ to be of the same order in N. The probability $P_t(\vec{\Omega})$ that the order parameters will have the value $\vec{\Omega} = (\Omega_1,...,\Omega_n)$ at time t is given by

$$P_t(\vec{\Omega}) = \sum_{\vec{s}} P_t(\vec{s}) \ \delta\left[\vec{\Omega} - \vec{\Omega}(\vec{s})\right]$$

Our purpose is to find a differential equation for $P_t(\vec{\Omega})$ using the microscopic master equation. This will be possible only if the local fields $h_i(\vec{s})$ depend on $\vec{s}$ only through the values of the order parameters: $h_i(\vec{s}) = h_i[\vec{\Omega}(\vec{s})]$. Using (2) one finds:

$$\frac{d}{dt} P_t(\vec{\Omega}) = \sum_{\vec{s}} \sum_j w_j(\vec{s}) p_t(\vec{s}) \left[\delta\left(\vec{\Omega} - \vec{\Omega}(\vec{s}) + 2s_j \vec{\omega}_j \right) - \delta\left(\vec{\Omega} - \vec{\Omega}(\vec{s})\right)\right]$$

where $\vec{\omega}_j = (\omega_j^{(1)}, \dots, \omega_j^{(n)})$. The average value of any function $\Phi[\vec{\Omega}]$ is: $\langle \Phi \rangle_t = \int d\vec{\Omega}\, P_t(\vec{\Omega}) \Phi[\vec{\Omega}]$. Its time derivative is given by

$$\frac{d}{dt} \langle \Phi \rangle_t = \sum_{\vec{s}} \sum_j w_j(\vec{s}) p_t(\vec{s}) \left[\Phi[\vec{\Omega}(\vec{s}) - 2s_j \vec{\omega}_j] - \Phi[\vec{\Omega}(\vec{s})]\right]$$

$$= \sum_{\vec{s}} \sum_j w_j(\vec{s}) p_t(\vec{s}) \left\{ \sum_{m \geq 1} \frac{2^{2m}}{(2m)!} \sum_{k_1=1}^{n} \dots \sum_{k_{2m}=1}^{n} \omega_j^{(k_1)} \dots \omega_j^{(k_{2m})} \frac{\partial^{2m} \Phi[\vec{\Omega}(\vec{s})]}{\partial \Omega_{k_1} \dots \partial \Omega_{k_{2m}}} \right.$$

$$\left. - s_j \sum_{m \geq 0} \frac{2^{2m+1}}{(2m+1)!} \sum_{k_1=1}^{n} \dots \sum_{k_{2m+1}=1}^{n} \omega_j^{(k_1)} \dots \omega_j^{(k_{2m+1})} \frac{\partial^{2m+1} \Phi[\vec{\Omega}(\vec{s})]}{\partial \Omega_{k_1} \dots \partial \Omega_{k_{2m+1}}} \right\}$$

Inserting the unit operator $\int d\vec{\Omega}\, \delta\left(\vec{\Omega} - \vec{\Omega}(\vec{s})\right)$ and performing partial integrations gives:

$$\frac{d}{dt} \langle \Phi \rangle_t =$$

$$\int d\vec{\Omega}\, \Phi[\vec{\Omega}] \sum_{m \geq 1} \frac{2^{2m}}{(2m)!} \sum_{k_1=1}^{n} \dots \sum_{k_{2m}=1}^{n} \omega_j^{(k_1)} \dots \omega_j^{(k_{2m})} \frac{\partial^{2m}}{\partial \Omega_{k_1} \dots \partial \Omega_{k_{2m}}} \sum_{\vec{s}} \sum_j w_j(\vec{s}) \delta\left(\vec{\Omega} - \vec{\Omega}(\vec{s})\right) p_t(\vec{s})$$

$$+ \int d\vec{\Omega}\, \Phi[\vec{\Omega}] \sum_{m \geq 0} \frac{2^{2m+1}}{(2m+1)!} \sum_{k_1=1}^{n} \dots \sum_{k_{2m+1}=1}^{n} \omega_j^{(k_1)} \dots \omega_j^{(k_{2m+1})} \frac{\partial^{2m+1}}{\partial \Omega_{k_1} \dots \partial \Omega_{k_{2m+1}}}$$

$$\sum_{\vec{s}} \sum_j s_j w_j(\vec{s}) \delta\left(\vec{\Omega} - \vec{\Omega}(\vec{s})\right) p_t(\vec{s})$$

From this, since $\langle \phi \rangle_t = \int d\vec{\Omega}\, P_t(\vec{\Omega}) \Phi[\vec{\Omega}]$, we deduce:

$$\frac{d}{dt} P_t(\vec{\Omega}) =$$

$$\sum_{m \geq 1} \frac{2^{2m-1}}{(2m)!} \sum_{k_1=1}^{n} \dots \sum_{k_{2m}=1}^{n} \sum_j \omega_j^{(k_1)} \dots \omega_j^{(k_{2m})} \frac{\partial^{2m}}{\partial \Omega_{k_1} \dots \partial \Omega_{k_{2m}}} P_t(\vec{\Omega}) \left(1 - \bar{s}_j(\vec{\Omega}) \cdot \tanh\left(\beta h_j[\vec{\Omega}]\right)\right) +$$

$$\sum_{m \geq 0} \frac{2^{2m}}{(2m+1)!} \sum_{k_1=1}^{n} \dots \sum_{k_{2m+1}=1}^{n} \sum_j \omega_j^{(k_1)} \dots \omega_j^{(k_{2m+1})} \frac{\partial^{2m+1}}{\partial \Omega_{k_1} \dots \partial \Omega_{k_{2m+1}}} P_t(\vec{\Omega}) \left(\bar{s}_j(\vec{\Omega}) - \tanh\left(\beta h_j[\vec{\Omega}]\right)\right)$$

where:

$$\bar{s}_j(\vec{\Omega}) = \sum_{\vec{s}} s_j \delta\left[\vec{\Omega}-\vec{\Omega}(\vec{s})\right] P_t(\vec{s}) \Big/ \sum_{\vec{s}} \delta\left[\vec{\Omega}-\vec{\Omega}(\vec{s})\right] P_t(\vec{s})$$

We may now conclude that for finite β:

$$\frac{d}{dt} P_t(\vec{\Omega}) = \sum_\mu \frac{\partial}{\partial\Omega_\mu}\left(P_t(\vec{\Omega})\left[\Omega_\mu - \sum_j \omega_j^{(\mu)} \tanh(\beta h_j[\vec{\Omega}]) \right] \right) + \sum_{m\geq2} O\left[V(\frac{n}{V})^m\right] \qquad (A1)$$

(where $\lim_{N\to\infty} |V_\mu|/V = O(1)$). For $N\to\infty$ we find deterministic evolution equations for the order parameters if $\lim_{N\to\infty} n/\sqrt{V} = 0$. In the latter case the solution of Eq. (A1) is given by $P_t(\vec{\Omega}) = \int d\vec{\Omega}_0 \, P_0(\vec{\Omega}_0) \, \delta\left[\vec{\Omega}-\vec{\Omega}(\vec{\Omega}_0,t)\right]$, in which $\vec{\Omega}(\vec{\Omega}_0,t)$ is the solution of

$$\vec{\Omega}(0) = \vec{\Omega}_0$$

$$\frac{d}{dt}\Omega_\mu(t) = \lim_{N\to\infty}\sum_j \omega_j^{(\mu)} \tanh(\beta h_j[\vec{\Omega}(t)]) - \Omega_\mu(t) \qquad (A2)$$

REFERENCES

[1]   W.A. Little, Math. Biosci. **19** (1974) 101

[2]   J.J. Hopfield, Proc. Natl. Acad. Sci. U.S.A. **79** (1982) 2554

[3]   D.O. Hebb, The organization of behaviour (1949). New York, Wiley

[4]   D.J. Amit, H. Gutfreund & H. Sompolinsky, Phys. Rev. **A32** (1985) 1007

[5]   D.J. Amit, H. Gutfreund & H. Sompolinsky, Phys. Rev. Lett. **55** (1985) 1530

[6]   D.J. Amit, H. Gutfreund & H. Sompolinsky, Phys. Rev. **A35** (1987) 2293

[7]   J.L. van Hemmen, D. Grensing, A. Huber & R. Kuehn, Z. Phys. **B65** (1986) 53

[8]   B. Derrida, E. Gardner & A. Zippelius, Europhys. Lett. **4** (1987) 167

[9]   A.C.C. Coolen & Th.W. Ruijgrok, Phys. Rev. **A38** (1988) 4253

[10] A.J. Noest, Phys. Rev. Lett. **63** (1989) 1739

[11] I. Kanter, Phys. Rev. Lett. **60** (1988) 1891

[12] A. Canning & E. Gardner, J. Phys. **A21** (1988) 3275

KINETICALLY DISORDERED LATTICE SYSTEMS[1]

J. Marro and P.L. Garrido

Departamentos de Fısica Aplicada y de Física Moderna,
Facultad de Ciencias, Universidad de Granada, E-18071-Granada, España.

1.- Introduction and some definitions.

This lecture deals with a series of Ising-like model systems in which a relatively complex kinetic process induces the presence of non-equilibrium steady states which are characterized, even dominated by a kind of (dynamic) disorder. It thus results a situation having some connection with the case of more familiar disordered systems [see, for instance, Ziman 1979, Chowdhury 1987], e.g. spin glasses, random-field systems and diluted magnets; some of our considerations may also bear some relevance in relation to the study of certain cellular automata, and perhaps to that of neural networks, which is the topic of this meeting. The material we are presenting has essentially been published before in two recent papers by Garrido and Marro [1989] and by López-Lacomba, Garrido and Marro [1990], although we shall also describe some unpublished results obtained in collaboration with J.J. Alonso, J.M. González-Miranda and A. Labarta. Further related efforts may be found in recent papers by Künsch [1984], Grinstein et al. [1985], Browne and Kleban [1989] and Droz et al. [1989], for instance.

Consider for simplicity a regular lattice $\Omega = \{r_i; i=1,2,\ldots,N\}$ in a d-dimensional space with spin variables $s_i \equiv s_{r_i} = \pm 1$ at each lattice site. Denote by $s \equiv \{s_r; r\in\Omega\}$ any of the (2^N) possible configurations, and by $P(s;t)$ the probability of s at time t. The latter evolves in time as implied by a Master Equation, namely

(1.1) $\quad dP(s;t)/dt = \sum_{s'} c(s|s')P(s';t)$

---

[1]Partially supported by Dirección General de Investigación Científica y Técnica, Project PB88-0487, Plan Andaluz de Investigación (Junta de Andalucía), and Commission of the European Communities.

where $c(s|s')$ are the elements of a matrix c of transition rates per unit time from s' to s . The matrix c involves a _series_ of competing, _local_ spin-flip mechanisms each producing stochastically, as in the so-called Glauber [1963] dynamics, the change $s_i \to -s_i$ of the variable at a site r_i . This generates a new configuration s' , to be denoted specifically either as s^{ri} or else as s^i , from s with a given probability per unit time, $c(s^r|s)$. That is, we shall assume in the following that

$$(1.2) \quad c(s'|s) = - m^{-1} \sum_{l=1}^{m} \sum_{r} \pi_{x \neq r} \delta_{s_x, s'_x} c_{\varphi_1}(s^r|s) s_r s'_r ,$$

where $r \in \Omega$, and φ_1 represents the value of a given parameter such as temperature, chemical potential, magnetic field, sign or strength of interactions, or any combination of these, etc.

The asymptotic behavior of $P(s;t)$ as implied by (1.1) and a given kinetic process (1.2) may be very varied in principle. This may be investigated by solving each specific model [see, for instance, Glauber 1963 and Garrido et al. 1987]. It sounds however more interesting a priori, in an effort to construct a theory of nonequilibrium phenomena for instance, to introduce the concept of an appropriate effective Hamiltonian [cf. Garrido and Marro 1989]. That will be our main approach here.

Let us denote by $P^{st}(s)$ the stationary solution or limit of $P(s;t)$ as $t \to \infty$, and assume this is positively defined for all s . When P^{st} exists, one may introduce an object $E(s)$ such that

$$(1.3) \quad P^{st}(s) = Z^{-1} \exp\{-E(s)\} , \quad Z \equiv \sum_{s} \exp\{-E(s)\} .$$

It follows that $E(s)$ is analytic, and one may write quite generally that

$$(1.4) \quad E(s) = \sum_{k=1}^{N} \sum_{(i_1 \ldots i_k)}' J^{(k)}_{i_1 \ldots i_k} s_{i_1} \ldots s_{i_k}$$

where Σ' sums over every set of k lattice sites in the system. We shall only be interested in the following on objects $E(s)$ such that

$$(1.5) \quad J^{(k)}_{i_1 \ldots i_k} = 0 \quad \text{for all} \quad k \geq k_o$$

where k_o is independent of N , at least for $N > N_o$. When a unique stationary distribution function P^{st} exists such that (1.3)-(1.5) hold, the resulting object E(s) will be termed the **effective Hamiltonian** of the system. While that short-ranged object E(s) will differ essentially from the actual Hamiltonian of the system under consideration, it in fact represents an effective Hamiltonian in the sense of equations (1.3), e.g. it may be used to study nonequilibrium stationary states and phase transitions by applying standard methods of equilibrium statistical mechanics.

The conceptually simplest realization of the general model defined above are the kinetic Ising models studied by Glauber [1963] and others. Those systems, which correspond to having m=1 in Eq. (1.2), consist of any lattice Ω whose configurations, assuming for the moment that there is no external magnetic field or chemical potential, have a potential or configurational energy given by

$$(1.6) \quad H(s) = - J \sum_{nn} s_{r_i} s_{r_j} \, ,$$

where the sum is over nearest neighbour (nn) pairs of sites. Moreover, the lattice is in contact with a thermal bath at temperature T which induces changes in s . Namely, the bath provokes spin-flips with a prescribed rate depending on the change of the energy (1.6) which would cause the flip:

$$(1.7) \quad c(s^r|s) = f_r(s) \, \exp[-K s_r \sum_q s_{r'}] = f_r(s) \exp[-\tfrac{1}{2}\delta H]$$

where $\delta H \equiv [H(s^r) - H(s)]/k_B T$, $K \equiv J/k_B T$, the sum is over the q nearest neighbours of site r and $f_r(s) = f_r(s^r)$ (>0) as required by the detailed balance condition:

$$(1.8) \quad c(s^r|s)/c(s|s^r) = \exp(-\delta H) \, .$$

The transition rates used before by Glauber [1963] and others [see, for instance, López-Lacomba et al. 1990] to deal with various problems correspond to different realizations for the function $f_r(s)$. Each of those choices drives the system asymptotically towards the same stationary state, the canonical or Gibbs equilibrium state characterized by the distribution $P^{st}(s) = \text{const.} \exp\{-H(s)/k_B T\}$ corresponding to the temperature T and to the energy H(s) . It follows simply in those (trivial) cases that $E(s) = H(s)/k_B T$, and the system properties may be obtained in principle, and some times also in practice, from the computation of Z in (1.3).

The situation is however more complex, and more interesting and challenging nowadays, for the dynamics (1.2) with m≥2 . The homoge-

neous Markov process (1.1) now results from a competition between
several stochastic spin-flip mechanisms, each acting as if it were
associated to a different value of a given parameter (e.g. corresponding to a different temperature), and occurring at a rate (1.7)
satisfying (1.8) which guarantees it would individually drive the
system to the corresponding Gibbs equilibrium state. Nevertheless, that
competition essentially complicates the system dynamics, e.g. the
resulting transition rate (1.2) will not satisfy (1.8) in general and
this may induce the presence of nonequilibrium steady states, as occurs
when a system is not isolated but acted on by some external agent.
Consequently, the existence of P^{st} , and that of an effective Hamiltonian $E(s)$, is so an open question in general.

It may be mentioned that the above concept of an effective
Hamiltonian may also be applied to the study of the so-called probabilistic (spin-flip) cellular automata [Lebowitz et al. 1990] which have
no configurational energy similar to (1.6) defined, but a certain
dynamical process. For instance, the one-dimensional case $(\Omega \equiv Z)$

$$(1.9) \quad P(s|s';\delta t) = \pi_{i\in\Omega} \text{const.}[1+s_{r_i} w(s'_{r_i})]$$

where w is any function such that $1\pm w \geq 0$. We shall not consider that
possibility explicitly in the following but refer, for instance, to the
paper by Grinstein et al. [1985].

2.- On the existence of an effective Hamiltonian.

The conditions for the existence of an effective Hamiltonian
(EH) for the general system defined above were investigated before by
Garrido and Marro [1989] and by López-Lacomba et al. [1990]. The main
general results in those studies are summarized in this section and in
the next one; they provide a framework to study those systems which is
simultaneously simple, systematic and powerful.

One may easily prove that, under some mild conditions [see, for
instance, Ligget 1985], a system with a finite number of sites, N ,
and any dynamics c starting from any configuration may reach any
other configuration in a finite number of steps. That is, there exists
a unique, non-zero stationary probability distribution $P^{st}(s)$, one for
each rate, and the system will tend asymptotically to it independently
of the initial distribution. Thus, the system admits a unique object
$E(s)$, independent of the initial probability distribution, for each
choice of transition rates $c(s^r|s)$.

Now, in order to solve the problem of computing the coefficients $J^{(k)}$ in $E(s)$ efficiently, it is convenient (see, however, section 4) to consider those cases in which the resulting transition rate (1.2), say

$$(2.1) \quad c(s^r|s) = \sum_{l=1}^{m} p_l\, c_l(s^r|s) , \qquad \sum_{l=1}^{m} p_l = 1 ,$$

satisfies a global detailed balance (GDB). That is, unless otherwise indicated we shall assume that (2.1) satisfies that

$$(2.2) \quad c(s|s^r)\, P^{st}(s^r) = c(s^r|s)\, P^{st}(s) \quad \text{for all} \quad s \quad \text{and} \quad s^r$$

or $c(s^r|s) = f_r(s) \exp(-\tfrac{1}{2}\delta E)$ where $\delta E \equiv E(s^r) - E(s)$ and $f_r(s) = f_r(s^r) > 0$, the latter implying that $f_r(s)$ has no dependence on the value of s_r. When this holds, we just have:

$$(2.3) \quad \ln[c(s^r|s)/c(s|s^r)] = -\delta E = 2s_r \sum_{k=1}^{N} \sum_{(r,i_2\ldots i_k)}' J^{(k)}_{r,i_2\ldots i_k} s_{i_2}\cdots s_{i_k}$$

and the unknowns $J^{(k)}$ follow by identifying coefficients. Moreover, it also follows that the necessary and sufficient condition for any dynamics c to fulfil the GDB condition (2.2) is that any set of k spin variables satisfies

$$(2.4) \quad \sum_{s} s_{i_1} s_{i_2} \cdots s_{i_j} s_{i_a} \cdots s_{i_k} \ln[c(s^{i_j}|s)c(s|s^{i_a})/c(s|s^{i_j})c(s^{i_a}|s)] = 0$$

for all i_j and i_a.

It also follows from above that, in the infinite-volume limit ($N \to \infty$), $E(s)$ may not be unique. Nevertheless, in order for $E(s)$ to be useful it needs to have some short-ranged nature such as the one specified in (1.5). That is, one is only interested in cases in which the number of coefficients $J^{(k)}$ in $E(s)$ remains constant, independent of N, and their expressions are also independent of the system size, and one may show this will certainly occur in many cases of interest. As a matter of fact, there is a large and interesting class of situations having that property when the system dynamics satisfies GDB. Namely, the object $E(s)$ is the EH of the system when the resulting transition rate (1.2) or (2.1) both satisfies the condition (2.2) of GDB **and** depends only on a constant number of sites that is independent of N.

3.- General application to one-dimensional systems.

The previous facts indeed suggest an efficient formalism to investigate the existence of an EH for one-dimensional systems and, when it exists, to find the corresponding explicit expression [López-Lacomba et al. 1990]. Consider the system defined in section 1 with $d=1$ whose dynamics consist of a competition as in (2.1) with

$$(3.1) \quad c_i(s^i|s) = f_i^{(1)}(s) \, \exp\{-\tfrac{1}{2}[H_i(s^i)-H_i(s)]\} \ .$$

Here, the functions f_i are assumed to be analytic, positive defined and independent of the variable s_i , so that each individual transition rate $c_i(s^i|s)$ satisfies the condition of detailed balance (1.8). It is further assumed that the individual physical Hamiltonians involved by the rates contain a term which corresponds to the action of an external magnetic field, i.e.

$$(3.2) \quad H_i(s) = - K^{(1)}_1 \sum_{i=1}^{N} s_i - K^{(1)}_2 \sum_{i=1}^{N} s_i s_{i+1} \ ,$$

and that the functions f_i have the following properties:

(i) They are invariant under the interchange of s_{i-1} and s_{i+1} ; consequently, they may be written as

$$(3.3) \quad f_i^{(1)}(s) = q_0^{(1)}(s_{11}) + g_1^{(1)}(s_{11})\sigma_1^i + g_2^{(1)}(s_{11})\sigma_2^i$$

where $\sigma_1^i \equiv \tfrac{1}{2}(s_{i-1}+s_{i+1})$, $\sigma_2^i \equiv s_{i-1}s_{i+1}$, and s_{11} represents the set of occupation variables in $f_i^{(1)}$ excluding s_{i-1} and s_{i+1} .

(ii) They are homogeneous in the sense that the coefficients g in (3.3) are independent of i , as is reflected in our notation.

(iii) They are symmetrical in the sense that $s_{i+m}^i \in s_{11}$, where $m^i < \tfrac{1}{2}(N-1)$, implies that $s_{i-m}^i \in s_{11}$, and $f_i^{(1)}$ has the same dependence on both, s_{i+m}^i and s_{i-m}^i .

(iv) Each individual dynamics has a few-body nature in the sense that it does not involve "many" neighbours of site i, i.e. $\max_{(1)}\{m^i|s_{i+m}^i \in s_{11}\} \equiv M \ll \tfrac{1}{2}(N-1)$ as certainly occurs in the familiar rates.

It follows from the results in the previous section that the object $E(s)$ always exists for the one-dimensional system just defined. In order to show under what conditions GDB is satisfied so that $E(s)$ is the system EH, one first notices that

$$(3.4) \quad \ln[c(s|s^i)/c(s^i|s)] = s_i[D_0(s_i^*) + 2D_1(s_i^*)\sigma_1^i + D_2(s_i^*)\sigma_2^i]$$

where

(3.5) $D_0(s_1^{*})\equiv\frac{1}{4}(b+d+2a)$, $D_1(s_1^{*})\equiv\frac{1}{4}(b-d)$, $D_2(s_1^{*})\equiv\frac{1}{4}(b+d-2a)$

with

(3.6a) $a(s_1^{*}) \equiv \ln\{\Sigma_1[g_0^{(1)}-g_2^{(1)}]\exp[K_1^{(1)}]/\Sigma_1[g_0^{(1)}-g_2^{(1)}]\exp[-K_1^{(1)}]\}$,

(3.6b) $b(s_1^{*}) \equiv \ln\left[\dfrac{\Sigma_1[g_0^{(1)}+g_1^{(1)}+g_2^{(1)}]\exp[K_1^{(1)}+2K_2^{(1)}]}{\Sigma_1[g_0^{(1)}+g_1^{(1)}+g_2^{(1)}]\exp[-K_1^{(1)}-2K_2^{(1)}]}\right]$,

(3.6c) $d(s_1^{*}) \equiv \ln\left[\dfrac{\Sigma_1[g_0^{(1)}-g_1^{(1)}+g_2^{(1)}]\exp[K_1^{(1)}-2K_2^{(1)}]}{\Sigma_1[g_0^{(1)}-g_1^{(1)}+g_2^{(1)}]\exp[-K_1^{(1)}+2K_2^{(1)}]}\right]$,

$1 = 1,2,...m$. Notice that s_1^{*} represents the set of spin variables appearing in a , b and d , and that s_1^{*} may differ from the set s_{11} ; in any case, one still has the properties (i)-(iv) above. Thus, a necessary and sufficient condition in order to have GDB here is that, for all j and $k \neq j$:

(3.7) $\underset{s}{\Sigma} s_{i_1}...s_{i_m} G^{j,k}(s) = 0$, $i_1 \neq j,k$, $1=1,...,m$, $m \geq 0$,

where the functions $G^{j,k}(s) \equiv s_j s_k \ln[c(s^j|s)c(s|s^k)/c(s|s^j)c(s^k|s)]$ are given for the model in this section as

(3.8) $G^{j,k}(s) = s_j[D_0(s_k^{*})+2D_1(s_k^{*})\sigma_1^{k}+D_2(s_k^{*})\sigma_2^{k}]-$
$-s_k[D_0(s_j^{*})+2D_1(s_j^{*})\sigma_1^{j}+D_2(s_j^{*})\sigma_2^{j}]$.

Then, taking k as a nn of j , e.g. $k=j-1$, the result is that the condition of GDB implies that

(3.9) $D_1(s_{j-1}^{*})+D_2(s_{j-1}^{*})s_{j-2}=D_1(s_j^{*})+D_2(s_j^{*})s_{j+1}$,

and the only way to satisfy (3.8) is by requiring that

(3.10) $D_1(s_{j-1}^{*})=D_1(s_j^{*})=$const. , $D_2(s_{j-1}^{*})=D_2(s_j^{*})=0$.

The consequences of those facts are noteworthy: The necessary and sufficient condition for having GBD in the one-dimensional model is that

(3.11) $\qquad \ln[c(s|s^i)/c(s^i|s)] = s_i[D_0(s_i^\bullet) + 2D_1\sigma_i^i]$,

where D_1 is a constant and

(3.12) $D_0(s_j^\bullet) = \sum\limits_{m\geq 0} \quad \sum\limits_{(i_1...i_m)}' A^m_{i_1...i_m} s_{j+i_1}...s_{j+i_m}$,

and that the coefficients A^m satisfy

(3.13) $A^m_{i_1...i_l...i_m} = A^m_{i_1-i_1...-i_l...i_m-i_1}$.

Thus, the coefficients in (1.4) for the EH of the one-dimensional system are given by

(3.14) $J^{(k)}_{i_1..i_j..i_k} = (2^{N+1})^{-1} \sum\limits_{s} s_{i_1}..s_{i_j}..s_{i_k} [D_0(s_{i_j}^\bullet) + 2D_1\sigma_1^{ij}]$.

For instance, $J^{(2)}_{i,i\pm 1}=-\frac{1}{2}D_1$.

Summing up, when a system satisfies the condition of GDB, a situation which sounds appealing and is also very frequent in practice, there is always a unique short-ranged object $E(s)$, and we found the necessary and sufficient conditions for GDB to hold when $d=1$. Namely, a one-dimensional system has an EH which is given by (3.14) if and only if D_0 fulfils (3.12) and (3.13), D_1 is a constant, and D_2 is zero. Moreover, $D_2=0$ implies among other facts that the EH for a system satisfying GDB will have the familiar nearest-neighbour Ising structure with only pair interactions. The latter fact may also be concluded for $d=2$ [Garrido and Marro 1989]. As a matter of fact, although the applications become very involved as one increases the system dimension, the formalism in this section may in principle be generalized for $d>1$.

4.- Steady states near equilibrium.

Notice that in the absence of the GDB condition the system may still have an EH and one may devise alternative methods, usually more indirect and specific ones, to determine the unique short-ranged object $E(s)$. For instance, one may relate in some cases the coefficients $J^{(k)}$ to some relevant correlation functions to be computed independently [see, for instance, Browne and Kleban 1989]. Alternatively, one may follow the philosophy and formalism we described before; e.g. when $D_2\neq 0$ is small enough one may attempt an expansion around $D_2=0$. Such procedures, however, will usually show a strong dependence on the

details of the system of interest or involve lengthy algebraic
manipulations. Instead, we shall focus in this section on a general
method which demonstrates the existence of a "linear regime" where a
nonequilibrium system may still retain most canonical features.

Consider the competing dynamics (2.1) where each $c_i(s^r|s)$ is
in the form (3.1) so that it satisfies the (individual) detailed
balance condition (2.2) with respect to a given Hamiltonian $H(s;\varphi_i)$.
Here φ_i represents the set of parameters characterizing a class of
Hamiltonians; for instance, one may imagine $H(s;\varphi_i)$ of the nn Ising
type with an external field, (3.2), i.e. $\varphi_i=\{K_1^{(1)},K_2^{(1)}\}$. Moreover, we
shall assume that each set of values φ_i differ by a small enough
amount from some reference values, φ, say $\varphi_i=\varphi+\delta_i$ for every i, so
that one may expand the Hamiltonian as

$$(4.1) \quad H(s;\varphi_i) = H(s;\varphi) + (dH/d\varphi)\cdot\delta_i + \tfrac{1}{2}(d^2H/d\varphi d\varphi):\delta_i\delta_i + \cdots$$

By using this expansion into condition (2.2), one obtains:

$$(4.2) \quad c_i(s^r|s) = c^0(s^r|s) + c_i^{(1)}(s^r|s) + c_i^{(2)}(s^r|s) + \cdots$$

where the contributions at each order, $c^0(s^r|s)$ and $c_i^{(n)}(s^r|s)$ with
$n=1,2,\ldots$, satisfy a kind of detailed balance condition, namely that

$$(4.3) \quad c^0(s^r|s)\exp\{-H(s;\varphi)\} = c^0(s|s^r)\exp\{-H(s^r;\varphi)\} ,$$

$$(4.4) \quad c^0(s^r|s)\exp\{-H(s;\varphi)\} [\Omega_i(s)-\Omega_i(s^r)] + c_i^{(1)}(s^r|s)\exp\{-H(s;\varphi)\} =$$
$$= c_i^{(1)}(s|s^r)\exp\{-H(s^r;\varphi)\} ,$$

$$(4.5) \quad c_i^{(2)}(s^r|s)\exp\{-H(s;\varphi)\} [\Gamma_i(s)-\Gamma_i(s^r)-\Omega_i(s^r)\{\Omega_i(s)-\Omega_i(s^r)\}] +$$
$$+ c_i^{(1)}(s^r|s)\exp\{-H(s;\varphi)\} [\Omega_i(s)-\Omega_i(s^r)] +$$
$$+ c_i^{(2)}(s^r|s)\exp\{-H(s;\varphi)\} = c_i^{(2)}(s|s^r)\exp\{-H(s^r;\varphi)\} ,$$

where $\Omega_i(s) \equiv -[dH(s;\varphi_i)/d\varphi]\cdot\delta_i$ and $\Gamma_i(s) \equiv \tfrac{1}{2}[(dH(s;\varphi_i)/d\varphi)\cdot\delta_i]^2 -$
$\tfrac{1}{2}[d^2H(s;\varphi_i)/d\varphi d\varphi]:\delta_i\delta_i$. Then, after using the result (4.2) in the
expression defining the effective Hamiltonian, one obtains

$$(4.6) \quad E(s) = H(s;\varphi) + [dH(s;\varphi)/d\varphi]\cdot[\Sigma_i p_i\delta_i] + O(\delta^2) .$$

That is, one may define a "linear regime", where the relevant parameters (temperature, interaction strength, external magnetic field, etc.)
appearing in the individual Hamiltonians which are associated to each

transition rate (via a detailed balance condition) have values close enough to a given set of values, such that one has both i) that GDB is satisfied by the system, and ii) that an EH always exists which is given by (4.6). Therefore, a system defined in such a linear regime, while being capable of a full nonequilibrium behavior which may differ qualitatively from the equilibrium one, is expected to suffer "small" departures from the canonical equilibrium associated with the reference Hamiltonian $H(s;\varphi)$ in the sense that it fits most relevant canonical qualities, say i) and ii).

5.- Several one-dimensional examples.

In this section we shall initiate the application of the formalism to several one-dimensional systems with a practical interest. Consider first the (trivial) case in which every rate is defined with respect to the same Hamiltonian, say (3.2). That is, $H_i(s) \equiv H(s)$, $K_1^{(1)} \equiv h/k_B T$ and $K_2^{(1)} \equiv K$. It follows that $\ln[c(s|s^i)/c(s^i|s)] = 2s_i[h/k_B T + 2K\sigma_i^i]$, so that GDB is (trivially) satisfied, $J^{(1)}_i = -h/k_B T$, $J^{(2)}_{i,i\pm 1} = -K$, and the rest of coefficients are zero. Consequently, $E(s) = H(s)$ independent of the specific choice for the rate function.

More interesting is the situation in which the dynamics is a mixture of two or more rates each defined with respect to a different "Hamiltonian", $H_i(s)$. As an illustration, López-Lacomba et al.[1990] considered within the present formalism a case solved before by Garrido et al. [1987] which may be characterized by $K_1^{(1)} \equiv 0$, $K_2^{(1)} \equiv \varphi_1 K$ with $\varphi_1 = [k_B(T-\delta T)]^{-1}$ and $\varphi_2 = [k_B(T+\delta T)]^{-1}$, and $c_i(s^i|s) = \{\cosh[2K_2^{(1)}\sigma_i^i]\}^{-1}\exp\{2K_2^{(1)}\sigma_i^i s_i\}$. It follows rather straightforwardly from the results in section 3 that $D_0 = D_2 = 0$ and $D_1 = \frac{1}{2}b$, with $b = \ln(\{[1+p\widetilde{\varphi}_1+(1-p)\widetilde{\varphi}_2]/[1-p\widetilde{\varphi}_1-(1-p)\widetilde{\varphi}_2]\}$, $\widetilde{\varphi}_i \equiv \tanh(2K\varphi_i)$, and, consequently, that $J^{(2)}_{1,i\pm 1} = -\frac{1}{4}b$ is the only non-zero coefficient. Thus, by defining an "effective temperature" T_{eff} such that $J^{(2)}_{1,i\pm 1} = -J/k_B T_{eff}$, one obtains $\tanh(2J/k_B T_{eff}) = p\widetilde{\varphi}_1 + (1-p)\widetilde{\varphi}_2$. This is precisely the result which was only obtained before [Garrido et al. 1987] after solving the model explicitly. The relative simplicity of the present method also becomes evident by considering different choices for $f_i^{(1)}(s)$.

The same formalism may be applied to the study of a system in which the dynamics involves a competition between several interactions, a case which may be more relevant in relation to the topic of this meeting given that it may include a kinetic version of the familiar spin glass systems. Consider, for instance, the case characterized by $K_1^{(1)} \equiv 0$ and $K_2^{(1)} \equiv J_i/k_B T$, $J_1 = K$ and $J_2 = K + \delta K$. Clearly, one obtains the same formal results as for the two-temperatures model except that the

need is now for an "effective interaction" $J_{eff}=J^{(2)}{}_{i,i\pm1}=-\frac{1}{4}b$. In spite
of that formal similarity, the present case bears a novel physical
significance. This becomes evident when one considers the possibility
of an external magnetic field h , i.e. $K_1^{(1)}\neq0$. The two-temperatures
model would then require $K_1^{(1)}=h/(T-\delta T)\neq K_1^{(2)}=h/(T+\delta T)$, implying that GDB
is not satisfied, while we now have $K_1^{(1)}=K_1^{(2)}=h/T$, so that GDB holds and
an EH exists whose only non-zero coefficients are $J^{(1)}{}_i=-h/k_BT$ and
$J^{(2)}{}_{i,i\pm1}$, the latter being the same as for $h=0$.

One may consider in the same way kinetic versions of random
field systems, e.g. the simple case of a mixture of two dynamics such
that the associated "Hamiltonians" are characterized instead by
$K_1^{(1)}=h/k_BT$, $K_1^{(2)}=0$, and $K_2^{(1)}=K_2^{(2)}=K$, i.e. one acts with probability
p as if the external magnetic field were h , and the other acts with
probability $1-p$ as if there were no external field. The study of this
problem for different rate functions evidences in particular the
outstanding role played by the details of the dynamics on the qualitative features of the (nonequilibrium) steady state.

One may also consider with little effort the so-called voter
model [see, for instance, Kernstein 1986] which belongs to the class of
systems whose definition does not involve any Hamiltonian but a certain
dynamical process. Namely, any configuration s evolves via spin-flips
with a rate $c(s^i|s)$ satisfying, for $d=1$:

$$(5.1) \quad \ln[c(s|s^i)/c(s^i|s)] = s_i \ln \left|\frac{1+(l-1)\sigma_i{}^i-l(1-2p)}{1-(l-1)\sigma_i{}^i+l(1-2p)}\right|$$

where $0\le l,p\le1$. It follows that GDB is only fulfilled either for $l=1$,
when $E(s) = -\frac{1}{2}\ln[p/(1-p)]\Sigma_i s_i$, or else for $p = 1/2$, when $E(s) = -\frac{1}{4}\ln[(2-l)/l]\Sigma_i s_i s_{i+1}$.

Notice that the above four simple models depict an interesting
situation. In particular, each example satisfies GDB only for some
range of values of the system parameters or for certain families of
transition rates. When that is the case, we obtain explicit expressions
for the EH, and it follows that the nonequilibrium system can be mapped
onto an equivalent equilibrium one with some "effective" value for the
relevant parameter, say T , J or h .

Finally, once the general method has been illustrated, we turn
to the detailed study, including critical behavior, of a one-dimensional system having a considerable physical interest. Namely, we are
interested now in a kinetic random-field model with rates $c_i(s;x)$
depending on $\delta H_i=H_i(s^x)-H_i(s)$ with $H_i(s)=-K\Sigma_x s_x s_{x+1}-h_i\Sigma_x s_x$ where h_i is a

random variable. Given that it follows from above that there is an EH with the familiar Ising structure, we write from the start $H_{ef}=-K_{ef}\Sigma_i s_i s_{i+1}-h_{ef}\Sigma_i s_i$. We first notice that different rates may imply different expressions for both, K_{ef} and h_{ef}. For instance, $c(s;x)=a\cdot\exp(\delta H/2)$ produces $K_{ef}=J/k_B T$ and $h_{ef}=\frac{1}{2}\ln[<<e^{J/k_B T}>>/<<e^{-J/k_B T}>>]$ where $<<\cdot>>$ represents an average with respect to the distribution $g(h)$ for the random variable h_i, which we shall assume to have the symmetry $g(h)=g(-h)$, whereas $c(s;x)=a[2\cosh(\delta H/2)]^{-1}\cdot\exp(\delta H/2)$ produces $K_{ef}=J+\frac{1}{4}\ln(<<c_{>}>>/<<c_{<}>>)$, where $c_{\pm}\equiv e^{h/k_B T}/\cosh[(2J\pm h)/k_B T]$, and $h_{ef}=0$.

In order to go further, we select the latter rate for simplicity and the distribution $g(h) = \frac{1}{2}p[\delta(h-h_o)+\delta(h+h_o)]+(1-p)\delta(h)$ which, in addition to an obvious intrinsic interest, has the merit of generalizing a case previously solved by Grinstein and Mukamel [1983]. It follows that $K_{ef}=bJ+\frac{1}{4}\ln\{[\frac{1}{2}p(c_+^{+}+c_-^{-})+p']/[\frac{1}{2}p(c_-^{+}+c_+^{-})+p']$ where we write $c_{\pm}^{+}\equiv\exp(bh_o)/\cosh(2bJ\pm bh_o)$, $c_{\pm}^{-}\equiv\exp(-bh_o)/\cosh(2bJ\pm bh_o)$, $p'\equiv(1-p)/\cosh(2bJ)$ and $b\equiv1/k_B T$. Consequently, by making $J=1$ and $b_{ef}\equiv K_{ef}/J$ we have $b_{ef}\approx b+O(b^3)$ for $b\to0$, i.e. the system tends to present Ising equilibrium behavior when T is high enough. Now, if one defines $t\equiv e^{-b}$, $b_{ef}\equiv-\ln t_{ef}$, it follows that $t_{ef}\approx t\{[1+t^{4ho}+2t^{4+2ho}]/[t^4+2t^{2bo}+t^{4+4ho}]\}^{\frac{1}{4}}$ for $p=1$ implying four different behaviors:

1) When $h_o=0$, we have $t_{ef}=t$ and $T_{ef}=T$, i.e. equilibrium behavior, as one should expect.

2) When $0 < h_o < 2$, we have $t_{ef}\approx2^{-\frac{1}{4}}t^{1-\frac{1}{2}ho}+O(t^n)$ with $n>0$, or either $b_{ef}\approx\frac{1}{4}\ln2+(1-\frac{1}{2}h_o)b+O(t^n)$ as $b\to\infty$, i.e. there is a simple linear relation near $T=0$.

3) When $h_o = 2$, we have $b_{ef}\approx\frac{1}{4}\ln3+O(t^n)$ for $b\to\infty$, i.e. the system is "hot", and it remains so even as $T\to0$, as a consequence of the competition between local random fields and thermal fluctuations.

4) When $h_o > 2$, we have $b_{ef}=O(t^n)\approx0$ for $b\to\infty$, i.e. the randomness induced by the fields is so strong that the system is completely disordered.

Also interesting is the case $p\neq1$ with $h_o=\infty$ which is characterized by $t_{ef}=\{[1+p+2pt^{-4}]/[1+p+2pt^4]\}^{\frac{1}{4}}$.

The behavior of the system energy follows as $e = N^{-1}\Sigma_i<s_i s_{i+1}> = -N^{-1}(d/dK_{ef})\Sigma_s\exp[-H_{ef}(s)] = -\tanh(b_{ef}) = (t_{ef}^2-1)/(t_{ef}^2+1)$, and the mean square fluctuations of the energy are given by $(\delta E)^2 = <E^2>-<E>^2 = -(d/db_{ef})<E> = -N(db/db_{ef})(de/db)$. In the present nonequilibrium situation one may define two different "specific heats": $C_1 \equiv de/dT = b^2(db_{ef}/db)[\cosh b_{ef}]^{-2}$ and $C_2 \equiv b^2 N^{-1}\delta E^2 = -b^2(de/db_{ef})=^2[\cosh b_{ef}]^{-2}$. It thus follows that $C_1/C_2 = db_{ef}/db=(t/t_{ef})(dt_{ef}/dt)$, i.e. there is no fluctuation-dissipation theorem in general.

Now let us describe the critical behavior of the kinetic random-field model of interest; this results quite novel. For instance, the correlations are $C(r) \equiv \langle s_i s_{i+r} \rangle = (\tanh b_{ef})^r$ and, when one defines a correlation length, say α , such that $C(r) \equiv \exp(-r/\alpha)$, it follows that $\alpha = -1/\ln \tanh b_{ef}$. Now, define a critical index v such that $\alpha \equiv t^{-v}$, one has a critical behavior for $b \to \infty$ which depends on the value of h_o :

a) When $h_o = 0$, one has $\alpha \approx \frac{1}{2}\exp(2b)$ diverging with $v=2$, i.e. the equilibrium behavior, as expected.

b) When $0 < h_o < 2$, one has instead $\alpha \approx 2\exp[b(2h_o)]$ whose divergence depends on the value of h_o , namely $v=2-h_o$.

c) When $h_o = 2$ and $h_o > 2$ one has $\alpha \approx \frac{1}{2}3^b$ and $\alpha \approx \frac{1}{2}$ respectively, i.e. $v=0$ corresponding to the absence of a critical point, even for $b=\infty$, due to the randomness induced by (relatively) strong local fields. It also follows that, for $h_o=\infty$, one has $\alpha \approx \frac{1}{2}[(1+p)/2p]^b$ which only diverges as $p \to 0$, the equilibrium case.

The same models are also being studied at present by means of mean-field and Monte Carlo techniques. Some interesting, preliminary results will be presented graphically which we hope will convince you, together with the above facts, that such a series of nonequilibrium systems with competing dynamics has an amazingly rich and interesting physical behavior which deserves a more detailed and systematic study.

Acknowledgements

We acknowledge very useful discussions with Juán José Alonso, Julio F. Fernández, Jesús M. González-Miranda, Amílcar Labarta and Antonio I. López-Lacomba.

References

Browne D.A. and P. Kleban 1989: Phys.Rev.A 40, 1615.
Chowdhury D. 1987: Spin Glasses and Other Frustrated Systems, Princeton Univ. Press, Princeton, N.J.
Droz M., Z. Rácz and J. Schmidt 1989: Phys.Rev.A 39, 2141.
Garrido P.L., A. Labarta and J. Marro 1987: J.Stat.Phys. 49, 551.
Garrido P.L. and J. Marro 1989: Phys.Rev.Lett. 62, 1929.
Glauber R.J. 1963: J.Math.Phys. 4, 294.
Grinstein G., C. Jayaprakash and Yu He 1985: Phys.Rev.Lett. 55, 2527.
Grinstein G. and Mukamel 1983: Phys.Rev.B 27, 4503.
Kerstein A.R. 1986: J. Stat. Phys. 45, 921.
Künsch H. 1984: Z.Wahrscheinlichkeitstheorie und verwandte Gebiete 66, 407.
Lebowitz J.L., E. Speer and C. Maes 1990: "Rigorous Results on Probabilistic Cellular Automaton", J.Stat.Phys., and references therein.
Ligget T.M. 1985: Interacting Particle Systems, Springer-Verlag, Berlin, and references therein.
López-Lacomba A.I., P.L. Garrido and J. Marro 1990: J.Phys.A, in press.
Ziman J. 1979: Models of Disorder, Cambridge Univ. Press, New York.

A Programming System for Implementing Neural Nets

A. S. Bavan

School of Computing and Information Technology

Thames Polytechnic, Wellington Street, London SE 18.

Abstract

This paper presents a programming system called NPS, for implementing neural networks, which is portable and have the ability to deal with network problems in general. The main aim of the system is to support portability and model independence by facilitating the implementation of a range of neural network models on a range of hardware. NPS is based on a specialised neural network language called NIL [Bava89]. NIL is a machine independent network specification language designed to map a spectrum of neural models onto a range of architectures. As part of the overall project, a neurocomputer architecture [Pach88] was also proposed, which is currently being implemented in CMOS.

1. Introduction

Artificial neural networks is a rapidly expanding field of research based on algorithms and hardware which is inspired by the neural structure of the brain to solve recognition, planning, optimisation and other pattern processing and classification problems. Although a growing number of neural network models have been developed to support a variety of applications, neural network programming is still mainly done using conventional languages like C, LISP and OCCAM [MayD87]. However, specialised neural programming systems like P3 [Zips86] and CONE [Hans87] are already available and commercial environments like SAIC [SAIC88], and ANZA [Anza87], have appeared on the scene during the last few years. The majority of these systems are PC-based and designed to operate with propriety accelerator boards.

Typically a neural network programming environment (NNPE) consists of an integrated set of software and hardware tools for specifying and executing neural network models and applications. Most of the currently available systems suffer from a major draw back which is that they are associated with a particular execution hardware. Thus limiting the use of applications built on these systems to their respective hardware.

The major goal of the work described in this paper is to achieve portability of a spectrum of neural network models over a range of hardware. Theis resulted in the

1- design of a neural network programming system called NPS, which is portable and have the ability to deal with network problems in general;

2- specification of a primitive processing element for building a parallel MIMD neurocomputer, configured from an array of these elements[Pach88] which is currently being implemented in CMOS.

NPS uses a network implementation language, NIL as its kernel to provide the user with a full programming system consisting of: *A programming language*, to specify network models; *A utility*, to save partially trained networks for further training; *Libraries of functions* and *algorithms*, to aid the network construction and run standard models.

Having given an overview of the system, the rest of the paper deals with the detailed aspects of the system beginning with the network programming system in section 2. In this section, each component of the NPS system is described. This includes the description of the Ucl Neurocomputer architecture, NPS command interpreter, and the systems view of the NIL language system. Sub section 2.1 describes the NIL language in greater detail with suitable examples. Finally in section 3 conclusions are drawn.

2. A Network Programming System (NPS)

The NPS is a set of integrated tools for building and executing neural network models and applications. This network programming system as shown in Figure 1 consists of:

1. a network implementation language (NIL) system;

2. an algorithms library;

3. a utility for saving partially trained network for further training and recall;

4. a main module comprising a command interpreter and a set of mappers;

5. a neurocomputer architecture simulator based on a primitive processing element.

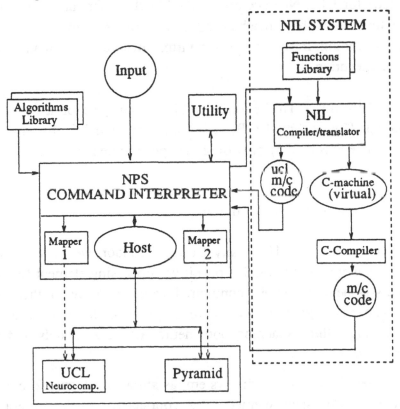

Figure 1. NPS

In this system, the user inputs control commands to perform a particular task such as compiling and executing a source network program on a desired hardware, executing a neural network model from the algorithm library or saving a partially run network in a file etc. In the case of a NIL program, it calls the NIL translation system to translate into the appropriate target machine code. Then the command interpreter takes the file containing the target machine code and passes it to the relevant mapping system for mapping the code on the target machine and setting up of the network ready for running. Once this is done the command interpreter initiates and facilitates the running of the network by providing I/O transfer facilities between the user and the network. Apart from compiling and executing NIL programs, the user can load and save partially run programs, stop and abort a running network, execute one of the standard models available in the library. The functions library contains function definitions for various types of nodes for use in a NIL program.

NIL System - This sub system consists of a NIL compiler and a translator, and a library of useful function definitions for nodes. NIL program can be either translated into the virtual machine(C-Machine) based on "C" language and then compiled into object code using a standard "C" compiler or compiled into target machine code for UCL Neurocomputer Architecture simulator. The command interpreter then takes these machine codes and maps them on the appropriate hardware for running. In the case of the virtual machine, it is loaded on a Unix machine for execution.

Algorithms Library - The algorithms library consists of a set of popular models such as Hopfield model, Back-Propagation model, Boltzmann machine, etc. These can be executed by providing the appropriate parameters.

Utility - This is a single function utility package containing software for saving a partially run network for further training at a later time.

Ucl Neurocomputer - At University College London we have designed and currently implementing in CMOS, a primitive processing element for building a parallel MIMD neurocomputer, configured from an array of these elements [Pach88]. The main goal of the neurocomputer is to support a range of connectionist algorithms spanning both neural network models and semantic network models.

The overall structure of the system, as shown by Figure 2, consists of a set of arrays of processing elements(PE) connected to a host computer that controls the activities of the network. A PE in an array is linked by a bi-directional, point-to-point connection to its two neighbours and communicates by sending message packets (see Figure 3). Each message consists of three fields: a destination address (ie-the identity of the destination PE), a logical input channel number, and a data value. When a PE receives a packet, it compares the address field with its own address and if it matches then the processor is interrupted and the packet is passed to the application process. If the address is different then it is passed to its other neighbour. In this system both the host and the individual PE can broadcast messages by placing a special destination address and input number in their packets. In the case of a broadcast packet, each PE takes a copy of the packet and passes the original to its other neighbour.

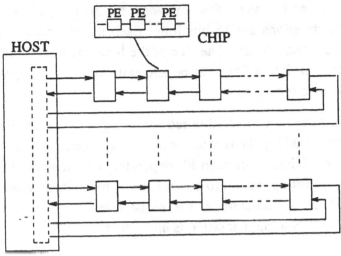

Figure 2. UCL Neurocomputer

Each PE as shown in Figure 3, consists of three units: a communication unit, a processor and a local memory. The communication unit has two I/O Buffers and a name register to hold its address.

A Packet:

16 bit	16 bit	16 bit
destination	input channel	data value

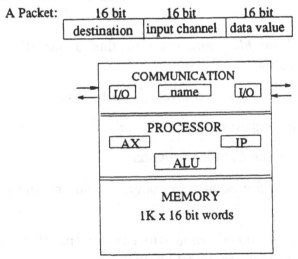

Figure 3. UCL Neuro-Chip

The processor consists of a primitive ALU, supporting a reduced instruction set of 16. This instruction set enable the processor to perform all the necessary functions required by a network program by providing instructions such as *load, store, add, sub, mult and, xor*, etc. There is only one working register, the accumulator AX, and a very few memory mapped registers. All the data addresses and instructions are 16-bits long. Each instruction consists of a 4-bit opcode and 12-bit data address. The size of the local memory can be up to 1K X 16-bit words long which is found to be adequate enough to support a range of neural network node functions.

The neurocomputer is configured by initialising each PE with a unique address and then loading them with appropriate codes. These codes can be either identical or different for each PE depending on the overall task. Each PE also has a local operating system to control and manage its activities. Currently, a simulation of the architecture at the register level is running on a UNIX machine and a CMOS implementation is under way.

Command interpreter - The command interpreter in the programming system is like a Unix shell. It accepts commands from the user and interprets it and calls the appropriate function/sub-system. This means that it shares some of its commands with the NIL system. That is - some of the commands can be used within a NIL program.

There are all together seven commands available to user at this level. They are

1. **load** -*m file_name* - for loading a partially run network for execution.

2. **save** *file_name* - to save a partially run program for running later.

3. **stop** - to stop a running network.

4. **go** - to run a loaded network.

5. **run** -*m file_name* - to compile, load and run a source program file.

6. **exec** -*m model_name* - to execute one of the standard models available in the model library.

7. **abort** - to abandon the execution of the network.

2.1 Network Implementation Language (NIL)

The main motivations behind the design of NIL is to produce a neural network specification language which is

1. **machine independent** - The language must be independent of any hardware, especially neurocomputers.

2. **neural network independent** - The language must be capable of representing a wide range of neural network models and applications.

Using these motivations as a basis, NIL was designed to exhibit some features [Ange88, Bahr87, MayD87] which can be considered desirable in a network language. That is - *Small and specialised; Readable; General; Implicitly Parallel; Reusable; Supports graphical representations.*

Using the above mentioned properties as the general guidelines, NIL was designed. NIL consists of two sub-languages :

1. **a network specification sub-language** which specifies the connections between nodes and the functions performed by the nodes in the network;

2. **a manipulation sub-language** which allows the user to observe and modify the network.

A program written in this language has the following syntax.

begin *network specification* **begin** *manipulation part* **end end**

The Specification Sub-Language - the specification sub-language consists of link statements for linking nodes in the network; **function definitions** for specifying the computational behaviour of the nodes.

In NIL, the network is built using three types of link statements. These statements, namely, a **link** statement which is the most basic form, a **rep** statement, and a **construct** statement implement a series of mappings using appropriate functions with specified lists of input and output parameters.

link statement - a link statement has the following syntax.

name(input_list : wt_list) -> (output_list)

The *name* refers to the specific function performed by that node. The *input_list* and *output_list* represent the set of input and output values associated with that node. The *wt_list* represents the list of initial values such as initial weights and status values. To illustrate the use of the link statement, consider following three link statements.

$$fl([a, b, c] :[1, 3, 5]) \rightarrow ([d, e, f])$$
$$fl([d, e] : [7, 9]) \rightarrow ([g, h])$$
$$f2([f, g, h]) \rightarrow ([i])$$

These implements the network in Figure 4.

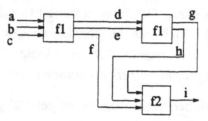

Figure 4. A Simple Network

rep statement - There are two forms of rep statements. A **rep** statement of the form

$$rep[3] \; ff([X[I], Y[I]]) \rightarrow ([X[I+1], Y[I+1]])$$

would replicate the function ff three times as follows(Figure 5).

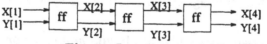

Figure 5. A Sequential Replication

A **rep** statement of the following form

$$rep[2] \; ff([X[I], Y[I]]) \rightarrow ([A[I], B[I]])$$

would produce the following effect(Figure 6).

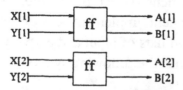

Figure 6. A Parallel Replication

construct statement - This allows the programmer to build a complete network structure with regular connections. For example, the following **construct** statement

> **construct** ([i=2] in(iv[i], sig[p]) -> (ou[i])
> :[p=1] hi(ou[i], erbk[k]) -> (out[p], sig[p])
> :[k=1] op(ou[i], out[p], eop[1], rcl[1]) -> (erbk[k], result[k])

implements the network in Figure 7.

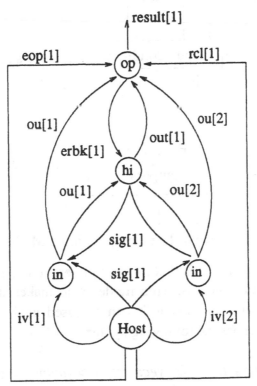

Figure 7. A Regular Network

The above construct statement is equivalent to the following set of statements.

> **rep**[2] in(iv[i], sig[1]) -> (ou[i])
> hi([ou[1],ou[2]], [erbk[1]]) -> ([out[1]], [sig[1]])
> op([ou[1], ou[2]], [out[1]], [eop[1]], [rcl[1]]) -> ([erbk[1]], [result[1]])

Weights can also be introduced into these statements using a random function.

Definition of functions
Once the topology of the network is specified, the computational behaviour of each node(or a group of nodes in the case of a number of nodes performing the

same type of computation) is described in the form of function definitions. The computational model of a function that represents a neuron in NIL tries to capture the overall properties of a neuron. This is based on the popular view that a neuron only starts firing when the inputs that arrive at the dendrites satisfy certain conditions. This generalisation led to the computational model shown in Figure 8.

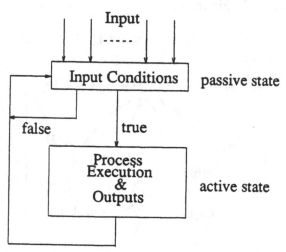

Figure 8. The Computational Model

We believe this model is flexible and general enough to accommodate a wide range of artificial neurons. This model also makes it possible to define functions in a general way so that it can be used again. A function which implements this model has the following syntax.

> **fun** *name* (*input_vectors* {*:wt_vectors* }) -> (*output_vectors*)
> "{"
>   *body*
> "}"

In this function, the parameter list consists of three types of vectors, they are input, output, and weight vectors. The weights vectors are sets of constant values used to initialise variables. Each element of the I/O vectors acts like a *virtual* I/O channel. The function body consists of a set of *GUARDED PROCESSES* with input conditions as its head. A guarded process has the following form.

input condition => { process description }

A guarded process becomes eligible for execution when the corresponding input condition becomes true.

Input Condition - The input condition is represented by an array with an index-range. If new input values are present in all the elements of the input array as specified by the index range then the corresponding process is executed. This input condition can be expressed as an "ALT" statement in Occam [MayD87]. An input condition has the following syntax.

*vector_1[index_range] {,vector_i[index_range] }**

The above syntax should produce a family of input condition that would cater for every eventuality as illustrated below.

x[3]   one specific channel input.
x[k..l]   sub range.
x[k1..m1,k2..m2]   set of sub range.
x[1,12], y[2,5,9]   subset of x and y.

A general structure of a function definition and a link statement which uses it are as follows.

sum([a, b, c] : [2, 5]) -> ([k, l, m])

fun sum (x[n1] : w[n2]) -> (y[n1])
"{"
  body
"}"

In the above definition *sum* is defined to be a function that takes an input vector x and a weight vector w each with n1 and n2 elements respectively and produces an output vector y. When sum is called n1 takes a value 3 and n2 a value 2 along with initialisation x[1] <- a, x[2] <- b, x[3] <- c, and w[1] <- 2, w[2] <- 5 and produces outputs y[1], y[2], and y[3] which are passed to other nodes in the network in k, l and m. This mechanism makes it possible for user to build general purpose functions that can be used by a number of nodes wishing to perform same type of computation but with different number of input elements.

Statements - The statements available for describing a process consists of "if", "do", "assignment" and "skip".

Output Statements - There is no special statement to perform output in this sub-language. This is achieved by assigning values to the elements of the output vectors.

A general function to perform summation and thresholding can be coded in NIL as follows.

```
fun lay2(x[n]) -> (y[i])
int    i, n, j;
real   x[10], y[10], netj;
{
    x[1..n] => { j := 1; netj := 0.0;
            do (j<=n) -> netj := netj + x[j];
                    j := j + 1;
                    od
            if (netj>=0) -> y[1] := 1.0;
                    ->-> y[1] := 0.0;
                    fi

    }
}
```

The Manipulation Sub-Language - Once a network is built, the MANIPULATION sub-language is used to observe and manipulate the network. This sub-language consists of control commands for stopping and starting the network; feeding and receiving data from individual nodes; looking at the current inputs and outputs of a node or nodes; inspection and modification of state variables of a node(s); creating and deleting links and nodes; and saving and reloading incomplete process(network).

Control Commands
Following control commands are used to observe and manipulate the network.

- **readst all** - reads I/O and weights of all nodes;

- **readst** node[3], node[5] - reads I/O and weights of nodes 3 and 5;

- **delete** *link_1(src_node, dstn_node)* - deletes link_1 between two nodes;

- **join** *link_1(src_node, dstn_node)* - establishes a link, link_1, between two nodes;

- **add** node[25] = name([a, b,..]) -> ([x]) - adds a new node with the named function;

- **go** and **stop** - executes and halts the network;

- **input** *link_1 = value_1,.. , link_n = value_n* - loads input values on to links;

- **output** *[text] op_list* - prints outputs;

- **ldconst** *node[5].w[2] = value_1,..,node[6].w[9]* - load weights into nodes;

- **save** *filename* - save the current network configuration;

- **load** *filename* - loads a previously saved or a new network.

Statements such as "do", "if", "assignment" and "skip" are also available for use in this part.

3. Conclusions

Current system generates code for UCL Neuro-computer simulator, and "C" code for execution on a Unix machine. So far most of the popular models have been coded in NIL and satisfactorily executed on both of these machines. Main strengths of this system lies in its hardware independence, ability to define very complex neurons with its tried and tested statements, and its amenability to be translated into high and lower level target languages. Currently work is under way to build a graphic system that can be interfaced to the language.

Acknowledgement

The author wishes to thank Marco Pacheco of University College London, for his advice and assistance during the preparation of this paper.

References

[Ange88] Angeniol. B, Texier. J, Mateu. J, "SLOGAN : An Object-Oriented Language for Neural Network Specification", nEuro'88, France, June,1988.

[ANZA87] ANZA User's Guide, Hect-Nielsen Corporation, Release 1.00, 1987.

[Bahr87] C. Bahr and D. Hammerstrom, ANNE - Another Neural Network Emulator

[Bava89] Bavan. A. S, "NIL-PLUS: A Neural Network Implementation Language", Proc. First Neural Computing Meeting, London, April 1989, pp 171-178, The Institute of Physics.

[Hans87] W. Hanson , C. Cruz , J. Tam, "CONE - Computational Network Environment", Proceeding of the first IEEE International Conference on Neural Networks, 1987. Vol3 pp 531-538

[MayD87] May D, "Occam2 language definition", INMOS, February 1987.

[Pach88] Pacheco M, Bavan S, Lee M & Treleaven P, "A Simple VLSI Architecture for Neurocomputing", Proceedings of the International Neural Network Society, First Annual Meeting, Boston, Massachusetts., September 1988, pp 398.

[SAIC88] "DELTA/SIGMA/ANSim", editorial, Journal of Neurocomputers, Vol 2, Number 1, 1988.

[Zips86] D. Zipser, D. Rabin, "P3 : A Parallel Network Simulating System", in Parallel Distributed Processing by D.E. Rumellhart, et al, Vol 1 Chapter 13, 1986.

AN AUTO-AUGMENTING NEURAL NETWORK ARCHITECTURE FOR DIAGNOSTIC REASONING

Sanjeev B. Ahuja
AI Research & Development Laboratory
Swiss Life Insurance & Pension Company
CH-8022 Zürich, Switzerland.

Woo-Young Soh
Department of Computer Science
University of Maryland, Baltimore
Maryland, U.S.A.

1. Background

During the past two decades, there have been tremendous efforts in the field of artificial intelligence(AI) to develop diagnostic reasoning models, including *statistical pattern classifiers* based on Bayes' theorem, *rule-based systems* employing deductive reasoning, and *associative network systems* which are built on inductive reasoning paradigms. Although each of those models has its strengths within relatively well suited application domains, none provide adequate mechanisms to support the interactions among knowledge structures which are necessary to capture the generative capacities of human diagnosticians in novel situations. Achieving such interactions has been one of the greatest difficulties associated with implementing models of diagnostic decision-making that can reason in the presence of imprecise or incomplete information, using traditional AI representations like frames [Minsky, 1975] and production rules [Newell & Simon, 1972]. Throughout this period, however, there has been an unabated interest in developing radically different models of human memory and cognition. This impetus for a new perspective to cognition can be largely attributed to a desire to overcome the shortcomings of existing symbol processing models to emulate human decision-making in terms of its underlying structure, representation and processing of knowledge, but more importantly, to better understand its evolution with experience over time. Not surprisingly, there has been a resurrection of a previously abandoned approach to studying the mind, called "Parallel Distributed Processing", "Connectionism", or "Neural Networks". Unlike classical approaches which associate intelligence with the explicit manipulation of structured symbolic expressions, this new approach proposes that intelligence emerges as a result of the transmission of activation levels within large networks of highly interconnected nodes, at times without specific semantic content. Implicit in the structure and behavior of neural models, they exhibit many of the interesting phenomena that are commonly associated with intelligence but which have eluded successful explication using traditional symbol-processing techniques. Research efforts over the past, however, have almost exclusively concentrated on low-level perceptual processing and/or pattern classification tasks. For high-level cognitive tasks which require knowledge-based decision-making such as does diagnostic reasoning, symbol processing approaches have continued to be the technique of choice.

In this paper, an alternative approach to diagnostic problem solving is presented, which has its knowledge representation basis in the symbol processing models of AI(an associative or semantic network), and an underlying processing metaphor reflective of a neuron-like or neurally inspired microstructure (a parallel spreading activation paradigm). Whereas, a semantic conceptual network provides a means of incorporating the functional knowledge, such as the causal-associative relationships that are required for diagnostic reasoning, a neural processing paradigm extends the possibility of emulating the interactions of domain concepts by providing a physical structure to the knowledge that enables the model to depict both, an ability to respond in novel situations and to enhance its diagnostic reasoning capabilities based on past experience. The proposed diagnostic model, LIBRA/Dx, makes the hypothesis that simultaneous satisfaction of multiple local constraints among domain concepts is sufficient to construct a plausible global explanation for a set of observed signs and symptoms. Such local constraints are dynamic in the sense that they can adapt to a given context, and that prior episodes can be captured as "experience" by an incremental modification of these constraints. Through the formation of conceptual clusters among observed signs and symptoms, this latter phenomenon, called episodic learning, characterizes the transition of a human diagnostician from a novice to an expert. Furthermore, it is suggestive of the process through which diagnosticians successfully identify "syndromes" or constellations of signs and symptoms that might share a common pathology, even though no such explicit relationships are known a priori within the domain knowledge. The remainder of this paper is organized as follows. Section 2 describes aspects that are characteristic of diagnostic problems in general, followed by an introduction to the fundamentals of our proposed neurally-inspired approach to diagnostic problem-solving. In the context of such a neural network model, Section 3 elucidates the basic concepts of learning as an evolutionary process and describes the specifics of an auto-augmenting neural network architecture which is intrinsic to the phenomenon of experience-based learning for diagnostic decision-making, as proposed within our approach.

2. A Connectionist Approach to Diagnostic Reasoning

Diagnostic reasoning is a high-level cognitive task that requires the inference of an explanation for observed signs and symptoms, or "manifestations", by accounting for the presence of each observed symptom with the hypotheses or "disorders" which constitute the explanation. In most diagnostic domains, causal associations between disorders and manifestations are generally imprecise. For example, that a disorder d <u>may cause</u> manifestations $m_1, m_2, \ldots, m_j$ or that a manifestation m <u>may be caused by</u> disorders $d_1, d_2, \ldots, d_i$, merely categorizes a set of possible manifestations of a disorder and a possible set of causative disorders for a manifestation, respectively. In general, of all the possible causative disorders for an observed manifestation, only one is necessary to account for its presence. Moreover, despite the fact that multiple disorders may have occurred simultaneously in a specific case, input information is usually incomplete since the expected manifestations associated with these disorders may not all be observed. These factors contribute to the ambiguity that is characteristic of any complex diagnostic reasoning task, since more than one plausible explanation might exist for any observed set of manifestations. In such circumstances, one needs a means of arriving at the *most plausible* explanation that configures itself dynamically with only the most relevant disorders, to form a tailor-made explanation in each context.

A diagnostician typically starts off with overlapping "descriptions" of disorders which specify sets of manifestations that are likely to be observed in the event that any of the corresponding disorders occur. Somehow the appropriate disorders are assembled to form a context-sensitive explanation that is both coherent and also supports inferencing. The combination of such a representational scheme together with a spreading activation processing mechanism constitutes our diagnostic engine for activating coherent assemblies of disorders based on observed manifestations, that draws inferences which are consistent with the knowledge represented by the activated disorders [Ahuja, Soh & Schwartz, 1989]. In this model, one can define a *connectivity matrix* between a node and all of its neighbors to which it is directly connected. Each element of the matrix specifies an association of that strength between a node and one of its neighbors. For example, if n_j is a node which is directly connected with a node n_i, then a bidirectional link exists between the two nodes with the strengths of association w_{ij} and w_{ji} respectively, from node n_j to n_i and vice versa. Weights may be highly asymmetric in the two directions and lie in the interval $[-1.0, +1.0]$ with a negative weight implying a *preventive* association or an *inhibitory* link and a positive weight implying a *supportive* association or an *excitatory* link. A simple bipartite (2-layer) causal network would consist of two distinguished sets D and M. Each disorder node d_i in set D is connected through *causal* links to a set of manifestation nodes $man(d_i)$ in set M, which d_i can cause. Similarly, each manifestation node m_j in set M is connected through *evoking* links to a set of disorder nodes $evokes(m_j)$ in set D, which the presence of m_j suggests. The cardinality of the set $evokes(m_j)$, represents the *fanout* of the node m_j [Anderson, 1983].

In the context of LIBRA/Dx the following features are to be noted:

- for any given manifestation node m_j that is said to be present, all plausible disorders that can account for its presence, $evokes(m_j)$, get activated initially;

- of the activated disorders that form the set $evokes(m_j)$ for a manifestation m_j, only one disorder stays active at certain level during (and possibly at the end of) the problem solving by inhibiting all the other alternative disorders; (of course, it may be that some of the inhibited disorders get reactivated subsequently, in order to account for other manifestations that are observed);

- each observed manifestation is accounted for by at least one disorder that stays active - the activated disorders provide a *cover* for M^+;

- disorders that are found to be active at the end of a problem solving session are those that present a globally most plausible explanation for the observed manifestations - parsimony is equated to the *highest degree of plausibility*.

Given a manifestation m_j which evokes a set of alternative disorders as its likely causes, each of these disorders are viewed as *competitors* for explaining the presence of m_j. In this competitive notion, each disorder within an $evokes(m_j)$ set for an observed manifestation m_j inhibits every other disorder in that set from being the cause of m_j. Each $evokes(m_j)$ set therefore, denotes a potential cluster of mutually inhibitory disorders, that compete for the activation of the commonly shared manifestation. Of course, this inhibition can occur if and only if that manifestation is observed to be present. In this sense the inhibitory clusters are dynamic and not known a priori [Rumelhart and Zipser, 1986]. In LIBRA/Dx, when a node assumes an activity level above its normal (threshold) resting level, its neighbors actively compete for the *energy* possessed by that node. The ability of the neighboring nodes to compete for shared node's activity is proportional to their own respective activation level. Consequently, the activation that this node acquires is at the expense of activation that is made available to its competitors and vice versa. Two separate aspects of contention are identified between competitors in an inhibitory cluster: weighted contributions made to each competitor from manifestations that are shared; and a mutually attenuating phenomenon among competitors via active manifestations that are shared. The ability of disorder node d_i to compete for m_j's output activity $a_j(t)$ is proportional to its own activation level $a_i(t)$ and to the weight of association w_{ji}. Thus, the activation transferred from m_j to d_i at time t, representing the contribution of node m_j to the belief in node d_i, is equal to:

$$OUT_{ij}^+(t) = \frac{|a_i(t)| \cdot w_{ij} \cdot a_j(t)}{\sum_k |a_k(t)| \cdot |w_{kj}|} \quad \forall \; d_k \; \epsilon \; evokes(m_j) \tag{2.1}$$

if d_i has a competitor, and $w_{ij} \cdot a_j(t)$ otherwise. This enables highly activated nodes to extract a proportionately larger part of m_j's finite activation, leaving a successively smaller portion for competitors with lower activation. In diagnostic reasoning, whenever supporting evidence is found for one disorder among several plausible ones that are alternatives, then not only does it contribute to a higher belief that the disorder is present, but it also takes away from the belief that one of its alternatives is present. Therefore, each node in an inhibitory cluster also seeks to *drain* its competitors of their respective activations. Since competing disorders can influence each other only via the active manifestation that they all share, this drain of activation among competitors can be expressed as a correction of the original support that is received by each competing disorder d_i from the common manifestation node m_j, by an amount due to the influence of its competitors:

$$OUT_{ji}^-(t) = (1 - \frac{|a_i(t)| \cdot (1 + w_{ji})}{\sum_k |a_k(t)| \cdot (1 + w_{jk})}) \cdot |a_i(t)| \cdot a_j(t) \qquad (2.2)$$

An intuitive justification for this attenuating phenomenon is that, of the belief in the presence of manifestation node m_j that originally contributes to the activation of node d_i, the amount $OUT_{ij}(t)$ is due to the likelihood of the *competitors* of d_i being present. It is by an amount that is proportionate to the strength of its competitors, therefore, that the activation of node d_i should get adjusted. However, the significance of an observed manifestation is diminished if other manifestations which share its cause are not observed, which reflects as a proportionately lower belief in the presence of that cause. To accommodate this effect of non-independence among manifestations, one can compute the active fanout $ACTFO_i$: a weighted measure of the portion of the manifestations of d_i that are in fact found to be present:

$$ACTFO_i = \frac{\sum_p w_{pi}}{\sum_k w_{ki}}, \quad \forall \ active \ m_p \ \epsilon \ man(d_i) \qquad (2.3)$$

Eq. 2.1 above denotes the contribution of an individual manifestation m_j to the belief in the presence of its associated disorders $\{d_i\}$. Since each of the observed manifestations of d_i supports its presence, contributions from all observed manifestation is accumulated by the disorder d_i. The accumulated belief is denoted by the measure of belief(MB) in the presence of a disorder d_i and is represented as a net transfer of activation that gets *injected* into a disorder from each of its manifestations m_j in the set $man(d_i)$ at time t:

$$MB_i(t) = \sum_j OUT_{ij}^+(t), \quad \forall \ m_j \ \epsilon \ man(d_i). \qquad (2.4)$$

Similarly the attenuation of the activation of disorder d_i by its competitors in the inhibitory cluster via their shared manifestation m_j is accumulated over all the manifestations which the disorder d_i can cause. This denotes a measure of disbelief(MD) in the presence of a disorder d_i and is represented as a net transfer of activation that gets *ejected* out of a disorder at time t:

$$MD_i(t) = \sum_j OUT_{ji}^-(t), \quad \forall \ m_j \ \epsilon \ man(d_i). \qquad (2.5)$$

Using these definitions one can compute the net input into a disorder node d_i at time t as:

$$\Delta F_i(t) = tanh(MB_i(t) \cdot ACTFO_i - MD_i(t))$$ (2.6)

The hyperbolic tangent function $tanh$ maps activation values in the interval $[-\infty, +\infty]$ into the interval $[-1.0, +1.0]$ and is used here as an approximation of the net input $\Delta F_i(t)$ that ranges in the interval $[-1.0, +1.0]$. Given this competition-based parallel activation approach to satisfying local constraints among domain concepts in a causal network, a global interpretation of the effects is possible by allowing the activation levels of each accumulator node to reach a stable state. The activation of node d_i at time $(t + \Delta t)$ is equal to the activation at time t, plus the net flow of activation $\Delta F_i(t)$ into the node, minus the decay:

$$a_i(t + \Delta t) = a_i(t) + \mu \cdot [\Delta F_i(t) \cdot \{1 - a_i(t)\} - \delta_i \cdot \{a_i(t) - \theta_i\}]$$ (2.7)

where μ is a constant that is used to control the size of a time slice during numerical simulation; δ_i is the rate at which the activation of node d_i decays in the absence of stimulus; and θ_i is the threshold or the natural resting level of activation for node d_i.

3. Learning for Diagnostic Problem-Solving

An equally important research issue when building cognitive models is that of making them learn or "evolve" over time. Learning is described as changes made to a system, which are adaptive in the sense that they enable the system to do the same tasks drawn from the same population more effectively the next time. This description of learning, however, acknowledges only one aspect of the evolution of a system over time: that of skills refinement, or the modification of representations based on repeated encounters and practice. Another significant component of learning in intelligent systems is their ability to construct new representations based on the discovery of new concepts and establish new associations with existing concepts in the same application, or on the generalization of existing concepts which share certain regularities. The latter process of learning is also called *conceptual clustering*: formation of object classes describable by simpler concepts. It is generally observed, that there is a transition period during which a novice diagnostician gathers experience with unknown or atypical case scenarios while gradually transcending into the more valuable category of an "expert". It is not obvious how this experience-based phenomena can be modeled as an automated learning paradigm. The aspects of a population of episodes that are characteristic of the domain, and therefore also observable within each case, are generally not known a priori and get discovered only during problem-solving. At the heart of neural networks research therefore, has been the development of learning strategies, specifying how to present the stimuli and change some network parameters in accordance with the response of the system. A learning scheme effectively, allows the network response to change depending upon the nature of the input stimuli. This is the only means through which the system is allowed to adapt itself, in order to produce desired outputs by re-organizing its previously encoded knowledge.

There are two general classes within which learning schemes can be categorized [Rumelhart & McClelland, 1986]; supervised schemes which use some sort of error propagation for learning and unsupervised schemes which use competition among alternatives for learning. *Supervised* learning is a tutoring process, in which the system evolves with repeated practice on typical input patterns, by gradually modifying its weight matrix to successively reduce its error in mapping them onto an expected set of output patterns over all known case-scenarios. *Unsupervised* learning does not make use of any tutoring inputs. Instead, the system adapts itself to model successively more and more of the implicit characteristics found in the input patterns that it encounters. Typically, in such schemes one ensures that directionality of modifications is such that its aggregate effect on the weight matrix gradually transforms it to model the statistical properties of the input population. Both paradigms are relevant for diagnostic reasoning, depending upon the kind of information to be assimilated. On the one hand, it makes logical sense to devise an unsupervised learning scheme to detect regularities in manifestation patterns while gradually adjusting connection strengths with each case-scenario, since one does not a priori know which patterns are likely to appear. On the other hand, it should be possible to mould the behavior of the network through intentional changes in connection strengths, in such a way that the domain model performs correctly in all known case-scenarios. Since a diagnostic reasoning system is generally required to identify a global explanation (i.e., one which is most plausible for observed manifestations), the "degree of relevance" with which alternative disorders explain an observed manifestation can be used as a competitive criterion for adjusting their respective connection strengths to that manifestation within the causal network. This section describes the construction of an auto-augmenting neural network which not only uses the existing domain knowledge, but which also allows case-specific knowledge to be assimilated through a gradual unsupervised learning process, under a unified framework for diagnostic reasoning.

3.1 Architecture of a Causal Network

Having constructed a causal network from causal-associative knowledge about the domain, how is one to know whether this "semantic network" is capable of even learning the correct patterns of activation for the typical case scenarios? A method needs to be devised for constructing a network architecture that is not only representative of the functional relationships among concepts, but which is also capable of responding with the correct patterns of activation using a specific activation paradigm. This capability would require appropriate structural relationships to be established between the concepts so that the chosen activation paradigm can perform correctly. We describe here a network architecture and representation scheme from the perspective of its ability to provide an effective structural basis for capturing functional dependencies existing among domain concepts in a causal-associative network. This network structure is additionally, able to capture any higher-order relationships which might exist between domain concepts, but which are detected only during problem-solving with real-world case-scenarios. It is a known fact, for example, that a mutually-exclusive (i.e., a logical XOR) relationship among two parameters m_1 and m_2 cannot be captured within a simple network of the type shown in Figure 3.1a, without the presence of a "higher-level" concept that is sensitive to co-occurrence of these two parameters [Rumelhart, Hinton & Williams, 1986]. In the network of Figure 3.1b, 's' is such a "higher-level" node which can be made to detect the simultaneous presence of nodes m_1 and m_2 and send a strong inhibitory signal to suppress the output of node 'f', which would otherwise have been active because of excitatory inputs from m_1 and m_2.

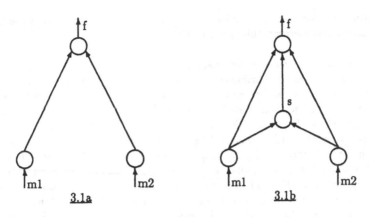

Figure 3.1: Learning the Exclusive-OR

In fact, whenever the representation provided by the outside (to the network) world is such that the similarity structure of the input and output patterns are very different, a network without internal representations (i.e., a network without hidden layers) is unable to perform all necessary classificatory mappings. Since one cannot always know a priori all possible input patterns, relationships which exist among input elements must get determined only by observing large populations of case scenarios. Consequently, a network with a fixed number of nodes may not be able to accommodate some of these unforeseen complex dependencies among input elements, as may be necessary to faithfully replicate the functional behavior mandated by the problem domain. A network that can add or delete nodes can dynamically reconfigure itself whenever it detects a new dependency, or when it determines an existing one to be spurious. LIBRA/Dx employs just such a dynamic network reconfiguration feature, which was adopted in order to allow the representation of both domain and case specific knowledge for the problem domain, with a clear delineation of the two types of knowledge and the role that is played by each during problem-solving.

The XOR problem is trivially solved through the introduction of an intermediate or "hidden" node s as shown in Figure 3.1. Co-occurrence of manifestations in general, happens to be an important characteristic of the diagnostic domain which is observable within every case scenario. It is a higher order relationship [Rumelhart & McClelland 1986] which requires a multi-layer network for detection. The more difficult problem in tailoring a network to learn such relationships, however, is that usually there exists no a priori information about higher-order relationships which might exist among different parameters of the case scenarios within a chosen application domain. These can normally only be detected over a period of time while processing several different case scenarios, or in other words through experience. Even when all known input patterns are available, the process of tailoring a network is time consuming and provides no guarantee of finding an optimal architectural solution. Without case specific or episodic knowledge therefore, it is difficult if not impossible, to predict ahead of time whether a network that one starts with is capable of identifying and learning the necessary relationships between input parameters. In the context of diagnostic problem solving, the intermediate node s can be thought of as a "syndrome" which denotes the $[m_1.m_2]$ constellation of manifestations. Medical diagnosticians often utilize such knowledge to make and/or substantiate a preliminary diagnosis. Whereas some co-occurrence patterns among the input manifestations may be known to the domain specialist, he is unlikely to be aware of all potential constellations that might be significant. In automating this process, one must ensure that only the relevant constellations get identified while the spurious ones are either ignored or, if at all identified, gradually suppressed. Moreover, since the introduction of intermediate nodes would warp the existing functional representation of causality that is encoded within the links connecting disorders to their manifestations, this is not a solution of choice.

Past attempts on using connectionist approaches for diagnostic problem-solving do not address any of these issues, since causal relationships or the functional knowledge of a domain is not explicitly encoded into the network. In earlier models, hidden layers with arbitrary numbers of nodes in each are constructed, without necessarily establishing first the number that might be adequate, or even needed in the first place. Supervised learning schemes are then employed to attempt associating different manifestation sets with corresponding disorders which might constitute acceptable explanations in each case [Bounds & Lloyd, 1988]. Inspite of the reported levels of performance, this approach makes a number of assumptions about the application domain which cannot easily be justified. This approach would also leave a black box between the set of disorder and manifestation nodes, which would make it impossible to generate any meaningful explanations about network behavior. Besides the fact that a possibly rich resource of existing domain knowledge is sacrificed, it needlessly forces an entire domain to be re-learned - a painfully slow and tedious process, not only at the beginning but also each time a new piece of knowledge is discovered and gets added to the domain.

The Augmented Causal Network

We propose an architecture that allows a domain network to be appended with a complementary case network, that is capable of learning any relevant high-order relationships among domain concepts from specific case-scenarios. Of course, the spreading activation paradigm has to be flexible enough to allow this knowledge to be simultaneously incorporated within the decision-making process in the future, as "experience". The domain(manifestation-disorder) and case (manifestation-syndrome) networks can be referred to as the domain and case "spaces" respectively, since they can be thought of as representing the two plausible diagnostic spaces within which the explanation for all observed manifestations is to be found.

If the domain space which is defined by domain experts is considered to be vertical, the case space is constructed orthogonal to it and shares all the input manifestations of the vertical (domain) space. The horizontal (case) space consists of syndrome nodes, each of which denotes a constellation of manifestations. The *case space* consists of several layers, each of which consists of syndrome nodes representing constellations of manifestations that are of the same size (i.e., co-occurrence of the same number of manifestations). In Figure 3.4, a node s_i^j represents the ith node of the jth layer. The first syndrome layer consists of nodes denoting constellations of two manifestations each, which are connected "below" to the corresponding manifestation nodes of the domain space. For example, the node s_2^1 denotes the co-occurrence pattern of manifestations m_1 and m_3, and is connected to those nodes in the domain space. Syndrome nodes within a specific layer of the case space are also connected to an adjacent layer "above" them, whose nodes denote constellations of manifestations that are larger in size by one. For example, the syndrome nodes s_1^1, s_3^1, and s_5^1 of size two are connected to the node s_1^2, which denotes the constellation of the three manifestations m_1, m_2, and m_4. Links between syndrome nodes of adjacent layers are similar to the causal-evoking links of the domain space, in the sense that the presence of a specific syndrome node would "cause" the co-occurrence of its corresponding smaller syndrome nodes also to be observed, which are implicitly connected through successively smaller syndrome nodes to the constellation of manifestations that they denote. Since the notion of a syndrome node is equivalent to that of a constellation of manifestations which are observed to co-occur, it is connected to the corresponding manifestation nodes whose co-occurrence it denotes. In this sense, the syndrome node as a feature extractor is functionally similar to the hidden node of connectionist models developed earlier. However, it differs from such a hidden node in the following sense: first, it is created and deleted dynamically on demand and second, it is created in a separate case space so as not to warp the existing conceptual relationships of the domain knowledge, as mentioned previously.

Every syndrome node is also connected through causal-evoking links to the disorder node in the domain space which can potentially cause the constellation of manifestations that it denotes, as shown in Figure 3.4. Since each syndrome node denotes the specific instance of the co-occurrence of its corresponding manifestations, links between the disorder nodes which can cause all these manifestations and the syndrome node are semantically justified.

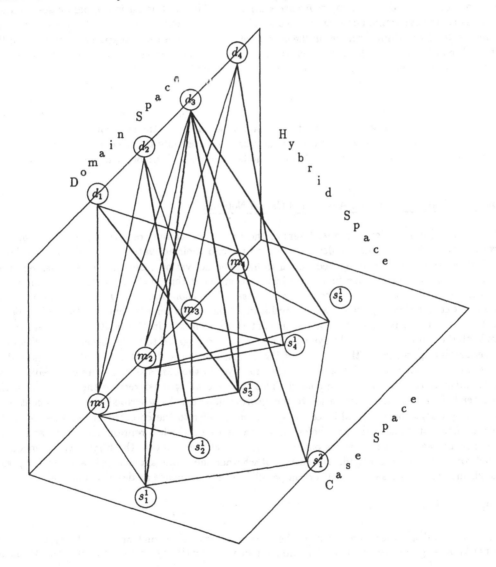

Figure 3.4 Augmented Causal Network

For example, the syndrome node s_2^1 representing the co-occurrence of manifestations m_1 and m_3 is connected to the disorder node that can cause those manifestations. Since links between syndrome nodes and disorder nodes connect the domain space to the case space, we call this as the "hybrid space". As explained below, through links contained in this hybrid space syndrome nodes can provide additional supporting evidence in favor of the presence of the disorders to which they are connected. If the connection strengths between different syndrome nodes can be made to assimilate the frequencies of co-occurrence among the input manifestations over past case scenarios, then the pattern (syndrome) which occurs more often has stronger connections linking it to the corresponding manifestation nodes than the one which occurs rarely. In any specific case-scenario, syndrome nodes which have a higher activation due to the input manifestations selectively propagate additional activation to the disorder nodes to which they are connected, thereby influencing the final decision process based on the co-occurrence statistics within the population of past case-scenarios. In this sense the syndrome nodes are treated as additional source of input to resolve the competition between alternative disorder nodes in the process of determining an explanation or diagnosis for the observed manifestations. Through the creation of such syndrome nodes, the augmented network architecture provides sufficient structural support to learn the high order relationships, while utilizing the existing conceptual relationships among the domain concepts to interpret the network results. Therefore, the creation of syndrome nodes and their links in the domain space is determined not only by the co-occurrence statistics within the input population but also by the other conceptual dependencies that already exist within the domain network. Since the syndrome nodes are created only as the need for them gets identified, the network captures only the necessary structural dependencies existing between the input and output nodes, while simultaneously avoiding the possibility of introducing conceptual inconsistencies.

Flow of Activation within the Augmented Causal Network

Since the syndrome nodes in higher layers of the case space cannot receive the activity from the lower layers by the time these nodes are expected to send their activation to the disorder nodes to which they are connected, it seemed logical to have the activation from the input nodes be propagated through the case space first. The justification being that such a process would allow the accumulation of additional evidence for the purposes of decision-making. One could interpret this as an *input completion* step, where an observable input pattern is complemented with the knowledge about co-occurrence patterns that are significant in its context. In the next step, the activation from both the input nodes and the syndrome nodes gets propogated to the disorder nodes. Consider this sequential approach for the network of Figure 3.4, where the disorder node d_3 is connected to s_1^2 in the second layer and s_1^1, s_3^1, and s_5^1 in the first layer in the case space. Before d_3 receives these inputs, their activations must be updated. This means that s_1^2 must receive the inputs from the first layer before sending its activation to the disorder node d_3. As described above, this control of activation propagation is logical since the system must refer to the co-occurrence statistics of the observed manifestations in the past before making a decision, which is represented by the activation pattern in the case space. It turns out, however, that at stabilization the final decision made by the network remains unchanged, irrespective of whether the input activation is allowed to spread through the three spaces of the network in parallel, or in a sequence as described above.

3.2 Proposed Learning Schemes

On the one hand, a supervised learning scheme is necessary to instruct or tutor the system about internal knowledge representations that are necessary to identify typical case scenarios. An unsupervised learning scheme on the other hand, must detect regularities among different case scenarios and capture these as "experience", which can then be used to modify the internal representations formed earlier to make increasingly plausible inferences in the future. The first scheme also serves as a validation mechanism which ensures that the assimilation of case knowledge together with the domain knowledge does not introduce conceptual inconsistencies. Whereas the former scheme addresses the skills acquisition aspect of learning, it is the latter which characterizes the transition from a novice to expert with experience. A modification of the generalized delta rule (GDR) [Rumelhart, Hinton & Williams, 1986] is used as the supervised learning paradigm to emulate the decision-making for typical case scenarios. A competition-based unsupervised learning paradigm was then developed for the model, to capture the co-occurrence statistics of real case-scenarios.

Supervised Learning within the Augmented Network

The initial network consists only of the domain space, denoting the set of causal associations between disorders and manifestations of an application domain. This configuration describes the declarative causal knowledge without considering the high order relationships that might be found to exist between input manifestations in the real-world. A modification of GDR was considered for LIBRA/Dx, as it is generally applicable to any network architecture and the model can be trained in an "off-line" mode after having primed it with the existing domain knowledge, rather than starting from scratch (no knowledge) as has been done in the past. Since activation spreads through three different spaces in the augmented network, supervised adjustments must occur within all three spaces. A variant of GDR is used in both the domain and case spaces, while a variant of the Delta rule is used for adjusting the hybrid space. In local representations like the one used for the proposed model where each node of the network corresponds to one concept, one could conceivably provide tutoring input for each of the intermediate (disorder) nodes also, but the model does not mandate it.

The network is given a set of tutoring inputs each of which consists of a pair of patterns: an input pattern $M_p = \{m_{p1}, m_{p2}, \cdots\}$, and a corresponding target output pattern $T_p = \{t_{p1}, t_{p2}, \cdots\}$, where p denotes the pth tutoring pattern. An input pattern M_p, is presented to the manifestation layer and propagated through the domain and hybrid space via the case space. This generates an actual output pattern, $A_p^o = \{a_{p1}^o, a_{p2}^o, \ldots\}$, which is compared with the target output to compute the error $E_p^o = \{e_{p1}^o, e_{p2}^o, \cdots\}$, where the superscript ^{o} stands for the top disorder layer that produces the network output. Then the weights on both the upward evoking and downward causal links are modified based on the error, since the weights on both types of links contribute to the error. The sum-squared error E can be expressed as,

$$ E = \sum_p E_p = \frac{1}{2} \sum_{pi} (t_{pi} - a_{pi}^o)^2 \,, \tag{3.1} $$

The goal is to find an algorithm determining the change in each weight w_{ij} from node j to node i in the network, in order to minimize the error in achieving the target outputs. This is proportional to the derivative of the error with respect to the weight and can be expressed as,

$$\Delta_p w_{ij} \propto -\frac{\partial E_p}{\partial w_{ij}} . \tag{3.2}$$

This can be rewritten as a product of two factors; the change in error as a function of the change in the net input to the node i, and the effect of changing the weight w_{ij} on the net input as,

$$\frac{\partial E_p}{\partial w_{ij}} = \frac{\partial E_p}{\partial \Delta f_{pi}} \frac{\partial \Delta f_{pi}}{\partial w_{ij}} , \tag{3.3}$$

where Δf_{pi} is the net input. Using the net input to a node and the weights on the links connecting the node its neighbors:

$$\frac{\partial \Delta f_{pi}}{\partial w_{ij}} = \frac{\partial}{\partial w_{ij}} \tanh(ACTFO_{pi} \sum_j OUT^+_{pij} - \sum_j OUT^-_{pji})$$

$$\frac{\partial \Delta f_{pi}}{\partial w_{ij}} = sech^2(ACTFO_{pi} \sum_j OUT^+_{pij} - \sum_j OUT^-_{pji})$$
$$\cdot \frac{ACTFO_{pi}OUT^+_{pij}}{w_{ij}}(1 - \frac{OUT^+_{pij}}{a_{pj}}) . \tag{3.4}$$

The error in the first part, however, is available only for the output nodes. In the intermediate and the syndrome layers the error should be computed by propagating the error in the output layer back to those layers. The error signal δ_{pi} for each node is computed first, which is defined using the chain rule as,

$$\delta_{pi} = -\frac{\partial E_p}{\partial \Delta f_{pi}} = -\frac{\partial E_p}{\partial a_{pi}} \frac{\partial a_{pi}}{\partial \Delta f_{pi}} , \tag{3.5}$$

where a_{pi} denotes the actual output of node i. From the activation function Eq. 2.7,

$$\frac{\partial a_{pi}}{\partial \Delta f_{pi}} = \frac{\partial}{\partial \Delta f_{pi}}(a_{pi} + \mu(1 - a_{pi})\Delta f_{pi}) = \mu(1 - a_{pi}). \tag{3.6}$$

The first part of Eq. 3.5 represents the error with respect to the output of node i. If the node i is the output node, from Eq. 3.1,

$$\frac{\partial E_p}{\partial a_{pi}} = -(t_{pi} - a_{pi}). \tag{3.7}$$

The error in the output layer $E^o = \{e^o_1, e^o_2, \cdots\}$ is computed by subtracting the target output T from the actual output D^o. Substituting Eq. 3.6 and Eq. 3.7 in Eq. 3.5, we get

$$\delta_{pi} = \mu(1 - a_{pi})(t_{pi} - a_{pi}). \tag{3.8}$$

There are two potential sources of error in the output nodes, since the activation in the input nodes propagates into the output nodes through two paths, the domain space and the hybrid space via the case space. Therefore, the weights on both paths should be adjusted to minimize the error in the output nodes. The weight change between the intermediate and syndrome nodes is determined based on the error for those nodes which is propagated back from the output nodes. The error for the intermediate nodes is computed by propagating the error in the output layer back to the domain space and the case space via the hybrid space, through the chain of links connecting the output nodes. In the domain space, the error for the node in an intermediate layer is propagated from the nodes in the layer above the layer that are connected to that node. In the case space, a node of syndrome layer is connected to the nodes in the layer directly above it and to the nodes in the domain space potentially causing that node. Therefore, the error for the syndrome node in a syndrome layer is the sum of the errors propagated from the nodes in the layer directly above the layer in the case space as well as from the nodes in the domain space connected to the node through the hybrid space. Thus the error signal for a node in the intermediate or syndrome layer should be computed by combining the error signal of the nodes receiving the activation from that node. From Equation (3.5), for a node i in the intermediate or syndrome layer,

$$
\begin{aligned}
\frac{\partial E_p}{\partial a_{pi}} &= \sum_k \frac{\partial E_p}{\partial \Delta f_{pk}} \frac{\partial \Delta f_{pk}}{\partial a_{pi}} \\
&= \sum_k \frac{\partial E_p}{\partial \Delta f_{pk}} \frac{\partial}{\partial a_{pi}} \tanh(ACTFO_{pk} \sum_j OUT^+_{pkj} - \sum_j OUT^-_{pjk}) \\
&= -\sum_k \delta_{pk} \cdot sech^2(ACTFO_{pk} \sum_j OUT^+_{pkj} - \sum_j OUT^-_{pjk}) \\
&\quad \cdot \frac{1}{a_{pi}}(ACTFO_{pk}OUT^+_{pki} - OUT^-_{pik}),
\end{aligned}
\tag{3.9}
$$

where $k \,\epsilon\, evokes(i)$ and $j \,\epsilon\, man(k)$. Substituting Eq. 3.6 and Eq. 3.9 in Eq. 3.5, we get

$$
\begin{aligned}
\delta_{pi} = \mu(1 - a_{pi}) \sum_k (\delta_{pk} \cdot sech^2(ACTFO_{pk} \sum_j OUT^+_{pkj} - \sum_j OUT^-_{pjk}) \\
\cdot \frac{1}{a_{pi}}(ACTFO_{pk}OUT^+_{pki} - OUT^-_{pik}))
\end{aligned}
\tag{3.10}
$$

From above, the weight change from j to i is made by substituting Eq. 3.4 in Eq. 3.2,

$$
\begin{aligned}
\Delta_p w_{ij} = \eta_1 \delta_{pi} \cdot sech^2(ACTFO_{pi} \sum_j OUT^+_{pij} - \sum_j OUT^-_{pji}) \\
\cdot \frac{ACTFO_{pi}OUT^+_{pij}}{w_{ij}}(1 - \frac{OUT^+_{pij}}{a_{pj}}),
\end{aligned}
\tag{3.11}
$$

where η_1 represents a learning rate. The weight change w_{ji} from node i to j on the downward causal link can be defined in the similar fashion to the upward links as,

$$\Delta_p w_{ji} \propto -\frac{\partial E_p}{\partial w_{ji}} . \tag{3.12}$$

This can be rewritten as a product of two factors; the change in error as a function of the change in the net input to the node i and the effect of changing the weight w_{ji} on the net input as,

$$\frac{\partial E_p}{\partial w_{ji}} = \frac{\partial E_p}{\partial \Delta f_{pi}} \frac{\partial \Delta f_{pi}}{\partial w_{ji}} . \tag{3.13}$$

The first part representing the change in error for the node i as a function of the change in the net input to the node i is same as in the computation of the upward weight change. Therefore, the error signal δ_{pi} defined above can be used again. From the net input Δ_{pi} the second part is computed as,

$$\frac{\partial \Delta f_{pi}}{\partial w_{ji}} = \frac{\partial}{\partial w_{ji}} \tanh(ACTFO_{pi} \sum_j OUT^+_{ij} - \sum_j OUT^-_{ji})$$
$$= sech^2(ACTFO_{pi} \sum_j OUT^+_{ij} - \sum_j OUT^-_{ji})$$
$$(\sum_j OUT^+_{ij} \cdot \frac{BOOL_{pj} - ACTFO_{pi}}{\sum_q w_{qi}} - \frac{a_{pi} OUT^-_{ji}}{\sum_q a_{pq} w_{jq}}) , \tag{3.14}$$

where $BOOL_{pj} = 0$ if the node j is absent, and 1 otherwise. Thus the downward weight change w_{ji} from node i to j is defined as,

$$\Delta_p w_{ji} = \eta_2 \delta_{pi} \cdot sech^2(ACTFO_{pi} \sum_j OUT^+_{ij} - \sum_j OUT^-_{ji})$$
$$\cdot (\sum_j OUT^+_{ij} \cdot \frac{BOOL_{pj} - ACTFO_{pi}}{\sum_q w_{qi}} - \frac{a_{pi} OUT^-_{ji}}{\sum_q a_{pq} w_{jq}}) , \tag{3.15}$$

where η_2 represents a learning rate.

Unsupervised Learning in Case Space

After the network has been constructed and trained under supervision with a set of typical case scenarios as tutoring inputs, the network can be used to process atypical (real-world) case scenarios. If we consider the process of supervised learning as "skills acquisition", or the accumulation of basic skills, then the process of unsupervised learning can be considered as "skills refinement", or the augmentation of domain knowledge with case knowledge from the real-world. A syndrome node in the augmented causal network represents a constellation of manifestations that have been observed to co-occur. The weights are adjusted in such a way that larger connection weights get assigned to links connecting a frequently occurring syndrome node and its corresponding sub-syndromes (manifestations), as opposed to the rare syndromes. The intention of the unsupervised learning is to gradually modify(increase or decrease) the weights of select links in the case space such that the co-occurrence statistics of the input nodes are captured within the composite representation of this space. In other words, the links are adjusted in accordance with the regularities which are detected among the population of input stimuli. For a pair of manifestation nodes m_j and m_k in the input layer(S^0), let w_{ij} and w_{ik}, respectively, denote the weights on the links that connect them to the ith syndrome node which signifies their co-occurrence, s_i^1 in the first syndrome layer(S^1). Then w_{ij} is representative of the frequency with which m_j is found to co-occur with m_k, as opposed to co-occurring with the other manifestations in the input layer. Similarly, w_{ik} is representative of the frequency with which m_k is found to co-occur with m_j, as opposed to co-occurring with the other manifestations in the input layer. In the reverse direction, the weight w_{ji} reflects the likelihood of manifestation m_j to be observed whenever the node s_i^1 is activated by the presence of manifestation m_k. Similarly, w_{ki} reflects the likelihood of manifestation m_k to be observed whenever the node s_i^1 is activated by the presence of m_j. At the time the syndrome node s_i^1 is created, the following weights are assigned to its links in the case space:

$$w_{ij} = w_{ik} = \frac{1.0}{||S^1||} \quad and \quad w_{ji} = 1.0,$$

where $||S^1||$ is the cardinality of the set of $_nC_2$ pali-syndromes. The same approach is adopted for initializing the higher layers in the case space.

From the point of view of spreading activation through a network, no distinction is made between a syndrome node and a disorder node. Consequently, within the first syndrome layer S^1, if s_p^1 and s_q^1 are two syndrome nodes that share an observed manifestation, then s_p^1 and s_q^1 are competitors which form an inhibitory cluster along with other syndrome nodes in the first layer that share the same manifestation. However, since syndrome nodes themselves do not form parts of an explanation, but rather help in the selection of the most plausible explanatory disorders instead, they do not compete with the disorders in seeking the activation of a share of manifestation. In this regard, the case nodes are thought of as "composite" manifestations and treated as such during the problem solving. On presenting a case scenario from the input population to the network and allowing the network to settle, nodes for the most frequently observed syndrome which are relevant to the case will end up with the most activation in the case space. At stabilization, the weights on links that connect these active nodes to the other nodes in the network are modified as follows. For any active node s_j^{n-1}, connected through a weight w_{ij} to the ith node s_i^n among the "k" nodes of the layer above, the weight w_{ij} is adjusted to:

$$w_{ij} = w_{ij} - g \cdot (w_{ij} - \Delta w_{ij}), \quad where \quad \Delta w_{ij} = \frac{e_{ij}}{\sum_k e_{kj}} .$$

The "error" $e_{kj} = r_{kj} - w_{kj}$, if $r_{kj} > w_{kj}$, and 0.0 otherwise; for all nodes s_i^n to which s_j^{n-1} is connected above. r_{ij} is the ratio of the activations of the syndrome node s_i^n and s_j^{n-1}, while the coefficient g is the "Learning Rate". Note that modification of the weight w_{ij} on the link which connects the syndrome nodes s_j^{n-1} and s_i^n is based upon a redistribution from a common pool of weight. This pool is formed by *proportionate* contributions from each of the weights w_{kj}, that connect the node s_j^{n-1} to the k nodes in the layer above. The coefficient "g" represents this contributing proportion, and since $\sum_k w_{kj} = 1.0$, it is precisely the re-distribution pool after each case that is processed. However, the downward weight w_{ji} in the reverse direction is not modified. This latter decision is justified since given the presence of a syndrome node s_i^n, none of the nodes s_j^{n-1} whose co-occurrence it represents are expected to be absent. A more detailed description of the behavior of this algorithm can be found in [Ahuja & Soh, 1989].

Conclusions

We have identified a practical approach for augmenting the domain knowledge represented within a causal network with context-sensitive case knowledge derived from a prior population of case scenarios. Our aim was to enable a diagnostic system to function within the collective characteristics of its environment through an experience-based process called episodic learning. Within the problem domain of diagnostic reasoning, the significant but largely ignored issue of whether a given connectionist network is even capable of learning is addressed. By providing an auto-augmenting architecture that can represent higher-order relationships without warping the original semantics that are associated with domain concepts, the proposed model effectively provides a mechanism for incorporating the co-occurence statistics of frequently occurring manifestation patterns as experience, under a unified framework for diagnostic decision-making.

References

Anderson, J. A. (1983): The Architecture of Cognition, Harvard University Press, Cambridge, MA.

Ahuja, S.B., Soh, W.-Y. & Schwartz, A.(1989): A Connectionist Processing Metaphor for Diagnostic Reasoning, Int. Journal of Intelligent Systems, Vol. 4, 155-180.

Ahuja, S.B. and Soh, W.-Y.(1989): Temporal Evolution of a Causal Network Through Conceptual Clustering, In Connectionism in Perspective, R. Pfeifer, et. al. (Eds.), Elsevier Science Publishing Company, North-Holland, 283-298.

Bounds, D. & Lloyd, P., (1988): A Multilayer Perceptron Network for the Diagnosis of Low Back Pain, In IEEE Proceedings of the International Conference on Neural Networks, Vol. 2, 481-489.

Minsky, M. (1975): A Framework for Representing Knowledge, In The Psychology of Computer Vision, P.H. Winston (Ed.), McGraw-Hill, New York, 211-277.

Rumelhart, Hinton & Williams, (1986): Learning Internal Representations by Error Propogation, In Parallel Distributed Processing, Volume 1: Foundations, MIT Press, Cambridge, MA., 318-362.

Rumelhart, & McClelland, (1986): PDP Models and General Issues in Cognitive Science, In Parallel Distributed Processing, Volume 1: Foundations, MIT Press, Cambridge, MA., 110-146.

Rumelhart & Zipser, (1986): Feature Discovery by Competitive Learning, In Parallel Distributed Processing, Volume 1: Foundations, MIT Press, Cambridge, MA., 151-193.

Formal Integrators and Neural Networks

José R. Dorronsoro and Vicente López
Instituto de Ingeniería del Conocimiento
Universidad Autónoma
28049 Madrid, Spain.

June 1990

1   Introduction

A vast amount of questions posed in different branches of Physics, Chemistry and Engineering require for their solution the numerical integration of large systems of Ordinary Differential Equations (ODEs).

Some of those questions can not be yet answered but for systems of reduced dimensionality, due to the very high number of computations and the limitations on the speed at which they are performed. Vectorial and parallel architectures in modern computers have allowed to face problems that were out of reach a decade ago, and it would be rather desirable to have available low cost hardware that could integrate systems of thousands of coupled differential equations in a reasonable time.

The object of this note is to show that simple, Neural Network–type architectures can be used for accurate and efficient numerical integration of a variety of large systems of Ordinary Differential Equations, namely those casted in the so called Universal Polynomial Formats, that is,

$$\dot{y}_i = a^i + \sum_j a^i_j y_j + \sum_{j_1 \leq j_2} a^i_{j_1 j_2} y_{j_1} y_{j_2} +$$

$$\cdots \tag{1}$$

$$\sum_{j_1 \leq j_2 \leq \cdots \leq j_P} a^i_{j_1 j_2 \ldots j_p} y_{j_1} y_{j_2} \cdots y_{j_P}, \quad 1 \leq i \leq N.$$

Well known examples of these systems are the Linear, Lotka–Volterra and Ricatti equations. Moreover [Kr], using relatively simple transformations, it is possible to bring into the Polynomial Format other ODE systems originally not in that form.

Many instances where the simplification introduced by these approaches can speed up tremendously the needed computations can be found in different branches of Physics. For the sake of illustration, the particular case of theoretical studies of Intramolecular Vibrational Energy Redistribution (IVR) in the scope of semiclassical mechanics will be discussed (see section 3 and [LLFVM,LFLM] for more details).

Artificial Neural Networks are viewed in this article not as a particular paradigm of Artificial Intelligence but as a collection of simple, efficient processors capable by the their connectivity of performing accurate massive parallel computations. The link with the Universal ODE formats discussed above arises from the particular numerical integrator we will consider, namely, formal Taylor series step integration. On the one hand, this integrator suits itself very naturally to the general Polynomial Format; on the other, the recurrent computation of the derivatives that this integrator requires and the partial summation approximation of the solutions at the next step lend themselves to be easily implemented as resulting from a feedforward, sigma–pi type Artificial Neural Network, with the connecting weights being dictated by the parameters of the ODE system to be integrated. These two issues, the formal integrator and its Neural Network implementation, will be discussed in the following section. In the third section we will consider in some detail their application to the Intramolecular Energy Distribution problem.

In any case, we want to point out that the connection presented in this note between Neural Networks and Taylor Series integrators of ODEs can have more far reaching consecuences when considered together with error correcting methods developped for learning from examples. In fact, we are presently studying the performance of these Neural Networks Architectures for learning differential equations from trajectories and preliminar results suggest that they can be quite useful not only for determining potential ODE models giving rise to specific time series, but also for nonlinear forecasting of time series with chaotic behaviour. The results of these analysis will be published elsewhere.

2   Numerical Features of the Integrator

In this section we will briefly discuss the mathematical aspects of the formal integration of an autonomous ODE system in the Polynomial Format introduced above. As already shown in [FLC], the key idea is to exploit the relatively simple structure of the system to approximate adequately its solutions by means of the partial summation of the corresponding Taylor series expansions. Of course, this assumes at least local analiticity for these solutions, but given the obvious analiticity of the right hand side of equation (1), this can be taken for granted. Under this assumption, expanding the solutions $y_1, \ldots, y_N$ around a given point T in a power series of $t - T$ as

$$y_i(t) = \sum_0^\infty \frac{y_i^{(n)}(T)}{n!} (t - T)^n, \quad i = 1, \ldots, N,$$

formal differentiation gives the corresponding series for the derivatives $\dot{y}_i$. Then, substituting these expansions on the Ricatti system and equating same order coefficients, we obtain the following recurrence relations for $n = 0, 1, \ldots$

$$y_i^{(n+1)}(T) = a^i \delta_{in} + \sum_j a_j^i y_j^{(n)} + \sum_{j_1 \leq j_2} a_{j_1 j_2}^i \left(\sum_{l=0}^n \frac{n!}{l! n - l!} y_{j_1}^{(l)} y_{j_2}^{(n-l)} \right) +$$

$$\cdots$$

$$\sum_{j_1 \leq j_2 \leq \cdots \leq j_P} a_{j_1 j_2 \ldots j_P}^i \left(\sum_{l_1 + l_2 + \cdots + l_P = n} \frac{n!}{l_1! l_2! \ldots l_P!} y_{j_1}^{(l_1)} y_{j_2}^{(l_2)} \cdots y_{j_P}^{(l_{P2})} \right),$$

where δ_{in} denotes Kronecker's delta. For the particular case of a Ricatti system only the constant, linear and quadratic terms appear, and we simply obtain

$$y_i^{(n+1)}(T) = y_i^{(n+1)} = a^i \delta_{in} + \sum_j a_j^i y_j^{(n)} +$$

$$\sum_{j_1 \leq j_2} a_{j_1 j_2}^i \left(\sum_{l=0}^n \frac{n!}{l! n - l!} y_{j_1}^{(l)} y_{j_2}^{(n-l)} \right)$$

Since $y_i^{(0)}(T) = y_i(T)$, the above formulae allows the recurrent computation of the power series coefficients of the desired solutions and, therefore, their partial summation up to any order. Hence, starting at initial time T_0, one can get the solution trajectories by step integration.

A useful feature for any numerical ODE integrator is the capability to provide prediction–correction estimates. This is more so when the integrator is to be used in equations whose solutions present a highly varying behavior. The autonomous Ricatti format can give rise to precisely this kind of solutions: well known examples are the Lorentz systems which can present random behavior (see for instance [BO], Chapter 4).

Most classical ODE integrators can be used to derive this type of estimates (see [SB], Chapter 7). For the proposed formal integrator the situation is particularly straightforward: the explicit computation of the derivatives allows easy estimates of the errors incurred when approximating the exact solution $y_i(t)$ by means of partial summation up to a, say, order K, namely

$$|y_i(T+\delta) - \sum_{n=0}^{K} \frac{y_i^{(n)}(T)}{n!}\delta^n| \simeq \frac{|y_i^{(K+1)}(T)|}{(K+1)!}\delta^{(K+1)}.$$

As a consequence, given a desired precision for the partial sum approximation, a choice can be made for a convenient integration step and, therefore, a simple predictor–corrector type algorithm can be added to this formal integration method. For instance, if the error is to be held below a given bound ϵ, the desired step value δ assuming approximation by order K partial sums is given by

$$\delta = \min_i \left(\epsilon \frac{(K+1)!}{|y_i^{(K+1)}(T)|} \right)^{\frac{1}{K+1}}.$$

It is very clear from the above considerations that the computational structure of the proposed integrator is extremely simple. In fact, both the recurrent computation of the derivatives of the y_i and their partial sum approximation lend themselves naturally to software vectorialization and, even, parallel implementation. These approaches to a concrete implementation would certainly result in adding computational efficiency to the integrator. A particularly attractive example arises when considering a Neural Network type implementation. As seen in the above formulae, the Taylor series algorithm only requires elementary operations, that is, in the language of the

Neural Networks community, it has a highly local nature. For the general polynomial case, the computational flow can be naturally mapped into a sigma-pi type, feedforward Neural Network architecture (see [RM]). For instance, the values of the solutions derivatives given by a Ricatti system integrator (which only involves linear and quadratic connections) can be seen as the activations of the following network

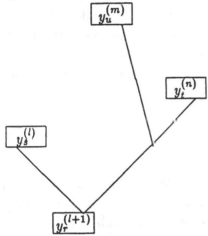

(we have $m + n = l$ and the connecting weights in this case are given as combinations of the system coefficients and numerical factors resulting from the recurrence relations between derivatives, and are therefore fixed).

These considerations suggest the possibility of a relatively simple hardware implementation of the Taylor series as an special purpose integrator for particular equations, as for instance those associated with the above mentioned IVR problem. However, a more compelling reason for the development of these hardware implementations can be derived from explicit complexity considerations.

The computation of the number of operations required to perform an integration over a complete time interval is fairly straightforward. Restricting again our attention to the Ricatti case and considering only multiplications, and assuming a time interval divided in a number S of steps and a full system of N equations to be solved by K-th order partial summation, the total number of operations becomes

$$\frac{N^3(K^2 + K)}{2}S + O(N^2, K^2, S)$$

The contributions from K and S come from precision considerations and we will not take them into account in what follows. In a serial computer, the time needed to perform an integration using the Taylor series algorithm would be of the order of N^3, preventing therefore its applicability in moderate–to–large systems. However, the paralelization inherent in the network architecture outlined above allows to drop a power of N from the contribution of the number of equations in the system, resulting therefore a required time of order N^2.

3   Applications

We shall illustrate the ideas exposed in this note in a particular example relevant in molecular physics, the Intramolecular Vibrational Energy Redistribution problem. Let us first do a brief review of the problem.

Current theoretical studies are based on the semiclassical approach that amounts to integrate the Classical Hamilton equations of motion for sets of initial conditions selected by semiclassical arguments. Approximations to the time variation of quantum expectation values are then calculated as suitable averages over the selected set of trajectories. In general, the model Hamiltonians used for Intramolecular Vibrational Energy Redistribution in molecules have usually the following form:

$$H = \frac{1}{2}\sum_{i=1}^{N}\sum_{j=1}^{N} p_i G_{ij} p_j + \sum_{i=1}^{N} D_i(1 - \exp\{-\alpha_i q_i\})^2$$

where q_i are the internal coordinates and p_i their conjugated momenta. Typically, the Wilson G matrix elements, G_{ij}, are taken to be constant and equal to the value at equilibrium internuclear distances and angles. D_i and α_i are the parameters of the usual Morse Potential. For this model Hamiltonian, the classical equations of motion to be integrated in time are the following $2N$ coupled ordinary differential equations

$$\dot{q}_k = \qquad\qquad \sum_j G_{kj} p_j$$

$$\dot{p}_k = -2D_k\alpha_k(\exp\{-\alpha_k q_k\} - \exp\{-2\alpha_k q_k\})$$

$$k = 1\ldots N.$$

They can be brought into the desired Ricatti format for parallel integration by introducing the following set of N new variables

$$z_i = \exp\{-\alpha_i q_i\} \qquad i = 1\ldots N.$$

and the equations of motion are transformed to the following enlarged system of $3N$ equations

$$\dot{q}_k = \sum_j G_{kj} p_j$$
$$\dot{p}_k = -2D_k\alpha_k z_k(1 - z_k)$$
$$\dot{z}_k = -z_k P_k$$

$$k = 1\ldots N.$$

As can be seen by inspection, the resulting system is far from being a full Ricatti system. In fact, the Wilson G matrix elements G_{kj} are null for non connecting atoms and summation in the time derivative of the internal coordinate q runs only for a few terms. This results on the serial implementation requiring order N operations for the computation of the solution for a time step whereas in the neural network implementation integration time is $O(1)$ provided that N units per layer are available.

Intramolecular Energy Redistribution among vibrational degrees of freedom in molecules has received an enormous attention in the last decade, mainly due to advances in experimental techniques and theoretical interpretations that make more plausible the attainment of the elusive Laser Assisted Intramolecular Selective Chemistry [M]. Theoretical Studies are fundamental not only for interpretation of experimental data but as a guidance for families of chemicals and selective excitations that should be experimentally explored.

The aforementioned scheme, although more feasible than other quantum approaches, involves massive computation. More than a hundred degrees of freedom are relevant for molecular models of practical interest, and to characterize regions of phase space with chaotic or regular trajectories, an accurate numerical integration is needed; moreover, integration has to be carried for many different trajectories and for long time periods. Thus, the improvement introduced by the neural network implementation of the integrator can offer a crucial enlargement of the size of the molecules that can be studied.

References

[BO]   C. M. Bender and S. A. Orszag, *Advanced Mathematical Methods for Scientists and Engineers*, McGraw–Hill 1978.

[FLC]   V. Fairén, V. López, V. Conde, *Am. J. Phys.* **56** (1988), 57–61.

[Kr]   E. D. Kerner, *J. Math. Phys.* **22** (1981), 1366–1371.

[LLFVM] S. M. Lederman, V. López, V. Fairén, G. A. Voth and R. A. Marcus, *Chem. Phys.* **139** (1989), 171–184.

[LFLM]   V. López, V. Fairén, S. M. Lederman and R. A. Marcus, *J. Chem. Phys.* **84** (1986), 5494–5503.

[M]   R. A. Marcus, *Faraday Disc. Chem. Soc.* **75** (1983), 132.

[RM]   D. E. Rumelhart, J. L. McClelland and the PDP Research Group, *Parallel Distributed Processing*, MIT Press 1986.

[SB]   J. Stoer and R. Burlirsch, *Introduction to Numerical Analysis*, Springer–Verlag 1980.

DISORDERED MODELS OF ACQUIRED DYSLEXIA

M.A.Virasoro
Dipartimento di Fisica
Universita' di Roma "La Sapienza"

Abstract

We show that certain specific correlations in the probability of errors observed in dyslexic patients that are normally explained by introducing additional complexity in the model for the reading process are typical of any Neural Network system that has learned to deal with a quasiregular environment. On the other hand we show that in Neural Networks the more regular behavior does not become naturally the default behavior.

1. Introduction

During this workshop we have discussed a few times about possible confrontation between the ideas of Neural Networks and the reality of biological systems.

In his talk Erdos introduced an imaginary line along which animal species can be ordered in terms of the complexity of their neural system. On one end lie simple, primitive, hardwired nematods, while on the other end one can imagine the human species, with its large and to a certain extent disordered brain. On both ends confrontation between models and experimental facts is possible.

Both have their relative advantages and disadvantages. Primitive animals provide exciting testing ground because one can envisage the possibility of checking details at the microscopic level. On the other end, when studying the brain of higher animals, the large number of degrees of freedom and the intrinsic disorder, make it possible the use of statistical methods.

But how disordered is the brain? In the history of neural sciences this question has always been at the center of a heated debate. In the last 130 years starting from the fundamental work by Broca on language impairment as a consequence of a localized lesion in the cortex, an impressive amount of clinical cases have accumulated demonstrating a rather tight correlation between the area of the lesion and the type of impairment[1]. If two patients A and B suffer from lesions in two separated areas of the brain and as a consequence patient A has function a impaired and function b normal while patient B has function b impaired while a is normal then functions a and b are said to be anatomically dissociated. Many examples of this type have led to the hypothesis of *Localization* in brain organization: different regions of the brain would work almost independently of each other and would perform different specific functions. Undoubtedly from this perspective one gets the conviction that in the brain there is a lot of functionally relevant structure so that perhaps disorder must be looked for at a smaller scale. Otherwise a straightforward extrapolation of localization all the way down to the single neuron leads to assuming the existence of the so called "Grandmother" neurons, i.e. single neurons that fire exclusively in front of a unique stimulus.

Recent results coming from brain lesion studies push, at least superficially, in the direction of further localization. For instance there are the so called category specific impairments (as the famous JB patient[2] who suffered of a category specific anomia: he could not give the right name to fruits and vegetables which were presented visually to him). Or, for instance, acquired dyslexia (i.e. impairment in the reading ability as a consequence of a trauma) that together with data from normal readers have led neuropsychologists to the complicated model in fig 1[3]. What is the meaning of the boxes in this model? When a certain function can be analyzed, from a

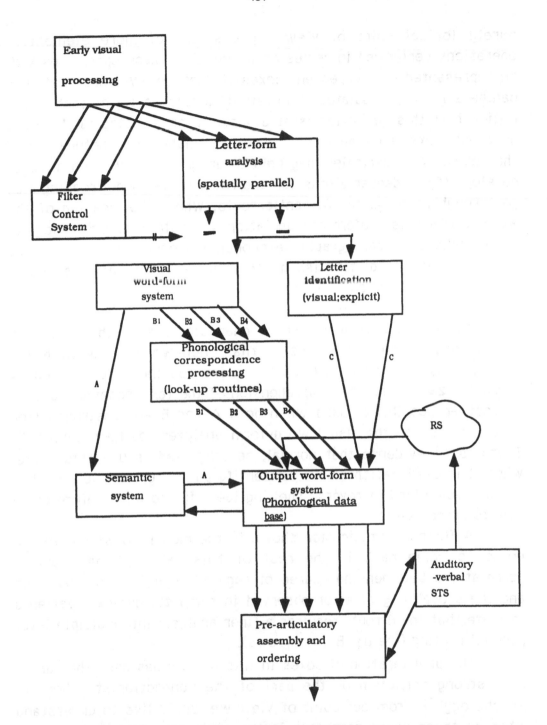

Fig.1: (From Ref 2). Model of the word-form multiple levels approach to the reading process. A=the semantic "route". B=the phenonmenological "route" and C= the compensatory strategy "route" .B,B2,... represent different levels of operation of the multiple level system. The "cloud" shape (rs) represents the operation of the hypo-thetical reversed spelling procedure

purely logical point of view, in a series of more elementary operations performed in series or in parallel, these operations will be represented by independent boxes if there is evidence that they can be affected, modulated or impaired independently of each other. Notice that this definition is at a purely phenomenological level. It does not make any assumptions about the material structure behind the boxes. Some examples may help to understand how such a model develops (for a deeper discussion see ref4. In the model reading may go through route A, the Semantic route, where the meaning of the word is identified before pronunciation, or route B, the phonological route, where certain grapheme-phoneme pronunciation rules are applied. They are distinguished because there are patients whose pattern of dysfunction looks as if one of the routes is much more damaged than the other one: *Phonological dyslexia*, where the phonological route is damaged, is characterized by the fact that words are much better read than non-words (meaningless neologisms); *Semantic dyslexia* where reading is preserved is characterized by lack of comprehension of the text meaning. Route C is another possibility used when both A and B are destroyed (for instance because the visual word-form analyzer has been damaged). There is no evidence that normal, healthy readers use this route: when it is used, patients have to identify painfully every letter, then through some kind of reverse spelling identify the word before being able to pronounce it.

A further complication shown in the model, to which we are going to come back in the rest of this talk, is that certain correlations between the degree of regularity of the pronunciation and the probability of error observed in semantic dyslexic patients indicate that the B route requires a finer analysis into multiple level parallel routing (the B_1, B_2, ... arrows).

The proliferation of boxes in this models has been the target of a strong criticism on the part of the Connectionist school in Psychology[7]. From our point of view we would like to understand what do these boxes represent in the brain architecture. Are they smaller, topographically separated, subareas of the cortex which can be roughly differentiated anatomically and/or histologically? Are they preprogrammed or rather the consequence of some kind of self-

These arguments can be refraced as saying that: 1) there is no grand design of the brain; 2) even if there was one it could not be stored in the DNA; 3) even if it was stored in the DNA there would be no way to put it into production.

We then prefer the most amorphous model capable of the specific behavioral response in question.

Here we will show, for instance, that the fine structure of the B route into multiple levels can be explained away as being typical of any amorphous system dedicated to the task of learning to read English where phonetic rules coexist with exceptions.

For this purpose we imagine a *quasiregular* environment. This concept was introduced in ref. 7 to indicate a body of knowledge that is systematic but admits irregularities. It is a rather common situation for a biosystem to be in front of similar stimuli most of which require a simple "regular" response but a small number characterized by some complicated features need a special, less regular response. In this case we should classify the reponse to a particular stimulus in terms of its *regularity* . The rigourous definition of regularity is far from trivial but we will show that a definition that is a natural extension of Chaitin s one in Information theory can be formulated in terms of how easy it is to induce the rule from examples.

When there is coexistence of more regular and less regular responses generalization can proceed in different ways. We will show that in some situations the more natural choice for a Neural Network may differ from the best strategy sometimes adopted by animals (the default rule-governed behavior).

The effects of a *lesion* are then investigated to check for the correlation between the probability of loosing one particular response and its regularity. It is observed in patients[8] that the larger the regularity the smaller is the probability of error. We prove that we should expect such correlation *in a large class of amorphous models*. We will also discuss the so called "regularization"[8] behavior in some patients.

2. How to make useful Statistical Inferences

In this talk we would like to discuss properties that are common to a large class of models rather than being specific to a few perhaps irrealistic ones. Specific models nicely exemplify these general arguments[9,10]. To deal with an ensemble of Neural Networks we use the methods of statistical inference. A statistical inference tells us what should we expect on the basis of what we know. In its Bayesian formulation the argument begins with the introduction of a prior probability measure, which normally states that in the absence of any knowledge about the system a certain number of choices are equally probable. It then proceeds through the restriction of this measure by whatever partial information one has about the system and finally concludes that the unknown observable takes its more probable value (the principle of maximum entropy). In the case of large systems, where the number of parameters that determine a configuration of the system is large, different measures can lead to identical statistical inferences for specific observables a fact that is explicitely used in most applications[11,12].

Neural Networks (NN) are prototypes of complex adaptive system and most of our analysis applies to any system in this class. A NN contains a structure that is supposed to be fixed and a large number N of parameters that are adjusted through a "learning" process. It is in this N dimensional space of parameter configurations that the probability mesure on the initial state of the system is defined[13,14,15]. We can consider this measure as part of the definition of the model. It is convenient to partition the continuum of configurations in sufficiently small cells with equal probability measure. In this way configuration space is discretized and an initial entropy equal to the logarithm of the number of configurations can be defined. In the simple case of the Perceptron, E. Gardner has given a convenient definition that has the advantage that averages on it can be easily calculated.

3. Entropy and generalization

The network has to learn the correct stimulus-response mapping but it proceeds, as usual, by learning instances of it. When learning one example the probability measure obviously evolves. It will become concentrated on those configurations that respond correctly on the new example while it is zero on those that do not. A new example implies an increase in the information contained in the system and a corresponding decrease in its entropy. As defined the entropy variation will satisfy:

$$\Delta S \leq \log N_2 - \log(N_1 + N_2) \tag{1}$$

where $N_2(N_1)$ is the number of configurations that do (do not) respond correctly.

Equation (1) becomes an equality if the learning protocol keeps the N_2 configurations equally probable, i.e. it does not introduce an arbitrary bias. This minimum amount of information can be conveniently assigned to the learned example (but for that particular network at that particular time). The information content of a mapping defined a' la Chaitin, is proportional to the minimum length of a program that implements the mapping on a Universal Computer. Because Universal Computers can be translated into one another, this definition is, to some extent, machine independent. In NN the emphasis changes because translating NN into one another might be rather unnatural. The corresponding definition defines the information content of the mapping as the sum total of the information content of all the examples, the latter defined as above. Finally a mapping is more or less regular if its information content is larger or smaller.

With this definition a more regular mapping is easier to generalize[14,15] in the sense that it can be induced from a fewer number of examples. The argument is very simple: the first learned instances of any mapping produce roughly the same decrease in entropy because it is impossible to know from the first few examples whether the mapping of which they are instances is more or less regular. But after a certain number of examples the decrease of entropy *per example* has to become smaller for more regular mappings because it has to add to a smaller total quantity. It then follows immediately from eq(1) that:

$$\frac{N_2}{N_1+N_2} \tag{2}$$

is larger. But this is just the probability that a system that has learned P examples automatically responds correctly to the P+1 one.

4. The role of the irregular examples in the generalization. The default rule-governed behavior is not typical of a Neural Network

In a quasiregular environment the situation is more complicated because there is coexistence of mappings with different degrees of regularity. For instance, machines that learn to read English[7] have to deal with examples that follow a simple phonetic rule but also less regular exceptions. In general we should expect (and so it is found in particular cases[9]) interaction between regular and irregular examples: the induction of the rule is affected by the presence of exceptions. When the frequency of irregular examples is small so that they represent a small perturbation on an otherwise regular world it is evident that the entropy decrease per regular example will, at least initially, behave as before while the one of irregular responses will remain permanently large. On the other hand, asymptotically, the entropy decrease per regular pattern will not decrease to zero in general.

To induce the rule behind a certain behavior a minimum number of examples are necessary. When there is coexistence between behaviors that differ a lot in regularity the NN may be in a situation in which the number of more regular examples is sufficiently large while the number of less regular ones is not enough. In this case the less regular behavior cannot be predicted and it plays the role of simply exceptions to the rule. For the sake of clarity we can then concentrate on the coexistence of a regular mapping and an infinitely irregular one. Then, by assumption, on a subset of arbitrary uncorrelated input stimuli the response will be exceptional and random and therefore it is impossible by definition to induce through generalization the next exceptional response. How does the system generalize this experience when it faces new stimuli? There are different possibilities realisable through the choice of different learning protocols, but there is one that is more natural because less biased. If f_I is the frequency of the irregular examples then the average entropy variation will be:

$$\Delta S = f_I \Delta S_{irr} + (1-f_I)\Delta S_{reg} \qquad (3)$$

If learning is *unbiased*, the residual entropy of the system is maximized, ΔS is minimized. This means choosing that particular arrangement that is realized by the largest number (exponential of ΔS) of equally probable configurations. In general this means that ΔS_{reg} is not minimal and therefore is different from zero. As a consequence p_R, the probability that on a new example the system assumes the more regular behavior, will be less than one. We can describe this behavior by saying that the rule has not become a *default* rule. Some interesting consequences follow because it is easy to realize that the number of errors that will come out will be roughly

Number of errors: $p_R(1-f_I)+f_I(1-p_R)+p_R f_I$

which is larger than f_I: the number of errors the system would make if it applied always the rule. It has been argued that in many cases the human being chooses the default rule behavior[16]. But it is important to notice that the derivation of such a better strategy depends strongly on the fact that while both the NN and the human being do not know the complicated rule behind the apparently irregular examples, only the human being **knows that he does not know**. Without this information the NN can not disregard the elss regular examples.

5. Regularity effects are well reproduced in most models

Let us now imagine a random, uncorrelated damage on the system to simulate a lesion.

The lesion is a random walk in the space of configurations starting from a particular region. Because, by definition, the lesion is not correlated with the previous learning we can assume that the probability distribution after the damage is identical to the distribution before the damage modulated by a decreasing function of the distance to the initial region. The definition of this distance is model dependent. In ref.10 the distance factor in the probability of error was discussed in the particular model. It was found that a convenient distance between two responses was given by the number of phonetic features in which they differ. Remarkably, similar results were found in patients.

If one considers the probability of making an error in one regular response against the same probability in an irregular one, independently of the actual outcome, then the distance is irrelevant and the relative probabilities can be calculated from (1):

$$\frac{\text{prob(error in one regular response)}}{\text{prob(error in one irreg. response)}} = \frac{\exp\Delta S_{reg}}{\exp\Delta S_{irr}} \tag{4}$$

It is, therefore, less than one. This result can be checked by preparing two lists of equal numbers of regular and irregular responses and counting the mistakes produced. Observations in patients and models verify (4) [8,9,10].

6. "Regularization" behavior in patients and in models

However, a discrepancy between certain patients behavior and numerical models in the *type* of error produced when mistaking an irregular response has also been observed. Patients[10] have a tendency to *regularize*, i.e. to prefer the regular pronounciation once the correct response has been lost. In models similar effect are much too small[8]. This discrepancy is a consequence of the fact discussed in Section 4 that the more regular behavior has not become the default behavior in the model. In fact eq.(3) allows for a probability of generalizing the regular behavior sensibly less than one.

Methodologically statistical inferences are extremely valuable to decide whether a specific aspect of the behavior of a system deserves a specific explanation or whether this is unnecessary because it is a consequence of differences in available phase space. In this talk we discussed examples of both types. Of course it is possible (though not probable) that rule governed behavior becomes the default rule behavior in a quasiregular environment. The challenge remains to find a sufficiently general learning mechanism that does just that.

Acknowledgments:

I thank D. Norris and K. Patterson for useful conversations

References

[1] Kandel E. R. and J.H. Schwartz *Principles of Neural Science*, Elsevier, New York, Amsterdam, Oxford, 1985. See Table 52-2 in page 701.
[2] Hart J., R.S. Berndt and A. Caramazza, Nature 316 (1985) 439-440

[3] Shallice, t. and R. McCarthy, Phonological reading: from patterns of impairment to possible procedures. In Surface Dyslexia: neuropsychological and cognitive studies of phonological reading (eds. K. Patterson, J. C. Marshall and M. Coltheart) Earlbaum, London 1985.

[4] Fodor J. A. and Z.W. Pylyshyn Connectionism and cognitive architecture: A critical analysis, Cognition 28 (1989)3-71

[5] Virasoro M.A. Disordered Models of the Brain. In New Developments in Hardware and Software for Computational Physics (eds A.M.J. Ferrero, R.P.J. Perazzo and S.L. Reich) North Holland 1989.

[6] Edelman G. Neural Darwinism: the theory of neuronal group selection, Basicc Books, New York, 1987.

[7] Seidenberg M.S. and McClelland J.S. A distributed developmental model of visual word recognition and pronunciation: acquisition, skilled performance and dyslexia. In From Reading to Neurons: Towards Theory and Methods for Research on Developmental Dyslexia. MIT Press/ Bradford Books, Cambridge Mass 1988.

[8] Shallice T., Warrington E.K. and McCarthy R. Reading without Semantics. Quartely Journal of Experimental Psychology, 35A (1983)111-138

[9] Virasoro M.A., Analysis of the effects of lesions on a Perceptron. Journal of Physics A: Math. Gen., 22 (1989) 2227-2232

[10] Patterson K., Seidenberg M.S. and McClelland J.S. Connections and Disconnections: Acquired Dyslexia in a Computational Model of Reading Processes. In Parallel Distributed Processing: Implications for Psychology and Neurobiology R.G.M. Morris (ed) Oxford University Press 1989.

[11] S. Geman and D. Geman "Bayesian Image Analysis" in Disordered Systems and Biological Organization", Ed. E. Bienenstock et al. Springer Verlag. Berlin, Heidelberg 1986.

[12] R.O. Duda et P.E. Hart "Pattern Classification and Scene Analysis", Wiley eds

[13] E.Gardner. Journal of Physics A: Math. Gen.,21A (1988) 257

[14] Carnevale P. and Patarnello S. Europhysics Letters 7 (1987)1199

[15] Denker J., Schwartz D., Wittner B., Solla S., Hopfield J.J., Howard R. and Jackel L. Complex Systems 1 (1987)877.

[16] A. Prince and S. Pinker "The nature of Human Concepts: Insight from an Unusual Source" MIT preprint.

HIGHER ORDER MEMORIES

IN OPTIMALLY STRUCTURED NEURAL NETWORKS

Karl E. Kürten

HLRZ c/o KFA Jülich, D-5170 Jülich, FRG
and
Institut für Theoretische Physik, Universität zu Köln,
D-5000 Köln, FRG

Abstract. Based on one-step Hebbian learning we incorporate higher order correlations into the flexible architecture of a neural network optimized in first order. Sparse connectivity as well as a suitable summation technique eliminate the proliferation problem of higher order terms.

1 Introduction

Most of the currently popular binary neural network models having their roots in the pioneering work of Mc Culloch and Pitts [1] and Caianiello's "neuronic equations" [2] are completely connected systems based on linear threshold dynamics. On the one hand, it is well known that first order networks are highly limited to linearly separable problems, whereas higher order networks have often been believed to be impractical due to the exponential increase of the total number of coupling coefficients favoured by a high degree of connectivity. On the other hand, computational and biological aspects suggest strongly the study of artificial networks equipped with higher order interactions. In fact, modelling the highly complex interactions of neighbouring synapses on the dendritic trees might require the incorporation of multi-linear weights which not only increase the complexity of the model but also lead to highly increased storage capacities as well as to improved discrimination properties [3].

2 The associative memory model

The transmission of information in a system of N formal neurons can be described in terms of the dynamics of a set of binary "firing" units σ_i which assume the value +1 when neuron i is "active" and -1 when it is not. We consider a sparsely connected model and specify that each neuron receives input from only K other units of the network. The total net internal stimulus h_i can then be modelled as a nonlinear irreducible superposition of input signals:

$$h_i(t) = c_{i0} + \sum_{j_1} c_{ij_1} \sigma_{j_1}(t) + .. + \sum_{j_1 < .. < j_K} c_{ij_1..j_K} \sigma_{j_1}(t)..\sigma_{j_K}(t) \quad . \quad (1)$$

The sums are only taken over those K neurons with which unit i is connected and self-coupling effects are excluded. The integer $Nx2^K$ matrix C defines the synaptic efficiencies between the model neurons. Here, the zero order couplings c_{i0} play the role of the individual activation thresholds. The state of the network at time t is then specified by one of the 2^N possible "firing patterns" $\{\sigma_1(t),...,\sigma_N(t)\}$ which evolve on a 2^N-dimensional hypercube according to the deterministic threshold rule

$$\sigma_i(t+1) = \text{sign}[h_i(t)] \qquad\qquad i = 1,...,N \quad . \qquad (2)$$

It is worth to mention that by allowing arbitrary coupling coefficients in (1) all possible $2^{2^{**K}}$ Boolean transition functions of K inputs can be assigned a sector of the 2^K dimensional hypbercube $|c_{ij(1)...j(K)}| \leq 1$. It can be shown further that a minimal space of couplings $c_{ij(1)...j(K)} \in \{0,+1,-1\}^{2^{**K}}$ already generates all possible Boolean functions of order K [4].

In order to achieve faithful storage of information encoded in a set of p arbitrary memory configurations $S^{(1)},...,S^{(p)}$ the coupling constants have to be chosen such that each component i of a memory pattern aligns with the corresponding induced field h_i such that the stability condition

$$S_i^{(\mu)} h_i(S_1^{(\mu)},...,S_N^{(\mu)}) = \kappa_{i\mu} > \kappa \qquad\qquad (3)$$

is satisfied. Furthermore, the magnitude of the positive global stability parameter κ is required to be as large as possible in order to ensure optimal stability and large basins of attraction for the individual patterns.

It is known that the 2^K synaptic coefficients describing the degrees of freedom of the system allow the network to store up to $p = \alpha 2^K$ randomly chosen patterns, where α depends on the nature of the patterns and the learning algorithm which specifies the coupling coefficients. Thus, the network offers an impressive storage capacity which however might become impractical due to the exponential increase of the higher order contributions. We will show that this problem can be eliminated by admitting a drastically reduced interconnectivity as well as by employing a suitable summation technique for the inclusion of the higher order terms.

3 Quasi-optimal network architecture

Recent analytical as well as numerical studies have convincingly shown that "optimally" diluted networks show a quantitatively similar or even better retrieval performance than their fully connected counterparts [5-7]. In contrast, random dilution of synaptic bonds usually decreases the performance of the network depending on the degree of dilution [8]. To be more specific, the retrieval performance as well as the storage capacity of a sparsely connected model can be largely improved if the connectivity is *not* chosen at random but adapted to the specific information the network is asked to memorize.

Here, we are confronted with a combinatorial optimization problem, where each cell i has $\binom{N}{K}$ different ways to choose its connecting neighbours in order to maximize the minimal polarization parameter $\kappa_{i,\mu}$ in eq.(3). This problem has been successfully attacked with suitable Monte-Carlo techniques disfavouring the choice of low-efficacy synapses [5,6], though, alternatively, genetic algorithms and deterministic optimization procedures may serve as well.

Moreover, work in progress suggests strongly that it is far more important to aim at an "optimal" network connectivity than to concentrate on optimal, time consuming learning procedures for the modification of the coupling coefficients. To be more precise, for small storage parameters α, widely used Hebbian learning prescriptions are able to stabilize uncorrelated information almost as well as the optimal perceptron rule.
Hence, the natural extension of the classical Hebbbian learning rule to order K

$$c_{ij_1j_2\ldots j_K} = \sum_{\mu=1}^{P} s_i^{(\mu)} s_{j_1}^{(\mu)} s_{j_2}^{(\mu)} \ldots s_{j_K}^{(\mu)} \tag{4}$$

simplifies the search for optimal network architectures appreciably, since the synaptic weights do not have to be learnt by an iterative process but may be explicitly evaluated.

4 Implementation of higher order interactions

In principle, the combinatorial optimization strategy adopted in refs. [5,6] could be extended to higher orders. However, we assume that the optimal connectivity in first order is also a good approximation for a quasi-optimal connectivity in the next orders. Thus, we restrict multi-cell interactions to the subspace containing the optimal neighbours which evolved within the first order approximation.

The technical evaluation of the s-order term in (1) is extremely simple if one includes reducible higher order interactions by adding symmetric and redundant terms, where all indices are not different. Note that the latter would already appear in lower-order contributions. In this special case it is straightforward to show by interchanging the order of the summations that such a modified s-order contribution in eq.(1) takes the simple form

$$\sum_{j_1 \cdots j_s} c_{j_1 \cdots j_s} \, \sigma_{j_1}(t) \cdots \sigma_{j_s}(t) = \sum_{r=1}^{p} S_i^{(r)} [\, m^{(r)}(t)\,]^s \qquad (5)$$

with

$$m^{(\mu)}(t) = \sum_j S_j^{(\mu)} \sigma_j(t) \qquad (6)$$

representing the overlap of a current spin configuration $\sigma_j(t)$ with a given pattern $S^{(\mu)}$. Thus the coupling matrix C needs neither be calculated explicitly nor be stored but the coupling coefficients are implicitly incorporated in the overlap [3]. Hence, for technical applications one does not need more storage capacity than for a first order network. Also the evaluation of the induced field of an s-order network requires essentially the same computational effort as that of a first order network.

Note that only odd orders enable the network to store also the complementary patterns since the field is not symmetric under sign inversion $S^{(\mu)} \to -S^{(\mu)}$ as is the case for even s. Consequently, one never retrieves the complementary pattern. Instead, the presentation of a complementary pattern leads to a one step convergence to the memorized pattern in an even order network.

On could indeed sum all modified higher order terms up to order K by a geometric series such that the total synaptic input (1) takes the simple closed form

$$h_i(t) = \sum_{r=1}^{p} S_i^{(r)} \left(\frac{1 - (m^{(r)})^{K+1}}{1 - m^{(r)}} \right) \qquad \text{for } m^{(r)} \neq 1 \qquad (7)$$

while the term in brackets takes the value K + 1 for $m^{(r)} = 1$. In this report, however, we restrict ourselves to the study of pure odd-order networks.

5 Computer simulations

In this section we demonstrate how the performance of a network improves with increasing order. We choose $p = \alpha K$ uncorrelated random patterns and optimize the stability condition (3) within first order. With the aid of a Monte Carlo procedure we determine optimal connections for each individual cell by maximizing the smallest plarization parameter $\kappa_{i\mu}$ in eq.(3). The search process is performed independently for each cell i resulting in asymmetric couplings as well as in widely varying numbers of connections for each individual cell. For the retrieval phase, where the patterns are degraded by random noise and presented as initial conditions we only admit one single *odd* s-order term in the dynamics (1). Figure 1 displays the recall performance of a network consisting of N=200 cells with <K> = 15 connections and 27 memorized patterns for several orders s.

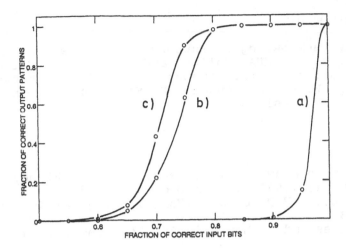

Fig.1: Mean fraction of correctly recalled
bits as a function of the fraction of
correctly presented input bits for
orders a) s=1, b) s=3 and c) s=5.

To emphasize the potential of the *optimized* architecture we point out that according to exact results in the thermodynamic limit for networks with *random* connectivity even imperfect storage is only possible up to the critical $\alpha_c = 2/\pi$ [7] with $p = \alpha K$. However, our first order results for $\alpha = 1.8$ demonstrate clearly that a pattern-specific connectivity allows the network not only to store the information perfectly but also leads to small basins of attraction.

Comparing first and third order results we observe that the size of the basins of attraction is drastically enlarged. Increasing the order further to s=5 still gives some improvement, whereas higher order networks for s=7, s=9 and s=11 show quantitatively the same behaviour as the fifth-order network. The expected saturation is due to the use of the Hebbian-like rule, which is by no means optimal. Only optimal learning algorithms such as generalized perceptron rules which can be naturally extended to any order will allow a further increase of the size of the basins. Our results indicate also that the network connectivity optimized in first order is presumably still a good approximation in third order but no longer optimal in higher orders.

6 Conclusion

The central result of this contribution is that linear threshold networks with pattern-specific connectivity clearly outperform networks with *random* architectures. Moreover, adding self-consistently higher order correlations improves the retrieval performance of the networks dramatically, particularly in view of the minor technical effort required.

Acknowledgements

The author acknowledges support from the Höchstleistungs-rechenzentrum at KFA Jülich and from the German Science Foundation under contract number Se 251/32-1. Thanks are due to G. Kohring, W. von Seelen and D. Stauffer for many helpful and informative discussions.

References

[1] W.W.McCulloch and W.Pitts,1943,Bull.Math.Biophys.5,115-133
[2] E.Caianiello,1961,J.theor.Biol.1,204-235
[3] D.Psalits, C.H.Parc and J.Hong,1988,Neural Networks 1, 149-163
   C.L. Giles and T. Maxwell,1987, Applied Optics 26,23, 2972-4978.
[4] K.E. Kürten, to be published
[5] K.E. Kürten,1990, J. Phys. France 51,1585-1594
[6] K.E.Kürten,1990, Parallel Processing in Neural Systems and Computers,World Scientific, ed. by R.Eckmiller, G.Hartmann and G. Hauske, 191-194
[7] M.Bouten, A.Komeda and R.Serneels,1990,J.Phys.A 23,2605-2612
[8] B.Derrida, E.Gardner and A.Zippelius,1987, Europhys.Lett.4, 167-173.

RANDOM BOOLEAN NETWORKS
FOR AUTOASSOCIATIVE MEMORY:
OPTIMIZATION AND SEQUENTIAL LEARNING

D Sherrington and K Y M Wong

Department of Theoretical Physics, Oxford University,
1 Keble Road, Oxford OX1 3NP, U.K.

Abstract:

Conventional neural networks are based on synaptic storage of information, even when the neural states are discrete and bounded. In general, the set of potential local operations is much greater. Here we discuss some aspects of the properties of networks of binary neurons with more general Boolean functions controlling the local dynamics. Two specific aspects are emphasised; (i) optimization in the presence of noise and (ii) a simple model for short-term memory exhibiting primacy and recency in the recall of sequentially taught patterns.

1. Introduction

Many simple models of neural networks follow McCulloch and Pitts [1] in idealising the neurons to have only two states, firing and non-firing. Such idealised neural states can be described by binary variables, Ising $\sigma_i(t) = \pm 1$ or Boolean $V_i(t) = 1, 0$, i denoting the neuron and t the time; to maintain uniformity with most conventional work in the Physics literature we shall employ Ising variables in this paper. With deterministic synchronous dynamics the local neuron update rule is therefore

$$\sigma_i(t+1) = F_i\big(\sigma_{i_1}(t), \ldots, \sigma_{i_C}(t)\big) \tag{1}$$

where i labels the neuron under consideration, $i_1, \ldots, i_C$ are the neurons feeding it, and F_i is a Boolean function taking only ± 1 as possible values. In general F_i is different for different i. There is a similar but probabilistic relationship for asynchronous dynamics [2].

Most of the work in the literature assumes that F_i belongs to the subset expressible as

$$F_i(\sigma_{i_1}, \ldots, \sigma_{i_C}) = \text{sign}\Big(\sum_{j=i_1}^{i_C} J_{ij}\sigma_j - W_i\Big) \tag{2}$$

where $\{J_{ij}\}$ are the synaptic efficacies and W_i is a threshold. This subset is based on a comparison with cortical neural networks in primates. More generally, however, the set of possible Boolean functions is much greater and indeed could be utilized in hardware implementations, for example based on random access memories (RAMs) as manifestations of the F_i. This talk will consider this more general situation [2].

Discussion will be restricted to recursive networks based on random interconnections of the neurons, and to dilute networks in which the number C of incoming (or outgoing) connections per neuron is much less than $\ln N$, where N is the total number of neurons; we shall concentrate on both C and N tending to infinity. The dilution restriction implies that correlations are effectively unimportant and permits the use of the annealed approximation [3] in analysing the retrieval properties. In consequence, if a system is started with a finite overlap $m(t = 0)$, where

$$m^\mu(t) = N^{-1} \sum_i \sigma_i(t)\xi_i^\mu , \tag{3}$$

with a single member of a set of stored random uncorrelated patterns $\{\xi_i^\mu\}$; then its temporal development can be described in terms of an iterative map

$$m(t + 1) = f(m(t)) . \tag{4}$$

The stable fixed points m^* of this iterative map give the possible asymptotic retrieval overlaps, whilst the unstable fixed points give the boundaries of the basins of attraction. The actual form of $f(m)$ depends on the choice of $\{F_i\}$.

If the $\{F_i\}$ are viewed as look-up tables, synaptic storage such as given in eqn (2) distributes information in an extended fashion throughout the whole look-up space of the $\{F_i\}$ on each neuron i; this follows immediately from the observation that the pattern information is stored in the $\{J_{ij}\}$. An alternative complementary storage localises information about pattern $\{\xi^\mu\}$ around the sites $(\xi_{i_1}^\mu, \ldots, \xi_{i_C}^\mu)$ in each look-up table F_i; distance (and therefore the concept of locality) is measured in terms of the fractional Hamming distance between site labels (the fractional number of different bit-entries).

In this paper we concentrate on two issues concerning Boolean networks, (i) optimal storage in the presence of training noise, and (ii) a simple model for short-term memory, utilizing the possibility of a third entry in the look-up tables and exhibiting the psychological features of primacy and recency in the recall of sequentially presented patterns.

2. Optimization

It is straightforward to demonstrate that if a single pattern is merely stabilised by choosing

$$F_i(\xi_{i_1}, \ldots, \xi_{i_C}) = \xi_i ; \text{ all } i ; \tag{5}$$

with all other entries in the Boolean functions completely random, then the resultant network is non-retrieving [4,2]. On the other hand, if the information is 'diffused' to sites in look-up table F_i close to $(\xi_{i_1}, \ldots, \xi_{i_C})$ then retrieval becomes possible [4]. This diffusion can be achieved by the use of 'training noise' interposed between the output of neurons $(i_1, \ldots, i_C)$ and the actual input to neuron i during the training process. A relatively small amount of diffusion suffices for retrieval to be possible, leaving 'space' for the storage of many other patterns.

A more precise measure of the role of training noise can be ascertained by considering the optimization of an appropriate cost function [5]. In particular consider the cost function

$$C = - \sum_{\mu,i} \xi_i^\mu F_i(\eta_{i_1}^\mu, \ldots, \eta_{i_C}^\mu) P_{d_t}(\eta|\mu) \tag{6}$$

where

$$P_{d_t}(\eta|\mu) = \prod_i [d_t \delta(\eta_i^\mu + \xi_i^\mu) + (1 - d_t)\delta(\eta_i^\mu - \xi_i^\mu)] ; \tag{7}$$

i.e. the cost is minus the sum (over patterns and sites) of the correct bit at neuron i for pattern μ, times the output bit at neuron i when a noisy version of pattern μ is presented at the input, times the probability of getting the noisy version from the clean at a training noise d_t. The objective is to choose a set of $F_i(\{\sigma\})$ such as to minimize C over an ensemble of random patterns and to monitor the retrieval properties of such an optimized network. Details of this optimization are presented elsewhere [4]. Here we note only some results.

The network capacity is given by the greatest p at which the stable fixed point m^* is non-zero. This is maximum for $d_t = 0^+$; i.e. infinitesimal training noise. It yields a filling of the sites of F_i according to the nearest neighbour majority rule which states that the content of any site addressed by one of the $(\xi_{i_1}^\mu, \ldots, \xi_{i_C}^\mu)$ follows the majority of the corresponding $\{\xi_i^\mu\}$ if there is a majority or is randomly chosen ± 1 if there is no majority, whilst unaddressed adjacent sites follow the majority bit of the patterns addressing their neighbours, and so on. [1] Such a network exhibits a high capacity $p_c = 1.1165(2/C^2)2^C$ for uncorrelated patterns, but retrieves only with errors scaling as $(2/C^2)$ and from narrow basins within $O(2/C^2)$ from perfect overlap [2]. At p_c the system has a first order transition from retrieval with a finite overlap (for $p < p_c$) to non-retrieval (for $p > p_c$).

When the training set is made up of complementary pairs, $\{\xi_i^\mu\}$ and $\{-\xi_i^\mu\}$, and the training noise d_t tends to the maximum of $1/2$, then the above procedure yields an emulation of synaptic storage with the synapses given by the Hebb rule

$$J_{ij} = C^{-1} \sum_\mu \xi_i^\mu \xi_j^\mu . \tag{8}$$

The storage is of course now extended in the look-up space and the network has the well-known properties found by Derrida et al [3], namely a capacity scaling as C, a wide attractor basin, and a retrieval overlap going continuously to zero as the critical storage is approached.

---

[1] This can be viewed as the creation of Voronoid cells of influence.

In fact, at any finite training noise, one has wide retrieval for p less than a value of order C. For d_t less than a critical value, increasing p leads to a transition to a regime of narrow retrieval before the eventual first order transition to non-retrieval at a larger p. For d_t greater than this critical value the transition is direct to non-retrieval via a second-order transition. The critical value d_c is given by [5]

$$d_c/(1 - d_c) = \exp(-\sqrt{2\pi}/C) \tag{9}$$

Fig. 1 shows the phase diagram in the high training noise limit.

So far we have considered only a network with deterministic dynamics. An analogy with synaptic rounding noise is given by choosing to operate according to eqn (1) or its converse, according to some 'retrieval noise' or temperature. Rigorous pursuance of eqn (1) corresponds to no retrieval noise (or zero temperature). Increasing noise (or temperature) yields more trangression from eqn (1). Such retrieval noise shrinks the retrieval error/basin boundary curve as a function of storage to lower p, thereby reducing the critical capacity [6].

Finally, in this section, we draw attention to our companion paper in this proceedings, wherein is considered the role of noise in training synaptic networks [7]. The phase diagram for synaptic networks has broadly similar features, but variations in detail. Note also that in such networks the limit of high training noise again yields Hebbian synapses.

The high training-noise network which is optimal over all Boolean functions therefore lies on the boundary of the synaptic subset. Over any training noise the optimum over all Boolean nets lies outside the synaptic subset and away from the boundary. The optimal synaptic net also lies away from the boundary but on the other side. This is illustrated schematically in Fig. 2.

3. 3-state functions

Thus far we have assumed that F_i takes only two values ± 1. For some purposes it is advantageous to allow also a third possibility [8], u or 0, such that the output is chosen randomly each time a u site is addressed. In an earlier paper [2] we have discussed the role of u-states on storage and retrieval in a network trained by the random presentation of samples drawn from a random ensemble of uncorrelated patterns. In this work [2] we allowed updates from u to ± 1 if a pattern requiring ± 1 addresses a u-site, at a rate we denoted the registration probability p_r, and from ± 1 to u if a pattern requiring the opposite bit addresses a site, with a correction probability p_c.

4. Primacy and recency in short-term memory

In this section we comment on a different application of 3-state local functions. This is to producing a simple model of working (short-term) memory which exhibits

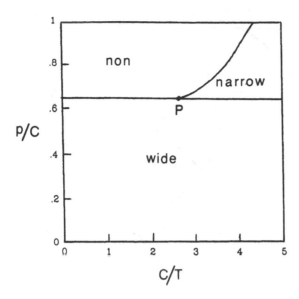

Fig. 1: The phase diagram of Boolean networks in the high training noise limit. The training noise temperature is defined as the inverse logarithm of the signal-to-noise ratio, i.e. $T = [\ln((1 - d_t)/d_t)]^{-1}$. Here both the number p of stored patterns and the training noise temperature T scale as C. The triple point P is $(p/C, C/T) = (2/\pi, \sqrt{2\pi})$.

primacy and recency in recalling a sequential list of patterns [9]. Recency refers to the phenomenon of recalling the patterns which have been learned most recently. Primacy refers to recalling the first patterns to be learned. There is experimental evidence of primacy and recency occuring whilst patterns in between are forgotten [10]. In ref. 9 we present a model based on a synaptic network with bounded synapses. Here we comment on an analgous Boolean network.

Consider a network in which initially all F_i are in the u state. Uncorrelated patterns are now learned serially. We denote the serial number by s. Each pattern is then presented in a sequence of n noisy versions and the addressed sites are updated with probability p_r if the addressed site is in a u-state, a change from ± 1 to u is performed with a probability p_c if the input bit is opposite to the existing bit, and no change is made if the addressing bit is identical to the addressed bit. For each pattern s a 'lifetime' $\Lambda(s)$ can be defined as the maximum number of further patterns which can be presented such that pattern s can still be retrieved.

There is clearly a minimum value of n to produce non-zero $\Lambda(s)$. This critical n varies as a function of s, being smallest for small s and reaching an asymptotic value as s is increased. The asymptotic $\Lambda(s \to \infty) \equiv \Lambda_\infty$ gives the asymptotic capacity for recall. A second (higher) n gives the greatest Λ_∞; too small n gives insufficient implanation of

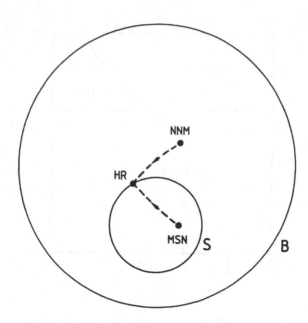

Fig. 2: A schematic illustration of optimal networks in the set of all Boolean nets (B) and its synaptic subset (S). The dotted lines indicate the trajectories of the optimal net when d_t increases from 0^+ and tends to 1/2 in the two sets. Here NNM, HR and MSN represent the nearest neighbour majority rule, Hebb rule and maximally stable network [7] respectively.

information for retrieval, too large n gives excessive dominance to the latest presented patterns at the expense of earlier ones.

Measured in terms of the total number of learned patterns, there is a window during which recall of pattern s is possible. This window is $s \leq t \leq s + \Lambda(s)$. After t patterns have been learned, this inequality gives the positions s of the patterns which can be recalled. Beyond the critical n one always has recency. Primacy and recency with forgetting of intermediate patterns occurs only for p_r/p_c greater than a critical value.

After presentation of many patterns, the distribution of individual $F_i(\{\sigma\})$ among $(\pm 1, u)$ settles down to an asymptotic value. For primacy to occur in a new learning schedule, the distribution of $\pm 1, u$ sites must deviate from their asymptotic form; above we have discussed only a 'tabula rasa' start but the result is the same, if less dramatic, for a starting situation with a sufficient excess of u states over that in the asymptotic limit. Thus, for this model to be potentially applicable to the description of primacy in short-term memory we must include not only sequence presentation 'time' but also a second 'real-time' during which resetting of the distribution occurs. A more extensive discussion, couched in terms of bounded synaptic networks, is discussed in ref. 9.

5. Conclusion

In this paper we have considered two aspects of simple Boolean networks; the first illustrating complementarity with conventional synaptic networks [2] and the role of noise in optimizing storage and in bridging the transition to a Hebbian network; the second a simple model of short-term memory with the potential for exhibiting primacy and recency. Many extensions can be envisaged, for example to different architectures [8] and different pattern correlations, as well as to other discrete models with more states.

Acknowledgements

We would like to thank Prof. I. Aleksander for stimulating our interest in Boolean storage and in the use of u-states, the late Dr. E. Gardner for stimulating our interest and that of much of the neural network community in the application of statistical mechanical techniques to the optimization of neural networks, Dr. J. Shapiro for many stimulating discussions on primacy and recency, and Messrs R. Blackburn, P. Kahn and A. Rau for assistance in part of the above analysis.

We would also like to thank the Science and Engineering Research Council of the United Kingdom for financial support and the Aspen Center for Physics for its hospitality to one of us (DS) during the writing of this paper.

References

[1] W.S. McCulloch and W. Pitts, BULL. Math. Biophys. **5**, 115 (1943)
[2] K.Y.M. Wong and D. Sherrington, J. Phys **A22**, 2233 (1989)
[3] B. Derrida, E. Gardner and A. Zippelius, Europhys.Lett. **4**, 167 (1987)
[4] K.Y.M. Wong and D. Sherrington, Europhys. Lett **7**, 197 (1988)
[5] K.Y.M. Wong and A. Rau (unpublished)
[6] R.A. Blackburn and D. Sherrington (unpublished)
[7] K.Y.M. Wong and D. Sherrington (these proceedings)
[8] I. Aleksander, "Neural Computing Architectures" ed. I. Aleksander (Kogan Page: London) p133 (1988)
[9] cf also K.Y.M. Wong, P. Kahn and D. Sherrington (to be published)
[10] R. Atkinson and R. Shiffrin, Sci. Am. **225**, 85 (1971)

---

[2] There are other complementary aspects which we have not discussed here – for example in the character of attractors (see refs 2,4).

5. Concluding Remarks

In this paper we have considered two aspects of multiple Boolean networks...

Acknowledgements

We would like to thank Prof. I. Aleksander for stimulating our interest in Boolean storage and in the use of models...

References

[1]
[2]
[3]
[4]
[5]
[6]
[7]
[8]
[9]
[10]

PARTICIPANTS

Prof. ABBOTT, L., Physics Dept., Brandeis University, Waltham, Mass.
02254, USA

Prof. AGULLÓ, F., Facultad de Ciencias, Universidad Autónoma de Madrid,
Cantoblanco 28049 Madrid, Spain

Prof. AHUJA, S.B. Rentenanstalt, General-Guisan-Quai 40, and AI R & D
Laboratory, B-484, Zürich, Switzerland

Mr. ANDRADE, M.A., Dpto. de Bioquímica, Fac. de Ciencias Químicas,
Universidad Complutense, 28040 Madrid, Spain

Mr. ARNAUS, M., Urgel 259, Atco. A, 08036 Barcelona, Spain

Mr. AVELLANA, N., Dpto. de Informática, Universidad Autónoma de Barcelona, 08193 Bellaterra, Barcelona, Spain

Dr. BADDELEY, R., Centre for Computational Cognitive Neuroscience,
University of Stirling, Scotland

Mr. BAFALUY, Dpto. de Física, Universidad Autónoma de Barcelona,
08193 Bellaterra, Barcelona, Spain

Prof. BAVAN, A.S., Thames Polytechnic, School of Comput. and Inform.
Tech. Wellington St., London, S.E. 18, England

Miss BAUER, K., Institut f. Physik III, Universität Regensburg,
D 8400 Regensburg, Germany

Prof. BATALLAN, F., Dpto. Física de Materiales, Universidad Complutense
Ciudad Universitaria, 28003 Madrid, Spain

Prof. BERNASCONI, J.,Asea Brown Boveri, Corporate Research,
CH 5405 Baden, Switzerland

Dr. BOCHEREAU, L., ENGREF/CEMAGRAF, 26 rue St. Louis, F 78000 Versailles,
France

Prof. BOLLE, D., Institute for Theoretical Physics, University of Leuven, Celestijnenlaan 200 D, B 3030 Leuven, Belgium

Mr. BONET, J., Dpto. de Física, Universidad Autónoma de Barcelona,
08193 Bellaterra, Barcelona, Spain

Dr. BÖS, S., Institut f. Theoretische Physik III, JLU Giessen, Hein-
rich-Buff-Ring 16, 63 Giessen, Germany

Prof. BOUTEN, M., Limburgs Universitair Centrum, Universitaire Campus,
B 3610 Diepenbeek, Belgium

Prof. BOURGINE, P., ENGREF/CEMAGREF, 26 Rue St. Louis, 78000 Versailles,
France

Prof. CAMPBELL, C., Kingston Polytechnic, Dept. of Applied Physics,
Kingston-upon-Thames, Surrey, KT1 2EE, England

Dr. CANNING, A., Dept. de Physique, Université de Genève, CH-1211
Genève, Switzerland

Mr. CATALA, A., Facultad de Informática, Pau Gargallo 5, 08028 Barcelona, Spain

Mr. CODINA, J., F.I.A., U.P.C. (ESAII0) Pau Gargallo 5, 08028 Barcelona,
Spain

Dr. COOLEN, A.C.C., Dept. of Medical and Phys. Physics, University of
Utrecht, NL 3584 CC Utrecht, Holland

Mr. DIAZ, A., Depto. de Física Fundamental, Universidad de Barcelona,
Diagonal 647, 08028 Barcelona, Spain

Dr. DUPONT,P. Inst. voor Theor. Fysica, Celestijnenlaan 200D,
B 3030 Heverlee, Belgium

Prof. ELIZALDE, E., Dept. E.C.M., Facultad de Física, Universidad de
Barcelona, Diagonal 647, 08028 Barcelona, Spain

Miss ETXEBERRIA, A., Dpto. de Lógica y Filosofía de la Ciencia, (UPV/
EHU), San Sebastian, Spain

Prof. ERDÖS, P. Institute of Theoretical Physics, University of Lausanne,
BSP, 1015 Lausanne, Switzerland

Mr. FUERTES, J.M., F I B- U P C (ESAII), 5 Pau Gargallo, 08023 Barcelona, Spain

Mrs. GARCIA, J., Dpto. de Electrónica, Universidad de Extremadura,
06071 Badajoz, Spain

Prof. GARRIDO, L., Dpto. de Física Fundamental, Facultad de Física,
    Universidad de Barcelona, Diagonal 647, 08028 Barcelona, Spain
Mr. GARRIDO, F., Centro Nacional de Microelectrónica, Universidad
    Autónoma de Barcelona, 08193 Bellaterra, Spain
Prof. GESZTI, T., Atomic Physics Dept., Eötvös University, Puskin u.
    5/7, H 1088 Budapest, Hungary
Dr. GONZALEZ, S., Instituto de Ingeniería del Conocimiento, U.A.M.
    Módulo C-XVI, 28049 Madrid,Spain
Prof. GONZALEZ MIRANDA, J.M. Dpto. de Física Fundamental, Universidad
    de Barcelona, Diagonal 647, 08028 Barcelona, Spain
Prof. GUSTAFSON, K. Dept. of Mathematics, University of Colorado,
    Campus Box 426, Boulder, CO 80309, USA
Prof. HERTZ, J.A., Nordita, Blegdamsvej 17, D 2100 Copenhagen 0,
    Denmark
Dr. HERZ, A., SFB 123, Universität Heidelberg, Im Neuenheimer Feld
    294, D 69 Heidelberg, Germany
Dr. HOLZHEY, C., Institute for Medical Psychology, Goethestr. 31,
    D 8000 Munich 2, Germany
Dr. JAIN, S., Dept. of Chemistry, Durham University, South Rd. Durham
    DH1 3LE, England
Prof. JARAMILLO, M.A., Dpto. de Electrónica, Universidad de Extrema-
    dura, 06071 Badajoz, Spain
Prof. KARKHECK, J. GMI Engineering and Management Institute, 1700 W.
    Third Ave. Flint MI 48504, USA
Prof. KINZEL, W., Institute of Theoretical Physics II, JLU, D 6300
    Giessen, Germany
Dr. KOMODA, A., Institute of Theoretical Physics, Limburgs Universi-
    tair Centrum, B 3610 Diepenbeek, Belgium
Mr. KURRER, C., Beckman Institute, University of Illinois, 405 N.
    Mathews Ave., Urbana, IL 61801, USA
Dr. KÜHN, R., SFB 123, Universität Heidelberg, Im Neuenheimer Feld
    294, D 6900 Heidelberg, Germany
Prof. KÜRTEN, K.E., Institut f. Theoretische Physik, Zülpicher Str.
    77, D 5000 Köln 41, Germany
Prof. LOPEZ, V., Instituto de Ingeniería del Conocimiento, U.A.M.,
    Módulo C-XVI, 28049 Madrid, Spain
Prof. MARRO, J., Facultad de Ciencias, Universidad de Granada, Campus
    de Fuentenueva, 18071 Granada, Spain
Mr. MARTIN, M., Facultad de Informática (UPC), Pau Gargallo 5,
    08028 Barcelona, Spain
Prof. MARTIN, P., Dpto. de Electrónica y Tecnología de Computadores,
    Campus Fuentenueva, 18071 Granada, Spain
Prof. MAS, F., Dpto. de Química Física, Fac. de Químicas, Universidad
    de Barcelona, Diagonal 647, 08028 Barcelona, Spain
Dr. MERELO, J.J., Dpto. de Electrónica, Facultad de Ciencias, Campus
    Fuentenueva, 18071 Granada, Spain
Mr. MERTENS, S., Institute for Theoretical Physics, Bunsenstr. 9,
    D 3400 Göttingen, Germany
Prof. MORAN, F., Dpto. de Bioquímica, Universidad Complutense, Ciudad
    Universitaria, 28040 Madrid, Spain
Dr. NICOLIS, S., INFN Sezione di Roma, Piazzale Aldo Moro 2, 00185
    Roma, Italy
Dr. NOEST, A.J.,Physics Dept., University of Utrecht, Princetonplein 5,
    3508 TA Utrecht, Holland
Mr. PAGONABARRAGA, I. Depto. de Física Fundamental, Universidad de
    Barcelona, Diagonal 647, 08028 Barcelona, Spain
Dr. PASCUAL, M.F., Universidad de Sao Paolo, Instituto de Física,
    Brasil, C.Postal 2.0516, CEP 01498 Sao Paolo, Brasil
Dr. PEREZ, A., Dpto. de Física Fundamental, Universidad de Barcelona,
    Diagonal 647, 08028 Barcelona, Spain
Mr. PEREZ, C.J., Centro Nacional de Microelectrónica, Universidad
    Autónoma de Barcelona 08193 Bellaterra, Spain

Miss POVILL, A., Química Física, Facultad de Químicas, Diagonal 647,
    08028 Barcelona, Spain
Dr. PUY, J. Escuela Técnica Sup. de Ingenieros Agrarios (UPC)
    Av. Alcalde Rovira Roure 177, 25006 Lérida, Spain
Mr. RITORT, F., Dpto. de Física Fundamental, Universidad de Barcelona,
    Diagonal 647, 08028 Barcelona, Spain
Dr. RIEGER, H., Institut für Theoretische Physik, Universität Köln,
    Zülpicher Str. 77, D 5000 Köln 41, Germany
Prof. RUBÍ, M. Dpto. de Física Fundamental, Facultad de Física, Univer-
    sidad de Barcelona, Diagonal 647, 08028 Barcelona, Spain
Prof. RUJÁN, P., Fachbereich Physik, Universität Oldenburg, D 2900
    Oldenburg, Germany
Miss SACRISTÁN, A., Dpto. de Informática, Universidad Autónoma de
    Barcelona, 08193 Bellaterra, Barcelona, Spain
Mr. SANTOS, E., Telefónica Investigación, Emilio Vargas 6, 28043
    Madrid, Spain
Mr. SANZ, J.C. , U.N.E.D., Dpto. de Matemáticas, EISIM, Univ. Poli-
    técnica, Madrid, Spain
Prof. SCHULTEN, K., Beckman Institute, University of Illinois,
    405 N. Mathews Ave. Urbana, IL 61801, USA
Dr. SCHURMANN, B., Corporate Research and Development, Siemens,
    ZFE IS INF 23, Otto-Hahn Ring 6, D 8 Munich, Germany
Prof. SERNEELS, R., Limburgs Universitair Centrum, Universitaire
    Campus, 3610 Diepenbeek, Belgium
Prof. SHERRINGTON, D., Dept. of Theoretical Physics, University of
    Oxford, 1 Keble Rd. Oxford OX1 3NP, England
Dr. SOMMERS, H.J., FB Physik, Universität GSH Essen, Pf 103 764,
    D 43 Essen-1, Germany
Prof. SOURLAS, N., Ecole Normale Supérieure, 24 rue Lhomond, F 75231
    Paris Cedex 05, France
Miss SOUSA, M.C. Facultad de Químicas, Diagonal 647, 08028 Barcelona,
    Spain
Prof. TAHIR-KHELI, R.A., Dept. of Physics, Temple University, Barton
    Hall 009-00, Philadelphia 19122, PA, USA
Dr. TALLET, A., Laboratoire de Photophysique Moléculaire, Bât. 213,
    F 91405 Orsay, France
Mrs. TORRAS, C., Instituto de Cibernética (UPC-CSIC), Diagonal 647,
    08028 Barcelona, Spain
Dr. TORRENT, M.C. , Depto. ECM, Facultad de Física, Universidad de
    Barcelona, Diagonal 647, 08028 Barcelona, Spain
Dr. TREVES, A. Dept. of Experimental Psychology, University of Oxford,
    South Parks Rd., Oxford OX1 3UA, England
Mr. TRIGUEROS, P.P., Facultad de Químicas, Diagonal 647, 08028 Barce-
    lona, Spain
Dr. VIANA, L., Laboratorio de Ensenada, Instituto de Física, UNAM,
    Apdo. Postal 2681, Ensenada, B.C. Mexico
PROF. VIRASORO, M.A., Dipto. di Física dell'Università di Roma
    "La Sapienza", Pia. Aldo Moro, 4, Roma, Italy
Dr. VIVES, E. Dpto. ECM, Facultad de Física, Universidad de Barcelona,
    Diagonal 647, 08028 Barcleona, Spain
Dr. WONG, K.Y.M., Dept. of Theoretical Physics, University of Oxford,
    1 Keble Rd., Oxford OX1 3NP, England
Dr. YAU, H.W., Dept. of Physics, University of Edinburgh, JCM Bldg.
    Mayfield Rd. Edinburgh EH9 3JZ, Scotland